INDUSTRIAL AND ENGINEERING APPLICATIONS OF ARTIFICIAL INTELLIGENCE AND EXPERT SYSTEMS

INDUSTRIAL AND ENGINEERING APPLICATIONS OF ARTIFICIAL INTELLIGENCE AND EXPERT SYSTEMS

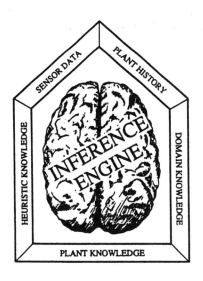

IEA/AIE
97

Proceedings of the Tenth International Conference
Atlanta, Georgia, USA, June 10-13, 1997

Edited by

Don Potter
University of Georgia
Athens, Georgia, USA

Manton Matthews
University of South Carolina
Columbia, South Carolina, USA

Moonis Ali
Southwest Texas State University
San Marcos, Texas, USA

GORDON AND BREACH SCIENCE PUBLISHERS
Australia • Canada • China • France • Germany • India • Japan • Luxembourg • Malaysia •
The Netherlands • Russia • Singapore • Switzerland • Thailand • United Kingdom

Amsteldijk 166
1st Floor
1079 LH Amsterdam
The Netherlands

British Library Cataloguing in Publication Data

Industrial and engineering applications of artificial
    intelligence and expert systems : proceedings of the tenth
    international conference
    1. Artificial intelligence – Industrial applications –
    Congresses 2. Expert systems (Computer science) – Industrial
    applications – Congresses
    I. Potter, Don II. Matthews, Manton III. Ali, Moonis
    006.3

ISBN 90-5699-615-0

CONTENTS

SPATIAL AND TEMPORAL REASONING

NATURAL LANGUAGE UNDERSTANDING

PLANNING AND SCHEDULING

GENETIC ALGORITHMS II

MACHINE LEARNING

EXPERT SYSTEMS

INTELLIGENT USER-INTERFACES / TUTORING SYSTEMS

CASE/MODEL-BASED REASONING

NEURAL NETWORKS II

FUZZY LOGIC II

PREFACE

The editors present the Proceedings of the Tenth International Conference on Industrial and Engineering Applications of Artificial Intelligence and Expert Systems (IEA/AIE-97) with a great deal of pride and pleasure. This international meeting of applied intelligence researchers and practitioners attempts to maintain a balance between applied and theoretical aspects of intelligent systems research as well as extensions of tools and theory of intelligent thinking machines to solve real-life problems. Conference participants share their research and expertise on intelligent problem solving in various areas of real life.

The proceedings' papers present a broad spectrum of research and development including genetic algorithms, computer vision, knowledge management, spatial and temporal reasoning, natural language understanding, planning and scheduling, machine learning, expert systems, intelligent user-interface/ tutoring systems, fuzzy logic, case/model based reasoning, neural networks, and practical applications. We are proud to report contributions from many distinguished colleagues within academia, government, and industry. The contributors of these papers come from across the globe; we received more than 90 papers from more than 25 countries. Approximately 60% were selected. The selected papers were presented at the conference; they cover a wide area of new ideas on numerous topics that reflect the focus and theme of the conference. The proceedings' objective is to disseminate information not only to the participants of this conference, but also to researchers around the world.

We would like to thank the Organizing Committee for its tireless efforts in putting the conference together, from preparing the call for papers to final closing in Atlanta. The Program Committee did a wonderful job reviewing papers and returning reviews in a timely manner. The external reviewers deserve recognition as well for their efforts; and special thanks go to Angie Paul for her help with processing the reviews, submissions and notifications.

Don Potter, *Program Chair*
Manton Matthews, *Program Chair*
Moonis Ali, *General Chair*

CONFERENCE ORGANIZATION

Organizing Committee

Moonis Ali, South West Texas State University — General Chair

Manton Matthews, University of South Carolina — Program Chair

Don Potter, University of Georgia — Program Chair

Marco Valtorta, University of South Carolina — Tutorial Chair

John Miller, University of Georgia — Local Arrangements Chair

Wuxu Peng, South West Texas State University — Publicity Chair

Khosrow Kaikhah, South West Texas State University — Exhibit Chair

Cheryl Morriss, South West Texas State University — Registration Chair

Program Committee

Frank Anger
NSF, USA

B. el. Ayeb
Univ. of Sherbrooke
CANADA

Suchi Bhandarkar
Univ. of Georgia, USA

David Billington
Griffith Univ.
AUSTRALIA

Gautam Biswas
Vanderbilt Univ., USA

Kai H. Chang
Auburn Univ., USA

Paul Chung
Loughborough Univ., UK

Michael Covington
Univ. of Georgia, USA

Ken Ford
Univ. of West Florida
USA

Graham Forsyth
DSTO AMRL
AUSTRALIA

Ashok Goel
Georgia Inst. of Tech.
USA

T. Govindaraj
Georgia Inst. of Tech.
USA

Mehdi T. Harandi
Univ. of Illinois, USA

Tim Hendtlass
Swinburne Univ. of Tech.
AUSTRALIA

Lakhmi C. Jain
Univ. of South Australia
AUSTRALIA

Mark Juric
Platinum Technologies, Inc.
USA

Kazuhiko Kawamura
Vanderbilt Univ., USA

Larry Kerschberg
George Mason Univ., USA

Steve Kimbrough
Univ. of Pennsylvania
USA

Michael Magee
Southwest Res. Inst., USA

Mark Maloof
George Mason Univ., USA

Esme Manandise
Exploration Resources
USA

Ron McClendon
Univ. of Georgia, USA

Drew McDermott
Yale Univ., USA

Jose Mira-Mira
UNED, SPAIN

John Mitchiner
Sandia Natl. Lab., USA

Riichiro Mizoguchi
Osaka Univ., JAPAN

Laszlo Monostori
Hungarian Academy of
Sciences
HUNGARY

Yi Lu Murphey
Univ. of Michigan, USA

Gordon Novak
Univ. of Texas, USA

Dick Peacocke
Nortel, CANADA

. . . continued

Program Committee ... *continued*

Fred Petry
Tulane Univ., USA

Francois Pin
Oak Ridge Natl. Lab.
USA

Ashwin Ram
Georgia Inst. of Tech.
USA

Anil Rewari
Verity Inc., USA

Rita Rodriguez
NSF, USA

John Rose
Univ. of South Carolina
USA

Abdul Sattar
Griffith Univ.
AUSTRALIA

Nigel Shadbolt
Univ. of Nottingham, UK

Mildred L.G. Shaw
Univ. of Calgary
CANADA

Reid Simmons
Carnegie Mellon Univ.
USA

Robert E. Smith
Univ. of Alabama, USA

Motoi Suwa
ETL, JAPAN

Takushi Tanaka
Fukuoka Inst. of Tech.
JAPAN

Bruce Tonn
Oak Ridge Natl. Lab.
USA

Spyros Tzafestas
NTU, Athens, GREECE

Marco Valtorta
Univ. of South Carolina
USA

Elizabeth Whitaker
NCR, USA

Ian H. Witten
Univ. of Waikato
NEW ZEALAND

Xin Yao
UNSW & ADFA
AUSTRALIA

Wai-Kiang Yeap
Univ. of Otago
NEW ZEALAND

SPONSORS AND COOPERATIVE ORGANIZATIONS

International Society of Applied Intelligence

ACM/SIGART

American Association for Artificial Intelligence

Canadian Society for Computational Studies of Intelligence

European Coordinating Committee for Artificial Intelligence

Institute of Measurement and Control

Institution of Electrical Engineers

International Neural Network Society

Japanese Society of Artificial Intelligence

Southwest Texas State University

University of Georgia

University of South Carolina

AI Associates, Inc.

External Reviewers

Jon Hamlin	Tim W. Dollar
Chris Henderson	Ashraf Saad
Gordon Shippey	Isao Hayashi
William A. Stubblefield	L.B. Romdhane
Shigeyoshi Tsutsui	John Thornton
Harold Dale	S.S. Liao
Stephen D. Kleban	Kevin Novins
Andy Walker	Rattana Wetprasit

Xudong William Yu

AN EVOLUTIONARY OPTIMIZATION SYSTEM FOR SPACECRAFT DESIGN

Alex S. Fukunaga and Andre D. Stechert
Jet Propulsion Laboratory, MS 525-3660
California Institute of Technology
4800 Oak Grove Drive
Pasadena, CA 91109-8099
Email: {alex.fukunaga,andre.stechert}@jpl.nasa.gov

Abstract

Spacecraft design optimization is a domain that can benefit from the application of optimization algorithms such as genetic algorithms. However, there are a number of practical issues that make the application of these algorithms to real-world spacecraft design optimization problems difficult in practice. In this paper, we describe DEVO, an evolutionary optimization system that addresses these issues and provides a tool that can be applied to a number of real-world spacecraft design applications. We describe two current applications of DEVO: physical design of a Mars Microprobe Soil Penetrator, and system configuration optimization for a Neptune Orbiter.

1 Introduction

In theory, many aspects of spacecraft design can be viewed as constrained optimization problems. Given a set of decision variables X and a set of constraints C on X, constrained optimization is the problem of assigning values to X that minimize or maximize an objective function F defined on X subject to the constraints C. In practice, there are a number of theoretical and practical obstacles that make constrained optimization of spacecraft designs difficult.

First, while optimization of smooth, convex objective functions is well understood (and efficient algorithms are known to exist, see, for example, [Fle87]), global optimization on surfaces with many local optima is not. Traditional approaches to optimization usually fare poorly on these so-called "rugged" surfaces, which are often characteristic of real-world optimization problems.

Second, many real-world optimization problems are *black-box optimization problems*, in which the structure of the cost function is opaque. That is, it is not feasible to analyze the cost surface by analytic means in order to guide an optimization algorithm. Often, this occurs when $F(X)$ is computed by a complex simulation about which the op-

timization algorithm has no information (e.g., to evaluate a candidate spacecraft design, we could simulate its operations using a suite of legacy FORTRAN code about which very little is known except for its I/O specifications). Black-box optimization problems of this kind are challenging from a practical point of view for two reasons: 1) Executing a black-box simulation in order to evaluate a candidate solution is usually very expensive relative to, for example, evaluating a cost function that is expressed as a system of equations, and can take on the order of several *minutes*. This is particularly problematic because optimization algorithms for black-box problems are necessarily *blind* search algorithms that must repeatedly choose sample points from the solution space, evaluate them by running the simulation, and then apply various heuristics in order to choose the next points to sample, 2) Interfacing optimization tools to black-box simulations can be difficult, particularly when the black box is a complex software system that involves various components written in different languages, possibly running on a distributed environment running on a number of different platforms.

Finally, many spacecraft design engineers do not have the optimization expertise required to apply state of the art algorithms to their problems. Obtaining this expertise is often prohibitively expensive; as a consequence, optimization using algorithmic methods is sometimes not even attempted, because of the perception that it is not worth the effort and expense.

In this paper, we describe Design Evolver (**DEVO**), an optimization system developed in an effort to address these issues. To address the problem of optimization of difficult cost surfaces, DEVO implements a generic, reconfigurable implementation of an evolutionary optimization algorithm. To overcome the practical difficulties described above that arise when designing tools for black-box optimization problems, DEVO is integrated with MIDAS [GPS95], a recently developed integrated spacecraft design environment, making it possible to apply automated optimization to any spacecraft design model specified in this environment.

The rest of this paper is organized as follows. Section 2 describes the architecture of the DEVO system, focusing on the practical issues that arise in the integration of an evolutionary optimization algorithm into a spacecraft design environment, and the automated reconfiguration of the optimization algorithm for a particular problem instance. In Section 3, we describe two spacecraft design optimization problems which are currently being used as testbed applications for DEVO: the NASA New Millennium DS-2 Mars Microprobe, and the Neptune Orbiter spacecraft.

2 The Design Evolver (DEVO) System

DEVO is a system for spacecraft design optimization currently being developed at the Jet Propulsion Laboratory (JPL). The goal of DEVO is to provide an optimization tool that is seamlessly integrated into an existing computer-aided design (CAD) environment for spacecraft, which enables users to apply optimization algorithms, including evolutionary algorithms, with a minimal amount of human effort. A fundamental assumption in the DEVO design is that CPU cycles are plentiful and cheap relative to the cost of an engineer hand-tuning an optimization engine. In this section, we describe the DEVO system. We first describe the evolutionary optimization algorithms implemented in DEVO. We then describe MIDAS, the design environment into which DEVO has been integrated. Then, we describe various features of DEVO that address the practical issues that arise in the application of evolutionary optimization techniques to real-world, spacecraft design problems.

2.1 Reconfigurable Evolutionary Algorithms

The central component of DEVO is the Reconfigurable Evolutionary Algorithm (**REAL**). The REAL is an implementation of a generic evolutionary optimization strategy that can be reconfigured at runtime to behave as one of the various classes of evolutionary algorithms. Figure 1 (following [BS96]) shows the general schema for evolutionary algorithms which the REAL implements.

Briefly, an evolutionary algorithm works as follows: a population of sample points from the cost surface is generated. In a process analogous to biological evolution, this population is evolved by repeatedly selecting (based on relative optimality) members of the population for reproduction, and recombining/mutating to generate a new population.

By providing different implementations of functions such as *initialize*, *recombine*, *mutate*, and *select*, and a selection of encodings (representations) of solutions (e.g., bit-string encoding, possibly with Gray coding, floating point representations, etc.) that can be chosen at runtime, it is possible to reconfigure the REAL to simulate a wide variety of evo-

$$
\begin{aligned}
&t := 0 \\
&\textbf{initialize } P(t); \\
&\textbf{evaluate } P(t); \\
&\textbf{while not terminate do} \\
&\quad P'(t) := \textbf{recombine } P(t); \\
&\quad P''(t) := \textbf{mutate } P'(t); \\
&\quad \textbf{evaluate } P(t); \\
&\quad P(t+1) := \textbf{select } (P''(t) \cup Q); \\
&\quad t := t + 1; \\
&\textbf{end while}
\end{aligned}
$$

Figure 1: Algorithm schema for an evolutionary algorithm P is a population of candidate solutions; Q is a special set of individuals that has to be considered for selection, e.g., $Q = P(t)$.

lutionary algorithms. For example, using a null *recombine* function and implementing a *mutate* function that applies Gaussian mutation, we achieve the canonical Evolutionary Programming (cf., [Fog95]) algorithm.

Currently, the REAL supports bit-string representations of numerical parameters, as well as floating point number representations. Various mutation, recombination, and selection operators are available. Furthermore, the REAL supports a number of different population structures, including the traditional generational population structure [Gol89], a steady-state population structure [Sys89], and a distributed population structure [Tan89]. Thus, the REAL can be configured to simulate a wide range of common Genetic Algorithm (cf. [Gol89]) and Evolutionary Programming (cf. [Fog95]) variants.

2.2 Spacecraft Design Model

A *spacecraft design model* is a software simulation of a spacecraft design. The design model takes as input decision variables to be optimized, and outputs an objective function value, which is assigned as the result of an arbitrarily complex computation (i.e., the simulator is a black-box simulation).

Thus, the design model is domain-specific, and is provided by the end users, i.e., spacecraft designers. In order for an optimization tool such as DEVO to be useful in practice, it must support a wide range of design models, which may consist of models implemented using various languages on different platforms. It is not feasible to expect spacecraft designers to implement their models in a particular language on a particular platform – if such inconvenient constraints were imposed, the optimization system will not be used by spacecraft designers.

The Multidisciplinary Integrated Design Assistant for Spacecraft (**MIDAS**) [GPS95] is a computer-aided design environment developed at JPL that allows a user to integrate a system of (possibly distributed) design model components

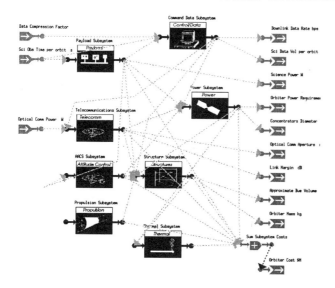

Figure 2:   Screen shot of a MIDAS methogram (part of the Neptune Orbiter model).

using a *methogram*, a graphical diagram depicting the data flow of the system. Each node in the methogram corresponds to a design model component, which may be one of 1) a model in a commercial design tool such as IDEAS, NASTRAN, or SPICE, 2) a program written in C, C++, or FORTRAN, 3) a MIDAS built-in tool, or 4) an embedded methogram (i.e., methograms can be hierarchical). Inputs to nodes in the methogram correspond to input parameters for the component represented by the node, and outputs from a methogram node correspond to output values computed by the component. Since it was implemented as a distributed object system and since an output node can be used to compute an arbitrary function of the parameters in the model, MIDAS provides a uniform interface to a wide variety of design models without requiring optimization algorithms to have strong dependencies on the target simulation. Figure 2 shows a screen shot of the MIDAS methogram for the Neptune Orbiter model.

2.3   Automated Configuration of the REAL

As we mentioned in Section 1, a significant practical obstacle to applying optimization algorithms to spacecraft design is the optimization knowledge and effort required of spacecraft designers. We therefore designed DEVO to be usable with as little input as possible beyond what is already provided by the user in the MIDAS spacecraft design model.

In order to run DEVO, the user is required to input the following:

- A MIDAS methogram that encapsulates the design model;

- A list of decision variables, as well as ranges of their

possible values. These may be continuous, discrete, or enumerated types;[1]

- An output from a methogram node that corresponds to the user's objective function value; and

- A termination condition for a run of the optimization algorithm. This can be either 1) a time limit, 2) a maximum number of simulation runs, or 3) a simple check for convergence of the algorithm, i.e., no improvement is made for some number of simulation runs (a default value is provided).

The user input listed above is sufficient for DEVO to automatically configure the REAL to an appropriate default configuration and run the evolutionary algorithm. Based on the decision variable types and ranges specified by the user, the genome evolved by the REAL is appropriately configured, and the evolutionary algorithm is executed. A user can, of course, manually configure the REAL using either a command-line or graphical user interface.

Thus, the effort and knowledge required by the spacecraft designer to run DEVO is minimal, since essentially all that is required for a user to use DEVO to optimize a design is to specify the decision variables and the constraints on them, and to specify an objective function.[2]

We should make it clear that by no means are we claiming that DEVO can provide a default configuration that works well for all spacecraft design problems. In fact, in the absence of any knowledge of the cost surface structure, it is quite possible that any default configuration of the REAL may be no better than random search.[3] *However, even if the default configuration chosen by DEVO performs relatively poorly, we argue that applying some optimization algorithm is better than not applying any optimization at all* (particularly when computational resources are readily available).

When a run of a particular configuration has terminated, DEVO continues the optimization process by saving the best solution found so far, and restarting the optimization process using another configuration.[4] This process is repeated until terminated by the user. The strategy currently used by DEVO to choose the next REAL configuration is a simple randomized strategy: generate the next configuration randomly, the only constraint being that the configuration is compatible with the decision variable types.[5] The general problem of *adaptive problem solving*, i.e., reconfiguring a

---

[1]Enumerated types are mapped onto discrete values by DEVO.

[2]In many cases, this objective is already available in the methogram.

[3]See [WM94] for a recent theoretical perspective on this issue.

[4]We are currently investigating whether it is better to start the next optimization run from scratch, or to seed the initial population using solutions found in previous runs.

[5]It is acceptable to repeat configurations, since the performance of evolutionary algorithms is stochastic.

problem solver to perform well on a particular problem instance, poses many difficult theoretical challenges, and is beyond the scope of this paper (see [FCM+97] for a more detailed discussion of this problem, as well as more sophisticated solutions that we are currently investigating).

2.4   Optimization as an Interactive Process

Most of the work on evolutionary optimization focuses on optimization as a fully automated process, in which the *initialize* step in Figure 1 is accomplished by random initialization. Design optimization, however, is often an interactive process, since the designers who developed the model have sufficient domain expertise to suggest some "reasonable guesses" as to what good decision variable parameters might be. Indeed, in many cases, it is difficult for a completely automated optimization process to produce designs of better quality than a human engineer.

A reasonable alternative is to attempt *local optimization* of a design that is initially specified by a human engineer. Therefore, in addition to the usual random initialization functions, we have implemented an initialization function for the REAL that generates the initial population based on random perturbations of decision variable parameters which the user has specified in the model. The perturbations apply random noise with Gaussian or uniform noise applied at the user's discretion.

2.5   Additional Implementation Details

2.5.1   Handling Instabilities in Black-Box Simulation

A common problem when trying to apply evolutionary algorithms to black-box simulations is the possibility of instability in the simulations. Spacecraft design models are often one-of-a-kind prototype systems, designed and implemented to provide proof of concept in the hands of an experienced engineer. Consequently, they are not necessarily robust enough to be executed with the thousands of different assignments of decision variable values that an evolutionary algorithm attempts. A particular input parameter combination could, for example, cause an arithmetic exception, causing the simulation to crash. If the optimization system is not designed to anticipate such failures, the whole optimization system could fail as a result. It could be argued that such instability could be symptomatic of an unreliable simulator whose results should not be trusted at all (and therefore needs to be fixed immediately). However, we take the view that it is not feasible to demand that the simulation software be made completely robust for the sake of the optimization system and designed DEVO so that it would circumvent simulation instability as much as possible.

DEVO protects the optimization system from simulation software instability by exploiting fault detection features built into MIDAS, and by separating the optimization process from the simulation process as much as possible. MIDAS is capable of detecting common failures (e.g., core dumps, arithmetic exceptions) that occur when executing a process that corresponds to a node in a methogram. DEVO monitors this information, and immediately aborts the execution of the simulation upon detecting a failure. Some failures can actually cause MIDAS itself to crash. Since DEVO is implemented as a separate process which invokes MIDAS and manipulates it through its CORBA interface, DEVO can detect MIDAS crashes and abandon the candidate solution evaluation that called the failed MIDAS simulation. It is, of course, not possible to detect with certainty if a simulation has entered an infinite loop; however, if DEVO has been waiting for the result of a simulation for an abnormally long time (e.g., if a simulation run is taking n standard deviations more time than an average simulation run to date), then it is assumed that the simulation is trapped in a loop, and DEVO will terminate the simulation.

An interesting issue is what to do with the evaluation of solutions that cause failures that are detected as described above. We currently apply the simple policy of assigning the worst possible fitness values to these solutions. On one hand, this has the effect of causing the evolutionary algorithm to avoid solutions that are very similar to the offending solutions. Assuming that the simulation is unstable in *regions* of the solution space, rather than isolated points, this is a reasonable policy. On the other hand, if this is not the case, then this policy could cause an undesirable bias in the evolutionary search. We are currently investigating the significance of this bias.

2.5.2   Parallelization

Executing a complex spacecraft design model simulation often takes a significant amount of time. For example, a single execution of our current Neptune Orbiter (see 3.2) takes several minutes on a Sun Ultra workstation. Given that a single run of an evolutionary algorithm requires hundreds to thousands of candidate solution evaluations, this poses a serious problem, if we want to complete the optimization processes in a reasonable amount of time. Fortunately, evolutionary algorithms are particularly well-suited to parallelization, since each candidate solution can be evaluated independently of the other solutions, i.e., the *evaluate* step in Figure 1 can be parallelized with near-linear efficiency. Therefore, DEVO distributes simulations over a network of workstations using Parallel Virtual Machines (PVM) [GBD+94].

3   Spacecraft Optimization Problems

In this section, we describe two specific spacecraft design optimization problems to which we are currently applying the DEVO system. The first is a low-level optimization of the physical dimensions of a soil penetrator microprobe. The second is a system-level optimization of the configuration of the communication system of an orbiter spacecraft. These examples are illustrative of the wide range of different spacecraft design optimization problems to which DEVO can be applied.

3.1 The Mars Soil Penetrator Microprobe

As part of the NASA New Millennium program, two microprobes, each consisting of a very low-mass aeroshell and penetrator system, are planned to launch in January, 1999 (attached to the Mars Surveyor lander), to arrive at Mars in December, 1999. The 3kg probes will enter the Martian atmosphere and orient themselves to meet heating and impact requirements. Upon impacting the Martian surface, the probes will punch through the entry aeroshell and separate into fore- and aftbody systems. The forebody will reach a depth of 0.5 to 2 meters, while the aftbody will remain on the surface for communications.

Each penetrator system includes a suite of highly miniaturized components needed for future micro-penetrator networks: ultra low temperature batteries, power microelectronics, and advanced microcontroller, a microtelecommunications system and a science payload package (a microlaser system for detecting subsurface water).

The optimization of physical design parameters for a soil penetrator based on these Mars microprobes is the first testbed for the DEVO system. The microprobe optimization domain in its entirety is very complex, involving three stages of simulation: separation from the Mars Surveyor, aerodynamical simulation, and soil impact and penetration. The complete design model for the penetrator is currently under development. Below, we describe the current model, which implements the simulation of stage 3 (impact/penetration).

Given a distribution on parameters describing the initial conditions including the angle of attack of the penetrator, the impact velocity, and the hardness of the target surface, the optimization problem is to select the total length and outer diameter of the penetrator, so as to maximize the expected ratio of the depth of penetration to the length of the penetrator. We maximize this ratio, rather than simply maximizing the depth of penetration, since for the Mars microprobe science mission, the depth of penetration should be at least as large as the overall length of the penetrator).

Using the default REAL configuration generated by DEVO (a canonical generational GA using bit-string encodings, one-point crossover, bit-flip mutation, population size of 50), it is consistently possible to generate a near-optimal design after about 50 generations.

3.2 The Neptune Orbiter

Neptune Orbiter is a mission concept currently being studied under the Outer Planet Orbital Express program at the Jet Propulsion Laboratory. The goals of the mission are to put a spacecraft in orbit around Neptune using state-of-the-art technologies in the areas of telecommunications, propulsion, orbit insertion, and autonomous operations. The spacecraft is expected to arrive at Neptune (30 a.u.) 5 years after launch in 2005 using a Delta launch vehicle. The subsystem requirements include 100 kbps data rate, solar electric propulsion, solar concentrator power source and a cost of less than $400 million (in FY 94 dollars).

For the initial phase of the optimization effort, the focus is on the orbital operations of Neptune Orbiter. The launch and cruise phases of the mission will be included in the optimization once the orbiter problem is well understood. The driving constraints of the orbiter problem are the optical communication aperture, transmit power and spacecraft mass. The transmit power is a direct input into the integrated spacecraft design model. The other inputs include the science observation time per orbit and the data compression factor. The output of the model that is being maximized is the science data volume per orbit. For designs in which the spacecraft mass is greater than 260 Kg, the data volume output is zero. A spacecraft with a dry mass of greater than 260 Kg is too heavy to lift on the target launch vehicle. Thus the mass limit constrains the optimization problem. Currently, we are using cost models in conjunction with the simulation of the orbiter as described above to obtain our cost function - a quantitative estimate of the science return (measured in, e.g., volume of science data obtained per dollar cost of the spacecraft).

4 Conclusions

Designing a widely applicable tool for black-box design optimization poses a significant technical challenge. In this paper, we have described DEVO, an evolutionary optimization system for spacecraft design that provides a design optimization tool that can be applied to real-world spacecraft design optimization problems with minimal human effort.

Much of the recent work in the evolutionary algorithm literature focuses on development of specialized representations and techniques to customize an evolutionary algorithm for a particular application. While we agree that developing specialized algorithms for particular applications is the best methodology for obtaining the best performance for any particular domain, this approach is often infeasible in practice, due to the human expertise and effort required to develop a specialized algorithm.

For problems that are unique in nature, a promising approach is to provide tools that make it possible to apply very general methods with little overhead. As long as the application of the method is virtually free of human effort, it is worthwhile to use available computational resources to approach the problem in a "brute-force" manner off-line, since the potential benefits of improving design quality can be quite substantial. The development of DEVO is a first step in this direction.

So far, we have found that the default behavior of DEVO has been sufficient for finding near-optimal solutions to the Mars soil penetrator microprobe problem.[6] However, we believe that in order for reconfigurable systems such as DEVO to become more useful, techniques for intelligently, automatically configuring the system to suit a particular problem instance must be developed and integrated into the system. We are currently investigating this problem [FCM+97].

Finally, we note the utility of system development efforts such as DEVO to the evolutionary algorithm research community. A myriad of promising approaches to evolutionary optimization have been proposed in the literature. However, the success of a particular technique for a given problem depends largely on the match between the technique and the problem [WM94], and thus, assessing the utility of a particular approach is mostly an empirical, problem-specific issue. Since applying a new technique to a real-world problem is often difficult and time-consuming, evolutionary algorithm researchers often restrict their evaluation of new approaches to synthetic cost functions (e.g., [DeJ75], [WMRD95]) or other easily implemented problems (e.g., the Traveling Salesperson Problem) whose relationship to most real-world problems is tenuous. By providing a stable, uniform interface to a wide variety of black-box optimization problems, DEVO provides evolutionary algorithm researchers with a framework into which many new techniques can be easily integrated, enabling the evaluation of new approaches on real-world problems. As an example, we have recently integrated an *incremental evolution* technique [FK95] into DEVO, and demonstrated its utility (compared to standard genetic algorithms) on the DS-2 probe design problem (see [FCM+97] for details).

Acknowledgments

The research described in this paper was performed by the Jet Propulsion Laboratory, California Institute of Technology, under contract with the National Aeronautics and Space Administration. Thanks to Julia Dunphy and Jose Salcedo for assistance with MIDAS. Bob Glaser and Celeste Satter provided the domain models for the Mars microprobe and Neptune Orbiter, respectively.

References

[BS96]   T. Back and H-P Schwefel. Evolutionary computation: An overview. In *Proc. IEEE International Conf. Evolutionary Computation*, 1996.

[DeJ75]   K. DeJong. *An Analysis of the Behavior of a Class of Genetic Adaptive Systems*. PhD thesis, University of Michigan, Department of Computer and Communication Sciences, Ann Arbor, Michigan, 1975.

[FCM+97]   A.S. Fukunaga, S. Chien, D. Mutz, R. Sherwood, and A. Stechert. Automating the process of optimization in spacecraft design. In *Proc. IEEE Aerospace Conf. (to appear)*, 1997.

[FK95]   A.S. Fukunaga and A.B. Kahng. Improving the performance of evolutionary optimization by dynamically scaling the evaluation function. In *Proc. IEEE International Conf. on Evolutionary Computation (ICEC)*, 1995.

[Fle87]   R. Fletcher. *Practical methods of optimization, 2nd ed.* Wiley, 1987.

[Fog95]   D.B. Fogel. *Evolutionary computation: Toward a New Philosophy of Machine Intelligence.* IEEE Press, 1995.

[GBD+94]   A. Geist, A. Beguelin, J. Dongarra, W. Jiang, R. Manchek, and V. Sunderam. *PVM: Parallel Virtual Machine - a user's guide and tutorial for networked parallel computing.* MIT Press, 1994.

[Gol89]   D.E. Goldberg. *Genetic Algorithms in Search, Optimization and Machine Learning.* Addison-Wesley, 1989.

[GPS95]   J. George, J. Peterson, and S. Southard. Multidisciplinary integrated design assistant for spacecraft (midas). In *Proceedings of American Institute of Aeronautics and Astronautics (AIAA)*, 1995.

[Sys89]   G. Syswerda. Uniform crossover in genetic algorithms. In *Proc. International Conf. on Genetic Algorithms (ICGA)*, 1989.

[Tan89]   R. Tanese. Distributed genetic algorithms. In *Proc. International Conf. on Genetic Algorithms (ICGA)*, 1989.

[WM94]   D.H. Wolpert and W.G. Macready. The mathematics of search. SFI-TR-95-02-010, 1994.

[WMRD95]   D. Whitley, K. Mathias, S. Rana, and J. Dsubera. Building better test functions. In *Proc. International Conf. on Genetic Algorithms (ICGA)*, 1995.

---

[6]The randomized reconfiguration described in 2.1 has not been necessary so far; this is due to the fact that the default configurations of the REAL were calibrated using the Mars microprobe and Neptune Orbiter domains (the first problems to which the REAL has been applied). The default configurations may not be appropriate for future problem instances.

ON THE USE OF VARIABLE MUTATION IN AN EVOLUTIONARY ALGORITHM

Tim Hendtlass
Centre for Intelligent Systems
School of Biophysical Sciences and Electrical Engineering
Swinburne University of Technology
P.O.Box 218 Hawthorne 3122
AUSTRALIA
Email: thendtlass@swin.edu.au

Keywords. Evolutionary algorithm, optimization, mutation.

ABSTRACT.

When an evolutionary algorithm is used to optimize a function or process it performs a broad search of solution space. To find the probable optimum point the evolutionary algorithm must be complimented by a local search technique. Creep mutation is one such possible technique. Specifying suitable mutation control parameter values that suit both the evolutionary algorithm and the local search requirements is not easy, especially if the solution space is complex and little detail is known about it. This paper describes the use of cyclically varying mutation control parameter values to allow creep mutation to be successful over a wide range of parameter values.

1   INTRODUCTION.

It is often necessary to find the minimum of some function. The function can represent an industrial process, a chemical reaction or an optimization problem. Complex functions may be of considerable practical importance and yet not be amenable to formal mathematical analysis. In addition, the solution space may have a number of local minima that thwart the use of a gradient descent technique. In this case, an evolutionary algorithm may be the most suitable technique.

Evolutionary algorithms use an initial population of (probably) poor solutions and combine aspects of (usually two) parent solutions to produce one or more new solutions. The aim is to produce offspring which combine the best features of, and so out perform, their parents. The offspring are then used as parents to produce another, hopefully better, generation still. The better a solution in a generation performs, the more part it plays in parenting the next generation. In this way a bias towards better performing individuals is introduced. As a result beneficial traits are assimilated into the population while unproductive traits tend to die out.

As well as combining parts of the parents in the offspring, mutation of the offspring may also occur. This has the effect of introducing new material into the gene pool. The mutation amount is controlled by two factors, the probability that a mutation will occur (the mutation probability) and the maximum magnitude that such a mutation may take (the mutation radius).

Evolutionary algorithms are very computationally expensive as solutions tend to be slow to evolve. The larger the population size, the richer the pool of genetic material from which to construct offspring but the more processing is required for each generation and the longer the time taken to find the solution. The smaller the population size, the more important a role mutation plays in providing diversity in the genetic material from which the new solutions are built.

Evolutionary algorithms are very efficient at performing a broad search of a vast solution space but they are not, in general, able to directly find the optimum solution. To find the optimum solution the evolutionary algorithm must be teamed with a local search technique. A gradient based local heuristic is often used [see, for example, P&H95]. However, a suitable local heuristic is not always available. In this case reliance must be placed on a mutation driven creep to the optimum solution.

Using a mutation radius that is inversely proportional to the genetic diversity in the population is advantageous when the whole population is close to a local minimum. The mutation is increased as the population approaches the local minimum thus helping it to escape. However, it is disadvantageous if the population is close to the global minimum as the increased mutation will tend to make it harder for the population to converge to the desired solution. The likelihood of such a situation occurring increases as the population size decreases.

Using the ratio of successful to unsuccessful mutations as a mutation amount control parameter has been suggested [TB96]. It is assumed that if this ratio is less than some value (often 0.2), then the population is too diverse and the mutation amount should be decreased. Conversely, if the ratio is greater than this value, the population is too focused and the mutation amount should be increased. While there is no doubt that this approach will sometimes help, it raises the problem of choosing (another) problem dependent parameter value.

For a review of performance based mutation strategies, see [PJA95].

Some work has also been done [TF89] with mutation levels which exponentially decrease over generations. The problem with such an approach is knowing the rate at which to decrease the mutation amount.

Creep mutation relies on low mutation radii to find the final solution. Bridging local minima and providing adequate genetic diversity in a small population often require higher mutation radii. Using mutation to achieve two goals can make the choice of suitable mutation control parameters difficult, especially when small populations are used. This paper considers the use of cyclic mutation to allow both of these goals to be realized.

2   The function used.

To test the effects of cyclic mutation a number of tests were performed on a function. This function defined by

$$Z = 1 + \sqrt{R} - \cos(5 * pi * R)$$

where

$$R = \sqrt{x^2 + y^2}$$

This function produces the three dimensional surface shown in figure one. This surface is symmetrical about a global minimum at 0,0 which is surrounded by deep concentric local minima.

This function was developed as the deep local minima around the global maximum make this a harder problem to solve than, say, the F7 function [JDS et al 89] in which the

local minima become smaller and closer together as the global minimum is approached. The regular size of the minima permits an easier interpretation of the role mutation plays in crossing them than if the minima size varies.

Only the section of the surface shown in figure one was used for the tests, representing the region around the desired solution. The local minima preclude the use of any gradient descent algorithm and the restricted range of solution space ensures that crossover is rarely beneficial. The inter local minima spacing is about 20% of the search range along either axis.

Population members are started at random places on the x,y plane. Except in the few cases in which an early crossover leads almost directly to a solution, the population rapidly contracts to a few (often one) positions on the plane. Although crossover between similar parents does still play a part in local exploration, progress is mainly reliant on creep mutation to find the global minimum.

3   The evolutionary algorithm used.

Each chromosome consists of two genes, one which holds the (real) x coordinate and the other the (real) y coordinate that collectively specified the position of this individual in the solution space. After the members of each generation have been scored by calculating the value of the function at the position defined by the values in the individual's genes, the population is sorted into descending order of score. The probability of an individual being a parent for the next generation is score based. Let A_n be the score for the nth individual and S be the total of these scores for all the population. Then the probability that the ith individual will be

used as a parent is $\dfrac{\dfrac{100}{\sqrt{(score(i)+1)}}}{\displaystyle\sum_{x=1}^{N} \dfrac{1}{\sqrt{(score(x)+1)}}}$ percent.

An offspring is produced by the following two steps.
- Crossover is applied. This consists of independently

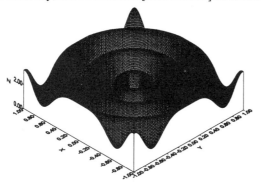

Figure 1a A plot of the function used.

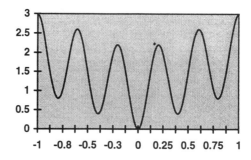

Figure 1b A cross section through the origin.

choosing each gene value in the offspring from the corresponding gene in a randomly chosen parent. An offspring could be identical to one of it's parents – in this case low levels of mutation were applied continuously until the offspring no longer matched either of it's parents.

- With a probability *mutation probability* the offspring are then mutated by adding random amounts in the range from 0 to *mutation radius* to each of the gene values.

Mutation is applied independently to each of the genes in the chromosome. As a result creep mutation tends to predominantly occur along the axes owing to the low probability of mutation occurring simultaneously to both genes. The mutation probability was limited to a maximum of 62.5%, as at mutation probabilities much higher than this the population was effectively being re-initialized each generation. Since small mutation in an individual could cause a large difference in performance, copying (at minimum) the best performing individual from each generation directly into the next generation was found to be essential. Without such copying, the most desirable gene pairs would have too little chance to influence the population before being lost as a result of mutation. Increasing the number of individuals copied had little effect until extreme numbers close to the population size were reached when exploration effectively ceased. Failure to preserve the best individual results in a predominantly random search.

4   RESULTS.

Each of the tests described below was repeated 500 times for each combination of mutation probability and radius in order to obtain a reasonable average performance estimate. An attempt was aborted and declared a failure if the optimum solution (0,0) had not been found after 5000 generations. Preliminary investigation showed that the number of attempts completed after 3000 generations was very low and that these tended to take tens of thousands of generations: an impractical time for realistic problems.

4.1 Constant mutation.

The first test used all combinations of a wide range of constant mutation probabilities and radii. The results are shown in figure two, with a more detailed map of the low probability and radii section of figure two in figure three. These figures adopt the common standard for all result graphs, with the success rate shown on the left in the 'a' graph and the average number of generations to solution (for those attempts that did find the global minimum) in

the 'b' graph on the right. The mutation probability is the percentage probability that mutation will occur to a gene value. The mutation radius is the maximum possible mutation expressed as a percentage of the search range, from -1 to 1. Typically at very low mutation radii and probabilities both the percentage correct and the average number of generations is very low. These correspond to the few cases in which the initial distribution is such that a fortunate crossover leads almost directly to the solution. Figure three shows some increase in the success percentage when the mutation probability and radius become large enough that the likelihood of bridging the local minima in one or two generations becomes significant. However at higher mutation radii, although bridging the local minima is easier, the size of the mutation steps makes it harder for the global minimum to be found, as evidenced by the increasing number of generations to solution. The net effect, however, is that the success rate with high mutation is poor.

4.2 Cyclic mutation probability, constant mutation radius.

Figure four shows the results of cyclically varying the mutation probability. The probability value used is the product of a base value and a boost factor. The boost factor is 1+(generation number mod 10). This is shown in figure six above. The mutation probability is thus increased every generation in equal steps from a base value to ten times the base value. The following generation it drops back from the extreme value to the base value again. The mutation probability axis in figure four shows the maximum value.

Performance improved slowly with increased mutation probability for all values of mutation radius. This improvement results from the higher likelihood of a series of small steps resulting in a net movement from one local minimum to the next. With a small mutation radius the number of consecutive steps required is large and the likelihood of this occurring is small. The success rate does increase slowly as the mutation radius is increased, as is to be expected. The improvement compared to the constant mutation case shown in figures 2 and 3 is marginal.

4.3 Constant mutation probability, cyclic mutation radius.

Figure five shows the results of cyclically varying the mutation radius. The radius value is increased in equal steps every generation from a base value to ten times the base value. The following generation it drops back from

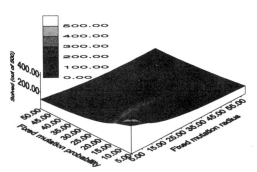

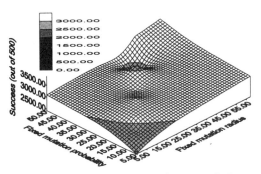

Figure 2a Number of solutions, mutation
probability and radius kept constant.

Figure 2b Average generations to solution,
mutation probability and radius kept constant.

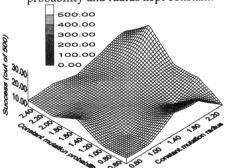

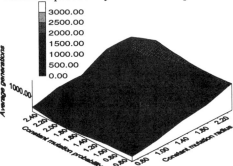

Figure 3a Number of solutions, mutation
probability and radius kept constant.

Figure 3b Average generations to solution,
mutation probability and radius kept constant.

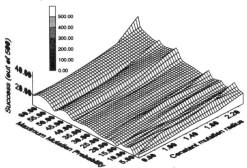

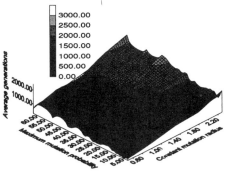

Figure 4a Number of solutions, cyclic mutation
probability but mutation radius kept constant.

Figure 4b Average generations to solution,
cyclic mutation probability but mutation radius
kept constant.

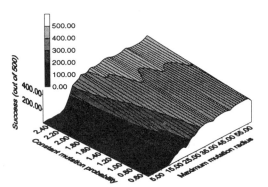

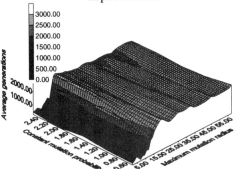

Figure 5a Number of solutions, constant mutation
probability but cyclic mutation radius.

Figure 5b Average generations to solution,
constant mutation probability, cyclic mutation
radius.

Boost factor = 1+ (Generation mod 10)

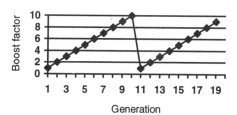

Figure 6. The boost factor when cycling mutation probability and / or radius.

the extreme value to the base value again as shown in figure six. The mutation radius axis in figure five shows the maximum value. Performance improved as soon as the radius used somewhere in the cycle was sufficient to allow direct movement from one minimum to another. This first occurred at a maximum radius of approximately 20% as expected. Note that the average generations to solution is highest at this point as there is a heavy reliance on - mutation almost equal to the maximum possible—a rare event. Movement by a sequence of steps is naturally possible but again with a low likelihood. Once the radius increases beyond this critical point, performance remains roughly constant with a slight improvement as the

mutation probability increases.

4.4 Cycling both mutation probability and radius.

Figure seven shows the results of cyclically varying both the mutation probability and the mutation radius in phase (both are boosted by the same factor between 1 and 10). Again the axes show the maximum values in the cycle. Note that performance is further enhanced when compared to cycling only probability or only radius both in terms of the number of successes achieved and also in terms of the number of generations taken. This is especially true for low mutation radii. This is to be expected, for at low mutation radii a fortuitous sequence of steps is required to move from one minimum to the next. The likelihood of this happening increases with the probability that mutation will occur each generation.

This is confirmed by the results in figure eight which shows the results of cyclically varying both the mutation probability and the mutation radius in anti-phase (the radius boost factor is 10 minus the probability boost factor). It will be noted that the performance at low mutation radii but high mutation probability has degraded significantly.

Note that performance in figure seven is good over a

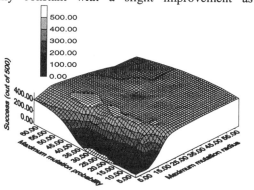

Figure 7a Number of solutions, both mutation probability and radius cycled in phase.

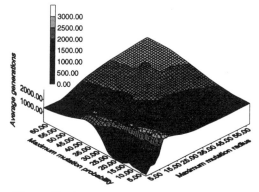

Figure 7b Average generations to solution, both mutation probability and radius cycled in phase.

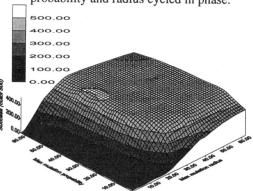

Figure 8a Number of solutions, both mutation probability and radius cycled out of phase.

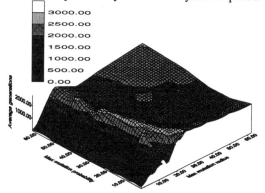

Figure 8b Average generations to solution, both mutation probability and radius cycled out of phase.

very wide range. The values required to solve the problem are not critical except that at least one of mutation probability and mutation radius must be above their problem dependent minimum value, which do not need to be accurately known.

5.   APPLICATION TO REAL LIFE PROBLEMS.

The work described in this paper concerns a single function specially designed so that creep mutation is of key importance in finding the optimum. The results obtained suggest that such an approach may be beneficial in the final stages of using an evolutionary algorithm to optimize many functions. A real life application will probably have more than two inputs and it will not be possible to simply plot the problem surface. Additionally, the local minima in the region of the global maximum are unlikely to be all the same size or regularly arranged. As a result it will be impossible to predict *a priori* the optimum combination of mutation probability and radius for the problem. Indeed the problem may not be solvable with any particular fixed combination owing to the variable terrain.

By using mutation probability and radius values which vary cyclically in phase a wide range of values occur and allow the different aspects of the problem to be addressed. While a simple variation is used in this paper, it is probably that any of a number of cyclic variations would be successful.

High levels of mutation are not beneficial during the part of the search when crossover is playing a major part. The benefits of combining desirable features from each parent are too likely to be lost by the extra change introduced by mutation. The low mutation part of the cycle enables progress to be made during the earlier part of the search. Using one strategy throughout relieves the user from having to specify the point at which to change strategies.

6. CONCLUSION.

Cyclic mutation variation together with the copying of elite members directly into the next population, as described in this paper, show promise that complex problems may be solved with smaller populations without requirement for *a priori* knowledge about the solution space. The values for the mutation control parameters are not critical.

If knowledge about the population space can be built up during evolution, for example by using a history record [P&H96], this can be used in addition to cyclic mutation variation.

7.   REFERENCES

[JDS et al 89]   J. David Schaffer, Richard A. Caruana, Larry J. Eshelman and Rajarshi Das, 'A Study of Control Parameters Affecting Online Performance of Genetic Algorithms for Function Optimisation', *Proceedings of the Third International Conference on Genetic Algorithms*, pp 51-60, California, June 1989.

[JP,TH95]   'Using a combination of Genetic Algorithm and Hill Climbing Techniques for the solution of a difficult letter recognition problem' John R Podlena and Tim Hendtlass. *Proceedings of the Eighth IEA/AEI International Conference*, pp 55-60, 1995.

[JP,TH96]   'Using the Baldwin effect to accelerate a genetic algorithm' John R Podlena and Tim Hendtlass. *Proceedings of the Ninth IEA/AEI International Conference AEI EAI-96* pp 253-258 1996.

[PJA95]   Peter J. Angeline, Chapter 11 in *'Computational Intelligence: A Dynamic Systems Perspective'* IEEE Press 1995.

[TB96]   'Evolutionary Algorithms in Theory and Practice' Thomas Back, Oxford University Press, 1996.

[TF89]   'Varying the probability of mutation in the genetic algorithm' Terence C Fogarty. *Proceedings of the Third International Conference on Genetic Algorithms.* pp 104-109 California, June 1989.

8.   ACKNOWLEDGMENTS.

It is a pleasure to acknowledge Howard Copland and John Podlena for their contributions during discussions about this work.

A GENETIC ALGORITHM FOR A HAMILTONIAN PATH PROBLEM

Eunice Ponce de León, Alberto Ochoa and Roberto Santana
Centro de Inteligencia Artificial
Instituto de Cibernética, Matemática y Física
Calle 15 #551, entre C y D Vedado
CP 10 400, Ciudad de la Habana, CUBA
eunice@cidet.icmf.inf.cu

ABSTRACT

The genetic algorithm (GA) is one of the stochastic search techniques with application to a wide variety of combinatorial optimization problems.

A conjecture of I. Hàvel, also attributed to P. Erdös, asserts that the simple graph G(k+1) (k>0), whose vertices are the subsets of cardinalities k and k+1 of the set {0,...,2k} and whose adjacency is given by subset inclusion, is Hamiltonian. The search of a Hamiltonian cycle in this graph is reduced to find a Hamiltonian path in the multi-level graph. In this paper we describe a GA approach to this conjecture. We present theoretical and computational results which show that this GA approach finds the optimal solutions when we search path for each level of the graph. We introduce an evolutive fitness function, and discuss the impact of using non standard crossover operators. Seeding and adaptive mutation rate are used. Hamiltonian cycles in G(6) and G(7) are constructed.

Key words: Genetic algorithms, NP-complete, Hamiltonian circuit, cycle or path, conjecture of I. Hàvel, graph theory.

INTRODUCTION

There is a kind of problems where GA has shown special results. This is the case of sequencing problems, the area of our interest. Sequencing problems are increasingly encountered in various applied sciences, including physics, biology, chemometrics, etceteras.

Sequencing problems, e.g. traveling salesman problems, routing problems, knapsack problems, graph coloring problem, Hamiltonian circuit problem, etceteras, are known to be NP- complete [DS89]. The number of all legal candidate solutions, ie the size of the search space, often grows non-polynomially (that is, faster than any polynomial of finite degree - often exponentially) with the number of constructional elements - the units with which candidate solutions are built. This kind of difficulty added to a multi-modal cost function in sequencing problems makes it useful to apply GA.

In this article we report on a GA for the Hamiltonian path problem in multi-level graphs. We present theoretical and computational results which show that this GA approach finds the optimal solutions when we search paths for each level of the graph. We use a standard representation but we introduce an evolutive fitness function, and discuss the impact of using non standard crossover operators. Seeding and adaptative mutation rate are used. Two Hamiltonian cycles are constructed with this method.

GA FOR HAMILTONIAN PATH PROBLEM IN MULTI-LEVEL GRAPHS

The Hamiltonian path problem (HPP) consists of finding a path through a directed graph that touches all nodes exactly once. Clearly, if a graph is fully connected this is an easy task. However, as edges are removed the problem becomes much more difficult, and the general problem is known as NP- complete.

For this article a family of multi-level graph was used. Today, in terms of its complexity and structure, it constitutes a serious challenge in finding a Hamiltonian path. This graph family is the result of attempts solve the conjecture of I. Havel, sometimes attributed to P. Erdös, which asserts that the simple graph G(k+1) (k > 0) whose vertices are the subsets of cardinalities k and k+1 of the set {0,...,2k} and whose adjacency is given by subset inclusion is Hamiltonian [DCQ88]. As you can see the size of the search space grows exponentially when k is increased.

Attempting to solve the Hamiltonian path problem directly with GAs rises many of same representation issues as in the case of traveling salesman problems (TSP) [DeJ85], [GGR⁺85]. They use the permutation representation. For example, in a 5-city TSP the string (b a c e d) means that city b is visited first, then city a and so on. However, the genetic operators used in the classical GA do not give valid offspring tours so new operators must be designed.

Crossover operators which have been suggested in the literature include [GL85], [OSH87], [WSF89], [SM90], [Sys91], [FM91] and [PC95]. In our approach we use standard representation using the advantage of the little proportion between edge and vertices that exists in the graph subjected to study.

MULTI - LEVEL GRAPHS (H_{K+1}) [Pon96]

Let $H_{k+1} = (V_h \, E_h)$ be a graph whose vertex set is V_h and whose edge set is E_h. Each vertex $v \in V_h$ is representing a circular vector of Hamming weight k+1 that has no endpoints[Pon96]. Let $|V_h|$ be a cardinal of set V_h, then $|V_h| = \binom{2k+1}{k} / (2k+1)$, which is a catalan number. This number represents all possible binary strings represented in circular form with exactly k coordinates in 0 and k+1 coordinates in 1.

For example: for k = 3, we have five vertices

The adjacency relation in H_{k+1} goes as follows:

Consider for example the vertex  for H_4. Select a coordinate with value 1. There is an adjacent vertex to 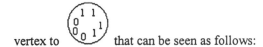 that can be seen as follows:

(1) write again 1 for the chosen coordinate

(2) write all the other values changed, for the remaining coordinates.

For example:

 is adjacent to

It will be defined the concept of the symmetrical vector of a circular vector.

Definition 1.- Let the symmetrical vector of a circular vector $(a_1,...a_n)$ be a resultant of doing a reading the circular vector counter clockwise and writing it clockwise.

That is $(a_n,...,a_1)$. Example, has as its symmetrical vector.

LEVELS OF H_{K+1} GRAPH

In [Pon96] we introduced a natural classification of the vertices of V_h, according to the number of unitary and zero blocks of the circular vectors that they represent.

Definition 2.- A block is the longest sequence of consecutive coordinates of a circular vector that has the same value. If all the coordinates are 1 the block is said to be a unitary block. The same happens with the zero blocks.

The circular vectors (respectively vertices) with the same number of unitary and zero blocks constitute one level of the graph H_{k+1}. There is a even number of blocks. We consider one level of graph H_{k+1} as a partial subgraph of graph H_{k+1}. We denote it $H_{k+1,b}$. where b is the level. When b=1, we have the first level and this represent that there is one unitary block and one zero block. When b=2 we are in the second level. There are two unitary blocks and two zero blocks, and so on. This multi-level graph is showed with an example of graph H_{k+1}, with k=4 as shown in Fig. 1.

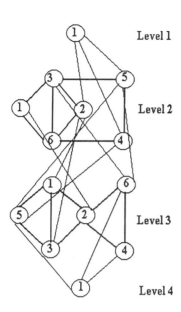

FIGURE 1 A multi-level graph.

In general, the graph H_{k+1} is a multigraph, but when we make the adjacency matrix, because we consider only the edges in our problem it is considered a simple graph. In this graph there are some vertices whose degree is two.

The GA works in each graph $H_{k+1,b}$ at all levels, in a separate way [POS96]. An algorithm that generates all vertices is given in [SP96].

GA COMPONENTS

Solution representation.- Candidate solutions are represented or encoded, as binary string or bitstrings. Each bitstring comprises M bit that mask the M edge in graph $H_{k+1,b}$: bit value 1 stands for "select the corresponding edge", whereas bit value 0 stands for "do not select the corresponding edge". This correspondence is given by adjacency matrix of graph $H_{k+1,b}$. If the graph have N vertices, then the bitstring have N bit put in 1.

An example of this encoding is the following 8-bit bitstring, when k=4, b=2, N=6.

(1 1 1 0 1 1 0 1)

This bitstring represent a subgraph of the graph

$H_{k+1,b}$. The size of the search space is $\binom{M}{N-1}$.

Objective function and fitness of a solution.- For calculating the objective function, first we count the degree of each vertex of the subgraph generated by the chromosome. This information is entered in the vector $W = (w_1,...,w_k)$.

Let $2 = (2,...,2)$ the vector whose coordinates are all 2, then

$D = |\ 2 - W\ |$ is the vector of difference coordinate to coordinate between 2 and W.

Thus, objective function F is defined as $F = 1000\ /\ D$.

When two coordinates of vector W are 1 and the rest zeros, F will reach its maximum value, that is 500. This function determine a necessary condition for the existence of the path, but this is not sufficient.

If two vertices have a degree of one and the rest a degree of two in the chromosome, it start to take on the form of a path. In this moment, the necessary condition can be only satisfied for those chromosomes that cycles and a subpath or no cycles and Hamiltonian path. We will 5allow a period between the beginning of the genetic algorithm and the appearance of this type of chromosome as the first evolutionary stage.

Once this quality among individuals is reached in the population it is then convenient to verify by the sufficient condition. This condition is given by a measure of the length of the path and of the cycles that are codified in the chromosome.

We suppose that we have p cycles and each one connects $t_1,...t_p$ vertices respectively and the subpath has the length L, then

$$F = 500 + t_1\ (t_1 - 1) + ... + t_p\ (t_p - 1) + L\ (L - 1),$$

as $t_1 + ... + t_p + L = k$, it can be seen that the maximum is reached for L = k.

A second evolutive stage starts when the first chromosome appears that fulfills the necessary condition. This stage is characterized by the both conditions checking simultaneously.

The evolutive objective function was introduced in natural way at the start of the GA, the graph generated by chromosomes are very complex. Many vertices could have a degree greater than two. This gave us the possibility of evaluating the first stage with the necessary condition and afterwards a second stage with sufficient condition. In this way the computational cost of the objective function is reduced.

Initial population: We tested two different initialization schemes. One is random and the other is seeding subpaths randomly in the initial population.

Restricted edge crossover operator.- This crossover follows from the fact that offsprings should preserve the edges that are in both its parents and they should not have those that are not present in either of its parents. On the sites where the values of both parents coincide the offsprings will keep this value. In the other case (When parents sites have different values) one of the offsprings is chosen randomly and it is assigned 1 or 0 values alternatively.

Example:

C1 = 1 0 0 1 0 0 1 1 1 0 1 0 1 1 0 1
C2 = 0 0 0 1 1 1 1 0 0 1 1 1 1 0 0 1

The total number of edges is 16, the path is of 10 vertices, therefore 9 edges are set to 1. H1 was chosen as the starting chromosome and given alternating. The underlined values are the positions where the values do not coincide.

Example:

H1 = 1 0 0 1 0 1 1 0 1 0 1 1 1 0 0 1
H2 = 0 0 0 1 1 0 1 1 0 1 1 0 1 1 0 1

It can be seen that this crossover guarantees that the restriction is adhered to, since both offspring have the same number sites for each value.

Mutation : Two randomly chosen chromosome sites with different values are exchanged. Adaptive mutation rate is used.

Selection operator.- In our GA each bitstring is chosen with a probability proportional to its fitness. Optionally, we also make use of heuristic selection. It is based in following property:

"For each vertex of the graph $H_{k+1,b}$, exists a symmetrical vertex. A vertex and his symmetrical vertex have the property of having the same neighboring [Pon96]. Departing of a Hamiltonian path can be obtained different other, substituting each vertex by his symmetrical".

Software.- All software needed for this study was programmed in the computer language C++. The GA was programmed using the software library GAL [ORP95], comprising domain-independent routines.

EXPERIMENTAL RESULTS

Figure 2 shows the crossover-mutation interaction. Varying the crossover probabilities from 0.1, 0.3, 0.5, 0.7 until 1, convergency curves are very close to each other. When the crossover probability is lower, the population is more homogeneous, mutation rate increases proportionally to the number of individual with the best fitness value. When the crossover probability is higher, happens contrariwise. In this way, the premature convergence is avoided. The values shown are for the average of 40 runs each with a different seed for random subpaths.

Utilizing a genetic algorithm with a adaptive probability of mutation, seeding in the initial population, standard selection and restricted edge crossover, two levels of 50 vertices were ordered. The Hamiltonian paths for the graph H_6 and H_7 were obtained ordering 42 and 132 vertices respectively.

Figure 3 shows the on-line convergence for different GA operators. We are searching a Hamiltonian path in the graph $H_{7,3}$. Initial population have 500 chromosomes and the number of generations is 250. The three start from an initial non random population, we used the seeding. We also used restricted edge crossover operator. The probability of mutation was variable.

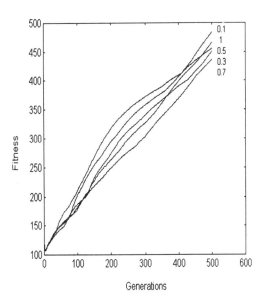

FIGURE 2 The crossover - mutation interaction

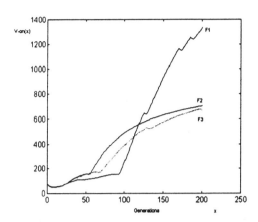

FIGURE 3 The on-line convergence for different GA operators

F1 : Use proportional selection. The optimum was reached and its value was 2960. F2 : The best fitness in the population was 1256. In this case we used heuristic selection. F3 : The best fitness in the population was 1040. Here standard and heuristic selection where alternated.

PRACTICAL IMPORTANCE

INTERCONNECTION TOPOLOGY OF PARALLEL COMPUTERS BASED IN HYPERCUBES

Nowadays, n-dimensional hypercubes are one of the most popular interconnection topologies of parallel computers. A parallel computer can be seen as a graph in which the processors correspond to the nodes and the communication links correspond to the edges.

One of the more important aspects concerning the parallel computers is how they can work in the presence of failures in the processors or in the links [BC91], [BCS92] and [BCH95]. Designing   alternative strategies that diminish the effect of the possible failures in the designed algorithms for use with parallel topology, is one direction in which much effort is being dedicated in this branch of computer technology.

The study of the characteristics of the n-dimensional hypercube, in our case, the regularities which it manifests, is of practicai importance. The problem of the existence of Hamiltonian paths in hypercubes with edge with failures is NP-complete.

The previous comments indicate some arguments that suggest the convenience of utilizing the GA to give solution to this type of problem. The solution

representation and the operators here introduced in the paper are conceived to search for Hamiltonian path in a certain levels of the hypercube.

FROM THE SEARCH OF THE PATHS TO THE SEARCH OF SUBSTRUCTURES IN THE GRAPHS

The objective function that was used could be also generalized for those problems in which we should attained the identification of other type of substructures over graphs. These substructures could be described in terms of a characteristic vector of the degree of the vertices. In these cases it would be necessary only to define the sufficient conditions, the ones that would be evaluated when the individuals attain a quality that justifies the calculus of the objective function.

CONCLUSIONS

The formulation of the problem of finding a Hamiltonian path in multi-level graphs is presented. The practical importance of this problem is explained. The way to extend this formulation to finding more complex structures or subgraphs is pointed out.

A new crossover operation and evolutive objective function are introduced.   One form of taking advantage of the heuristic knowledge of the problem was introduced in the seeding strategy in chromosomes in the initial population.

The experimental results reveal the efficacy of the GA paradigm for the important problem of finding a Hamiltonian path in very complex graphs.

REFERENCES

[BC91]   Bruck, J. and Cypher, R. (1991): Running   ascend-Descend algorithms on faulty hypercubes. *Research Report* September 9, 1991. I BM Research Division.

[BCS92]   Bruck, J.; Cypher, R. and Soroker, D. (1992): Tolerating faults in hypercubes   using   subcube partitioning. *IEEE transactions on computers*, Vol 41 No 5. May 1992.

[BCH95] Bruck, J.; Cypher, R. and Ho, Ch., T. (1995): On the construction of fault-tolerant cube-connected cycles networks. *Journal of parallel and distributed computing*, 25, pp.98-106.

[DeJ85] De Jong, K.A. (1985): Genetic Algorithms: a 10 year perspective. *Proc. First Int. Conf. Genetic Algorithms and their Applications.* Erlbaum.

[DS89] De Jong, K. A. and Spears, W.M. (1989): Using genetic algorithms to solve NP-complete problems. *Proc. Third Inter. Conf. on Genetic Algorithms and their Applications*, pp. 124-132. Kaufmann.

[DCQ88] Dejter, I. J.; Cordova, J. and Quintana, J. A. (1988): Two hamilton cycles in bipartite reflective kneser graphs. *Discrete Mathematics* 72, pp. 63-70.

[FM91] Fox, B.R. and McMahon, M.B. (1991): Genetic operators for sequencing problems. In *Foundations of Genetic Algorithms* (edited By G.J.E. Rawling)., pp. 284-300. Kaufmann.

[GL85] Goldberg, D. E. and Lingle, R. (1985): Alleles, loci and the traveling salesman problem. *Proc. First Int. Conf. Genetic Algorithms and their Applications*, pp.154-159. Erlbaum.

[GGR⁺85] Grefenstette, J.J; Gopal, R.; Rosmaita, B. and Van Gucht, D.(1985): Genetic algorithms for the traveling salesman problem. *Proc. First Int. Conf. Genetic Algorithms and their Applications*, pp. 160-168. Erlbaum.

[ORP95] Ochoa, A.; Rossete, A. and Ponce de León, E. (1995): Programación de un algoritmo genetico: un primer paso para la solución del problema. *Revista Ingeniería, Electrónica, Automática y Telecomunicaciones.* No. 2. Facultad Ingeniería Eléctrica. ISPJAE.

[OSH87] Oliver, I. M., Smith, D. J. and Holland, J. R. C. (1987): A study of permutation crossover operators on the traveling salesman problem. *Proc. Second Int. Conf. on Genetic Algorithms and their Applications*, pp 224-230. Erlbaum.

[Pon96] Ponce de León, E.(in press): La conjetura de Erdös y Havel: acciones de grupo, vectores circulares y grafos multiniveles. *Reporte de Investigación ICIMAF*, 1996.

[POS96] Ponce de León, E.; Ochoa, A. and Santana, R. (in press): La conjetura de Erdös sobre grafos Hamiltonianos: un experimento usando algoritmos geneticos. *Reporte de Investigación ICIMAF*, 1996.

[PC95] Poon, P.W. and Carter, J.N. (1995): Genetic algorithm crossover operators for ordering applications. *Computers Ops Res.* Vol 22, No.1, pp. 135-147.

[SP96] Santana, R. and Ponce de León, E. (1996): Algoritmos geneticos para un problema de busqueda de caminos Hamiltonianos. *Tesis de diploma.* Facultad de Matemática. Universidad de la Habana.

[SM90] Shahookar, K. and Majumder, P. (1990): A genetic approach to standard cell placement using meta-genetic parameter optimization. *IEEE Trans. Comput. Aided Design* 9 (5) may, pp. 500-512.

[Sys91]   Syswerda, G. (1991): Schedule optimization using genetic algorithms. In *Handbook of Genetic Algorithms* (edited by L. Davis), pp. 332-349. Van Nostrand Reinhold, Amsterdam.

[WSF89]   Whitley, D., Starkweather, T. and Fuquay, D. (1989): Scheduling problems and the traveling salesman: The genetic edge recombination operator. *Proc. Third Int. Conf. on Genetic Algorithms and their Applications*, pp. 133-140. Kaufmann.

GA-HARD OR GA-EASY: A CRITICISM TO EPISTASIS ANALYSIS

Ting Kuo
Department of International Trade
Takming Junior College of Commerce
No. 56, Huan Shan Road Sec. 1, Ney Hwu, Taipei, Taiwan
Email: tkuo@mailsv.tmjcc.edu.tw

Tel: (8862) 6585801-311

ABSTRACT

It is necessary to consider what makes problems easy or hard for genetic algorithms. Davidor [Dav91] has examined this question by applying Epistasis Analysis. In this paper, we first give a review of Epistasis Analysis. Then, we apply Epistasis Analysis to certain functions and construct five functions with differing degrees of difficulty for genetic algorithms, some GA-easy and some GA-hard. Finally, experimental results and conclusion are described. Unfortunately, Epistasis Analysis can not accurately estimate the GA-hardness of a function.

Keywords: Genetic Algorithms, GA-hard, GA-easy, Epistasis Analysis.

1 INTRODUCTION

Genetic algorithms have been applied in many diverse areas, such as function optimization [Jon75], the traveling salesman problem [GGRG85], scheduling [CS89], neural network design [HSG89], vision [BLM91], control [Kar91], and machine learning [Jon88]. Goldberg's book [Gol89] provides a detailed review of these applications. The fundamental theory behind genetic algorithms was presented in Holland's pioneering book [Hol75]. GAs are population-based search techniques that maintain populations of potential solutions during searches. A potential solution usually is represented by a string with a fixed bit-length. In order to evaluate each potential solution, GAs need a payoff (or objective) function that assigns a scalar payoff to any particular solution. Once the representation scheme and evaluation function are determined, a GA can start searching. Initially, often at random, GAs create a certain number (called the population size) of strings to form the first generation. Next, the payoff function is used to evaluate each solution in this first generation. Better solutions obtain higher payoffs. Then, on the basis of these evaluations, some genetic operators are employed to generate the next generation. The procedures of evaluation and generation are iteratively performed until the optimal solution(s) is (are) found or the time alloted for computation ends.

2 AN OVERVIEW OF GENETIC ALGORITHMS

Genetic algorithms are an adaptive search technique based on the principles of population genetics. The metaphor underlying genetic algorithms is that of natural evolution. John Holland is the founder of the field of genetic algorithms. The appearance of his book *Adaptation in Natural and Artificial Systems* in 1975 produced both the formalism and much of the theory for the field's subsequent expansion and acceptance. He and his students have had the most influence on this field over the past 20 years. With the advantages of great robustness and problem independence, genetic algorithms have proven to be a promising technique for machine learning, search, and optimization.

To gain a general understanding of genetic algorithms, it is useful to examine its components. Before a GA can be run, it must have the five components:

1. A chromosomal representation of solutions to the problem.

2. A function that evaluates the performances of solutions.

3. A population of initialized solutions.

4. Genetic operators that can used to evolve the population.

5. Parameters that specify the probabilities by which these genetic operators are applied.

These components are described below.

2.1   Representation

Usually, only two components of a GA are problem dependent: the representation and evaluation functions. Representation is a key genetic algorithm issue because genetic algorithms directly manipulate coded representations of problems. In principle, any character set and coding scheme can be used. However, binary character set is preferred because it yield the largest number of schemata for any given parameter resolution, thereby enhancing the implicit parallelism of genetic searches. Note that, in most GAs, the individuals are represented by fixed-length binary strings that express a schema as a pattern defined over alphabet $\{0, 1, *\}$, and describe a set of binary strings in the search space. Thus, each string belongs to 2^L schemata (L is the length of binary string).

2.2   Evaluation Function

Along with the representation scheme, the evaluation function is problem dependent. GAs are search techniques based on feedback received from their exploration of solutions. The judge of the GA's exploration is called an *evaluation function*. The notion of evaluation and fitness are sometimes used interchangeably. However, it is important to distinguish between the evaluation function and the fitness function. While evaluation functions provide a measure of an individual's performance, fitness functions provide a measure of an individual's reproduction opportunities. In fact, evaluation of an individual is independent of other individuals, while an individual's fitness is always dependent on other individuals.

2.3   Initial Population

Choosing an appropriate population size for a genetic algorithm is a necessary but difficult task for all GA users. On the one hand, if the population size is too small, the genetic algorithm will converge too quickly to find the optimum. On the other hand, if the population size is too large, the computation cost may be prohibitive. The initial population for a genetic algorithm is usually chosen at random.

2.4   Operators

Most research has focussed on the three primary operators: selection, crossover, and mutation. While selection according to fitness is an exploitative resource, the crossover and mutation operators are exploratory resources. The GA combines the exploitation of past results with the exploration of new areas of the search space. The effectiveness of a GA depends on an appropriate mix of exploration and exploitation. The three primary genetic operators are described below.

1. *Selection* (or *Reproduction*): The population of the next generation is first formed by using a probabilistic reproduction process, of which there are two general types: generational reproduction and steady-state reproduction. Generational reproduction replaces the entire population with a new one in each generation. By contrast, steady-state reproduction [WK88, Sys89] replaces only a few individuals in each generation. The resulting population, sometimes called the intermediate population, is then processed using crossover and mutation to form the next generation.

2. *Crossover*: A crossover operator manipulates a pair of individuals (called parents) to produce two new individuals (called offspring) by exchanging segments from the parents' coding. By exchanging information between two parents, the crossover operator provides a powerful exploration capability. A commonly used method for crossover is called one-point crossover. Assume that the individuals are represented as binary strings. In one-point crossover, a point, called the crossover point, is chosen at random and the segments to the right of this point are exchanged. For example, let x1=101010 and x2=010100, and suppose that the crossover point is between bits 4 and 5 (where the bits are numbered from left to right starting at 1). Then the children are y1=101000 and y2=010110.

3. *Mutation*: By modifying one or more of an existing individual's gene values, mutation creates new individuals to increase variety in the population. The mutation operator ensures that the probability of reaching any point in the search space is never zero.

2.5   Parameters

Running a genetic algorithm entails setting a number of parameter values. However, finding good settings that work well on one's problem is not a trivial task. Grefenstette [Gre86] used a *metalevel* GA to determine robust parameters settings for population size, operator probabilities, and evaluation normalization techniques. He described the genetic algorithm space in the following six parameters: population size, crossover rate, mutation rate, generation gap, scaling window, and selection strategy.

3   A REVIEW OF EPISTASIS ANALYSIS

It is well known that genetic algorithms typically deal with problems that are nonlinear. Here, nonlinear means it is not possible to treat each gene as an independent variable that can be solved in isolation from other genes. Geneticist call the interaction between genes as *epistasis*. Epistasis, derived from the Greek words *epis* and *stasis* (behind and stand), means stoppage or masking. There are two extremes of behavior of gene interaction [Raw91]:

- There is *Zero epistasis*: In this case every gene is independent of every other gene. That is, there is a fixed ordering of fitness (contribution to the overall objective value) of the alleles of each gene.

- There is *Maximum epistasis*: In this case no proper subset of genes is independent of any other gene. That is, there is *no* possible fixed ordering of fitness of the alleles of *any* gene.

A well-fitted GA problem should exhibit some, but not a very high degree of, epistasis. If genes in a chromosome have high epistasis, GAs will not be effective. Vose and Liepins [VL91] indicated that it is always possible to represent any problem with little or no epistasis. However, designing such a representation may be a hard work. Beasley, Bull, and Martin [BBM93] proposed an *expansive coding* for combinatorial problems. Despite enlarging the search space, this coding scheme did reduce the epistasis effectively. Davidor [Dav91] suggested a simple statistic, *epistasis variance*, to measure the degree of epistasis in a representation. He also indicated that present-day GAs are only suitable for solving problems that with medium epistasis. The following notation are adopted from Davidor's study. Without loss of generality, we may assume that a binary representation with length L is used. The function value of a string S is denoted by $v(S)$, and the average function value of all strings (2^L) is denoted by $\bar{V}$. First, the excess value of a string is defined as

$$E(S) = v(S) - \bar{V}, \qquad (1)$$

and the average allele value is denoted by

$$A_i(a) = \frac{1}{N_i(a)} \sum v(S). \qquad (2)$$

Where $a \in 0, 1$, and $N_i(a)$ is the number of all strings with a allele a in bit position i.

Then, the excess allele value can be described as

$$E_i(a) = A_i(a) - \bar{V}, \qquad (3)$$

and the excess genic value as

$$E(A) = \sum_{i=1}^{L} E_i(a). \qquad (4)$$

The genic value of a string S, i.e., the predicted string value, is defined as

$$A(S) = E(A) + \bar{V}, \qquad (5)$$

Thus, the difference $\epsilon(S) = v(S) - A(S)$ supposed to be a measure of the eepistasis of a string S. The epistasis variance is defined as

$$\sigma_\epsilon^2 = \frac{1}{N} \sum (v(S) - A(S))^2. \qquad (6)$$

To compute the epistasis variance, we can use the following two statistics: the fitness variance

$$\sigma_v^2 = \frac{1}{N} \sum (E(S))^2, \qquad (7)$$

and the genic variance

$$\sigma_A^2 = \frac{1}{N} \sum (E(A))^2. \qquad (8)$$

Finally, the epistasis variance (EV) can be derived from the following equation:

$$EV = \sigma_v^2 - \sigma_A^2. \qquad (9)$$

Davidor analyzed three epistatically different functions: zero epistasis, semi-epistasis, and total epistasis. Table 1 presents these three functions. Table 2 shows the results of the computation of epistasis variances.

4   EPISTASIS ANALYSIS GA-HARD AND GA-EASY FUNCTIONS

In the jargon of genetic algorithms, a function is called GA-easy or GA-hard depending on whether the genetic algorithms can find the optimum (optima) of the function. There are several approaches to studying whether a function is GA-easy or GA-hard. The most widely known approach is to study the deceptiveness of the function. Based on earlier work by Bethke [Bet80], Goldberg [GS87] introduced the concept of "deception". Although there is no well accepted definition of the term "deception", it is generally agreed that a function is deceptive if low-order schemata guide the search away from global optima. Formally, deceptive functions are defined in terms of schema partition. A *schema partition* of order-n involves a set of schemata having the same n significant

Table 1: The Zero, Semi-, and Total Epistasis Functions

S	Zero epistasis			Semi-epistasis			Total epistasis		
	$v(S)$	$E(S)$	$E(A)$	$v(S)$	$E(S)$	$E(A)$	$v(S)$	$E(S)$	$E(A)$
000	0	-3.5	-3.5	0	-3.5	-7	0	-3.5	-10.5
001	1	-2.5	-2.5	0.5	-3	-3	0	-3.5	-3.5
010	2	-1.5	-1.5	1	-2.5	-2.5	0	-3.5	-3.5
011	3	-0.5	-0.5	1.5	-2	1.5	0	-3.5	3.5
100	4	0.5	0.5	2	-1.5	-1.5	0	-3.5	-3.5
101	5	1.5	1.5	2.5	-1	2.5	0	-3.5	3.5
110	6	2.5	2.5	3.0	-0.5	3	0	-3.5	3.5
111	7	3.5	3.5	17.5	14	7	28	24.5	10.5

Table 2: Epistasis Variances of Zero, Semi-, and Total Epistasis Functions

Zero epistasis			Semi-epistasis			Total epistasis		
σ_v^2	σ_A^2	EV	σ_v^2	σ_A^2	EV	σ_v^2	σ_A^2	EV
5.25	5.25	0	28.875	16.625	12.25	85.75	36.75	49

bit locations, but in which the combination of n bits are different for each schemata. For example, 0*0*, 0*1*, 1*0*, and 1*1* are the set of schemata in a schema partition of order-2. The "global winner" of a schema partition is the schema that has the highest fitness value, where the fitness value of a schema is the average fitness of all strings contained in that schema. A schema partition is deceptive if the global winner of that schema partition is the schema containing the deceptive optimum. A function is order-n deceptive if all schema partitions of less than order-n are deceptive.

Deceptive functions are useful in analysis, testing, and design of GAs because they are of bounded difficulty (order-n). GAs that solve deceptive functions should also be able to solve other problems of up to the same level of difficulty. We will focus on functions of unitation: functions for which the function value of a string depends only on the total number of 1's in that string and not on the positions of those 1's. As opposed to the 2^l values in a general function over l bits, only $l + 1$ different function values are needed to consider. Such simplification makes functions of unitation particularly attractive for use as test functions and for theoretical work.

4.1 Symmetric Bipolar Function and Trap Function

A bipolar function is defined as a function that has two global optima that are maximally far away from each other and a number of deceptive attractors that are maximally far away from the global optima. Here the distance is measured in Hamming space. A symmetric bipolar function of unitation is a function that has two global optima of unitation $ut = 0$ and $ut = l$ (l, an even integer number, is the length of the bit string), respectively, a number of deceptive attractors of unitation $ut = l/2$, and function values that are symmetrical with respect to unitation $ut = l/2$. Since there are $\binom{l}{l/2}$ strings of unitation $u = l/2$, the total number of deceptive attractors is $\binom{l}{l/2}$. In our study, the test case was a six-bit symmetric bipolar deceptive function of unitation. This function was constructed by satisfying the sufficient conditions for a bipolar deceptive function [DHG92]. A bipolar deceptive function is a function where every schema partition is deceptive. In a bipolar function, a schema partition is deceptive if the schema (or schemata) containing the maximum number of deceptive attractors is (are) no worse than other competing schemata.

Like the bipolar function, trap functions are functions of unitation and deceptive functions too [DG91]. Ackley [Ack87] showed that a two-peak trap function was very hard to optimize. The test case we chose was an 8-bit trap function defined as follows:

$$f(ut) = \begin{cases} 160/7 * (7 - ut) & \text{if } ut < 7 \\ 200/1 * (ut - 7) & \text{otherwise,} \end{cases} \quad (10)$$

where ut is the number of 1's in a string. The global maximum is at $ut = 8$ and has a value of 200, but there is a local maximum at $ut = 0$, with a value of 160.

Table 3 presents these two functions. Where $order(x)$ refer to the number of 1's in x. These two

Table 3: Symmetric Bipolar Function (SBF) and Trap Function (TF)

$order(x)$	SBF(x)	TF(x)
0	1	160
1	0.2	960/7
2	0.6	800/7
3	0.9	640/7
4	0.6	480/7
5	0.2	320/7
6	1	160/7
7		0
8		200

functions are both deceptive and hard for GAs to optimize.

4.2 Non-Deceptive but GA-Hard Function

We choose a class of problems that are "easy" in the sense of being non-deceptive but which are, in fact, hard for conventional GAs to optimize [Gre93].

Let f be defined as

$$f(x) = \begin{cases} 2^{L+1} & \text{if } x = 0 \\ x^2 & \text{otherwise,} \end{cases} \quad (11)$$

where x is an L-bit binary string representing the interval [0, 1]. Clearly, for any schema H such that the optimum is in H, the average fitness of H is higher than all other schemata that do not cover the optimal solution. Because they pose no deception at any order of schema partition, functions such as Eq. (11) are often called "GA-easy" [Wil91]. However, the optimum of this function will probably never be found by a GA unless by a lucky crossover or mutation. This is because the schemata that contain the optimum have function values that vary widely, so the observed average fitnesses of the schemata do not reflect their true average fitnesses. In other words, large sampling errors are inevitable. Grefenstette [Gre93] called this a type of "needle-in-a-haystack" function (NIAH) because the global optimum of the function is isolated from the relatively good areas of the search space.

4.3 GA-Hard and GA-Easy Functions

To further examine the relation between the epistasis variance of a function and GA-hardness, we constructed five functions with differing degrees of difficulty for GAs. Table 4 presents these five functions and their epistasis variances. Where, $order(x)$ refers to the number of 1 in the x.

5 EXPERIMENTAL RESULTS AND CONCLUSION

It is well known that genetic algorithms typically deal with problems that are nonlinear. A well-fitted GA problem should exhibit some, but not a very high degree of, epistasis. If genes in a chromosome have high epistasis, GAs will not be effective. For the needle-in-a-haystack function stated above, we calculate its epistasis variance for the case $L = 10$. The epistasis variance is 4052.6778. This value explain why it is a nearly GA-impossible function. While the symmetric bipolar function and the trap function are both deceptive, their epistasis variances are quite different. The epistasis variances of the symmetric bipolar function and the trap function are 0.06214, and 187.190 respectively. This is because the function values associated with the trap function are bigger (200 times) than those for symmetric bipolar function. Furthermore, epistasis variance is not normalized, making comparisons between problems difficult. So, after normalizing the function values of the trap function by a factor of 200, we recompute the epistasis variance of the trap function. The epistasis variance is 0.004679 ($= 187.190/200^2$) now. We also normalize the needle-in-a-haysatck function such that the optimal solution is 1 and recompute the epistasis variance. The new value of epistasis variance is 0.00096578. Furthermore, if we normalize the needle-in-a-haysatck function such that the optimal solution is 10 and recompute its epistasis variance, the new value of epistasis variance is 0.096578, even less than the epistasis variance of F2 (0.174826). According to Davidor's view, function F2 should be GA-harder than the "needle-in-a-haystack" function.

Since the behavior of genetic algorithms is stochastic, their performance usually varies from run to run. Consequently, we replicated twenty runs of these functions for the following GA parameter settings: $p_c = 0.35$ and $p_m = 0.01$. Here p_c and p_m repre-

Table 4: The Epistasis Variances of Five Functions

$order(x)$	$F1(x)$	$F2(x)$	$F3(x)$	$F4(x)$	$F5(x)$
0	10	10	0	9	1
1	0	3.3332204	1	8	1
2	3.333333	1.1109606	2	7	1
3	2.222222	0.3702072	3	6	1
4	2.5925926	0.1232895	4	5	1
5	2.4691358	0.0409836	5	4	1
6	2.5102881	0.0135483	6	3	1
7	2.4965706	0.0044032	7	2	1
8	2.5011431	0.00135483	8	1	1
9	2.499619	0.0003387	9	0	1
10	2.500127	0	10	10.	10
EV	0.1575417	0.174826	0.0	0.1169	0.07825

sent the crossover and mutation rates, respectively. In all cases a population size of 20 was used. The performance of a single run was taken to be the best individual in the population at the end (i.e., after 100 generations) of that run. Table 5 summarizes the experimental results. For function $F1$, only 4 runs found the optimal solution. For function $F2$, all 20 runs found the optimal solution. These results show that function $F2$ is easy for GAs to optimize, and function $F1$ is GA-harder than function $F2$. Similarly, we replicated twenty runs of functions $F3$, $F4$, and $F5$. All parameter settings were same as stated above. For function $F3$, all 20 runs found the optimal solution. For function $F4$, all 20 runs failed to find the optimal solution. These results confirm that function $F3$ is easy for GAs to optimize and that function $F4$ is hard for GAs to optimize.

For function $F5$, the landscape is a flat area of the search space in which the whole set of points, except for the optimal point, have a value of 1. On such a landscape, it is not possible, by making local comparisons, to determine which direction is the best way to move. Consequently, selection does not provide any benefit since no information can be used. Using crossover and mutation operators, GAs may have a chance to find the optimal solution. There were four lucky runs in which the optimal solution was found.

The data presented in Table 4 imply that function $F3$ will be GA-easy and function $F2$ will be relatively GA-harder than the other four functions. However, experiment results show that this conjecture is partially correct: function $F3$ is a GA-easy function as we guessed, but function $F2$ is accurately a GA-easy function despite functions $F1$, $F4$, and $F5$ being GA-hard. Thus, epistasis variance can not accurately estimate the GA-hardness of a function. Furthermore, epistasis variance is not normalized, making compar-

isons between problems difficult. Moreover, the epistasis variance is dependent on the function values.

References

[Ack87]   D. H. Ackley. *A Connectionist Machine for Genetic Hillclimbing*. Kluwer Academic Publishers, Boston, MA, 1987.

[BBM93]   D. Beasley, D. R. Bull, and R. R. Martin. An overview of genetic algorithms: Part 1, fundamentals. *University Computing*, 15(2):58–69, 1993.

[Bet80]   A. D. Bethke. *Genetic Algorithms as Function Optimizers*. PhD thesis, Univ. of Michigan, 1980.

[BLM91]   B. Bhanu, S. Lee, and J. Ming. Self-optimizing image segmentation system using a genetic algorithm. In R. K. Belew and L. B. Booker, editors, *Proceedings of the Fourth International Conference on Genetic Algorithms and Their Applications*, pages 362–369, San Mateo, CA, July 1991. Morgan Kaufmann.

[CS89]   G. A. Cleveland and S. F. Smith. Using genetic algorithms to schedule flow shop releases. In J. D. Schaffer, editor, *Proceedings of the Third International Conference on Genetic Algorithms and Their Applications*, pages 160–169, San Mateo, CA, June 1989. Morgan Kaufmann.

[Dav91]   Y. Davidor. Epistasis variance: A viewpoint on ga-hardness. In G. J. E. Rawlins, editor, *Foundations of Genetic Algorithms*, pages 23–35. Morgan Kaufmann, San Mateo, CA, 1991.

Table 5: Number of Successful Runs out of 20 Runs

F1	F2	F3	F4	F5
4	20	20	0	4

[DG91] K. Deb and D. E. Goldberg. Analyzing deception in trap functions. Technical Report IlliGAL Report No. 91009, University of Illinois at Urbana-Champaign, 1991.

[DHG92] K. Deb, J. Horn, and D. E. Goldberg. Multimodal deceptive functions. Technical Report IlliGAL Report No. 92003, University of Illinois at Urbana-Champaign, 1992.

[GGRG85] J. J. Grefenstette, R. Gopal, B. J. Rosmaita, and D. V. Gucht. Genetic algorithms for the traveling salesman problem. In J. J. Grefenstette, editor, *Proceedings of the First International Conference on Genetic Algorithms and Their Applications*, pages 160–168, Hillsdale, NJ, July 1985. Lawrence Erlbaum Associates.

[Gol89] D. E. Goldberg. *Genetic Algorithms in Search, Optimiztion and Machine Learning*. Addison-Wesley, Reading, MA, 1989.

[Gre86] J. J. Grefenstette. Optimization of control parameters for genetic algorithms. *IEEE Transactions on System, Man, and Cybernetics*, SMC-16(1):122–128, 1986.

[Gre93] J. J. Grefensette. Deception considered harmful. In L. D. Whitley, editor, *Foundations of Genetic Algorithms, Volume 2*, pages 75–91. Morgan Kaufmann, San Mateo, CA, 1993.

[GS87] D. E. Goldberg and P. Segrest. Finite markov chain analysis of genetic algorithms. In J. J. Grefenstette, editor, *Proceedings of the Second International Conference on Genetic Algorithms and Their Applications*, pages 1–8, Hillsdale, NJ, July 1987. Lawrence Erlbaum Associates.

[Hol75] J. H. Holland. *Adaptation in Natural and Artificial Systems*. The University of Michigan Press, Ann Arbor, MI, 1975.

[HSG89] S. A. Harp, T. Samad, and A. Guha. Towards the genetic synthesis of neural networks. In J. D. Schaffer, editor, *Proceedings of the Third International Conference on Genetic Algorithms and Their Applications*, pages 360–369, San Mateo, CA, June 1989. Morgan Kaufmann.

[Jon75] K. A. De Jong. *An Analysis of the Behavior of a Class of Genetic Adaptive Systems*. PhD thesis, Univ. of Michigan, 1975.

[Jon88] K. A. De Jong. Learning with genetic algorithms: an overview. *Machine Learning*, 3(2/3):121–138, 1988.

[Kar91] C. L. Karr. Design of an adaptive fuzzy logic controller using a genetic algorithm. In R. K. Belew and L. B. Booker, editors, *Proceedings of the Fourth International Conference on Genetic Algorithms and Their Applications*, pages 450–457, San Mateo, CA, July 1991. Morgan Kaufmann.

[Raw91] G. J. E. Rawlins. Introduction. In G. J. E. Rawlins, editor, *Foundations of Genetic Algorithms*, pages 1–10. Morgan Kaufmann, San Mateo, CA, 1991.

[Sys89] G. Syswerda. Uniform crossover in genetic algorithms. In J. D. Schaffer, editor, *Proceedings of the Third International Conference on Genetic Algorithms and Their Applications*, pages 2–9, San Mateo, CA, June 1989. Morgan Kaufmann.

[VL91] M. D. Vose and G. E. Liepins. Schema disruption. In R. K. Belew and L. B. Booker, editors, *Proceedings of the Fourth International Conference on Genetic Algorithms and Their Applications*, pages 237–242, San Mateo, CA, July 1991. Morgan Kaufmann.

[Wil91] S. W. Wilson. Ga-easy does not imply steepest-ascent optimizable. In R. K. Belew and L. B. Booker, editors, *Proceedings of the Fourth International Conference on Genetic Algorithms and Their Applications*, pages 85–89, San Mateo, CA, July 1991. Morgan Kaufmann.

[WK88]   D. Whitley and J. Kauth. Genitor: A different genetic algorithm. In *Proc. of Rocky Mountain Conference on Artificial Intelligence*, pages 118–130, 1988.

CHARACTERIZATION OF SEEDS BY A FUZZY CLUSTERING ALGORITHM

Younes Chtioui[a], Dominique Bertrand[a], and Dominique Barba[b]

[a] Institut National de la Recherche Agronomique, Laboratoire de Technologie Appliquée à la Nutrition, BP 1627, Rue de la Géraudière, F-44316 Nantes Cedex 03, France.
[b] Institut de Recherche et d'Enseignement Supérieur aux Techniques de l'Electronique, Laboratoire des Systèmes Electroniques et Informatiques, La Chantrerie, C.P. 3003, 44087 Nantes Cedex 03, France.
Email: chtioui@nantes.inra.fr

ABSTRACT

Artificial vision was applied to the discrimination between 4 seed species (2 cultivated and 2 adventitious seed species). The main scope of this investigation was to study the strengths and the limitations of a fuzzy clustering technique, which was the fuzzy c-means algorithm (FCMA). Since the performances of FCMA depend on the initial configuration of the cluster centers, this study reports on a method for carrying out fuzzy classification with a non-random initialization of the cluster centers. A set of 58 size, shape, and texture features were extracted from color images in order to characterize each seed. The data was reduced by applying a principal component analysis and selecting the first 10 components. The first 10 principal components represented 93.16 percent of the total sum of squares. FCMA was applied, with the Euclidean distance, for the classification of learning and test data. For the present work, simulations showed that the classification performances depended on some parameters of FCMA. It was observed that there is no need to apply FCMA with a very high number of iterations. The best classification performances were 96.85 percent and 97.89 percent of the training and the test sets, respectively. FCMA may therefore be used in an on-line device for seed discrimination. However, further improvements are to be performed such as the use of a more sophisticated distance measure instead of the Euclidean distance.

Key words: artificial vision, fuzzy c-means clustering, color image analysis, seed.

INTRODUCTION

A pattern recognition system may use a supervised or an unsupervised learning approach. The unsupervised learning (also called cluster analysis) approach consists of partitioning data into a number of subgroups, such that the whole patterns of each subgroup have a certain degree of similarity. The clustering process plays an important role in the performance of any pattern recognition system. There are two major approaches to unsupervised classification: hierarchical and non-hierarchical [Spa80]. Hierarchical methods perform, in an iterative way, a partition of the data by merging (or splitting) patterns until the fixed number of clusters is obtained. The hierarchical methods suffer from the defect that erroneous decisions cannot be corrected. Once 2 patterns have been merged, they cannot be separated. Non-hierarchical methods are based on the minimization of a specified objective function, such as the within variance of each cluster. They use heuristic models to define an optimal partition of the data. Non-hierarchical methods are the most widely used methods for cluster analysis. This is due to their simplicity, and computational efficiency. The c-means algorithm is probably the most well-known method for non-hierarchical clustering. It is based on the minimization of the sum of the squared Euclidean distance between the patterns and the cluster centers. However, the c-means algorithm exclusively attributes each pattern to a single class. This « hard clustering » does not take into account the fact that the boundaries between the real clusters may be fuzzy. Therefore, the Fuzzy c-means algorithm (FCMA) has been developed [Bez80]. Here, each pattern is no longer attributed to a single class, but to all the available classes. The class to which a pattern is most likely to belong is then given a high membership value.

The major difficulty encountered during fuzzy clustering by FCMA is the definition of the location of the initial cluster centers.

Seed discrimination consists of identifying the species of each seed. Seed batches cannot be commercialized before passing a control test. Today, seed discrimination is still a manual task in many seed stations around the world. Visual seed identification is tedious and time consuming. There have been attempts for automated seed identification by near-infrared spectroscopy [Che95], and artificial vision [Pet92]. Near-infrared spectroscopy is a destructive method. However, artificial vision is a non-destructive method, flexible, and the analyses may give real-time results. Artificial vision makes it possible to discriminate between seed species according to their physical appearance. In seed discrimination by artificial vision, supervised classification techniques have usually been used, such as linear discriminant analysis [Zay90] and neural networks [Cht96]. The main limitation of any supervised classification method is that it requires the knowledge of the labels of the learning patterns. However, the labeled learning patterns can be unavailable such as in the case where the foreign materials which are present in a seed batch have never been previously observed. In this situation, the labels of the foreign materials must be found without supervised learning. In the present paper, we explored the potential of FCMA for the identification of 4 seed species. The usual practice is to apply FCMA with a few random initial configurations of cluster centers, and select the configuration which gives the best performances. However, the computational requirements of this approach can be prohibitive. In the present work, we investigated the use of a novel initialization method which avoids the need for « guessing » the initial location of the cluster centers. This novel initialization method was first introduced for vector quantizer design [Lin80]. We therefore report the improvement of FCMA by introducing a deterministic initialization in the place of a random initialization.

FUZZY CLUSTERING THEORY

Suppose we have n patterns in a p-dimensional space (x_1, x_2, ..., and x_n), and assume these patterns are to be clustered into c classes. In fuzzy clustering, W denotes the cluster membership matrix, $W=[w_{ij}]$, where w_{ij} represents the degree of membership of the pattern x_i with respect to the class j, ($0 < w_{ij} < 1$ and $\sum_{j=1}^{c} w_{ij} = 1$). In FCMA, we assess the degrees of membership of a pattern from its distances to the currently available cluster centers. Each element w_{ij} is a function of the

inverse of the distance of the pattern x_i to the class j. All the cluster centers are gathered in a c x p matrix noted Z, and each element of Z is noted z_{lj} ($l=1,2,...,c$ and $j=1,2,...,p$). The distance between the pattern x_i and the cluster center z_j is noted $d(x_i,z_j)$. In the following, k refers to the iteration number. The parameter ε, which is called the « tolerance parameter », is a scalar related to the ending criterion of the algorithm. FCMA proceeds as follows:

step 1:
Set $k = 1$, and select the value of the tolerance parameter $\varepsilon > 0$. Choose initial cluster centers $Z(k)$ (The way for choosing initial cluster centers is developed later in the text).

Step 2:
Compute the matrix of membership grades $W(k)$ using the formula

$$ w_{ij}(k) = \frac{\left[\dfrac{1}{d^2(x_i, z_j)} \right]^{\frac{1}{m-1}}}{\sum_{l=1}^{c} \left[\dfrac{1}{d^2(x_i, z_l)} \right]^{\frac{1}{m-1}}} $$

$$ \forall\, i \in \{1,2,...,n\}, \text{ and } \forall\, j \in \{1,2,...,c\} $$

if $\exists\ i \in \{1,2,...,n\}$ and $j \in \{1,2,...,c\}$ such that $d(x_i,z_j) = 0$, then $w_{ij}(k)=1$ and $w_{il}(k)=0 \ \forall\ l \neq j$
where m is a parameter which represents the fuzziness ($m > 1$). When m is equal to 1, a hard clustering is therefore on hand.

Step 3:
Compute the matrix of cluster centers $Z(k)$

$$ z_j(k+1) = \frac{\sum_{i=1}^{n} w_{ij}(k)^m x_i}{\sum_{i=1}^{n} w_{ij}(k)^m} $$

Step 4:
if $\left\| Z(k) - Z(k+1) \right\| < \varepsilon$ stop, otherwise set $k = k+1$ and go to step 2.

$\left\| Z(k) - Z(k+1) \right\|$ refers to any matrix norm and is defined by

$$ \left(\sum_{j=1}^{c} \sum_{l=1}^{p} \left| z_{jl}(k) - z_{jl}(k-1) \right|^{v} \right)^{\frac{1}{v}}, \ v = 1, 2, \infty $$

If $v = \infty$, the matrix norm simplifies to

$$\max_{j,l}\left|z_{jl}(k) - z_{jl}(k-1)\right|$$

When FCMA is used in order to classify learning and test data, the algorithm is first applied on the learning data. The membership grades of each test pattern are therefore assessed, as in step 2, by using the cluster centers of the learning data.

The choice of the initial cluster centers has a crucial importance on the performances of FCMA. Some authors have applied FCMA with a random initialization, but the clustering is highly dependent on the location of the initial cluster centers. Other authors proposed to apply the algorithm many times on the basis of different random initializations. In this case, among the several produced solutions, the one yielding the best performances is selected. The main drawback of this approach is that it is very time consuming. In this paper, we investigate a novel initialization method which is called « initial guess by splitting » [Lin80], and which proceeds as follows,

1)

Compute the average (z_i) and the standard deviation (σ_i) of the whole data set.

2)

« Split » each cluster center z_i into 2 close cluster centers $z_i-\phi$ and $z_i+\phi$, where ϕ is a fixed perturbation vector. Each component of ϕ may be a ratio of the standard deviation of z_i (i.e. $\forall\ j\in\{1,2,...,p\}$, $\phi_j = \dfrac{\sigma i_j}{\lambda}$, where λ is a constant).

3)

Calculate a new partition of the data set according to steps 2, 3, and 4 of FCMA.

4)

Go to step 2 until the number of clusters is equal to the number of initial clusters (c in FCMA).
It can be noticed that if the number of initial clusters (c) is odd, only the necessary number of clusters which present the highest standard deviations are split at the last iteration.

MATERIALS AND METHODS

Sample collection

Four seed species were obtained from the national seed testing station of France (Station Nationale d'Essais de Semences, Beaucouzé, 49071, France). The collection included 2 cultivated seed species which were lucerne (*Medicago sativa* L.) and vescue (*Vicia sativa* L.), and 2 adventitious seed species which were wild oat (mixture of 3 wild oat varieties: *Avena fatua* L., *Avena pubescens* L., and *Avena sterilis* L.) and rumex (mixture of 3 rumex varieties: *Rumex crispus* L., *Rumex longifolius* L., and *Rumex obtusifolius* L.). Each seed species was represented by a set of 588 seeds.

Image acquisition

The image acquisition system was formed by a high resolution RGB 768 (H) x 512 (V) color camera (KY-F55B, JVC Corp., Japan) linked to a frame grabber (VP1300-768-E-AT, Imaging Technology Inc., Bedford, USA). The camera unit included a C-mount adapter to support a Nikkor 50 mm lens (Nikkor AF, Nikon, Japan). The lighting was provided by two 18-watt neon lamps operated at 12 V DC which provided a color temperature of 950 K (OSRAM L 18w/12-950, LUMILUX de Luxe, Tageslicht-Daylight, Germany). The lamps were placed at each side of the camera. The seeds were placed individually on a white background, and in random positions and orientations under the camera. The distance from the camera lens to the seed surface was constantly maintained at 190 mm. This made it possible to acquire images of size 4 cm x 6 cm. The images were therefore acquired with a spatial resolution of 7.81 x 10^{-2} mm/pixel. Each pixel was represented by 3 bytes of data allowing up to 2^{24} colors to be represented. The number of seeds in a single image depended on the seed species. A single image of lucerne or rumex contained around 80 seeds, whilst an image of vescue or wild oat included about 20 seeds. A set of 42 color images were therefore acquired for the 4 seed species.

Data measurement

Each pixel in the color image has three components. Each component refers to the gray level value, of the corresponding pixel, in a particular color channel (red, or green, or blue channels). All the acquired color images were processed in order to achieve feature measurement. The processing of the images consisted successively in their filtering, binarization, and labeling. A set of 25 size and shape features were extracted from the binary images in order to characterize the morphology of the seeds (Table 1).

Table 1: Size and shape features used for seed characterization. All these features were measured from the binary images.

Feature	Interpretation
Area	The number of pixels in the region of a seed.
Perimeter	The number of pixels in the boundary of a seed.
Length	Maximum length of the seed through the centroid.
Width	
Thinness ratio (or circularity)	The degree of circularity of a seed.
Elongation	Ratio of the width to the length of a seed.
First ten magnitude Fourier descriptors	The low-frequency Fourier components describe the general shape properties of the contour, and the high-frequency ones describe the finer detail.
Seven invariant moments	Describes the shape of the seed.
Eccentricity	Describes how elongated the seed is.
Spread	Describes the spread of the seed.

Moreover, textural features were extracted because the texture, which represents the gray level distribution within the regions of the seeds, may play an important role for the discrimination between different seed species. Two kinds of methods were used for texture analysis : the local histogram method and the co-occurrence matrix method [Har73]. The local histogram method allowed the extraction of features such as the mean and the standard deviation of the gray level. The co-occurrence matrix method made it possible to characterize the distribution of the gray level within the region of each seed. Each texture feature had 3 components which were extracted from each color channel. Eleven texture features were extracted (Table 2). A set of 58 (33 + 25) morphological and textural features were therefore extracted to characterize each seed. All the measured features were invariant to the orientations and the positions of the seeds.

The whole data were gathered in a matrix of size 2352 rows (representing seeds) and 58 columns (representing features). This matrix of measured data was randomly partitioned into 2 matrices: the matrix of learning data which was of size 1592 rows x 58 columns (398 learning seeds for each seed species) and a matrix of test data of size 760 rows x 58 columns (190 test seeds for each seed species).

Table 2: Texture features used for seed characterization. Each of these features is a 3-dimensional vector.

	Feature
Local histogram features	mean gray level
	variance
	energy
	entropy
	kurtosis
	skewness
Grey level co-occurrence matrix features	Energy (or Angular Second Moment)
	Contrast
	Correlation
	entropy
	inverse different moment

Data reduction

The measured features were correlated. It was necessary to reduce the redundancy by extracting (or selecting) the key features which were effective in discriminating between pattern classes. Principal component analysis (PCA), which is a well-known method in multivariate analysis, achieves a dimensionality reduction of a data set of correlated variables. This is achieved by creating new un-correlated variables, called « principal components », while maintaining as much as possible the diversity present in the original data set. These principal components are ranked so that the first few components describe the largest part of the variation of the original data set [Jol86]. PCA was applied here on the learning data as active observations, and on the test data as supplementary ones.

Application of fuzzy c-means algorithm

The scope of the present study was to analyze the strengths and the limitations of FCMA for the discrimination of seeds. The performances of this unsupervised classification technique were studied pertaining to the values of different parameters, such as the number of iterations and the tolerance parameter ε. In previous studies, the fuzziness parameter m was usually set to the value 2.0. In this investigation, the fuzziness parameter m was varied from 1.0 to 9.0 in order to analyze its importance upon the clustering performances. The distance between a pattern and a cluster center was the Euclidean distance. The main advantage of the Euclidean distance is that it is very easily assessed. The parameter ν in the matrix norm assessment was set to 2. This matrix norm is called the « Frobenius » norm. The parameter λ defined in the initialization method was set to 10.

In the present work, the « de-fuzzification » step was based on the « maximum degree of membership ». This means that each pattern was finally attributed to the class which gave the maximum membership grade.

RESULTS AND DISCUSSION

Seed appearance

FCMA constructs decision frontiers on the basis of the physical characteristics of the seeds. The higher the difference between the seed species to be discriminated, the easier the identification by artificial vision. Seeds of lucerne are small, and have a dark yellowish color (Figure 1). Some seeds of lucerne however are almost brown in color. Seeds of vescue are circular and dark brown. Wild oat seeds, which are adventitious seeds, present a large variation both in size and color. There are small and light yellow wild oat seeds, as well as large and dark brown ones. Rumex seeds are quite homogeneous in both their size and color (brown). They have approximately the same size and shape as the seeds of lucerne. Thus, it might be anticipated that a perfect classification would not be possible. In the blue channel, most seeds appeared very dark, even the light-colored seeds of wild oat.

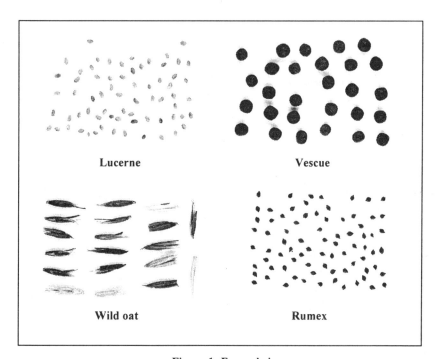

Lucerne

Vescue

Wild oat

Rumex

Figure 1: Example images.

PCA was applied in order to reduce the redundancy between the measured features. In the present work, the first 10 components were selected because they represented a large proportion of the total sum of squares (93.16 percent). PCA made it possible to achieve a selection of features that are combinations of the initial measurements, but this operation did not allow the number of measured features to be reduced. Nevertheless, the search for a classifier, in a low-dimensional space, was greatly simplified.

Clustering analysis

The sample collection was selected in order to include seeds which must « naturally » create 4 subgroups representing each seed species. The relevance of FCMA was estimated by its ability to cluster the sample collection into these 4 qualitative groups.

Since there is no theoretical method for choosing the optimal value of the fuzziness parameter m, its value was varied in order to study its effect on the classification performances of both the training and the test sets. This parameter was varied from 1.0 to 9.0. It was found that about 60 percent of the seeds were misclassified when this parameter had a value less than 1.5.

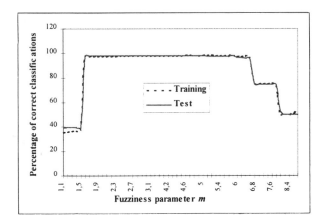

Figure 2: Evolution of the classification performances as a function of the fuzziness parameter m.

The best performances were obtained when m ranged from 1.5 to 6.4. Then, the performances decreased for values of m greater than 6.4 (Figure 2). These results show that the value of the fuzziness parameter m had a great effect on the clustering performances. The optimal value of m is dependent on the nature of the data to be clustered. The higher the value of m, the fuzzier the clustering. In the present study, a high fuzziness was not recommended. As FCMA gave good clustering performances when the value of m was 2.0, this parameter was set at 2.0 in all of the following experiments.

The performances of FCMA were then analyzed as a function of the value of the tolerance parameter ε. This parameter, which describes the accuracy of the location of the cluster centers, was varied from 10^{-1} to 10^{-10}. The percentage of correct classifications of both the training and the test sets were assessed for each value of ε. When ε was set to 0.1, only 60 percent and 73 percent of the seeds were correctly classified for the training and the test sets, respectively. Furthermore, it was observed that the performances were independent of ε when it was greater than 0.02 (Figure 3).

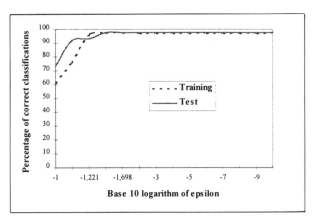

Fgure 3 : Evolution of the classification performances as a function of the base 10 logarithm of the tolerance parameter ε.

The number of iterations of FCMA, which were necessary to reach the convergence, was assessed as a function of different values of the tolerance parameter ε (Figure 4). FCMA needed only 11 iterations to converge when ε was equal to 10^{-4}, and as many as 32 iterations when ε was equal to 10^{-10}. As the number of iterations increased, the computation time of FCMA obviously increased. For this particular problem, it was not worthwhile to apply FCMA with a very small value of ε.

Figure 4 : The number of iterations of FCMA as a function of the base 10 logarithm of the tolerance parameter ε.

The confusion table, which represents the actual seed species (rows) and the predicted seed species (columns), is displayed when m and ε were set to 2.0 and 10^{-4}, respectively (Table 3).

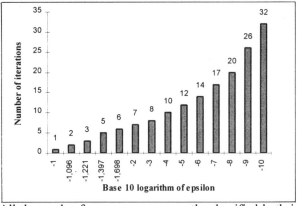

All the seeds of vescue were correctly classified both in the training and the test sets. The main confusions were between seeds of lucerne and rumex. This is due to the fact that these 2 seed species have close morphological features, and some seeds of lucerne have the same color as seeds of rumex. The classification results were 96.85 percent for the training set and 97.89 percent for the test set.

Table 3: Classification results obtained for the training and the test sets. The parameters *m* and ε were respectively set to 2.0 and 1e-4.

		Predicted species							
		Training set				Test set			
		Lucerne	Vescue	Wild oat	Rumex	Lucerne	Vescue	Wild oat	Rumex
Actual species	Lucerne	353			45	176			14
	Vescue		398				190		
	Wild oat	1	4	393		1		189	
	Rumex				398	1			189

Fuzzy clustering by FCMA made it possible to identify seeds with a high recognition rates. This means that the fuzzy description of the affiliation of each pattern, to all the available classes, is efficient. However, further improvements are to be performed such the use of other distance measures. The Euclidean distance makes the assumption that the shapes of the clusters are hyperspheroidal. This assumption is not usually met, because clusters may present other shapes, such as hyperellipsoidal shapes. Furthermore, the Euclidean distance does not take into account the presence of variable cluster densities, or the possibility of unequal numbers of patterns in each cluster. Therefore, other distances should be used, such as the Mahalanobis, or the « exponential » distance [Gat89]. Moreover, the clustering performances may be improved by introducing an additional class which corresponds to the class of the rejected seeds. During the de-fuzzification phase, the pattern which belongs to all the available classes with approximately the same membership grades should be attributed to the class of the rejected seeds. Such improvements are under study in our laboratory.

CONCLUSION

Fuzzy clustering is useful for the description of the fuzziness of the frontiers between the subgroups which are to be clustered. It was concluded that FCMA performances depend greatly on the value of the fuzziness parameter *m*. When this parameter was set to 1.4, only 63.70 percent and 61.40 percent of the learning and the test seeds were correctly classified, respectively. On the contrary, 96.85 percent of the learning set and 97.89 percent of the test set were correctly identified when *m* was set to 2.0. The present investigation showed that artificial vision has great potential for the design of an automatic seed discrimination device. The performances of recognition of seeds are, admittedly, dependent on the classification method, but they also depend on the measured features. In this application, quantitative features were extracted to characterize each seed. In future work, we intend to measure qualitative features which allow the description of the structural information of the seeds. Structural information may be of great use especially when seeds present close physical appearances.

ACKNOWLEDGEMENTS

This work was supported by a grant from *Institut National de la Recherche Agronomique* and *Région des Pays de la Loire* (France). The authors take pleasure in acknowledging the help of Yvette Dattée, Didier Demilly, and Maria Mannino.

REFERENCES

[Bez80] J. C. Bezdek. A convergence theorem for the fuzzy ISODATA clustering algorithms. *IEEE Transactions on Pattern Analysis and achine Intelligence*, vol. PAMI-2, no. 1, 1980.

[Che95] Y. R. Chen, S. R. Delwiche, and W. R. Hruschka. Classification of hard red wheat by feedforward backpropagation neural etworks. *Cereal Chemistry*, vol. 72, no. 3, pp. 17-319, 1995.

[Cht96] Y. Chtioui, D. Bertrand, Y. Dattée, and M. F. Devaux. Identification of seeds by colour imaging. Comparison of discriminant analysis and artificial neural network. *Journal of the Science of Food and Agriculture*, vol. 71, pp. 433-441, 1996.

[Gat89] I. Gath. and A. B. Geva. Unsupervised optimal fuzzy clustering. *IEEE Transactions n Pattern Analysis and Machine Intelligence*, ol. 11, no. 7, pp. 773-781, 1989.

[Har73] R. M. Haralick, K. Shanmugan, I. Dinstein. Texture features for image classification. *IEEE Transactions on Systems, Man, and Cybernatics*. 3(6), pp. 610-621, 1973.

[Jol86] I. T. Jolliffe. *Principal component analysis*. Springer-Verlag, New-York, 1986.

[Lin80] Y. Linde, A. Buzo, and R. M. Gray. An algorithm for vector quantizer design. *IEEE Transactions on Communications*, vol. COM-28, pp. 84-95, 1980.

[Pet92] P. E. H. Petersen and G. W. Krutz. Automatic identification of weed seeds by colour machine vision. *Seed Science and Technology*. vol. 20, pp. 193-208, 1992.

[Spa80] H. Spath. *Cluster analysis algorithms*. Ellis Horwood, Chichester, 1980.

[Zay90] I. Zayas, H. Converse, and J. Steele. Discrimination of whole from broken corn kernels with image analysis. *Transactions of he ASAE*, vol. 33, no. 5, pp. 1642-1646, 1990.

EFFICIENT AND ACCURATE ALGORITHMS IN MACHINE VISION INSPECTION

Tie Qi Chen[1] and Yi Lu Murphey[2]

[1]Department of Physics
Fudan University, Shanghai 200433
The People's Republic of China
tqchen@umich.edu

[2]Department of Electrical and Computer Engineering
The University of Michigan-Dearborn
Dearborn, MI 48128-1491
yilu@umich.edu
Phone: 313-593-5420, Fax: 313-593-9967

ABSTRACT

Automatic visual inspection is one of the primary applications of computer vision. Machine vision inspection requires efficient processing time and accurate results. In order to achieve efficiency and accuracy, it is important to explore domain-specific knowledge and apply the knowledge to the development of highly efficient and accurate machine vision algorithms. In this paper, we present three effective algorithms in machine vision inspection.

Keywords: machine vision, machine inspection, industry automation, image binarization, tilt detection and correction

1. INTRODUCTION

Visual inspection has broad applications in industry automation and covers the full range of technical difficulty in computer vision[BKD95, DaJ88, SNW95]. Because of its diverse application environment, it has been recognized that there is no pervasive generic solution in machine vision, each application requires a careful study of alternatives and perhaps even the invention of a new technique[Mun88, DMW94] in order to satisfy the practical requirements.

In this paper, we describe three key algorithms developed for inspecting Vacuum Florescent (VF) Displaying boards. VF boards are widely used in the automobile industry to display information about vehicle's status. Figure 1 shows an image of a VF board.

The displays are illuminated by circuit boards specially designed to disclose specific information including speed, mileage, fuel level, compass heading, and etc. In general, a VF board should correctly display a number of different functions. Each function can be specified by a set of on-segments and a set of off-segments. Once the circuit board is complete, the entire board must be tested for its proper functions. VF inspection is to test various displaying patterns of each VS board on the production. VF displays are mounted directly onto the circuit board using pin through hole and surface mounting. The functional test forces the display to show pre-specified patterns which are currently verified by the test operator. The functional test lasts about 4 to 8 seconds per circuit board and in that time an operator is expected to inspect at least two display patterns of VF board.

In [LuT96], we described a system for inspecting Vacuum Florescent (VF) Displaying boards. The system is responsible for detecting the following types of VF faults:

- *Missing segment* — When a voltage is applied to a segment's pin, that segment fails to light.
- *Dim segment* — When an ON-segment is dimmer than the other ON-segments. It is caused by lack of sufficient florescent material or an external voltage drop.
- *Segment void* — This is a black or dim spot in an ON-segment. This is caused by missing internal florescent material to display.

Figure 1 A VF display of compass/temperature/trip board.

• *Extra Segment* — When a test pattern is turned on, an extra segment that does not belong to the test pattern is also turned on.

The system we developed consists of two major programs, off-line learning and on-line testing. The off-line learning program is used to generate test features for every type of new VF board, and the on-line testing program is part of the integrated VF board test system running on the manufacturing line. The manufacturing environment imposes the following requirement:

• The functional test should be extremely fast. Since the test speed is directly related to the productivity of the manufacturing, efficiency of the machine vision algorithms is our major concern.

• The machine vision algorithms must be robust to lighting changes. The lighting on the production floor will change constantly throughout the day. Objects such as cabinets and boxes are moved around causing shadows and reflections in images. These variations change the intensity values of the pixels on the image interfering with the outcome of the test. This is the most challenging barrier for this project and also one of the most crucial aspects of the test.

• The test program must be variant in scale, tilt and location. The camera can be moved to different positions for different types of VF boards. The focus and the aperture setting of the camera may change in day-to-day operation. Furthermore, the test board can be positioned differently on the production line. Therefore, the machine vision inspection must be invariant to scale, tilt and translation.

Based on the discussion above, the following three algorithms are essential to this machine vision inspection problem: image binarization, image tilt detection and segmentation, and bounding box calculation. The following sections describe the three algorithms we developed for this particular application.

2. VF BOARD IMAGE BINARIZATION

Image binarization is often an important process in many machine vision system, since a large number of high level computer vision algorithms are developed based on binary images. Image binarization is often accomplished through the thresholding of a gray scale image. The accuracy of the thresholds directly impacts the resulting binary image. Therefore, image binarization is to correctly compute the threshold points. In this application we need to find the threshold point that correctly separates the gray scale values of the on segments from the others.

The binarization threshold is found by analyzing the histogram of the image. For a typical image of a VF board, the number of the pixels belonging to the ON segments is small in comparison with the total pixel number of the image pixels. Therefore in the histogram of a typical VF image, the highest peak should appears near 0 which represent the background and OFF segments. Let us denote this peak as P_L. The bright pixels belonging to the ON segments result in a peak in the higher intensity area. Let us denote it as P_H. Because the dark pixels usually have different intensities (or pixel values) and the pixels which form the segments at OFF state are usually brighter than the pixels belonging to the background, the distribution near P_L is not uniformed. Let us denote the peaks around P_L as $\{P_{L'}\}$. Similarly, for the bright pixels, some of them are brighter than the others, which result in a number of peaks around P_H. Let us denote them as $\{P_{H'}\}$. As defined above, P_L is the highest peak among $\{P_{L'}\}$ and P_H is the highest peak among $\{P_{H'}\}$. In order to separate the pixels belonging to the foreground from the background, the binarization threshold must locate in a valley between $\{P_{L'}\}$ and $\{P_{H'}\}$. However, the border between $\{P_{L'}\}$ and $\{P_{H'}\}$ are hard to determine. Figure 1 shows two typical histograms of VF images. Based on these considerations, we developed an algorithm to compute an effective threshold point. The algorithm consists of the following steps:

[Step-1] Generate the histogram.

[Step-2] Smooth the histogram using an averaging filter. In most cases, the peaks around P_L and the peaks around P_H will not be removed.

[Step-3] Find the highest peak P_L. Suppose P_L locates at X_L.

[Step-4] Define a function VMIN(X): for any location $X \in (X_L, 255)$, $V_{MIN}(X)$ equals to the minimum value of the histogram within $(X_L, X]$.

[Step-5] Search from X_L to 255, find the location corresponding to the maximum of $[X/V_{MIN}(X)]$. We denote it as X_H. We consider X_H the location of P_H because this peak has the largest peak-to-valley ratio to P_L. By finding the peak according to $[X/V_{MIN}(X)]$, we can skip the peaks around P_L which are usually higher than P_H.

[Step-6] Find the location where the histogram equals to $V_{MIN}(X_H)$. We denote it as X_t. X_t is the binarization threshold we will use. The

reason is that it is the location of the deepest valley between P_L and P_H and should be the most suitable border between $\{P_{L'}\}$ and $\{P_{H'}\}$. But why don't we find the deepest valley directly? Because the valleys outside (X_L, X_H) are usually deeper.

3. ALGORITHM FOR DETECTING TILTING ANGLE

As we described in the introduction section, VF boards are placed directly onto the fixture using pin through hole and surface mounting during the test. The boards are not always placed properly therefore the VF images be tilted. It is important to detect the tilting angle of a test VF board image so the test features learnt during the learning process can be correctly mapped to this VF board.

The algorithm we developed for detecting the tilting angle first extracts the ON pixels belonging to the top edge of the object. For example, the object is a tilted quadrilateral in Figure 3(a). The top edge of this object should be Side A. Often in VF board images, the ON pixels belonging to Side A and the ON pixels belonging to Side B are not easy to separate, since —

• the object edges are not sharp enough in the image due to noise;

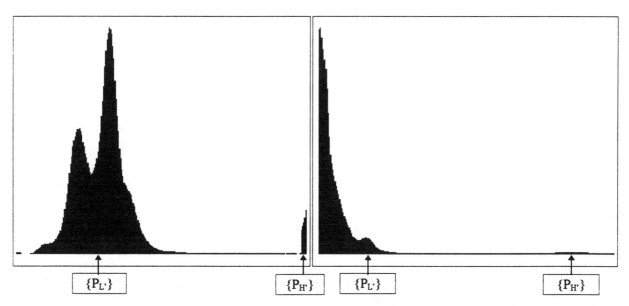

(a) The histogram with several significant peaks ∈ $\{P_{L'}\}$ existing in the left side of P_L.

(b) The histogram with a significant peak ∈ $\{P_{L'}\}$ existing between P_L and P_H.

Figure 2 Two types of histograms. Please note that small peaks are invisible in this figure.

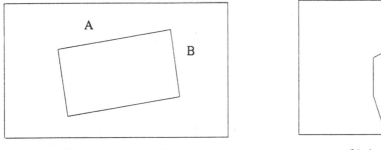

(a) A simple example.   (b) A complicated example.

Figure 3 Some examples of the tilting objects.

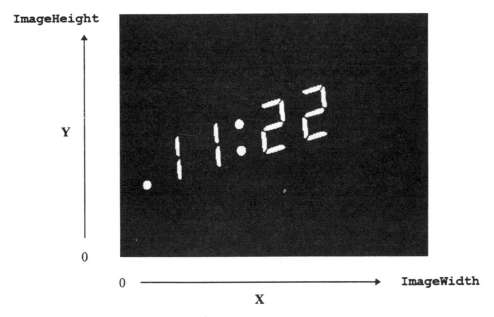

Figure 4 A typical image of a VFD board.

- the angle between Side A and Side B can be an obtuse angle close to 180°;
- Side B may be longer than Side A or Side A may be broken; etc..

A more complicated example is shown in Figure 3(b) in which the top edge is not a straight line. Figure 4 shows a typical image of a VFD board, in which the top edge of the object has unconnected short line segments

Here we introduce an effective algorithm for finding the tilting angle of a VF image. Let us imagine a rigid rod falling down from the top of the image. Initially the rod is hanging horizontally at the top of the image. No matter what a tilting angle it may have later, the length of the projection on the **X**-axis is equal to the image width, i.e. **ImageWidth**. Assume we let the left end (X = 0) fall down first. The rod will stop falling when it hits the first ON pixel. Suppose the location of this pixel is $(X_0=0, Y_0)$ and the rod is currently perpendicular to the X-axis. Let (X_0, Y_0) be the fulcrum and the rod rotate clockwise around this fulcrum until the rod hits another ON pixel. Suppose the location of this pixel is (X_1, Y_1). If $X_1 <$ **ImageWidth**$-1-X_1$, the mass center of the rod is at the right side of X_1. In this situation the rod is unstable, we take (X_1, Y_1) as the new fulcrum and continue to rotate the rod until we find the next ON pixel. This step repeats until we find an ON pixel (X, Y) such that $X \geq$ **ImageWidth**$-1-X$, namely, the mass center of the rod is between X_0 and X. In this situation the rod is stable and will not rotate further, and therefore the rod and the image have the same tilting angle.

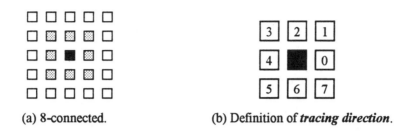

(a) 8-connected.   (b) Definition of *tracing direction*.

Figure 5 Border tracing.

This algorithm is fast and accurate and works on a broad range of tilting angle.

4. ALGORITHM FOR LOCATING SEGMENTS

In our VF board inspection, the test functions of each type of VF board are specified by ON and off segments. The location of a segment is specified by its bounding box. Once the bounding boxes of all the segments are obtained, we can compute a number of useful features including the occupancy ratio, the average pixel value, the standard deviation. These features will be used later on to detect the VF board defects. The bounding boxes of all the segments obtained during the learning program are then mapped to test images to detect defects.

One approach to find bounding boxes of segments is using connected component technique. Finding connected components in an image is a classic problem which has been intensively studied in the community of image processing and computer vision. We evaluated a number of connected component algorithms before we decided to develop our own algorithm to meet this requirement of this application. Rosenfeld and Pfaltz first introduced an algorithm that makes only 2 passes through the image and a large global table for recording equivalences. In 1981, Haralick proposed an iterative algorithm[]. This algorithm uses no auxiliary storage to produce the labeled image from the binary image, therefore it would be useful in environments whose storage is severely limited[HaS83]. However the number of scan times through the image is determined by the complexity of the connected components in this image. Lumia, Shapiro and Zuniga[LSZ83] developed a space-efficient two-pass algorithm that uses a local equivalence table.

All these algorithms are time consuming. Here we present a much more efficient algorithm to compute the connected components. This algorithm uses the technique of tracing the border of a complex polygon.,

therefore, not all the ON pixels are scanned and labeled belonging to a specific class, but the border of each class is found therefore the bounding box can be determined. In this algorithm we assume 8-connected *adjacency*(see Figure 4), so a ±45°-tilted line will not be considered as separate pixels, and furthermore, while tracing the border of an arbitrary polygon, less border pixels need to be covered. The tracing directions for each pixel is defined in Figure 5.

This algorithm does not use any auxiliary storage. The second definition is *tracing direction*. Each pixel has 8 neighboring pixels and each neighboring pixel represents a tracing direction. The 8 tracing directions which are used in our algorithm are defined in Figure 5. As mentioned above, some segments may be formed by connecting several components with thin lines, which means somewhere the width of a segment may be only 1 pixel. In this situation, some pixels must be scanned more than once while tracing the border. Therefore the tracing direction as well as the starting pixel of the border tracing must be used to determine when the border tracing should stop.

Our algorithm first scan the binary image from top to bottom to look for a border pixel. A border pixel is a ON pixel which has BLACK pixels in its neighborhood. Suppose the current scanned pixel P_s is (x_s, y_s). If P_s is a border pixel outside all the bounding boxes of the segments already found. Denote P_s as P_0, let $x_0 = x_s$ and $y_0 = y_s$, choose P_0 as the starting pixel of a new border tracing. We scan the neighboring pixels of P_0 counterclockwise until we find a BLACK pixel following by another border pixel P_{-1}, extract the tracing direction d_0 from P_{-1} to P_0 according to Figure 5. We initialize the new bounding box by letting $x_{min} = x_{max} = x_0$ and $y_{min} = y_{max} = y_0$, store the current border pixel P_c i.e. (x_c, y_c) and the current tracing direction d_c by letting $x_c = x_0$, $y_c = y_0$ and $d_c = d_0$. Label the segment as P_c. The next border pixel (x_n, y_n) and the next tracing direction d_n is searched as follows. Let $d_c = [(d_c+4) \textbf{ MOD } 8]$ which is the opposite direction of d_c (see Figure 5), scan the neighboring pixels of P_c clockwise from $[(d_c+1)$ **MOD** 8] until we find another border pixel P_n i.e. (x_n, y_n), extract the tracing direction d_n from P_c to P_n according to Figure 5. If $x_n = x_0$ and $y_n = y_0$ and $d_n = d_0$, then accelerate the scan by moving the searching point to the right side of the current bounding box, which can be done simply by setting $x_s = x_{max}$. Otherwise, we update the new bounding box by setting $x_{min} = x_n$ if $x_n < x_{min}$; $x_{max} = x_n$ if $x_n > x_{max}$; $y_{min} = y_n$ if $y_n < y_{min}$; and $y_{max} = y_n$ if $y_n > y_{max}$; store the current border pixel P_c i.e. (x_c, y_c) and the current tracing direction d_c by setting $x_c = x_n$, $y_c = y_n$ and $d_c = d_n$. Thses steps of operation repeat until we have traced all border points. If we also want to label the pixels inside the borders. Scan the binary image from top to bottom. Assign each unlabeled WHITE pixel the label of its labeled neighboring pixel. Because all the border pixels have been labeled, scanning the image one time is enough.

This algorithm guarantees to keep the interior of the segment at the right side of the tracer during the border tracing. According to the principle of topology, the border of any complicated polygon can be traced using this method as long as the starting pixel of the border tracing is at the outer border of the polygon (if the polygon also has inner borders). In our algorithm, the starting pixel of the border tracing is found by sequentially scanning the binary image, and therefore the first border pixel found is always at the outer border.

This algorithm is efficient in the sense that it only searches for new border pixels, therefore all the interior pixels and all the pixels inside the bounding boxes of the segments already found are skipped. After a new bounding box is found, the scan is accelerated by moving the scanning pointer to the right side of the bounding box. In many cases, this algorithm doesn't make even one entire pass through the image. There computational complexity of our algorithm is proportional to the length of the chain-code representation of the binary image.

5. CONCLUSION

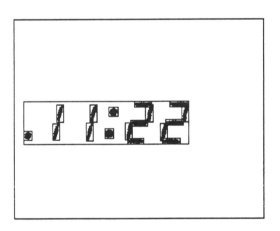

Figure 6 The result of Figure 4.

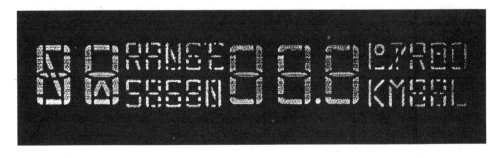

Figure 7(a) Another example of VF board image.

Figure 7(b) The result of Figure 7(a).

In this paper we presented three algorithms used in a machine vision inspection system, VF Learning and Detection system. The system has been designed and implemented for reliable inspection of various types of defects of VF boards. The system has two procedures, the learning and inspection procedures. The three algorithms presented in this paper provide operations critical to the system. The system currently has been integrated into the manufacturing line and has been proved to be robust to a plant environment where lighting condition can vary and shadow can occur in images, and the testing boards can be tilted and be placed in various positions.

6. ACKNOWLEDGMENT

This work is supported in part by Jabil Circuit, Inc. In particular, we would like to thank Mr. Anthony Tisler for introducing the problem to us and his contribution in the knowledge of VF manufacturing.

5. REFERENCES

[BKD95] Jerry Bowskill, Tim Katz and John Downie, "Solder Inspection using an Object-Oriented Approach to Machine Vision," SPIE Proceedings, Machine Vision Applications in Industrial inspection III, pp. 34 - 45, 1995.

[Chi88] R. T. Chin, "Automated Visual Inspection, 1981-1987," Journal CVGIP, Vol. 4, pp. 346-381, 1988.

[DaJ88] A. M. Darwish and A. K. Jain, "Machine Vision Techniques for Inspection of Printed Wiring Boards and Thick Circuit," Journal of Opt. Society, Am., A3, pp. 1465-1482, 1986.

[DMW94] A. D. Dorundo, J. R. Mandeville, and F. Y. Wu, "Reference-Based Automatic Visual Inspection of Electronic Packaging using a Parallel Image Processing System," SPIE Machine Vision Applications in Industrial Inspection II, pp. 38 - 57, February, 1994.

[HaS92] R. M. Haralick and L. G. Shapiro, "Computer and Robot Vision," Addison-Wesley Publishing Company, 1992.

[LuT95] Yi Lu and Anthony Tisler, "Gray Scale Filtering For Line And Word Segmentation," The Third International Conference on Document Analysis and Recognition, Canada, August, 1995.

[LuT96] Yi Lu and Anthony Tisler, "Machine Vision Inspection of VF Boards," 13th International Conference on Pattern Recognition, Vienna, Austria, Vol. C, pp. 839-843, August, 1996.

[Mun88] Joseph L. Mundy, "Industrial Machine Vision - - Is It practical?" Machine Vision -- Algorithms, Architectures, and Systems, Edited by Herbert Freeman, Academic Press, Inc., 1988

[SNW95] C. Sanby and L. Norton-Wayne, "Machine Vision Inspection of Lace using a Neural Network," SPIE Machine Vision Applications in Industrial Inspection III, Vol. 2423, pp. 315-322, 1995

Curvature-Symmetry Fusion in Planar Grasping Characterization from 2D Images

P.J. Sanz, A.P. del Pobil and J.M. Iñesta

Departamento de Informática, Universitat Jaume I, Campus de Penyeta
Roja, 12071-Castellón, Spain.
E-mail: sanzp,pobil,inesta@inf.uji.es

ABSTRACT

A new strategy is presented to simplify the real time determination of grasping points in unknown objects from 2D images. We work with a parallel-jaw gripper and assume point contact with friction, taking into account stability conditions.

This strategy is supported by a new tool that permits to establish a supervisor mechanism with the aim to seek grasping points from geometric reasoning on the contours extracted from 2D images captured by the system in execution time. This approach is named "curvature-symmetry fusion" (CSF) and its objective is to integrate curvature and symmetry knowledge in a single data structure to provide the necessary information to predict the suitable directions of the supervisor mechanism.

These algorithms have been implemented on a SCARA manipulator with one end point mounted camera. Visual feedback was used in the control system and the total time for the execution is about 2 or 3 seconds in our inexpensive prototype, making real applications feasible.

1. INTRODUCTION

The significance of the symmetry concept can be derived from the number of disciplines that make use of it, either for establishing behavior laws in certain physics domains [Jos91], or for characterizing planar shapes for recognition, handling, etc. [BA84], [VMU*95], [IBS96]. This latter application falls into our domain of interest: robotics.

From a mathematical point of view the existence of three basic types of transformation is stated [Led85]: rotations, reflections, and translations. They constitute the main categories of symmetries in the Euclidean space. An isometry can be constructed with combinations of those three transformations. Moreover, that kind of symmetries can be often found in nature and manufactured objects.

A number of authors have devised geometric methods for grasp determination. One approach [MNP90], particularly addressed to polygonal shapes, consists in determining all the regions that guarantee antipodal point grasps [Ngu88] through the use of inscribed circumferences. This approach is not applicable to a work universe of real manufactured objects, without shape restrictions, but the idea of taking distances from the centroid to opposite points in the outer contour is useful to deal with some kinds of symmetry, as it will be described below. There are other recent works in the same line for planar grasping characterization [Bla95] that make use of bitangential circles to seek for all potential pairs of candidate planar grasping points in the contour, guaranteeing force closure [Ngu88]. Note that these approaches are only applicable to contours with smooth curvatures, usually modeled with parametric curves.

In this work, we address a new approach dealing with the evaluation of the degree of proximity to shapes with different kinds of symmetry (other authors have dealt with the same problem, e.g. [ZPA95]) using curvature estimations, as a previous step to execute the geometric reasoning inherent to a SUpervisor MEchanism (SUME) that will analyze both global and local aspects of symmetry and curvature. We name this tool *Curvature-Symmetry Fusion* (CSF). More precisely, one important point is to find whether the minimum inertia axis, I_{min}, is a reflective symmetry axis or not. This fact makes easier the geometric reasoning, and it is commonly taken into account by other authors for avoiding difficulties and focusing in higher level grasp problems (e.g. learning [KFE96]).

The first stage to reach our objective in planar grasping determination is a careful segmentation that preserves shape [San96], and then to extract the shape features that will be used to obtain suitable grasping points for a given hand configuration. Objects are supposed nearly planar and homogeneous, and they are not known in advance: no attempt of shape identification is done. The only *a priori* knowledge is the geometry of the fingers of the parallel-jaw gripper.

The technical details of the strategy for grasping point determination are provided elsewhere [SDP*96], where a relation between the criteria that characterize a stable grasp are provided. This stability is defined in relation with statics and shape features provided by a real time imaging system. Including artificial vision in the grasp determination, as stated in that work, permits the model characterization through the definition of three thresholds:

> *Curvature threshold* (α). For taking into account smooth conditions on the grasping zones of the object, in order to guarantee the finger adaptation with a maximum contact.
> *Angular threshold* (β). For the non sliding condition, using the friction model.
> *Distance threshold* (γ). Permits a maximum distance from the centroid of the object to the grasping line $\overline{P_1 P_2}$ (the line that joins the grasping points P_1 and P_2).

Details on these thresholds are also found at [SDP*96], but the basic idea to preserve grasp stability is that the possible candidates to be grasping points need to hold three properties;

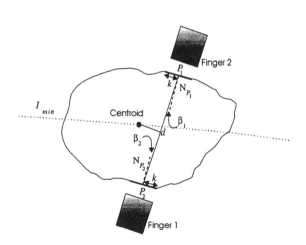

FIGURE 1. Geometric interpretation of the assumed stability conditions: smooth curvature in regions of radius k centered at P_1 and P_2; small angles β_1 and β_2 of the normals N_{P_1} and N_{P_2} with the grasping line; and a small distance, $d < \gamma$, from the centroid to that line.

1) Object-gripper finger adaptation. This implies that contour zones of radius k (the finger radius) must be found that, centered in the grasping points, present very low curvature conditions.
2) The grasping line is inside the friction cone; i.e. nearly normal to the contour at both points according with the "force closure condition" [Ngu88].
3) The grasping line is close to the centroid object.

There is a direct relation from each of these properties to the three previously defined thresholds. An important detail is that the knowledge used for their assessment and for all further geometric reasoning, is only derived from some few visual parameters captured in execution time: the object centroid, the direction of the minimum inertia axis (I_{min} - main axis), and a description of the object contour, **C** [SIP96]. The underlying idea in the former conditions would be in correspondence with the common sense reasoning [Dav90] used by people when they grasp an object on a planar surface, positioning both fingers on opposite sides of the object and close to its gravity center (e.g. a pencil on a table), and psycho-physic evidences exist that show this fact [OBB93].

The geometric interpretation of those conditions can be observed in Fig. 1, that shows a pair of grasping points, P_1 and P_2, under the assumption of a frictional point contact model, taking into account stability conditions. Note that d is the distance between the grasping line and the centroid, and β_1 and β_2 are the angles between the grasping line and the normals to **C** at the grasping points (N_{P_1} and N_{P_2}).

Here, we are going to focus only in those aspects of the proposed grasp model related with the geometric reasoning that leads to the development of the *Curvature-Symmetry Fusion* (CSF) scheme. This scheme arises from the need of providing knowledge from shape with the aim of make easier the task of the *SUpervisor MEchanism* (SUME) developed, aiming the grasping points determination in contour regions that satisfy all the proposed thresholds. Such a mechanism can be reduced to the following three steps:

> SUME_1.   To guarantee the invariance of the orientation of both normals (N_{P_1} and N_{P_2}) in a interval of radius k centered at P_1 and P_2, respectively, with a tolerance given by α; being $\alpha = 20°$.
>
> SUME_2.   To guarantee that the angles β_i hold

that $$\beta_1 = \left| \overline{P_1 P_2} \, \widehat{} \, N_{P_1} \right| \leq \beta \qquad \text{and}$$

$\beta_2 = \left| \overline{P_1 P_2} \stackrel{\wedge}{} N_{P_2} \right| \leq \beta$, being $\beta = 10°$ the angular threshold.

SUME_3. To guarantee that $d(\overline{P_1 P_2}, \text{centroid}) < \gamma$, where $\gamma = 0.03\ (2 \cdot I_a)$. Here I_a represents the main semilength of the main axis of the best fit ellipse.

The values for the three thresholds have been determined empirically.

The input of the SUME algorithm are two candidates to be grasping points, P_1^0 and P_2^0, obtained as intersections of the curve **C** with the axis of maximum inertia, I_{max}, through the GPCA algorithm (Grasping Points CAndidates) [SIP96]. Its output, P_1 and P_2, will be the aimed grasping points.

The rest of the paper is organized as follows: the CSF scheme is presented in section 2; experimental results are provided in section 3; followed by discussion and conclusions in section 4.

2. THE CURVATURE-SYMMETRY FUSION APPROACH, CSF

Some contour regions that satisfy certain conditions are needed to be found to perform the required geometric reasoning for grasp determination. These regions need to satisfy certain conditions related with the three thresholds described in section 1. For this, we need to:

1) Incorporate the gripper geometry to the model, translating the semilength of its fingers, k, to the points of the interval on the object contour.
2) Evaluate the curvature at each contour point.
3) Incorporate the information about the contour symmetry in the SUME algorithm in order to determine the possible directions for improving the positions of the grasping points.

Point 1) is trivial, so lets describe the curvature computation for point 2). Based on the k-vector notion [RJ73] (see Fig. 2) joining the considered point, p_i, with a point before, p_{i-k}, and after, p_{i+k}, it is defined as:

$$\vec{b}_{ki} = (x_{i-k} - x_i, y_{i-k} - y_i)$$
$$\vec{a}_{ki} = (x_{i+k} - x_i, y_{i+k} - y_i)_{,,,}$$

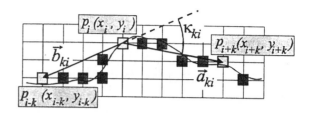

FIGURE 2. Geometric interpretation of the k-vectors ($k = 5$) for the computation of the discrete curvature.

the *k-angular bending*, κ_{ki}, is defined as the angle between both k-vectors (see Fig. 2):

$$\kappa_{ki} = -\vec{b}_{ki} \stackrel{\wedge}{} \vec{a}_{ki}$$

It is convenient to compute: κ_{ki}, $i = 1,...,N$, for every point in a contour. Note the need of smoothing that information with the aim to eliminate the quantization noise inherent to a digital contour, by means of a discrete convolution with a Gaussian kernel,

$$K_{ki}(w) = \kappa_{ki} \otimes G_i(w), \quad \text{being} \quad G_i(w) = \frac{1}{\sqrt{2\pi}w} e^{-\frac{i^2}{2w^2}},$$

where w has been empirically set to 2.5.

From the knowledge of symmetry a global characterization of the object's morphology is obtained, that can be fused with the knowledge of the local curvature at every point of the contour to get the CSF, with the aim to make the geometric reasoning involved easier. As a matter of fact, from the computation of the curvature at every point of the contour, by means of κ_{ki}, in addition to the point position in this contour, an adequate structure to visualize the symmetry and evaluate it is obtained.

The idea behind evaluating the symmetry degree associated with a digitized shape has been successful addressed, in a general way and without temporal constraints, in other works [ZPA95]. However, its computational effort is affected by an iterative search algorithm of the closest symmetry shape to the original one and an orientation symmetry axis optimization based on dynamic programming. Thus, it is not suitable in our context due to the real time robotics requirements. A different approach with the aim to reduce the computational effort is now provided, making use of the advantageous situation that represents *a priori* knowledge of the axis direction for the evaluation of symmetry.

For a planar shape two privileged directions exist related to its mass distribution (inertia), these directions are I_{min} and I_{max}, and they represent the eigenvectors of the moment of inertia covariance matrix [Pra91]. If mirror symmetry exists, these directions are the first candidates for that. As a result of that the next objective is evaluate the symmetry grade associated with these two directions.

From the curvature description K_{ki}, the symmetry degree is computed with respect to the I_{min} associated with the contour, using the intersection points $\{C \cap I_{min}\}$, computing Δ_i, such as:

$$\Delta_i = K_{k,P_3-i} - K_{k,P_3+i}; \quad i = 1,\ldots,N/2 \ ,$$

where $P_3 \in \{C \cap I_{min}\}$ traversing the contour clockwise from the initial point, PI, between P_1^0 and P_2^0. That is to say, Δ_i is computed as indicated, for each couple of points equidistant to P_3, covering all the contour. Using these quantities, we define the normalized global symmetric deficiency as:

$$\Phi = \frac{1}{N} \sum_{i=1}^{N/2} \Delta_i$$

where N is the total number of points in the contour. The same process is utilized to compute Φ for the I_{max} direction, but now, instead of P_3, we compute Δ_i with $P_1^0 \in \{C \cap I_{max}\}$, as introduced above. Note that $\Phi = 0$, i.e. perfect mirror symmetry with respect to the considered axis, exists only for an ideal mirror symmetric object. Some illustrative results in relation with the usefulness of Φ are shown in the results section (see Table 1).

With the aim to visualize the symmetry effect over the curvature evaluation/correction mechanism three different situations are considered, in a decreasing symmetry order. The main objective is to use this information to translate the intervals of radius k, with center at P_1^0 and P_2^0, searching for contour regions, taking into account the curvature smooth conditions (α) using K_{ki}.

Case_1) Mirror symmetry with respect to I_{min} and I_{max} directions (see Fig. 3). Once the high curvature level has been detected in the vicinity of P_1^0 and P_2^0, and taking into account the kind of symmetry detected, previously inferred from Φ, a clockwise turn is applied to seek for lower curvature regions. Note that if both points were not rotated in the same direction, the grasping line would get away from the centroid, which

involves instability. The stopping mechanism for this kind of symmetry is imposed by the $d(\overline{P_1 P_2}, \text{centroid}) < \gamma$ condition.

Case_2) Mirror symmetry with respect to the I_{min} direction (see Fig. 4). In this situation the improving directions for the movement of P_1^0 and P_2^0 are always in opposite directions for each other (towards P_3 or P_4) clockwise. The problem arises when during the seek for smooth curvature regions, the distance threshold γ is overpassed, because grasp instability is detected.

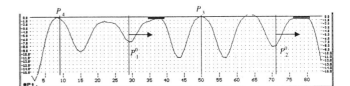

FIGURE 3. Nut. K_{ki} against i graph for that shape. The arrows indicate the possible improvement directions for decreasing the local curvature. The rectangles mark the selected grasp.

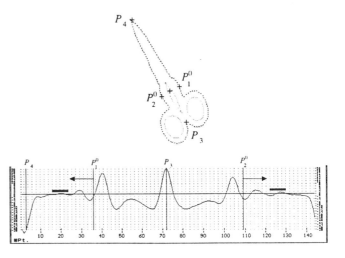

FIGURE 4. Scissors. K_{ki} against i graph for that shape. Arrows indicate the improvement directions for decreasing the local curvature. The rectangles mark the valid regions (according to local curvature) for the selected grasp.

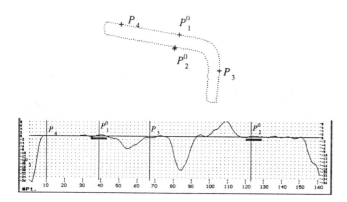

FIGURE 5. Allen key. K_{ki} against i graph for that shape. In this case the improvement is not needed.

Case_3) Symmetry does not exist (see Fig. 5). In this case, a systematic procedure is not possible because we do not know how the displacement of a point will affect the opposite one. This situation requires a more computationally expensive mechanism, that evaluates the impact on $\overline{P_1 P_2}$ each time that a point is moved, and then proceed according to that information.

The detailed procedure can be applicable for the angular evaluation/correction (β_i). The distance threshold is checked for each iteration in any case, acting as stopping condition. If any of the three thresholds are overpassed (α, β, γ) the grasp stability can not be assured.

3. EXPERIMENTAL RESULTS

A first experiment is presented to illustrate the normalized global symmetric deficiency concept and its applicability (see table 1). The data showed in that table are the mean and standard deviation computed from four digitizations for each object at different locations (position and orientation) over the work area. From that table some empirical results can be observed:

I_{min} direction. Looking at the table a gap between "pliers" and "pincers" is observed, $\Phi = 3$, and other gap between the latter and "caliper", $\Phi = 5$. After many trials we have followed a mirror symmetry approach for those images that present $\Phi < 5$, named $\Phi_c = 5$ to this critical value. In section 4 the results obtained in the intermediate

case of "pincers" appear, showing that the approach is suitable.

I_{max} direction. In this case the gap appear just between "nut" and the rest of shapes. So a value $\Phi_c = 3$ represents a good critical value.

A primary classification of the objects present in Table 1 would be the following: "mirror symmetry for I_{min} and I_{max} directions" {nut}; "mirror symmetry in I_{min} direction" {nut..pincers}; and "without symmetry" {caliper, allen key}. Note that only "nut" has mirror symmetry in both directions in correspondence with its inherent radial symmetry.

In order to visualize conditions to determine the grasping points described in section 1, we present some convenient results:

Case 1: Mirror symmetry in I_{min} and I_{max} directions exists. (e.g. nut; see Fig. 6). According to the approach when this kind of symmetry is presented, the predefined point translation movement is performed clockwise. The stopping condition is imposed simultaneously because the smooth curvature, and $d(\overline{P_1 P_2}, \text{centroid}) < \gamma$ conditions are both satisfied.

Case 2: Mirror symmetry in I_{min} direction exists (e.g. pincers in Fig. 7). From the points provided by GPCA, SUME firstly evaluates the curvature conditions (SUME_1). Fig. 6 shows that these points (P_1^0 and P_2^0) satisfy the necessary smooth curvature conditions. The problem lies on SUME_2. The trace of this algorithm is displayed in Table 3, in which the grasp instability prediction can be noted. Once again, the algorithm, as SUME_1 did in the case of Fig. 4, makes use of the existence of mirror symmetry and thus follows the same criterion to seek regions of good angular behavior. Firstly moves them toward PI (to the left in Fig. 7, dir-B):

$$P_1 \mapsto P_1 - n \wedge P_2 \mapsto P_2 + n ; \quad n = 1, 2, \cdots$$

and the stopping condition is reached with just one iteration (see Table 3: $d = 3.8 > \gamma = 3.0$), and then tries to the opposite direction (dir-A):

$$P_1 \mapsto P_1 + n \wedge P_2 \mapsto P_2 - n ; \quad n = 1, 2, \cdots$$

Two iterations are now performed with no success until the stopping condition is reached, so the proposed grasp can not be guaranteed. The effect of this lack of confidence for the considered shape when the gripper perform the grasp can be observed in Fig. 8.

Shape	I_{min}		I_{max}	
	$\overline{\Phi}$	σ_Φ	$\overline{\Phi}$	σ_Φ
Nut	1.86	0.74	1.99	0.69
Screwdriver	1.58	0.82	4.25	0.57
Scissors	1.59	0.92	4.85	0.32
Screw	1.69	0.62	5.23	0.75
Pliers	2.68	0.54	4.67	0.83
Pincers	4.42	0.61	5.60	0.87
Caliper	5.12	0.30	7.19	0.64
Allen key	6.38	0.58	5.46	0.78

Table 1. (See *Apendix* for corresponding shapes). Normalized global symmetric deficiency Φ computed for different images in both directions I_{min} and I_{max}. It shows the mean, $\overline{\Phi}$, and the standard deviation, σ_Φ, for each one.

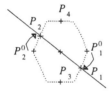

FIGURE 6. Nut: The choice of grasping points is done taking into account the prediction associated with its CSF degree.

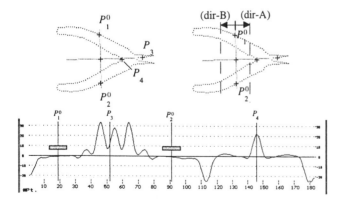

FIGURE 7. Pincers: In this figure, the smooth curvature conditions are satisfied (see the low values in the fingers adaptation zone). In the top right figure the directions taken by SUME_2 trying to correct the angular conditions (β_i) are displayed.

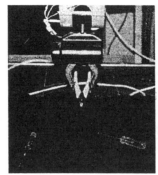

FIGURE 8. The gripper in action: "unstable" grasp of pincers (case-2). Here, the second condition of SUME is not satisfied, so stability is not guaranteed.

			(dir-B)	(dir-A)	(dir-A)'
	GPCA	SUME_1	SUME_2	SUME_2	SUME_2
P_1	(140,88)	(140,88)	(137,86)	(143,89)	(146,90)
P_2	(142,157)	(142,157)	(139,159)	(145,156)	(148,154)
β_1		24°	23°	24°	26°
β_2		24°	24°	24°	23°
d	0	0	3.8 > γ	2.2 < γ	5.2 > γ

Table 2. Related to Fig. 7. Here, the curvature conditions are satisfied (SUME_1), but the angular ones are not (SUME_2). Note the two stopping conditions, each in one direction, indicated by the condition $d > \gamma = 3$.

Case 3: Symmetry does not exist (e.g. Fig. 9). In this case SUME evaluates and corrects the grasping point candidates provided by GPCA. Finally, SUME finds that all conditions are held and therefore the grasp is stable. As shown above (Fig. 4), the intervals centered in the grasping points candidates hold the smooth curvature conditions imposed by SUME_1.

Following P_1^0 is evaluated/corrected through SUME_2, letting P_2^0 invariant to such translation. Once SUME_2 has been satisfied (so SUME_3 is also held by means of the algorithm implementation design), a suitable grasp is observed. The results of the evolution are presented in Table 3.

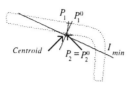

FIGURE 9. Allen key. Grasping points candidates (P_1^0, P_2^0), provided by GPCA and those corrected by SUME (P_1, P_2) are displayed.

	[1ª] GPCA	[2ª] SUME_1	[3ª] SUME_2
P_1	(156,75)	(156,75)	(150,74)
P_2	(147,92)	(147,92)	(147,92)
β_1		20°	0° < β
β_2		18°	0° < β
d	0		1 < γ

Table 3. Related to Fig. 9. A simplified trace of the stability seeking algorithm, from the points provided by GPCA. The left column shows the right direction of the translation performed. The angular threshold β is not overpassed and the same happens for γ, providing a stable grasp.

4. DISCUSSION AND CONCLUSIONS

A new tool that permits to establish a supervisor mechanism with the aim to seek grasping points based on geometric considerations over contours extracted from 2D images of shapes has been presented. This tool, named Curvature-Symmetry Fusion (CSF), permits the quantification of the symmetry degree associated with a shape (evaluated as a normalized global symmetric deficiency measure) from curvature estimations, in a simple and efficient manner, that has behaved well across the proposed work universe (mechanical tools). It is important to notice that, with that information we can characterize the degree of symmetry of unknown planar objects in a very simple, robust, and fast way like any application in robotics requires. Moreover, the system provides an open way for computing other grasps, relaxing the conditions imposed by some *a priori* thresholds, but taking into account the consequent impairment in the stability.

The global system has shown to work with a wide set of planar parts, and positioning algorithms have proven to be precise enough for the task, avoiding unnecessary expensive computacional means. In fact, the total time for the execution on a 486-DX33 computer and an inexpensive UMI RT-100 manipulator is between 2 and 3 seconds.

The method can be applied to simple industrial environments where parts are found at random positions and orientations, as long as a visual system is able to detect their shape features. This approach can be utilized in a number of possible application domains. One is the manipulation of hand tools, e.g. for supplying surgical instruments to the surgeon in an operating theater, where position and illumination conditions are highly controlled. Other domain would be the assistance to the handicapped, concerning to handling tasks. In these domains an effort in developing suitable computer-human interfaces for each use should be done.

ACKNOWLEDGMENTS: This work has been funded by a CICYT project (TAP 95-0710-C-01), Generalitat Valenciana (GV-2214/94), by Fundació Caixa-Castelló (P1A94-22), and a ESPRIT ("IOTA") project grants.

REFERENCES

[Bla95] Blake A. *A Symmetry Theory of Planar Grasp.* The International Journal of Robotics Research, Vol.14, No. 5, pp. 425-444. October, 1995.

[BA84] Brady M, Asada H. *Smoothed Local Symmetries and Their Implementation.* The International Journal of Robotics Research, Vol.3, No. 3, pp. 36-61. 1984.

[Dav90] Davis E. *Representations of Commonsense Knowledge.* Morgan Kaufmann Publishers. California. 1990.

[Jos91] Joshua SJ. *Symmetry principles and magnetic symmetry in solid state physics.* Adam Hilger. 1991.

[IBS96] Iñesta JM, Buendía M, and Sarti MA. *Local symmetries of digital contours from their chain codes.* Pattern Recognition, Vol. 29, No. 10, pp. 1737-1749, 1996

[KFE96] Kamon I, Flash T, and Edelman S. *Learning to Grasp Using Visual Information.* Proc. of the IEEE Int.*Conf.* on Robotics and Automation. Minneapolis, pp. 2470-2476. Minesota, April, 1996.

[Led85] Ledermann W. *Handbook of applicable mathematics.* Vol. V: Combinatorics and Geometry, Part B, chapt 11: "Symmetry". John Wiley. 1985.

[MNP90] Markenscoff X, Ni L, Papadimitriou CH. *The Geometry of Grasping*. The International Journal of Robotics Research. Vol.9, No 1, pp. 61-74. February 1990.

[Ngu88] Nguyen V-D. *Constructing Force-Closure Grasps*. The International Journal of Robotics Research. Vol.7, No 3, June 1988.

[OBB93] Opitz D, Bulthoff HH, Blake A. *Optimal Grasp Points: Computational Theory and Human Psychophysics*. Perception, pp. 22-123. 1993.

[Pra91] Pratt WK. *Digital Image Processing*. J. Wiley and Sons, New York. 1991.

[RJ73] Rosenfeld A, Johnston E. *Angle detection on digital curves*. IEEE Transactions on Computers. Vol. C-22, pp.875-878. September, 1973.

[SDP*96] Sanz PJ, Domingo J, del Pobil AP, and Pelechano J. *An Integrated Approach to Position a Robot Arm in a System for Planar Part Grasping*. Advanced Manufacturing Forum, Vol. 1, pp.137-148, Special issue on Applications of Artificial Intelligence. 1996.

[SIP96] Sanz PJ, Iñesta JM, del Pobil AP. *Towards an Automatic Determination of Grasping Points Through a Machine Vision Approach*. In *Proc.* of the Ninth International Conference on Industrial & Engineering Applications of Artificial Intelligence & Expert Systems (IEA/AIE), pp. 767-772. Fukuoka, Japan. 1996.

[San96] Sanz PJ. *Razonamiento Geométrico Basado en Visión para la Determinación y Ejecución del Agarre en Robots Manipuladores*. PhD Thesis, Jaume I Univ., Spain. 1996. (In spanish)

[VMU*95] Van Gool L, Moons T, Ungureanu D, Pauwels E. *Symmetry From Shape and Shape From Symmetry*. The International Journal of Robotics Research, Vol.14, No. 5, pp. 407-424. October, 1995.

[ZPA95] Zabrodsky H, Peleg S, and Avnir D. *Symmetry as a Continuous Feature*. IEEE Trans. Patt. Anal. Mach. Intell. Vol.(7) No. 12 , pp. 1154 - 1166, 1995.

Apendix

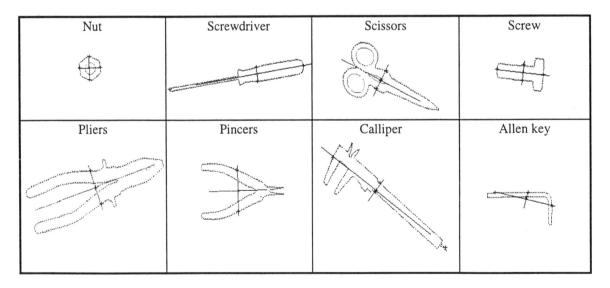

The collection of shapes for the experiment described in Table 1.

REQUIREMENTS SPECIFICATION FOR KNOWLEDGE-BASED SYSTEMS

John Debenham
Key Centre for Advanced Computing Sciences, University of Technology, Sydney,
PO Box 123, NSW 2007, Australia
debenham@socs.uts.edu.au

ABSTRACT

The requirements specification step in a knowledge-based systems design methodology is presented. Requirements specification is the first step and system analysis is the second step in this methodology. The product of requirements specification is the requirements model, and the product of system analysis is the conceptual model. Requirements specification is unified with system analysis in three distinct senses. The notation employed for the requirements model meshes neatly with the notation employed for the conceptual model. The structure of the requirements model is used to guide the construction of the conceptual model. Further, the requirements model and the conceptual model both employ uniform representations of each thing in the application.

INTRODUCTION

The terms 'data', 'information' and 'knowledge' are used in a rather idiosyncratic sense. The *data* in an application are those things which could be represented naturally as simple variables, the *information* is those things which could be represented naturally as relations, and the *knowledge* is those things which could be represented naturally in some rule language.

Requirements specification is the first step and system analysis is the second step in a knowledge-based systems design methodology. Unlike many current approaches to design, including KADS, this methodology is 'unified' in three distinct senses. First, the notation employed for the requirements model meshes neatly with the notation employed for the conceptual model. Second, the structure of the requirements model guides the construction of the conceptual model. Third, the requirements model and the conceptual model both employ uniform representations of each thing in the application. A single notion of "decomposition" [Deb96a] may be employed to simplify the representation of individual things in the requirements and conceptual models.

A knowledge-based systems design methodology should address maintenance. Maintenance is concerned with two key problems. First the problem of engineering the basic structure of the knowledge base so that it is inherently maintainable [Deb96a] [Deb95a]; matters to be considered here include the choice of structures for expressing knowledge and the way in which knowledge is represented in terms of those structures. Second, once the basic structure of the knowledge base has been engineered, the problem of presenting the represented expertise so that it can readily be modified within that structure in response to changes in circumstance; see for example [KGC96] [CBC92] [LHS92]. The methodology reported here addresses the first problem. This methodology employs a unified approach to modelling, it normalises the knowledge, it encourages the specification of constraints for knowledge and it supports a sound analysis of knowledge base maintenance [Deb95a].

To illustrate the first key maintenance problem identified above, consider the raw expertise [E1] "the profit on a part is the product of the cost price of the part and the mark-up factor of the part less 1". Consider also the raw expertise [E2] "the tax payable on a part is 10% of the product of the cost-price of the part and the mark-up factor of the part". Now both [E1] and [E2] may have buried within them the sub-rule [E3] "the selling price of a part is the cost price of that part multiplied by the mark-up factor of the part". If the expertise represented in rule [E3] should change then both rule [E1] and rule [E2] will have to be modified. Consider what happens if rule [E3] has not been explicitly identified; in this case rules [E1] and [E2] 'share an unstated sub-rule between them' and consequently constitute a hidden maintenance hazard.

The scenario just described concerned a sub-rule hidden inside two other rules. Even more complicated situations arise when rules are hidden inside relations and vice versa. The application of decomposition [Deb96a] helps to remove these hidden duplications from the requirements and conceptual models. When these duplications have been removed then this helps to ensure that those models can support maintenance [LHS92] [Deb89].

METHODOLOGY IN OUTLINE

The first two steps in a design methodology for knowledge-based systems are:

• *requirements specification*. This step focuses on the problem and proceeds top-down; it begins by constructing a rough view of the key requirements and ends with a 'fairly complete' specification called the "requirements model". The requirements model is functional and specifies *what* is required of the system in a broad sense. The requirements model will address system maintainability. The requirements model is expressed in terms of "r-schema".
• *system analysis*. This step focuses on the solution and proceeds bottom-up; using the structure of the requirements model as a guide it constructs a precise model of the system called the "conceptual model". The conceptual model employs a particular form of *knowledge representation* called "items". Items are expressed in terms of "i-schema". The conceptual model is declarative and specifies *how* the system will do what it is asked to do. The conceptual model inherits details of *what* the system should be able to do

by links from the requirements model.

These two steps have very different characteristics. They are contrasted in Figure 1 and Figure 2. Despite these different characteristics these two steps they are unified by the methodology [Deb96b] in three distinct senses.

• *vertical unification* means that a single schema is used to represent the *data* (individual data items), *information* ('relations') and *knowledge* ('rules') both during requirements specification (using r-schema) and during system analysis (using items).
• horizontal *model unification* means that the overall structure of the requirements model acts as a detailed guide for the construction of the conceptual model.
• horizontal *schema unification* means that the r-schema used in the requirements model translate naturally into the items used in the conceptual model.

These three forms of unification for requirements specification and system analysis are illustrated in Figure 3. Requirements specification consists of two tasks:

• *application representation*. This task is concerned with the construction of a rough view of key aspects of the application called the *application description*; this does not give all the details that would be required for implementation.
• *requirements identification*. This task is concerned

	Step	
	Requirements specification	System analysis
Product	requirements model	conceptual model
Representation	functional	declarative
Schema	r-schema	i-schema
Proceeds	top-down	bottom-up
Focus	the problem	the solution
Accent	what	how

Figure 1   Design step characteristics

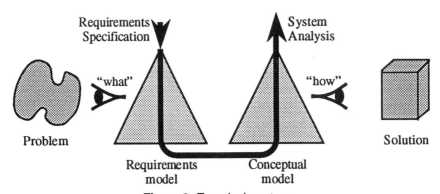

Figure 2 Two design steps

with the identification of those things in the application description which are relevant to the proposed system, and with the classification of those things as to how they should be implemented; this classified sub-set of the application description is called the *requirements model*. This task also addresses cost, scope and re-use.

The product of the requirements specification step is the requirements model. The requirements model is expressed in terms of the r-schema notation which is described below.

R-SCHEMA

Th term *chunk* refers to a real chunk of knowledge however large it may be.

A *declarative view* of a chunk is a representation that contains sufficient detail to fully specify how that chunk might be implemented. Items are a formal, declarative view of chunks. Items are described below. Items are presented in practice using i-schema. An item can represent a whole system, a whole knowledge base or a whole database. In other words, a set of items is itself an item.

A *functional view* of a chunk is a representation that specifies *what* that chunk should be able to do but not *how* it should do it. The "r-schema" notation gives a functional view of chunks. The r-schema notation is informal and compact. During the initial requirements specification step the degree of understanding of the application will necessarily be rather vague; r-schema are designed to accommodate this lack of detailed precision.

An r-schema describes *what* a chunk will be required to do. Part of the r-schema representation is the r-schema's "behaviour". When a chunk is implemented, the implementation will just "sit there" until it is asked to do something. A message which asks the implemented form of a chunk to do something is called a *request*. When the implementation of a chunk receives a request, it is expected to *do something* specific and to deliver an appropriate message in response; such a response is called a *response*. A request will be presented over some period of time. Some time, at or after the time that the request begins to arrive, the response will begin to appear. The request should have been fully presented by the time that the response has been completely delivered. From the time that the request begins to arrive to the time that the response is fully delivered the implementation of the chunk will be *active*. What the implemented chunk does internally while it is active is called the *activity*. There are three inter-related but separate *aspects* of behaviour:

- the *request presentation*;
- the *activity*, and
- the *response delivery*.

These three aspects are not considered to be sequential. That is, bits of the request may be presented after the implementation has become active, and bits of the response may be delivered before the implementation ceases to be active.

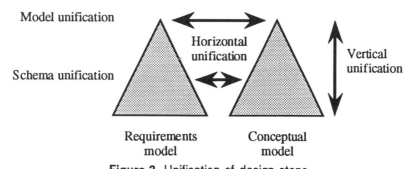

Figure 3 Unification of design steps

	Request presentation	Activity	Response delivery
What	What is the form of the request?	What is the form of the activity?	What is the form of the response?
Why	Why is the request being presented?	Why is the activity being performed?	Why is the response being delivered?
Who	By whom will the request be presented?	Who else will be contribute to the activity?	To whom should the response be delivered?
When	Will the request be presented at any particular time?	Is there any limit on the time taken for processing the activity?	Should the response be delivered at any particular time?
Where	Where will the request come from?	Does the activity need to take place anywhere in particular?	Where should the response be delivered?

Figure 4 Perspectives and aspects of behaviour

There are five features of each of the three above aspects. These five features are called *behaviour perspectives*:

- *what* is it?
- *why* is it happening?
- *who* else (or *what* else) is involved?
- *when* does it happen?
- *where* does it happen?

The three aspects and five perspectives of behaviour are summarised in Figure 4; together they provide a framework for describing behaviour. During requirements specification attention should only be given to those features of behaviour which appear to be important. Behaviour is only concerned with "what" the implementation is required to do and not with "how" it might do it.

An "r-schema" describes the function of a chunk. An r-schema has a *name*. An *optional* part of the r-schema view is the "mission"; the *r-schema mission* is a succinct statement of what that chunk is supposed to achieve within the general operation of the system. The r-schema view includes the specification of as many of that chunk's 'components', or sub-chunks, as can be readily identified. The "r-schema" view describes a chunk's "behaviour". An r-schema's constraints is an informal statement of any important constraint on the way in which the chunk should behave. An *r-schema* consists of:

- the r-schema *name*;
- the names of (at least some of) the *components* of that r-schema;
- (at least one of) the r-schema's *behaviours*;
- the r-schema's *constraints*, and
- (optional) the r-schema's *mission*.

During requirements specification, it is not necessarily required to quote *all* of the r-schema's components, behaviours or constraints. This degree of imprecision is intended to reflect the lack of understanding that necessarily pervades requirements specification during the design process. The r-schema notation is illustrated in Figure 5 which shows a sample r-schema for *"Tax Act*

name	Tax Act KB
components	*Individual Tax KB* *Corporation Tax KB*
behaviour	• Advise on tax payable by an individual on the presentation of a correctly prepared profile of that individual. • Advise on tax payable by a corporation on the presentation of a correctly prepared profile of that corporation. • Provide intelligent cross referencing to support requests to browse the knowledge base.
constraints	• If two different profiles can be prepared for a given individual or corporation then the advice given on the basis of those two different profiles should be the same.
mission	"To provide an effective implementation of the Tax Act to suit the general requirements of individuals, corporations and taxation specialists."

Figure 5 r-schema for item *"Tax Act KB"*

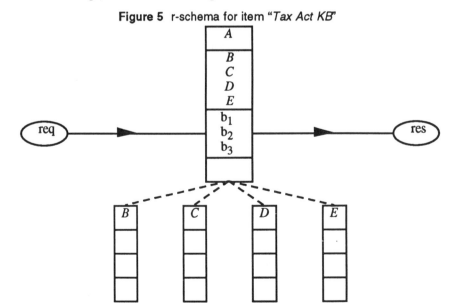

Figure 6 r-schema and four components

KB". r-schema can be used to describe data, information and complex knowledge or whole systems; thus **vertical unification is achieved for requirements specification**.

REQUIREMENTS SPECIFICATION

The requirements specification step begins with the "application representation" task. The application representation task begins with the construction of an overview of the system. This overview is then expanded into a fairly complete view of the system requirements called the application description.

Application Representation

Application representation is a top-down process that begins with the construction of a succinct view of the application called a "context diagram". The context diagram contains a single r-schema. Starting with the context diagram, a method of "r-schema decomposition" is continually applied to the r-schemas in the application description until the r-schemas derived represent either information or data.

A *context diagram* has an outer frame labelled "Application Description"; inside this outer frame there is an inner frame labelled "System Boundary"; inside the outer frame and outside the inner frame the participants who will interact with the system are shown, and inside the inner frame a single r-schema representing the system is shown. The context diagram is the starting point for the application representation task and thus is the starting point for the whole design process. The context diagram only shows those behaviours which the are of key importance to the operation of the system. In other words, the context diagram presents a deliberately summary view of the application.

The procedure for r-schema decomposition starts with any given r-schema and generates a set of new r-schema. In simple terms the procedure tries to allocate each behaviour to one of the r-schema's components; if this is not possible then that behaviour is replaced with a description of how the components can collectively contribute to that behaviour. For example, consider behaviour b_2 of r-schema A, shown in Figure 6. Suppose that b_2 in A may be performed entirely by component C; then the decomposition will be as shown in Figure 7. On the other hand, if b_2 in A can only be performed by components B, D and E collectively then the decomposition is as shown in Figure 8. In Figure 8 that behaviour b_2 has been replaced with behaviour b_7 in A, where b_7 is a rule which describes how behaviour b_2 in the new component F may be achieved by b_4, b_5 and b_6.

Requirements Identification

Having completed the application representation task, the business of requirements specification is completed with the second task of "requirements identification". The requirements identification task is concerned with:

- rejecting those things in the application description which are not relevant to the system;
- exploring the possible re-use of existing components;
- classifying those things as to be implemented 'by re-using an existing component', 'as a knowledge-based system component' or 'as a conventional programmed component';

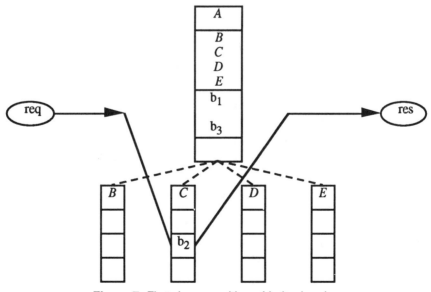

Figure 7 First decomposition of behaviour b_2

- estimating the cost of the system;
- deciding whether the scope of the initial system should be reduced, at least initially, and
- specifying those sections of the system which are considered to be highly volatile and which should be carefully engineered for maintainability.

The product of this task is called the *requirements model* The requirements model is an annotated extract of the application description. The requirements model will thus consist of a cascade decomposition of annotated r-schemas.

ITEMS

The r-schema notation has been introduced for describing the *function* of chunks. In this section items are introduced. Items describe what chunks actually *are*. Items have three important properties: items have a uniform format no matter whether they represent data, information or knowledge things [Ito91]; items incorporate two powerful classes of constraints, and a single rule of decomposition can be specified for items. The key to this uniform representation is the way in which the "meaning" of an item, called its *semantics*, is specified. Items can be defined either formally as λ-calculus expressions, or informally as i-schema.

The *semantics* of an item is a function which *recognises* the members of the "value set" of that item. The value set of an item will change in time τ, but the item's semantics should remain constant. The value set of an information item at a certain time τ is the set of tuples which are associated with a relational implementation of that item at that time. For example, the value set of the item named *part/cost-price* could be the set of tuples in the "part/cost-price" relation at time τ. Knowledge items have value sets too. Consider the rule "the sale price of parts is the cost price marked up by a universal mark-up factor". Suppose that this rule is represented by the item named *[part/sale-price, part/cost-price, mark-up]*, then this item's value set is the set of all quintuples:

(part-no., dollar-amount, part-no., dollar-amount, factor)

which satisfy this rule. This idea of defining the semantics of items as recognising functions for the members of their value set extends to complex, recursive knowledge items too [Deb96b].

An *item* consists of: an item name, item components, item semantics, item value constraints, and item set constraints. The *item name* is usually written in italics, eg *name*. The *item components* is a set of the names of items which this item represents some

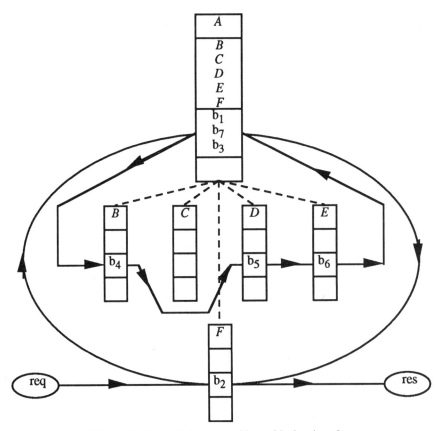

Figure 8 Second decomposition of behaviour b_2

association between. The *item semantics*, S_{name}, is an expression which recognises the members of the value set of this item. The *item value constraints*, V_{name}, is an expression which must be satisfied by all members of the value set of this item. The *item set constraints* [Deb95b] are constraints on the structure of this item's value set. The λ-calculus form for an item is:

<name> [<semantics>, <value constraints>,
<set constraints>]

Suppose that in an application each *part* is associated with a *cost-price*. This association could be subject to the "value constraint" that parts whose part-number is less that 1,999 will be associated with a cost price of no more than $300. This association could also be subject to the "set constraints" that every part must be in this association, and that each part is associated with a unique cost-price. This association could be represented by the information item named *part/cost-price*. The formal λ-calculus form for this item is:

$part/cost\text{-}price[\ \lambda xy\bullet[S_{part}(x) \wedge S_{cost\text{-}price}(y) \wedge$
$\quad costs(x, y)]\bullet,$
$\quad \lambda xy\bullet[\ V_{part}(x) \wedge V_{cost\text{-}price}(y) \wedge$
$\quad\quad ((x < 1999) \rightarrow (y \leq 300))\ ]\bullet,$
$\quad (Uni(part) \wedge Can(cost\text{-}price, \{part\}))_{part/cost\text{-}price}\ ]$

The λ-calculus form is not intended for practical use. In practice items are described using i-schema. The i-schema format and the i-schema for this item are shown in Figure 9. The meaning of the set constraints is as follows. "Uni(*a*)" is a *universal constraint* which means that "all members of the value set of item *a* must be in this association", "Can(*b*, A)" is a *candidate constraint* which means that "the value set of the set of items A functionally determines the value set of item *b*", and "Card" means "the number of things in the value set". The subscripts indicate the item's components to which that set constraint applies [KGC96]. The '∀' represents a universal constraint; the horizontal line and the '**o**' represents a candidate constraint. Data, information and knowledge can all readily be defined using items. Thus **vertical unification is achieved for system analysis**. For example, the *[part/sale-price, part/cost-price, mark-up]* knowledge item is:

name		name
name1	*name2*	components
x	y:	variables
meaning of item		semantics
value constraints		value constraints
set constraints		set constraints

part/cost-price	
part	*cost-price*
x	y
costs(x, y)	
x<1999 → y≤300	
∀	
-----------	**o**

Figure 9 i-schema format and the i-schema '*part/cost-price*'

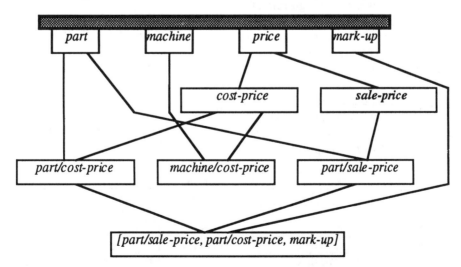

Figure 10 Simple conceptual diagram

[part/sale-price, part/cost-price, mark-up][

$$\lambda x_1 x_2 y_1 y_2 z \bullet [(\ S_{part/sale\text{-}price}(x_1, x_2)\ \wedge$$
$$S_{part/cost\text{-}price}(y_1, y_2) \wedge S_{mark\text{-}up}(z)\ ) \wedge$$
$$((x_1 = y_1) \rightarrow (x_2 = z \times y_2))] \bullet,$$

$$\lambda x_1 x_2 y_1 y_2 z \bullet [\ V_{part/sale\text{-}price}(x_1, x_2)\ \wedge$$
$$V_{part/cost\text{-}price}(y_1, y_2) \wedge V_{mark\text{-}up}(z)\ ) \wedge$$
$$((\ x_1 = y_1\ ) \rightarrow (\ x_2 > y_2\ ))] \bullet,$$

(Uni(*part/sale-price*) ∧ Uni(*part/cost-price*)

∧ Can(part/sale-price, {*part/cost-price, mark-up*})

∧ Can(*part/cost-price*, {*part/sale-price, mark-up*})

∧ Can(*mark-up*, {*part/sale-price, part/cost-price*})

$^{)}$*[part/sale-price, part/cost-price, mark-up]* $^{]}$

System Analysis

The conceptual model is derived during the system analysis step. The *conceptual model* consists of a specification of each thing in the application as an item. The conceptual model is constructed bottom-up by working up the requirements model as a guide. Each r-schema in the requirements model becomes one, or more, i-schema in the conceptual model. The structure of the requirements model is a cascade decomposition of annotated r-schemas. The structure of the requirements model provides a guide for the construction of the conceptual model. Thus **model unification is achieved** for both the requirements model and the conceptual model. A diagrammatic presentation of a simple conceptual model is shown in Figure 10.

From r-Schema to i-Schema

There are three important differences between r-schema and i-schema. First, r-schema are essentially informal and i-schema are essentially formal. Second, r-schema usually give a view of a set of items; as a very rough rule-of-thumb, the number of different behaviours in an r-schema can be indicative of the number of different items which that r-schema corresponds to. Third, r-schema describe *what* a chunk will be required to do. i-schema describe *how* an item will do what it is asked to do. Setting these differences aside for the moment, the specification of the name, components and constraints in both r-schema and i-schema has the same general intention. The mission statement in the r-schema notation has no parallel in the i-schema notation.

The behaviour specified in the r-schema is related to the semantics in the i-schema. There is a lot of useful information in the r-schema's behaviour which does not find its way into the i-schema's semantics. This useful information is imported to the conceptual model by using links. A rough correspondence between r-schema and i-schema is shown in Figure 11. Thus horizontal **schema unification is achieved** between r-schema and i-schema.

CONCLUSION

A unified approach to requirements specification and system analysis for knowledge-based systems has been introduced. A key feature of the approach is that all things are modelled in a completely uniform way using r-schema and i-schema. The characteristics of requirements specification and system analysis have been contrasted. Despite their different characteristics these two steps they are unified by the methodology [Deb96b] in three distinct senses. Vertical unification means that a single schema is used to represent the data (individual data items), information ('relations') and knowledge ('rules') both during requirements specification (using r-schema) and during system analysis (using items). Horizontal model unification means that the overall structure of the requirements model acts as a detailed guide for the construction of the conceptual model. Horizontal schema unification means that the r-schema used in the requirements model translate naturally into the items used in the conceptual model.

REFERENCES

[Deb96a] J.K. Debenham, "Knowledge Simplification", in *proceedings 9th International Symposium on Methodologies for Intelligent Systems ISMIS'96*, Zakopane, Poland, June 1996, pp305-314.

[Deb95a] J.K. Debenham, "Understanding Expert Systems Maintenance", in *proceedings Sixth International Conference on Database and*

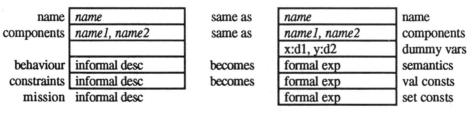

name	*name*	same as	*name*	name
components	*name1, name2*	same as	*name1, name2*	components
			x:d1, y:d2	dummy vars
behaviour	informal desc	becomes	formal exp	semantics
constraints	informal desc	becomes	formal exp	val consts
mission	informal desc		formal exp	set consts

Figure 11 Rough correspondence between r-schema and i-schema

Expert Systems Applications DEXA'95, London, September 1995.

[KGC96] B. Kang, W. Gambetta, and P. Compton, "Validation and Verification with Ripple Down Rules", *International Journal of Human Computer Studies* Vol 44 (2) pp257-270 (1996).

[CBC92] F. Coenen and T. Bench-Capon, "Building Knowledge Based Systems for Maintainability", in *proceedings Third International Conference on Database and Expert Systems Applications DEXA'92*, Valencia, Spain, September, 1992, pp415-420.

[LHS92] F. Lehner, H.F. Hofman, R. Setzer, and R. Maier, "Maintenance of Knowledge Bases", in *proceedings Fourth International Conference DEXA93*, Prague, September 1993, pp436-447.

[Deb89] J.K. Debenham, *"Knowledge Systems Design"*, Prentice Hall, 1989.

[Deb96b] J.K. Debenham, "Integrating Knowledge Base and Database", in *proceedings 10th ACM Annual Symposium on Applied Computing SAC'96*, Philadelphia, February 1996, pp28-32.

[Ito91] H. Ito "Interface for Integrating a Knowledge-Based System and Database Management System Using Frame-Based Knowledge Representation", in *proceedings Expert Systems World Congress*, J. Liebowitz (Ed), Pergamon Press 1991.

[Dat86] C.J. Date, *"An Introduction to Database Systems"* (4th edition) Addison-Wesley, 1986.

[Tay93] N. Tayar, "A Model for Developing Large Shared Knowledge Bases" in *proceedings Second International Conference on Information and Knowledge Management*, Washington, November 1993, pp717-719.

[Gra89] P.M.D. Gray, "Expert Systems and Object-Oriented Databases: Evolving a New Software Architecture", in *"Research and Development in Expert Systems V"*, Cambridge University Press, 1989, pp 284-295.

[Deb95b] J.K. Debenham, "Knowledge Constraints", in *proceedings Eighth International Conference on*

Industrial and Engineering Applications of Artificial Intelligence and Expert Systems IEA/AIE'95, Melbourne, June 1995, pp553-562.

[Deb96c] J.K. Debenham, "Characterising Maintenance Links", in *proceedings Third World Congress on Expert Systems*, Seoul, February 1996.

[KaM91] H. Katsuno and A.O. Mendelzon, "On the Difference between Updating a Knowledge Base and Revising It", in *proceedings Second International Conference on Principles of Knowledge Representation and Reasoning, KR'91*, Morgan Kaufmann, 1991.

KNOWLEDGE REQUIREMENTS AND ARCHITECTURE FOR AN INTELLIGENT MONITORING AID THAT FACILITATE INCREMENTAL KNOWLEDGE BASE DEVELOPMENT

Ellen J. Bass, Ronald L. Small, Samuel T. Ernst-Fortin
Search Technology, Inc.
4898 South Old Peachtree Road, #200
Norcross, Georgia 30071-4707
E-mail: ellen, rons, sfortin@searchtech.com

ABSTRACT

Being able to incrementally define and test knowledge bases for intelligent systems is desirable. However, as more knowledge is added, the knowledge engineer must ensure that undesirable interactions between the existing and additional knowledge do not occur. One intelligent monitoring system, the Hazard Monitor (HM), provides for the ability to add knowledge incrementally by including an arbitration process. This paper describes the knowledge requirements and the architecture for Hazard Monitor which facilitates incremental knowledge base development.

KEYWORDS

human-machine systems, intelligent interfaces, knowledge acquisition, knowledge-based systems, intelligent systems

INTRODUCTION

Intelligent automation can help operators of complex systems to monitor and control the systems for which they are responsible. How to design and validate the knowledge driving these intelligent systems remains an active research area. Many knowledge acquisition processes use incremental addition of knowledge by progressively widening the area of application. Usually verification of such systems has to cover the whole area of application at each step (and not just the additional area) making the knowledge verification process a major task.

Being able to define and test knowledge bases incrementally would be an improvement in the overall process. Such an incremental approach provides the side benefit that operators can use incrementally developed intelligent systems as new knowledge (and therefore new capability) is incorporated. In other words, the entire knowledge base does not have to be completely developed before the end users can take advantage of the functionality of the intelligent system.

Intelligent automation that provides for incremental development of knowledge bases must provide for the interaction of previously defined (and tested) knowledge with newly engineered knowledge. As more knowledge is added, the knowledge engineer must ensure that undesirable interactions between the existing and the additional knowledge do not occur. One intelligent monitoring system, the Hazard Monitor (HM), provides for the ability to add knowledge incrementally. In this paper, we describe the Hazard Monitor, its architecture, and its knowledge requirements. We describe how its architecture facilitates incremental knowledge base development by including an arbitration process.

WHAT IS A HAZARD MONITOR?

Hazard Monitor, is a knowledge-based technology that reduces the rate of preventable accidents regardless of the source of the problem (i.e., human, machine, or external environment). Traditional error prevention strategies (i.e., structured design, ergonomics, personnel selection, extensive training and automation) have been practiced for decades and may have reached their limits of effectiveness in further reducing human error rates. For example, in the past thirty-seven years, the percentage of serious U. S. commercial aircraft accidents due to pilot error has remained above 65% [Boe96].

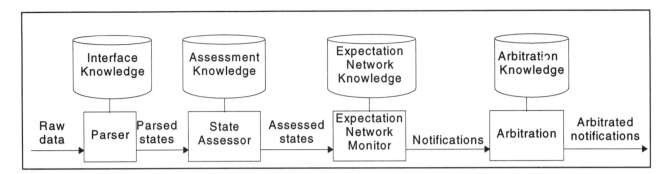

Figure 1 Hazard Monitor Architecture

HM complements designing for error prevention by helping the operator to avoid the negative consequences of potentially hazardous situations. The idea stems from the fact that in complex environments, hazards occur despite improvements in system design and advances in human-computer interaction. The literature describes many incidents (e.g., the Bhopal chemical plant disaster [Cas93] and the 1994 Nagoya airport Airbus crash [HD95]) which could have been avoided if the operators had recognized a deteriorating situation in time to avoid its adverse consequences [EB93]. HM aids in these situations by enhancing the problem recognition and identification process.

When tailored for a particular domain, HM aids in reducing the number of preventable accidents by alerting the human operator to discrepancies between actual and expected system states (a divergence that could lead to adverse consequences). Discrepancies are assessed as failed expectations in a situation within the context of goal-oriented activity. For example, during the approach to an airport, one expectation is for the aircraft landing gear to be down and locked. Defining expectations in this way allows discrepancy notification to be both timely and context-dependent as notification is tied to the operator's current situation, goals, and activities. Also notification can be based on the interaction of many system and environmental states instead of only one single state.

HM dispatches notifications that are available for presentation to the human operator so the operator can act to remedy the situation before unacceptable consequences occur. Because notifications have associated levels (e.g., warning, caution, advisory), HM also provides the ability to organize them (e.g., prioritize by severity) for presentation purposes.

Because human operators of complex systems can carry out several activities at the same time, HM monitors multiple activities concurrently. HM includes

functionality called *arbitration* to manage notifications generated during the process of monitoring multiple activities [SZS95]. HM suppresses duplicate notifications, arbitrates among similar notifications at different severity levels and among conflicting notifications, and provides the ability to condense sets of related notifications into more abstract information. With its arbitration capability, HM makes available for presentation the most relevant set of notifications at any given time.

HAZARD MONITOR ARCHITECTURE

The HM architecture consists of structured domain-specific knowledge, application-specific interface information, and a domain-independent engine which comprises four tailorable components: a parser, a state assessor, the expectation network monitor, and arbitration (see Figure 1).

The *parser* uses interface knowledge to format raw data coming from domain specific data sources. For example, when HM was integrated with a Boeing/Honeywell 747-400 Flight Management System (FMS) in a B747-400 aircraft simulator, the parser required interface knowledge for interpreting data coming from an IEEE 488 General Purpose Interface Bus (GPIB) [SZS+95].

The *state assessor*, using application and domain specific knowledge, integrates the parsed data into a higher level representation of the current system state. In effect, the state assessor computes value-added attributes from the parsed data and creates assessed states. The functionality required in the state assessor is driven by the needs of the expectation network monitor. For example, suppose the expectation network monitor needs the distance between the aircraft's present position and

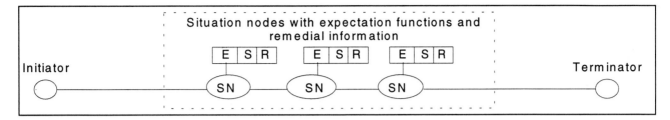

Figure 2 Expectation Network Architecture

its destination, but the FMS only provides individual leg length distances. The state assessor can determine how far the aircraft is from its destination by assessing this distance as the arithmetic sum of the remaining flight path leg lengths.

In addition to state assessment, the majority of the knowledge required by HM is for expectation network monitoring and for the arbitration process. These two components are described in the next two sections.

EXPECTATION NETWORK MONITOR AND ITS KNOWLEDGE REQUIREMENTS

The knowledge required for expectation network monitoring explicitly represents when to monitor system states that are relevant to the achievement of human operator activities within the domain of interest. The knowledge also explicitly represents the context in which notifications should be generated for presentation to the human operator. The knowledge for each activity is represented independently to allow for monitoring of multiple and sometimes competing goals or activities. This independence also allows the knowledge engineer to develop the knowledge for each activity separately.

HM employs *expectation networks* [GSZ+95] to structure the activity monitoring knowledge. An expectation network is a graph of correctly achieved activity, represented at the level of assessed states and their values. An expectation network also contains severity information related to falling outside the range of expected system state values. The knowledge in each expectation network is comprised of an initiator function, situation nodes with expectations and notification information, and a terminator function (see Figure 2).

INITIATOR

An *initiator*, illustrated on the left side of Figure 2, is used to identify the initial conditions for the start of monitoring for an activity. The initiator is composed of a set of initiating conditions. If the system state matches one of the initiator's conditions, HM begins monitoring the associated activity. When an expectation network

initiates, the running network is actively monitoring for goal accomplishment.

The knowledge engineer defines the initiator of the expectation network for an activity to be the precipitating conditions for that activity. In other words, the initiator for an expectation network is the set of conditions defining when the human operator is expected to consider applying the procedural knowledge for the activity. For example, when descending from cruise altitude on the way to the destination airport, pilots adjust the altimeter to the local barometric pressure at the transition altitude.[1] In the expectation network to monitor this activity, the initiator includes the procedure initiating condition (i.e., when the aircraft altitude is less than the transition altitude).

TERMINATOR

The *terminator* causes activity monitoring for the expectation network to cease. Terminators de-activate the network either because the activity is no longer relevant or because the goal of the activity has been achieved.

The knowledge engineer defines the terminator for an expectation network to be the set of conditions for which monitoring of the activity should terminate. The knowledge engineer includes conditions related to goal achievement as well as goal abandonment. For example, for the altimeter setting network, the terminator includes the goal achievement condition (i.e., the aircraft has landed) and the goal abandonment condition (i.e., the aircraft climbs above the transition altitude, implying that the pilot should no longer set the altimeter to the local barometric pressure but rather to the standard setting of 29.92 inches).

SITUATION NODES

Each expectation network has one or more *situation nodes*. Situation nodes (represented as ovals labeled "SN" in Figure 2) define the situation context along the way to achieving the goal of an activity. For an actively monitored expectation network, HM monitors only one

---

[1]The transition altitude in the United States is 18,000 feet MSL. Below this altitude, the altimeter is set by the pilot to the local barometric pressure.

situation node at a time. This situation node is called the *current situation node*.

The knowledge engineer defines for each situation node a *subcontext* within the context of the activity. The subcontext, called the *situation node constraint*, is an expression defining the applicability of the situation node. For example, for an aircraft takeoff activity, one situation node may have a situation node constraint of below 100 feet AGL[2] for expectations related to being near the runway (when the landing gear is expected to be extended). Another may have a situation node constraint of between 100 feet and 500 feet AGL for the initial climb (when the landing gear is expected to be retracted).

Since the reason to monitor is to provide feedback for the operator, the knowledge engineer defines the situation nodes in an expectation network to be circumstances in an activity when significant discrepancies between expected and actual states should be communicated to the operator. The knowledge engineer adds situation nodes to an expectation network based on changes in the expectations and the appropriateness for providing feedback.

Each situation node has an associated set of expectations. Each expectation is composed of three elements:

1. Expectation function ("E" in Figure 2),
2. Severity level ("S" in Figure 2), and
3. Remedy code ("R" in Figure 2).

The knowledge engineer defines each expectation function based on states and their expected values within the subcontext of the situation node. The expectation functions encode valid ranges for states based on factors such as the following:

- Safety,
- Regulations (e.g., [US96]),
- System performance,
- Standard operating procedures,
- Efficient use of resources, and
- Passenger comfort (in the case of transportation systems).

For each expectation function, the knowledge engineer creates a test defining what the expected state(s) should be.

When an expectation function evaluates false on the current situation node of an active expectation network, HM generates a notification. To generate a notification, HM requires two pieces of information to be defined. In Figure 2, this information is represented by two boxes adjacent to the expectation functions: the severity level and the remedy code.

The severity level informs the operator how serious the notification is. A common taxonomy for alert severity level in aviation systems is [RLW80]:

- Warning (i.e., requires immediate operator awareness and action),
- Caution (i.e., requires immediate operator awareness and subsequent corrective action), and
- Advisory (i.e., requires crew awareness and may require subsequent or future action).

Used in this way, the severity level encodes a priority order for operator action (i.e., handle warnings first, then cautions, then advisories).

In addition to severity level for each expectation, HM requires a remedy code used for presentation and for arbitration related purposes. Note that at the knowledge level, the remedy code and the expectation function are logically synonymous. That is, the remedy code is simply a short-hand way to reference the semantics of the expectation function. To keep the presentation technology separate from the monitoring process, the remedy code is also used to map a notification to its effect (e.g., a text string and/or an aural alert). The actual alerting method, tailored to the presentation system, must be defined. When HM is integrated with a text-based presentation system, the knowledge engineer defines a unique text message to display for each remedy code. The knowledge engineer must decide whether the presented information is prescriptive (i.e., presented as a command) or descriptive (i.e., a definition of the problem situation) or both. When HM is integrated with other presentation systems (e.g., one using synthetic voice), similar issues arise.

ARBITRATION CONFLICT RESOLUTION AND ITS KNOWLEDGE REQUIREMENTS

It is possible for notifications generated from unmet expectations of one network to conflict with the expectations of another. To provide unambiguous feedback to the human operator, HM should only dispatch notifications if the expectations of one monitored activity are still valid in the context of the other activity. Because of the need for notification conflict resolution, HM requires knowledge to determine when to suppress conflicting information and what to suppress. The knowledge engineer defines conflict resolution knowledge as suppressing notifications from a particular network while another network is either active or inactive.

For example, suppose a pilot of a transport aircraft is performing a normal descent to the destination airport when the cabin pressurization system fails. HM monitors the normal descent activity and expects the aircraft to descend at a safe and comfortable rate (e.g., at a rate of descent no steeper than 3,000 feet per minute). HM also monitors the emergency descent activity because of the cabin pressurization problem and expects the aircraft to descend at a faster rate (e.g., at a rate of descent steeper than 2,000 feet per minute). If the pilot decides to

---

[2]Altitude above ground level (AGL) is the altitude above the terrain.

descend at 3,500 feet per minute, HM would generate the rate of descent notification from the normal descent network. The knowledge engineer therefore defines conflict resolution knowledge to suppress the normal descent network rate of descent notification in the context of the emergency descent activity.

DISCUSSION

To test the architecture described in this paper, we have integrated HM with a NASA Langley PC-based aircraft simulation [Nas95] to which we have added an Advisory, Caution and Warning System (ACAWS) and other enhancements. We have implemented several expectation networks and we are in the process of developing the arbitration conflict resolution functionality and knowledge bases.

Based on our work to date, the conflict resolution function not only provides the ability to present only the relevant notifications to the human operator but also allows the knowledge engineer to define monitoring knowledge for each activity separately in an incremental manner. As each expectation network is added, the knowledge engineer defines the associated arbitration related knowledge.

While the architecture facilitates incremental knowledge development, the conflict resolution function puts a burden on the knowledge engineer to determine which expectation networks can be simultaneously active. It also forces the knowledge engineer to determine which of those networks have expectations for related states. However, this extra work is typically easier to do than de-conflicting an entire knowledge base in one pass.

We plan to investigate one strategy for improving the knowledge engineering process: designing knowledge acquisition tools for HM. Knowledge acquisition tools for HM should enable defining all of HM's knowledge (i.e., interface, assessment, expectation network, and arbitration), not just the conflict resolution knowledge. A subset of the requirements for these tools are based on the conflict resolution knowledge definition:

1. Identify networks that are potentially simultaneously active,
2. Identify networks with expectations for related states, and
3. Provide the ability for the knowledge engineer to define which notifications to suppress when expectation networks are active or inactive.

With such tools, the knowledge engineer will be better supported in the task.

ACKNOWLEDGMENTS

The research reported herein was supported in part by NASA contracts NAS1-19898 and NAS1-20210 (Terence Abbott, Contracting Officer Technical Representative) and by USAF contracts F33615-94-C-3802 (Capt. Anthony Moyers, Project Engineer) and F33615-95-C-3611 (Anthony Ayala, Project Engineer). The authors would like to thank Stacy L. Henrie for formatting help.

REFERENCES

[Boe96] Boeing Commercial Airplane Group (1996). *Statistical summary of commercial jet aircraft accidents: Worldwide Operations, 1959-95*, page 30. Seattle, WA.

[Cas93] Casey, S. M. (1993). *Set phasers on stun: And other true tales of design, technology, and human error.* Santa Barbara, CA: Aegean Publishing Company.

[EB93] Endsley, M. R. and Bolstad, C.A. (1993). Human capabilities and limitations in situation awareness. In *Combat automation for airborne weapon systems: Man/machine interface trends and technologies*. AGARD-CP-520. Neuilly Sur Seine, France: NATO-Advisory Group for Aerospace Research and Development,; pp. 19/1 -19/10.

[GSZ+95] Greenberg, A. D., Small, R. L., Zenyuh, J. P., and Skidmore, M. D. (1995). Monitoring for hazard in flight management systems. *European Journal of Operational Research*, 84, Amsterdam, Netherlands: Elsevier Science B. V.

[HD95] Hughes, D. and Dornheim, M. A. (January 30, 1995). Accidents direct focus on cockpit automation. *Aviation Week & Space Technology*. New York: McGraw Hill, pp. 52-5.

[Nas95] NASA (1995). *NASA Baseline FMS Documentation*. Last revision October 3, 1995.

[RLW80] Randle, R. J., Larsen, W. E., and Williams, D. H. (1980). *Some human factors issues in the development and evaluation of cockpit alerting and warning systems*. NASA Reference Publication 1055. Moffett Field, CA: NASA Ames Research Center.

[SZS95]   Skidmore M. D., Zenyuh, J. P., and Small, R. L. (1995). Hazard monitoring: Arbitrating among remedial directives in the cockpit. In R. S. Jensen and L. A. Rakovan (Eds.) *Proceedings of the Eighth International Symposium on Aviation Psychology.* Columbus, OH: The Ohio State University.

[SZS+95] Skidmore, M. D., Zenyuh, J. P., Small, R. L., Quinn, T. J., and Moyers, A. D. (1995). Dual use applications of hazard monitoring: Commercial and military aviation and beyond. In *Proceedings of the IEEE/AESS/SAE National Aerospace Electronics Conference (NAECON) 1995.*

[US96]   U. S. Department of Transportation Regulations (1996). *Federal Aviation Regulations and Airman's Information Manual.* Renton, WA: Aviation Supplies & Academics, Inc.

Analysing FORTRAN Codes for
Program Understanding and Reengineering

Ronald B. Finkbine, Ph.D. and Warren O. Mason

finkbine@babbage.sosu.edu and wmason@babbage.sosu.edu

Department of Computer Science
Southeastern Oklahoma State University
Durant, OK 74701
(405) 924-0121

Abstract

Legacy software represents a large investment for many organizations in the scientific community. The maintenance programmers, which keep these software systems operational are continually developing, modifying and updating code. In order to assist the maintenance programmer in understanding legacy software, it is desirable to have a software tool that can automate repetitive and computable tasks. Past research has often used the term "intelligent editor" to describe this function This research project takes this direction further, it attempts to build an expert system that is able to understand segments of legacy software. The field of *Program Understanding* attempts to determine the function of a code segment with or without programmer intervention. This research analyses FORTRAN numerical analysis programs for common algorithm usage and other software characteristics. This paper discusses the problems identified by a cursory analysis of FORTRAN programs from the *Collected Algorithms of the ACM*.

Introduction

The purpose of this research is to develop a general-purpose algorithm recognition system, capable of recognizing any well-defined and well-written algorithm. This project uses models, (*plans* Wills, 1990) to recognize common forms (code segments) within existing software in an attempt to gain knowledge about a legacy software system. To understand legacy software, maintenance programmers generally review all existing documentation; however, many software systems will have little up-to-date documentation. This general lack of accurate external documentation results in the situation where the best documentation for a legacy system is the code itself (Biggerstaff, 1990). Program understanding generally occurs by matching the code segment in question against a large, defined set of models, or common algorithms, rather than attempting to deduce what a code segment performs from its specific data flow (Rugabur, 1990).

The ARS (Algorithm Recognition System) is an expert system which performs three functions: 1) analyzes input source code input by the user against a library of model numerical functions, 2) contains a library of models of numerical functions for comparison and detection, and 3) allows a user to input source code that will become a model and then added to the existing library.

Background

Program understanding is a subfield of artificial intelligence and software engineering known by alternative names, such as software re-engineering and knowledge-based software engineering. The main thrust of research in program understanding is to discover general knowledge about a segment of source code. There are two possible uses for this type of software tool: 1) allow the developer of new software to compare the segment of code being tested to a known correct model within the expert system, and 2) allow the maintenance programmer an assistant for segmenting a large program under revision into known and unknown segments. It is hoped that the maintenance programmer will be able to make revisions, improvements, and enhancements to existing source code if an expert system is able to answer questions about the source code.

Targeted Problems

Legacy software, in general, exhibits a number of the following problems: 1) parameter identification, 2) identifying code segment that are replaceable by calls to commercial libraries (such as IMSL), 3) removing duplicate code to user library, 4) separation of intertwined codes, and 5) combining disparate codes into single equations.

The first problem, *parameter identification,* is the most simple. It involves searching the source code for variables that are assigned values within assignment statements (no reads) one time. Any usage, thereafter, is only on the right-hand side of assignment statements and is reference to the variable, not a modification to the variable. Therefore, these types of variables, or constants, can be identified by the parameter statement which indicates their true usage.

The second problem, *plan recognition,* is comprised of identifying code segments that are replaceable by calls to commercial libraries (such as IMSL). This will involve detecting codes similar to those used within commercial libraries.

The third problem, *duplicate removal,* consists of detecting and removing duplicate code to the user's library. This allows the user to designate a section of code as common and to look through their remaining programs searching for codes that are copies of the target.

The fourth problem, *algorithm separation,* involves detection/separation of overlapping algorithms within the same section of code. In **Figure 1** it can be seen that there are two initializations of arrays occurring within the same do-loop. This is good for optimizing computer resources, but not for optimizing the programmers' time for understanding and maintaining a program.

```
          DO 10 I = 1, N
              A(I) = 0
              B(I) = 0
   10         CONTINUE
```

Figure 1: Intertwined Algorithms

The fifth problem, *algorithm aggregation,* involves combining disparate codes into single equations. As displayed in **Figure 2**, an equation 1) can be coded in multiple ways. Though the computations are equivalent, the recognition of them must take these variations into account.

```
   1) distance = sqrt(side1*side1+side2*side2)
   2) a = side1 * side1
   3) b = side2 * side2
   4) distance = sqrt(a+b)
```

Figure 2: Equation Variations

Detailed Example

The example subroutine in Figure is a portion of Algorithm 423 from the *Collected Algorithms From ACM.* The statements are identified for this discussion and a number of the statements are of interest. Statements 8 and 9 are the saving of an array position to a scalar variable. Statements 9, 10 and 11 are integrally related and constitute a swap, the exchange of values within two positions of the same array. Statements 12, 13 and 14 are a summation of a column within a matrix. Statements 18, 19 and 20 are a summation of a product of a scalar and a column within a matrix.

```
[ 1]    SUBROUTINE SOLVE(N,NDIM,A,B,IP)
[ 2]    REAL A(NDIM,NDIM),B(NDIM),T
[ 3]    INTEGER IP(NDIM)
[ 4]    IF (N.EQ.1) GOTO 9
[ 5]    NM1=N-1
[ 6]    DO 7 K = 1, NM1
[ 7]       KP1=K+1
[ 8]       M=IP(K)
[ 9]       T=B(M)
[10]       B(M) = B(K)
[11]       B(K) = T
[12]       DO 7 I = KP1, N
[13]             B(I)=B(I) + A(I,K)
[14] 7CONTINUE
[15]    DO 8 KB = 1, NM1
[16]       KM1 = N - KB
[17]       B(K)=B(K) - A(K,K)
[18]       DO 8 I = 1, KH1
[19]             B(I)=B(I)+A(I,K) * T
[20] 8 CONTINUE
[21] 9 CONTINUE
[22]    B(1)=B(1) / A(1,1)
[23]    RETURN
[24]    END
```

Figure 3: Algorithm 423

ARS Overview

The ARS system is formed of a number of modules as shown in Figure 1. The processing of FORTRAN source programs begin with traditional compiler technology, lexical scanning and parsing. The input programs (whole programs, subroutines or functions) are transformed into an Internal Representation (IR) which resides within a Relational Database Management System (RDBMS). The use of a standard database allows for a correct and complete representation of data in much larger portions than traditional compilers. A 10,000 line program is going to produce an extremely large amount of data.

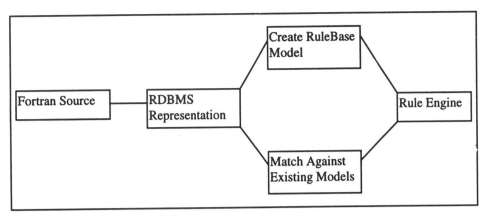

Figure 4: ARS System

From the IR one of two operations can be performed at user request. The most used operation will be plan recognition. This will analyze the user's source program, searching for models that have been pre-defined in the rule base system. The second operation available to the user is to be able to create a rule base for the specified input source segment. This allows the user to add a model to the models rule base and search source code for the designated model as well as the standard model library. This feature will allow the user to search for duplicate code within a single program or within an entire software system.

The rule system is divided into two subsystems: general-purpose and the user-specific library. The general-purpose library is a collection of models which are pre-inserted into the database. These include simple models such as an increment statement as well as models such as sorting algorithms.

Currently, this system is hosted on an IBM 80486 class machine with Windows 95 as an operating system. The ARS expert system itself is written in Delphi with a shareware RDBMS running in the background.

Current Status

This project currently has a number of its major components under revision. The RDBMS is complete and stable. The FORTRAN scanner and parser are complete and working properly. Elementary recognition of models has been performed. The expert system shell used for the preliminary recognition is being replaced by a custom written rule engine. This is necessary since usage of the DOS-based CLIPS expert system was not conducive to

repetitively executing for extremely small segments of code.

Acknowledgement

This project has been supported by the Faculty Research Fund at Southeastern Oklahoma State University.

Author Information

Mr. Mason is a undergraduate student in the B. S. program at SOSU. Dr. Finkbine is an Assistant Professor of Computer Science at Southeastern Oklahoma State University. He has a B.S. in Computer Science, 1985, and an M.S. in Computer Science, both from Wright State University. He has the Ph.D. in Computer Science from the New Mexico Institute of Mining and Technology.

References

Biggerstaff, Ted, *Design Recovery for Maintenance and Reuse*, in *IEEE Computer*, July, 1990.

Finkbine, Ronald B., *Recognition of High-Level Algorithms*, Ph.D. Dissertation, Department of Computer Science, New Mexico Institute of Mining and Technology, 1994.

Rugaber, Spencer, Stephen B. Ornburn, and Richard LeBlanc, Jr., *Recognizing Design Decisions in Programs*, in *IEEE Computer*, July, 1990.

Wills, Linda, *Automated Program Recognition by Graph Parsing*, Ph.D. Dissertation, MIT Artificial Intelligence Laboratory, July, 1992.

MODELING CAUSAL-EFFECT RELATIONSHIPS FOR ARTIFICIAL INTELLIGENCE APPLICATIONS

Kumbesan Sandrasegaran

Department of Electrical Engineering, University of Durban-Westville,
Private Bag X54001, Durban 4000, South Africa.
Email: kumbes@pixie.udw.ac.za

ABSTRACT

Cause-effect representations form the foundations for representing systems causality in a number of AI systems. Such a representation can be used for a number of different tasks. It can be used for simulating the behavior of a device, providing explanations about the operation of a device, generating diagnostic knowledge, etc. In this paper, a taxonomy of models of cause-effect relationships of device operation and how these models were used to generate diagnostic knowledge are presented.

INTRODUCTION

Research in causal reasoning can be classified into causal models and theories of causation. Causal models are computational models that apply AI and other techniques to predict or explain events in a particular domain. Theories of causation pertain to a domain independent, philosophical, and theoretical understanding of the nature of causality.

Causal models are further subdivided into explicit and implicit causal models. Explicit causal models use a graph of nodes and links to depict cause-effect knowledge explicitly. This can be used for a number of tasks such as diagnosis, explanations, etc. The nodes of a causal graph could represent states, events, actions, tendencies, and state changes [Rei76]. The links could represent causes, possibly causes, etc. Implicit causal models use structures that do not represent causality explicitly but can be used to predict the behavior or produce explanations of a system.

The cause-effect knowledge of a system can be explicitly specified or generated using one of the well known techniques such as causal analysis [DB84], QSIM [Kui84], method of comparative statics [IS86, IS94], stochastic simulation [Pea86], theory of causal structures [Ban90], CAUSIM [Fu91], etc.

Causal knowledge of devices have been represented in a number of different ways: mechanism graphs [Dek79], state diagrams [DB84], causal relation graphs [WHS84], causal process diagrams [SC86, Cha94], signed directed graphs [SKT91], transition spaces [Bor92], causal networks [TC89, BS92], etc. Most of these representations have a number of similar characteristics such as (a) nodes representing events or states, (b) directed arcs representing transitions between them, (c) abstractions of cause-effect knowledge at different levels of the functional hierarchy of the device, and (d) temporal relationships between events. However, there does not exist a general and domain independent representation formalism for describing the cause-effect knowledge of devices. Such a formalism will serve a number of purposes.

CAUSAL DEPENDENCY

Consider two adjacent events, A and B, in a causal graph of physical system. A number of possibilities may arise pertaining to causal dependency between A and B. These are:

- If A occurs, then B should occur immediately after A.
- If A occurs, then B may occur immediately after A.
- If B occurs, then A should have occurred immediately prior to B.
- If B occurs then A may have occurred immediately prior to B.

Each of the above statements address a different type of relationship between events A and B. These relationships between events are referred to as inter-event relationships. The first two inter-event relationship pertain causal dependence in the forward direction, i.e., from cause to effect. The last two inter-relationship pertain to causal dependence in the backward direction, i.e., from effect to cause. In this work, four primary type of inter-event relationship are used

- immediate dependent (id),
- immediate partial dependent (ipd),
- backward immediate dependent (bid),
- backward immediate partial dependent (bipd).

These inter-event relationships were formulated from a diagnostic perspective. However they can be applied to a number of different tasks. Each of the inter-event relationship will be defined next.

Immediate Dependent Relationship

The immediate dependent relationship, *id(x,y)*, means that if event x occurs then event y must occur immediately after event x. Thus two separate criteria must be satisfied:
(a) event y must occur if event x occurs, and
(b) event y must occur immediately after event x.

Immediate Partial Dependent Relationship

The immediate partial dependent relationship, *ipd(x,y)*, means that if event x occurs, then event y may occur but if event y does occur it will be soon after event x. This inter-event relationship expresses partial causal dependence in the forward direction.

Backward Immediate Dependent Relationship

The backward immediate dependent relationship, *bid(x,y)*, is used to denote the fact that if an effect event y is known to have occurred, then any one of its cause events, x, should have also occurred immediately before event y. The 'bid' relationship is used for inferencing in the opposite direction to the causality. It represents causality in the backward direction. If two events x and y hold an 'id' relationship, then it does not imply that there is a 'bid' relationship between them.

Backward Immediate Partial Dependent Relationship

The backward immediate partial dependent relationship, *bipd(x,y)*, means that if an effect event y is known to have occurred, then its cause event, x, may have occurred immediately before event y. It is used for inferencing in the opposite direction to the causality. The 'bipd' inter- relationship represents causality in the backward direction.

CAUSE-EFFECT RELATIONSHIPS

In this paper, cause-effect relationships are represented using the syntax:
sequence(cause(A), effect(B)).
It represents the fact that an event A causes an event B. A more general cause-effect relationship is,
sequence(cause($A_1,A_2,A_3,...,A_m$), effect($B_1,B_2,B_3,...,B_n$)),
since an event can have multiple causes and/or multiple effects. Events $A_1,A_2,A_3, ..., A_m$ represent cause events and events $B_1,B_2,B_3,...,B_n$ represent effect events. In this work, cause-effect relationships have been classified into five primitive types:

- 1-1 : representing one cause event resulting in one effect event,
- 1-m 'and' : representing one cause event resulting in m effect events,
- 1-m 'or' : representing one cause event resulting in one of m effects,
- m-1 'and': representing m cause events necessary for an effect event,
- m-1 'or' : representing one of m cause events resulting in an effect.

The first parameter (1 or m) represents the number of cause events and the second parameter (1 or m) represents the number of effect events. The third parameter ('and' or 'or') depicts the relationships amongst the sequences. Each type of causal sequence is described next.

The 1-1 Causal Sequence

A 1-1 causal sequence, sequence(cause(A), effect(B)) , contains one cause effect resulting in one effect event as shown in Figure 1. Event B is the only effect of event A and event A is the only cause of event B.

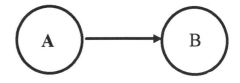

Figure 1. The 1-1 Causal Sequence

Amongst the four types of primary inter-event relationships, only the 'id' and 'bid' apply to the 1-1 causal sequence as shown below.

	'id'	'ipd'	'bid'	'bipd'
1-1	✓		✓	

Expression (1) is used to generate 'id' relationships in a 1-1 sequence. It states that if there exists an event x that has an effect y and there does not exist any other effect(s) of event x, then events x and y have an immediate dependent relationship. The notation $? denotes a multifield wildcard that matches to zero or more fields.

$$\text{sequence(cause(x),effect(y))} \land$$
$$\neg(\text{sequence(cause(x),effect(\$?:}\neg y))) \Rightarrow id(x,y) \ ...(1)$$
$$\text{sequence(cause(x),effect(y))} \land$$
$$\neg(\text{sequence(cause(?:}\neg x),\text{effect(y)))} \Rightarrow bid(x,y) \ ... (2)$$

The 1-m 'and' Causal Sequence

The 1-m 'and' causal sequence represents causality that contains one cause event resulting in many simultaneous effects, as shown in Figure 2. The syntax used is:

$$\text{sequence(cause(A), effect(} B_1 , B_2, B_3, , B_n))$$

Effect events B_1, B_2, B_3 and B_n , comprise all the simultaneous effect events resulting from the event A. Furthermore, each of the effect events, B_i, does not have any cause event but A. The distinguishing feature of the 1-m 'and' causal sequence syntax is that it contains two or more events in the effect slot of the sequence.

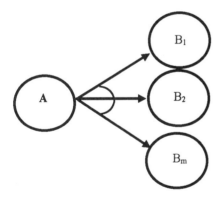

Figure 2. The 1-m 'and' Causal Sequence

	'id'	'ipd'	'bid'	'bipd'
1-m 'and'	✓		✓	

Expression (3) is used to generate 'id' relationships using the 1-m 'and' causal sequences and expression (4) is used to generate 'bid' relationships.

$$\text{sequence(cause(x),effect(}y_1, y_2, y_3, y_n))$$
$$\Rightarrow id(x,y_1) \land id(x,y_2) \land ... \land id(x,y_n) \ ...(3)$$
$$\text{sequence(cause(x),effect(}y_1 ,y_2 ,y_3 ,y_n))$$
$$\Rightarrow bid(x,y_1) \land bid(x,y_2) \land ... \land bid(x,y_n) \ ...(4)$$

The 1-m 'or' Causal Sequence

If an event results in one of many possible effects, the 1-m 'or' causal sequence is used, as shown in Figure 4.3. This type of causal sequence is represented as a collection of sequences:

$$\text{sequence(cause(A),effect(}B_1)),$$
$$\text{sequence(cause(A),effect(}B_2)),$$
$$...$$
$$\text{sequence(cause(A),effect(}B_n))$$

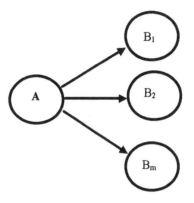

Figure 3. The 1-m 'or' Causal Sequence

Effect events B_1, B_2, B_3 and Bn, comprise all the possible effect events resulting from the event A. Furthermore, each of the effect events, B_i, does not have any cause event but A. Each of the cause-effect relationships is represented as a separate causal sequences. The difference between a 1-m 'or' causal sequence and m x 1-1 causal sequences is that the latter does not have a common cause event in the m causal sequences.

	'id'	'ipd'	'bid'	'bipd'
1-m 'or'		✓	✓	

Expression (4.31) pertains to the generation of 'ipd' inter-event relationships using 1-m 'or' sequences. Expression (4.32) pertains to the generation of 'bid' relationships. It states that if there exists an event y that has a cause x and there does not exist any other cause of event y then events x and y have a 'bid' relationship.

$$\text{sequence(cause}(x)\text{, effect}(y_1)) \wedge$$
$$\text{sequence (cause}(x)\text{,effect}(y_2)) \wedge ... \wedge$$
$$\text{sequence(cause}(x)\text{,effect}(y_n))$$
$$\Rightarrow \text{ipd}(x,y_1) \wedge \text{ipd}(x,y_2) \wedge ... \wedge \text{ipd}(x,y_n)...(5)$$

$$\text{sequence(cause}(x)\text{,effect}(y)) \wedge$$
$$\neg(\text{sequence(cause}(?:\neg x)\text{,effect}(y)))$$
$$\Rightarrow \text{bid}(x,y) \ ... \ (6)$$

The m-1 'and' Causal Sequence

When many concurrent events acting together result in another event then the causality displayed is of the type m-1 'and' causal sequence. This is shown in Figure 4. It is represented as:

$$\text{sequence(cause}(A_1,A_2,A_3,, A_m)\text{, effect}(B)).$$

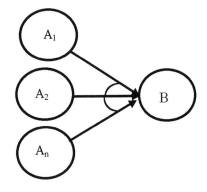

Figure 4. The m-1 'and' Causal Sequence

Events A_1, A_2, A_3, and A_n, comprise all the simultaneous cause events of event B. Furthermore, each of the cause events, A_i, does not have any effect event but B. The m-1 'and' sequence is the only sequence containing more than one event in the cause slot. Thus it is easily distinguished from the other sequences.

	'id'	'ipd'	'bid'	'bipd'
m-1 'and'		✓	✓	

Expression (7) pertains to generation 'ipd' relationships from the m-1 'and' sequence and Expression (8) pertains to generation of 'bid' relationships.

$$\text{sequence(cause}(x_1,x_2,x_3, \ ... \ ,x_m)\text{,effect}(y)) \Rightarrow$$
$$\text{ipd}(x_1,y) \wedge ... \wedge \text{ipd}(x_m,y) \ ...(7)$$

$$\text{sequence(cause}(x_1,x_2,...x_m)\text{,effect}(y)) \Rightarrow$$
$$\text{bid}(x_1,y) \wedge \text{bid}(x_2,y) \wedge ... \wedge \text{bid}(x_m,y)...(8)$$

The m-1 'or' Causal Sequence

In the m-1 'or' causal sequence, one of many possible cause events can result in an effect event, as shown in Figure 5. The syntax used is given below:

$$\text{sequence(cause}(A_1)\text{, effect}(B)),$$
$$\text{sequence(cause}(A_2)\text{, effect}(B)),$$
$$...$$
$$\text{sequence(cause}(A_m)\text{, effect}(B))$$

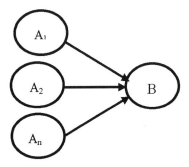

Figure 5. The m-1 'or' Causal Sequence

Events A_1, A_2, A_3 and A_n, comprise all the possible cause events of event B. Furthermore, each of the cause events, A_i, does not have any effect event but B.

	'id'	'ipd '	'bid'	'bpd'
m-1 'or'	✓			✓

Expression (9) is used to generate 'id' relationships in the m-1 'or' sequences. It states that if there exists an event x that has an effect y and there does not exist any other effect(s) of event x, then events x and y have an immediate dependent relationship.

$$\text{sequence(cause}(x)\text{,effect}(y)) \wedge$$
$$\neg(\text{sequence(cause}(x)\text{,effect}(?:\neg y))) \Rightarrow \text{id}(x,y)...(9)$$

COMPOSITE CAUSAL SEQUENCES

Composite causal sequences refer to those causal sequences that contain two or more primary causal sequences. An example of a composite causal sequence is shown in Figure 6. Event A causes event B and C and either event A or event D can cause event C. These sequences will be represented using the following syntax:

sequence (cause(A), effect(B, C)),
sequence (cause (A), effect(C)),
sequence (cause (D), effect (C)

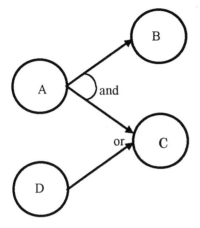

Figure 6 An example of a composite causal pattern

If event C is known to have occurred, what can be said about the occurrence of event A. Using the 1-m 'and' sequence, there is a 'bid' relationship between events A and C, i.e. bid(A,C). Using the 1-m 'or' sequence, there is a 'bipd' relationship between A and C, i.e. bipd(A,C). This is clearly inconsistent with the earlier inter-event relationship. However, by splitting the cause-effect graph as shown in Figure 7 and inserting a hypothetical event AC, the inconsistency can be solved.

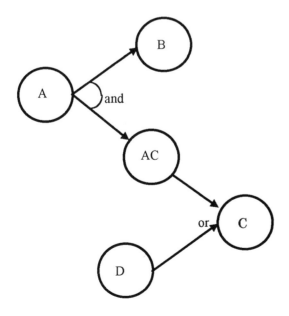

Figure 7. Modified Composite Cause-Effect Graph

For this graph in Figure 7, the following inter-event relationship can be obtained:

sequence(cause(A),effect(B,AC)):   id(A,B), id(A,AC), bid(A,AC), bid(A,B)
sequence(cause(AC,D),effect(C)): id(AC,C), id(D,C), bipd(AC,C), bipd(D,C)

There are no inconsistencies in the inter-event relationship above.

CONCLUSION

This paper presented a domain, application and task independent methodology for modeling cause-effect relationship. The inter-event relationships described are generated using cause-effect knowledge of a system. If the cause-effect graph of a device contains n nodes and if there are m types of inter-event relationships between events, then a maximum of m x n(n - 1) relationships are generated. Thus the worst case order of the generated inter-event relationships is $O(n^2)$.

REFERENCES

[Ban90]   V.R. Bandekar, "Causal Structures: Computation and Applications" in ECAI 90: Proceeding of the 9th European Conference on AI, pp.71-76, 1990.

[Bor92] G. C. Borchardt, "Understanding Causal Descriptions of Physical Systems", in Proceeding of the AAAI, pp.2 - 8, 1992.

[BS92] M. Botta and L. Saitta, "Use of causal models and abduction in learning diagnostic knowledge", International Journal of Man-Machine Studies, Vol. 36, pp.289-307, 1992.

[Cha94]   B.Chandrasekaran,   "Functional Representation and Causal Processes", to be published in Advances in Computers, ed. Marshall Yovits, Academic Press, 1994.

[Dek79] J. De Kleer, Causal and Teleological Reasoning in Circuit Recognition, Ph.D. thesis, AI Lab, MIT, 1979.

[DB84] J. de Kleer and J. S. Brown, "A qualitative physics based on confluences", in Artificial Intelligence, Vol. 24, pp. 7- 83, 1984.

[Fu91] L. M. Fu, "Causim: A rule based causal simulation system" in Simulation, Vol.56, pp.251-257, April 1991.

[IS86] Y. Iwasaki and H. A. Simon, "Causality in device behavior", Artificial Intelligence, vol.29, pp.3-32, 1986.

[IS94] Y. Iwasaki and H.A. Simon, "Causality and model abstraction" Artificial Intelligence, Vol.67, pp.143-194, 1994.

[Kui84] B. Kuipers, "Commonsense reasoning about causality: deriving behavior from structure" in Artificial Intelligence, Vol.24, pp.169-203, 1984.

[Pea86] J. Pearl, "Fusion, propagation, and structuring in belief networks" in Artificial Intelligence, vol.29, pp.241-288, 1986.

[Rei76]. C. Reiger, "An organization of knowledge for problem solving and language comprehension", Artificial Intelligence, Vol.7, pp.89--127, 1976.

[SC86] V. Sembugamoorthy & B.Chandrasekaran "Functional Representation of Devices and Compilation of Diagnostic Problem Solving Systems" in Experience, Memory and Learning edited by . J. Koldoner and C. Riesbeck, pp. 47-73., Lawrence Erlbaum, 1986.

[SKT91] J. Shiozaki, S. Karibe, and H. Tabuse, "Diagnostic rule generation from qualitative plant models" in Advances in Instrumentation and Control, Proceedings of the ISA/91 International Conference and Exhibition, pp.1405-1413, Vol. 2, ISA, 1991.

[TC89] P. Torasso and L. Console, Diagnostic Problem Solving: Combining Heuristic, Approximate and Causal reasoning,. London: North Oxford Academic, 1989.

[WHS84] M. Williams, J. Hollan, and A. Stevens, "An overview of steamer: an advanced computer-assisted instruction system for propulsion engineering", Behavior Research Methods and Instrumentation, Vol.13, no.2, pp.85-90, 1981.

An Extensible Frame Language for the Representation of Process Modeling Knowledge

Christian Rathke and Frank Tränkle
Institut für Informatik, Institut für Systemdynamik und Regelungstechnik
University of Stuttgart, D-70565 Stuttgart, Germany
rathke@informatik.uni-stuttgart.de, traenkle@isr.uni-stuttgart.de

Abstract

A knowledge-based process modeling tool (PROMOT) is being developed to assist chemical engineers in generating mathematical models of process units. For representing the process modeling knowledge, a language with concepts specific to the chemical engineering domain has been designed. The Model Definition Language (MDL) is an extension to a frame language called FRAMETALK, which has been specifically designed with the goals of adaptability and extensibility in mind.

The FRAMETALK system includes a meta level at which certain aspects of its functionality may be extended. Using the example of a continuously stirred tank reactor, it is shown how the frame representation is extended to support the formulation of process modeling knowledge with MDL. The ability to provide abstractions for domain specific concepts greatly reduces the complexity for the chemical engineer to express new modeling entities or to compose them to form new process models.

INTRODUCTION

The applications of mathematical modeling and computer-aided simulation in chemical engineering are manifold. Commercially available software tools are used for stationary design and optimization of process plants. Dynamical process simulation is applied to the design and the inspection of process control systems, the startup and shutdown of plants, their behavior in case of operation faults, as well as the design and operation of inherently dynamical processes. Generally speaking, model-based techniques gain more and more importance in chemical engineering in order to improve chemical processes with respect to growing economical and environmental demands.

However, the development and validation of adequate process models strongly limits the routine application of model-based techniques in process design as well as process operation. The problem arises from the complexity and variety of process units that constitute process plants. The mathematical models of these process units follow both from a rigorous application of the conservation laws of mass, momentum, and energy, and from the consideration of the occuring physico-chemical phenomena. These phenomena can be described by a large number of alternative formulations [MZ92], impeding the model development as well as the design of model libraries for process units. Only experienced modeling experts with a solid background in chemical engineering, general systems theory, and applied mathematics are capable of deriving process models of novel process units.

At present, process simulation environments such as DIVA [KHMG90] and SPEEDUP [Pan88] are widely used in chemical engineering. They provide powerful numerical methods to analyze and solve systems of differential-algebraic equations that constitute the process models. Their model libraries contain modeling entities in the granularity of individual process units that can be interconnected to process models by specifying the flowsheet of the considered plant. Unfortunately, these environments give only very limited support to the development of new process unit models as well as to the reuse and the documentation of existing models. To overcome this problem, considerable efforts have been initiated during the last decade to develop general as well as domain specific process modeling and simulation environments [Mar96, PB95] that support process model development, repository, analysis, and simulation. Despite the advances, the developed software systems still show some shortcomings. For instance, adequate concepts and language constructs to design, implement, handle, and document large chemical engineering specific knowledge bases are scarcely available.

The following section introduces a knowledge-based process modeling tool, whose knowledge base is implemented with the Model Definition Language (MDL). MDL is an extension to the frame representation language FRAMETALK [Rat96] with language concepts specific to the chemical engineering domain. In the subsequent sections, the extension of FRAMETALK for the implementation of MDL is illustrated by using the model development of a continuously stirred tank reactor as an example.

A PROCESS MODELING TOOL

A knowledge-based process modeling tool (PROMOT) [TGZG97a, TGZG97b] (Figure 1) is being developed to assist chemical engineers in generating mathematical models of novel process units. The object-oriented knowledge base of PROMOT contains general process modeling knowledge, from which the chemical engineer can interactively browse information and generate process unit models. This knowledge base is persistently stored in the *model library*, whose contents can be loaded to the temporary *workspace*. With either the graphical MDL-Editor or the textual MDL-Editor, the chemical engineer interactively develops *process unit models* by aggregating and specifying *modeling entities* in the workspace. Once a process unit model is completed, it can be saved to the model library for later reuse or exchange with collaborating chemical engineers. With its interface to the *pre-processing tool* and *code generator* [RGGZ94], PROMOT translates the internal symbolic representation of selected process unit models to DIVA simulation modules [KHMG90], which can be interconnected to simulation-ready plant models in the context of plant flowsheets. In summary, the major features of PROMOT are: an information system for process modeling knowledge, the interactive development of process unit models with the graphical MDL-Editor or the textual MDL-Editor, the repository and documentation of process unit models.

For the representation of the process modeling knowledge and its implementation in the knowledge base of PROMOT the Model Definition Language (MDL) has been developed. PROMOT's knowledge base contains elementary and composite modeling entities. These modeling entities have been obtained from a rigorous structuring of the process modeling knowledge by using concepts of general systems theory and object-oriented knowledge representation [TGZG97a]. Three distinct types of modeling entities are identified: *structural, behavioral,* and *material* modeling entities. Examples of *structural modeling entities* are thermodynamic phases, phase boundaries, and solid walls. Examples of *behavioral modeling entities* are balance equations, phenomenologi-

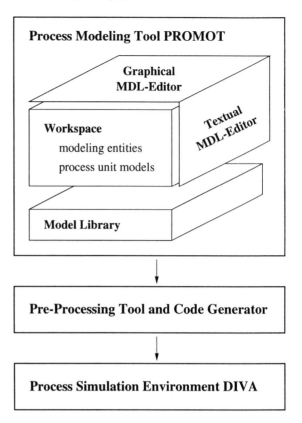

Figure 1: Architecture of the Modeling and Simulation Environment PROMOT/DIVA

cal relations, and physical property correlations that represent the physico-chemical and biological behavior of the structural modeling entities. *Material modeling entities* describe the pure chemical substances and mixtures being situated in the structural modeling entities by sets of parameters for the physical property correlations.

Both the modeling entities and the process unit models are defined as MDL structures, which are arranged in specialization and aggregation hierarchies. According to the three types of modeling entities, MDL provides expressions to define structural, behavioral, and material modeling entities. Composite structural modeling entities can be successively aggregated from more fine-grained structural modeling entities. Each structural modeling entity has a set of text attributes, which are informal representations of the physico-chemical assumptions and simplifications that lead to the mathematical model of the structural modeling entity. According to these attributes, the chemical engineer selects predefined behavioral and material modeling entities from the knowledge base and attaches them to the structural modeling entity.

Each MDL structure representing a structural modeling entity may be manipulated in both its graphical or textual form. The chemical engineer may develop models using both the graphical or the textual MDL-Editor. As soon as MDL structures in the graphical MDL-Editor are modified, the changes are propagated to the textual MDL-Editor and vice versa.

AN EXAMPLE

For illustrating the graphical and the textual representation of MDL structures, the model development of a continuously stirred tank reactor (CSTR) (Figure 2) is considered as an example.

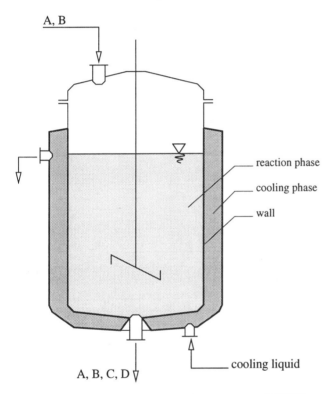

Figure 2: Continuously stirred tank reactor (CSTR) with a reaction and a cooling phase.

The CSTR is fed with the chemical substances A and B that react to the substances C and D in the liquid reaction phase. Because of the exothermal reaction the liquid reaction phase exchanges heat with the enclosing liquid cooling phase across the tank's wall.

For simplicity reasons, in the following we only consider the aggregation of the reactor model from elementary structural modeling entities and do not show its behavioral and material modeling entities. The process unit model of the reactor is aggregated from the elementary modeling entities reaction phase, cooling phase, and wall, where each modeling entity is given a unique name: "rph", "cph", and "w", respectively. Figure 3 depicts the graphical representation of the

reactor model in the graphical MDL-Editor, whereas Figure 4 shows its textual representation as a MDL expression in the textual MDL-Editor. The MDL structure of the reactor model is defined as a sub-class of the MDL structure "composite-device", which is the super-class of all process unit models in PROMOT. Its components are the MDL structures "reaction-phase", "wall", and "cooling-phase" with names "rph", "cph", and "w", which are interconnected via exchange terminals as defined by the double-sided arrows in Figure 3 and the construct ":links" in Figure 4.

The MDL structure of the reactor model itself has four pipe-line terminals with names "rph-inlet", "rph-outlet", "cph-inlet", and "cph-outlet". They define the interaction of the reactor with its surroundings and can be interconnected to pipe-lines for the reaction feed, the reaction outlet, the cooling liquid feed, and the cooling liquid outlet in the context of plant flowsheets. The terminal "rph-inlet" is identical to the terminal "in" of the reaction phase, i.e., the reaction feed is the inlet of the reaction phase. Similar identities exist for the terminals "rph-outlet", "cph-inlet", and "cph-outlet", which are expressed by the straight lines inbetween the terminals in the graphical MDL-Editor (Figure 3) and by the MDL construct ":is-eq-to" in the textual MDL-Editor (Figure 4).

A FRAME-BASED REPRESENTATION

The workspace of PROMOT (see Figure 1), which holds the current state of the process model being developed, and the model library, which holds the model knowledge base, are organized using a layered architecture. The upmost layer contains the modeling entities and process unit models as MDL structures. These structures are mapped to frame representations in the middle layer, which in turn are mapped to classes and instances of an object system at the base level.

The frame representation scheme – called FRAME-TALK [Rat96] – at the middle layer is specifically designed with the goals of adaptability and extensibility in mind. Usually, knowledge representation languages provide a fixed set of primitives to express knowledge about a domain. For instance, in rule-based languages such as OPS5 [For81] there are primitives for specifying rules, their condition and action parts, and data elements such as classes, integers and strings. In term description languages such as Loom [MB87] and Classic [BBMR89] there are primitives for specifying concepts, roles, and role descriptions. By combining the language primtives to more complex structures and giving them a name, programmers or "knowledge en-

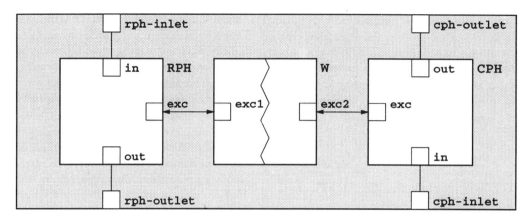

Figure 3: The MDL structure of the reactor model in the graphical MDL-Editor. The elementary modeling entities of the model are the reaction phase ("rph"), the cooling phase ("cph"), and the wall ("w").

```
(define-device :class "reactor"
   :super-classes ("composite-device")
   :documentation "CSTR with a homogeneous reaction phase"
   :components (("rph" :is-a "reaction-phase")
               ("w" :is-a "wall")
               ("cph" :is-a "cooling-phase"))
   :terminals (("rph-inlet" :is-eq-to "rph.in")
               ("rph-outlet" :is-eq-to "rph.out")
               ("cph-inlet" :is-eq-to "cph.in")
               ("cph-outlet" :is-eq-to "cph.out"))
   :links (("rph.exc" :is-linked-to "w.exc1")
           ("cph.exc" :is-linked-to "w.exc2")))
```

Figure 4: The MDL expression for defining the reactor model in the textual MDL-Editor.

gineers" define domain specific abstractions.

The number of compositional operators to define new abstractions is quite limited. For instance, it is usually impossible to adapt the general inferencing method to a more specific one and make the knowledge representation system use it in a specialized context. Also, there is usually only a restricted number of possible data types which may be used to encode the knowledge, e. g., there is no analogue to the abstract data type concept of programming languages, which allows to define arbitrarily complex data types.

In frame languages such as FRL [RG77] or KEE [KEE85] the situation is not much different. There are basically two different types of structured data: abstract and concrete frames. An abstract frame is a named structured type defined by its slots. A concrete frame is an instance of an abstract frame with specific values for all or some of the slots specified in the abstract frame. In abstract frames, slots may be further annotated by restrictions for their values, by default values, and by attached procedures. Attached procedures are activated under certain conditions such as when requesting or assigning a value. Abstract frames may be related to each other by specialization. A more specific frame inherits all the slots and the slot descriptions from the more general frames. Combining frames by inheritance and adding slots are the only way of generating new abstractions from previously existing ones.

A frame language is – like a programming language – a bare bones system which gives only little support for the knowledge engineer to represent the concepts of a specific application domain. The primitives of a representation language are like the atoms of the physical world and the physical laws which govern their behavior. They may be sufficient to describe every physical existence but in almost all cases using them is awkward and cumbersome.

A frame language may be adapted to the specific needs of the domain expert by providing predefined frames which represent basic conceptual entities of the application domain. For an architect, it is much easer to combine and extend representations for walls, windows, buildings and the important relations between them than to start with the "most general object" which subsumes everything.

Similar to architecture, the basic conceptual entities a chemical engineer may use to represent process models are just a starting point. To experiment with and build more complex models, the basic concepts must be extended, combined, and framed as new abstractions, which could serve as the building blocks of even higher level models.

Like the "conceptual atoms" of a frame system, the means for forming new abstractions (specialization and the definition of slot descriptions) are independent of and therefore semantically distant from any specific application domain. It is often necessary to not only extend the basic concepts but also the *mechanisms* for generating new abstractions from existing ones.

FRAMETALK

FRAMETALK is a frame language whose run time system includes a meta system at which certain aspects of the functionality of FRAMETALK may be extended. This meta system is implemented by a collection of classes, objects, generic functions and methods. Together they define how FRAMETALK frames are initialized, how slot descriptions are interpreted, and how and when attached procedures are activated. The FRAMETALK meta system is an extension to the meta system of CLOS, which is known as the Meta Object Protocol (MOP) [KdRB91] (Figure 5).

Each implementation layer supports a corresponding language. The CLOS language consists of macros and functions for defining classes, slots, generic functions and methods [Ste90, 770-864]. The FRAMETALK language consists of functions and macros for defining frames, slots, slot descriptions and attached procedures [RR96]. Like CLOS, FRAMETALK is embedded in LISP: FRAMETALK expressions are evaluated by the LISP interpreter and LISP forms may be used as arguments to, for instance, slot specifications.

CLOS plays an important role for FRAMETALK, because, first, it is FRAMETALK's implemen- tation language* and, second, a major part of FRAMETALK's functionality overlaps with the functionality of CLOS.

---

*The CLOS extensions to LISP are implemented as objects and generic functions in CLOS. In this sense, CLOS is implemented by itself.

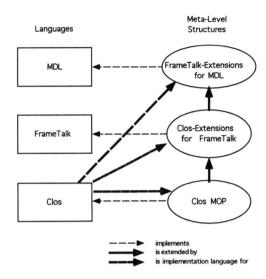

Figure 5: The Relationships between CLOS, FRAMETALK and MDL

The functionalities of CLOS, FRAMETALK and MDL are all specified using the CLOS language for defining classes, generic functions and methods as extensions to the MOP.

Viewing the progression of layers from CLOS to MDL as a way of reducing the semantic distance from a general programming language to an application specific language, an important step is taken when moving from a frame representation (FRAMETALK) to a domain specific representation (MDL). The computational paradigm of manipulating slots and thereby invoking attached procedures is replaced by a declarative approach of specifying relations from the chemical engineering domain between devices, phases and terminals.

The mapping from MDL to FRAMETALK is illustrated using the reactor example of section . In the chemical engineering domain, as in other technical domains, aggregate objects play an important role. The definition of the reactor device (Figure 4) mentions three components of a reactor: the reaction phase, the cooling phase and the wall. The device definition and the specification of the components relationship are transformed to a frame definition with slots and attached procedures. Figure 6 shows part of the definition of the reactor frame as a FRAMETALK expression. The frame has three slots: one for the reaction phase, one for the cooling phase and one for the wall between them. The components, which are also frames, and their container are linked by slots pointing to each other. In order to keep the relationships consistent, the slots have procedures attached. The procedures are activated *after* any of the slots are modified.

```
(defframe reactor (composite-device)
      (:slots (rph (:subframe reaction-phase
                        (:slots (inside (:frame reactor)
                                        (:if-set-after update-rph-slot)))))
                  (:if-set-after update-inside-slot))
            (cph (:subframe cooling-phase
                        (:slots (inside (:frame reactor)
                                        (:if-set-after update-cph-slot)))))
                  (:if-set-after update-inside-slot))
            (w (:subframe wall
                  (:slots (inside (:frame reactor)
                                  (:if-set-after update-w-slot)))))
                  (:if-set-after update-inside-slot))))
```

Figure 6: The Reactor Frame with Attached Procedures for Maintaining a Components Relationship.

Although the relationship between a container object and its parts could be defined using the means of the frame language, it is much easier for the chemical engineer to use the declarative approach and the domain specificy of MDL. Although in both cases, domain specific modeling entities such as "reaction-phase" and "wall" are used, the MDL expression reduces complexity by also providing language primitives and compositional operators from the chemical engineering domain. The FRAMETALK expression of Figure 6 has already been simplified by only mentioning the concepts which make up the components relationship. The "terminals" and "links" specifications of the MDL expression make the definition of the reactor device in FRAMETALK even more complicated.

The mapping from MDL to FRAMETALK is determined by the meta level structures of MDL which extend the meta level structures of FRAMETALK (see Figure 5). It is important to note that this is not a purely linguistic transformation from one language to the other. Instead, the domain specific elements of MDL cause the instantiation of MDL specific meta level structures.

The reduction of an MDL expression to a FRAMETALK expression corresponds to the generalization of a modeling unit to a frame, e.g., the modeling entity "reaction-phase" is an instance of "structural-modeling-entity" which is a specialization of a FRAMETALK-frame. This is the same kind of relationship which holds between a FRAMETALK frame and a CLOS class, i.e., a frame is a special type of class.

The FRAMETALK extensions for MDL are specified as an object-oriented program which contains MDL concepts as specializations of frames. In addition to the CLOS meta object protocol for extending the meta level concepts, the FRAMETALK meta systems provides various protocol extensions which allow to add new language primitives and to specify their semantics in terms of attached procedures. For instance, the MDL expression for the reactor device mentions "components", "terminals" and "links" as additions to the frame language.

EVALUATION

The representation of knowledge about a specific domain is a difficult and time consuming process. A language tool is necessary and can be helpful but is far less important than widely believed. Of much greater importance is the conceptual understanding of the domain and the identification and framing of the relevant entities and relations. A language's job is to allow the formulation of the knowledge without too much additional effort or representational virtuosity.

This supportive role is achieved if basic domain specific abstractions can be expressed and then used as starting points for specialization and composition in order to express more complex conceptual entities. In addition to the conceptual entities, the language glue for putting them together ought to be meaningful in the application domain as well.

A frame language provides a good basic set of mechanisms such as the abstract–concrete distinction, the specialization relationship and the possibility to define properties and relations. The attached procedures mechanism of most frame languages can be used as a backdoor to general programming facilities for implementing inference procedures which are based on the availability and modification of information.

For the process modeling domain, the basic mechanisms have been extended and shaped to fit more closely the characteristics of the domain. Common relations between modeling entities such as "components", "terminals" and "links" have been made an integral part of the modeling language MDL.

The ability to provide abstractions for these concepts greatly reduces the complexity for the chemical engineer to express new modeling entities or to compose them to form new process models.

A major feature of PROMOT is its knowledge base of modeling entities specific to the chemical engineering area, which is scarcely available in comparable process modeling tools and modeling languages [Mar96, PB95]. By aggregating modeling entities and by specifying the occuring physico-chemical phenomena as attributes, chemical engineers can derive process unit models on a more abstract level than on the equation level.

The current work concentrates on representing modeling entities for distillation processes, from which coarse as well as fine grained models of distillation columns and their boundary systems may be generated [TGZG97a]. A prototypical implementation of PROMOT is being used by chemical engineers to define the demands of its graphical user interface and the organization of its knowledge base for a proper and consistent model development of distillation processes [TGZG97b].

ACKNOWLEDGMENTS

The authors would like to thank Achim Gerstlauer, Ernst-Dieter Gilles, Bernd Raichle, and Michael Zeitz for many discussions and for commenting on earlier drafts of this paper. Bernd Raichle contributed significantly to the implementation of FRAMETALK. This work has been partially supported by the Deutsche Forschungsgemeinschaft (Ra 520/3-1) and the Sonderforschungsbereich 412.

References

[BBMR89] A. Borgida, R. Brachman, D. McGuinness, and L. Resnick. Classic: A structural data model for objects. In *Proceedings of the 1989 ACM SIGMOD International Conference on Management of Data*, pages 59–67. Association for Computing Machinery, 1989.

[For81] C.L. Forgy. Ops5 user's manual. Technical Report CS-81-135, Carnegie-Mellon University, 1981.

[KdRB91] Gregor Kiczales, Jim des Rivières, and Daniel G. Bobrow. *The Art of the Metaobject Protocol*. MIT Press, Cambridge, MA, 1991.

[KEE85] IntelliCorp. *KEE Software Development System User's Manual*, 1985.

[KHMG90] A. Kröner, P. Holl, W. Marquardt, and E.D. Gilles. DIVA – An Open System for Dynamic Process Simulation. *Comput. Chem. Engng.*, 14:1289–1295, 1990.

[Mar96] W. Marquardt. Trends in Computer-Aided Process Modeling. *Comput. Chem. Engng*, 20:591–609, 1996.

[MB87] R. MacGregor and R. Bates. The loom knowledge representation language. Technical Report ISI/RS-87-188, USC/Information Sciences Institute, Marina del Rey, Ca., 1987.

[MZ92] W. Marquardt and M. Zeitz. Rechnergestützte Modellbildung in der Verfahrenstechnik. In *I. Troch, editor; Modellbildung für Regelung und Simulation. VDI-Berichte 925, 307-341*, VDI-Verlag, Düsseldorf, 1992.

[Pan88] C.C. Pantelides. SPEEDUP – Recent Advances in Process Engineering. *Comput. Chem. Engng.*, 12:745–755, 1988.

[PB95] C.C. Pantelides and H.I. Britt. Multipurpose Process Modelling Environments. In L.T. Biegler and M.F. Doherty, editors, *Conference on Foundations of Computer-Aided Process Design*, pages 128–141. CACHE Publications, 1995.

[Rat96] Christian Rathke. *Using the CLOS Metaobject Protocol to Implement a Frame Language*, chapter 8, pages 129–143. CRC Press, Inc., Boca Raton, Fla., 1996.

[RG77] R.B. Roberts and Ira P. Goldstein. The frl-manual. Technical Report MIT-AI Memo 409, MIT, 1977. Cambridge, Ma.

[RGGZ94] S. Räumschüssel, A. Gerstlauer, E.D. Gilles, and M. Zeitz. Ein Präprozessor für den verfahrenstechnischen Simulator DIVA. In G. Kampe and M. Zeitz, editors, *Simulationstechnik, 9. ASIM-Symposium*, pages 177–182, Stuttgart, Germany, 1994. Vieweg Verlag.

[RR96] Christian Rathke and Bernd Raichle. FRAMETALK primer. Technical report, Institut für Informatik, 1996.

[Ste90] Guy L. Steele Jr. *Common LISP: The Language*. Digital Press, Digital Equipment Corporation, second edition, 1990.

[TGZG97a] F. Tränkle, A. Gerstlauer, M. Zeitz, and E.D. Gilles. Application of the Modeling and Simulation environment PROMOT/DIVA to the Modeling of Distillation Processes. Accepted as a contribution to PSE'97, ESCAPE–7, Trondheim, Norway, May 26–29, 1997.

[TGZG97b] F. Tränkle, A. Gerstlauer, M. Zeitz, and E.D. Gilles. PROMOT/DIVA: A Prototype of a Process Modeling and Simulation Environment. Accepted as a contribution to 2nd MATHMOD Vienna, Austria, Feb. 5–7, 1997.

EVALUATION AND USE OF CLIPS FOR DEVELOPING TEMPORAL EXPERT SYSTEMS

Susan J. Chinn
School of Business, Penn State Erie, The Behrend College, Erie, PA 16563
Gregory R. Madey
Graduate School of Management, Kent State University, Kent, OH 44242
Email: sjc6@psu.edu

ABSTRACT

Many problems in industry that lend themselves to an expert system solution include "time" as a problem dimension. Monitoring and control applications are two areas in which representing temporal elements and reasoning about those elements occur. For expert system developers, two special challenges exist in creating monitoring and control applications where time is involved. The first problem is the large volume of temporal relations that can occur among facts in the knowledge base. The second problem is that temporal applications impose an order on events, whereas an expert system is non-procedural. This research-in-progress evaluates an expert system shell, CLIPS, in terms of requirements that temporal expert systems should possess. We also use an extension to CLIPS that establishes and maintains temporal relations to create some typical monitoring and control scenarios for a workflow application.

INTRODUCTION

Many management problems that lend themselves to an expert system solution include "time" as a problem dimension. Planning and scheduling, diagnosing, monitoring and controlling applications are problems in which representing temporal elements and reasoning about those elements occur. Temporal expert systems have been developed for complex problems in aerospace, factory automation, and telemetry systems. Lockheed's Expert System shell (LES), for example, is a temporal expert system shell that has been used to monitor and diagnose problems with the Hubble space telescope [PA90]. REX is another temporal expert system shell developed for aerospace ground control operations [PPU+94]. Other expert systems using time have emerged from the academic realm, such as KNOWBEL, the Sootblowing Advisory Expert (SAX), and the Temporal Production System (TPS) [MWK93, Ste91, Mal92]. Real-time expert system tools such as G2 and RTworks also use temporal representation and reasoning [Ril96].

Developers of temporal expert systems for monitoring and control applications encounter two common problems 1) the number of temporal relationships that can occur, and 2) the problem of sequencing and ordering of activities in program execution. The first problem is especially critical in interval-based representations of time. Temporal relations formed among intervals constitute a set of constraints on the possible truthfulness of a state. With each state change, these temporal constraints must be checked to ensure that additional inferences do not create inconsistencies among already established relations. The problem is NP-complete if the temporal expert system explicitly stores all relations among intervals. As the number of intervals increases, the time needed to maintain consistency increases exponentially [All83, Bar93, VK86]. Performance aside, a practical limitation is that the amount of information may be overwhelming, producing many relations that may be redundant or are not "useful" for the application. Research on methods to limit constraint propagation has been an ongoing concern [AK83, Bar93, Dor92].

The second problem, of ordering and sequencing activities, is especially troublesome for rule-based expert systems. Time by its very nature implies that events are ordered, and planning and monitoring are time-oriented problems. A well-constructed rule-based program, however, uses declarative knowledge rather than procedural constructs to control rule activation. Rules that incorporate time constraints, however, require some procedural control mechanisms so that they can be fired in a prescribed sequence. In addition, functions that procedural programs easily perform, such as calculating intervals from sets of time points, loop endlessly because the derived information causes the assertion of a fact with new information, and yet the patterns still match on the new facts. Explicit program control mechanisms, which group rules based on when they should become available to be activated, often do not solve this problem.

These two problems often make writing temporal expert system applications cumbersome, with many additional constructs needed to capture and present temporal relationships. In a system where such relationships are needed, the solution may be to move some of the work to procedures called by the expert system shell.

This research-in-progress evaluates an expert system shell, CLIPS (C Language Integrated Production System), in terms of requirements that temporal expert systems should possess. We also use an extension to CLIPS that we developed that shifts some of the work in establishing and maintaining temporal relations from the developer to procedures called by the shell. We use the extension on some typical monitoring and planning scenarios in a workflow application.

REQUIREMENTS FOR A TEMPORAL EXPERT SYSTEM APPLIED TO CLIPS

We chose CLIPS as our expert system development tool for analysis because it is well documented, easily obtainable, and the source code is available. CLIPS was first developed in 1984 by the Artificial Intelligence

Table 1. CLIPS constructs.

Constructs	Definition
Deftemplate	A frame-like construct for defining a group of facts with the same set of attributes, similar to a record or structure definition
Deffact	A construct used to hold groups of facts. Deffacts are asserted at the beginning of the program in working memory
Defmodule	A blackboard-like construct used for partitioning a knowledge base. It allows other facts to be grouped together and controls access to constructs from other modules
Defrule	Rules which represent heuristic knowledge. Actions are defined on the right-hand side of a defrule that are subject to satisfaction of conditions on the left-hand side
Deffunction	A function with parameters that works directly in the CLIPS environment

Source: CLIPS Basic Programming Guide, 1993.

Table 2. Temporal requirements applied to CLIPS.

Requirement	Support Features of CLIPS
Knowledge Base	
Constructs to represent time	Defrules
	Attribute-value pairs
	Deftemplates
	COOL (inheritance)
Methods to represent temporal relationships	Deffunctions, user-defined
	Defrules, user-defined
Time granularity	User-defined only
Representation of absolute and relative time	Defrules, user-defined
Inference Engine	
Temporal inferencing	Conflict resolution to reduce search space
	Forward chaining
	Backward chaining present only in "research" versions
	Defmodules, control facts
	FuzzyCLIPS
Correctness of inferences	Truth maintenance through "logical" conditional element
Other Temporal Issues	
Time-line view of events	User-defined only
Historical data access	Attachments present to some databases (non-temporal)

Section of NASA's Johnson Space Center and is derived from the OPS5 family of languages [CLI93]. CLIPS's inferencing strategy uses forward chaining based on the Rete pattern-matching algorithm. Enhancements to the original version have included improved documentation, an architecture manual, utility programs to verify and validate rule-based programs, procedural and object-oriented constructs, and an enhanced user interface (for Windows, Macintosh, and X-Window). Table 1 presents some of the CLIPS constructs and their definitions that we use in the discussion to follow.

Table 2 lists requirements that a temporal expert system should possess and the constructs in CLIPS that support those requirements. Our examination of CLIPS in terms of the knowledge base requirements concerns how the shell can represent temporal relationships, varying time granularities, and the representation of both absolute and relative time. Research on temporal representation has focused in using *points* to represent *absolute* time, and *intervals* to represent *relative* time. In point-based representations of time, an event is expressed in relationship to a specific point of reference that can be quantitatively expressed [GD93]. Point representations lend themselves well to two ways in which we use time: 1) as *absolute* (or actual) dates and times from which we can describe related events (event *A* occurred before time point *t1*) and 2) as time-stamps, which associate a time point with an action or event (event *A* occurred at 14:20 hours on 5/1/96) [KG77]. Interval-based representations, in contrast, assert that events or actions do not occur instantaneously, but occupy an interval of time, however small, that can be measured [All83]. An interval-based approach allows for more complex expressions such as during, finishes, or overlaps. Thus we can talk about events or actions occurring in *relative* time without referring to specific time-stamps or points.

When we examine CLIPS in terms of knowledge base features, we find that although there is no direct support for constructs that allow the user to reason about events over time, deftemplate facts can be constructed to include slots with temporal information. Unfortunately, CLIPS does not support procedural attachments in deftemplate constructs; however, the object-oriented extension to CLIPS, COOL (CLIPS Object-Oriented Language) does provide message handlers acting as methods in class definitions. COOL also supports inheritance, as do deftemplate constructs by allowing multiple instances of facts with the same sets of attributes. The ability to define different time granularities is not supported directly by CLIPS except as a value in a slot defined by the user. CLIPS has conditional elements and predicate functions that can represent some temporally expressive concepts. For example, the *exists* predicate, which allows a match on at least one fact in working memory without regard to the total number of facts, can be used to express a concept such as "at any time."

Our requirements for inference engines that can support temporal representation and reasoning include the ability to use both forward and backward inferencing strategies and the ability to support truth maintenance. Forward chaining is well suited to data-driven applications, such as monitoring which uses temporal data. Backward chaining algorithms, often used in planning and diagnostic applications, use time as part of the consultative process to determine when events occurred, or the duration of certain states. Shells that combine both methods of inference could be extremely useful in planning problems, such as providing a temporal sequence in which goals could be solved [PA90]. Truth maintenance in temporal applications involves the "frame problem," where the system may need to determine which facts have not changed as events occur over time [MH69]. The assumption that facts will remain "true" unless explicitly changed, known as *persistence*, allows previous assertions to continue to be true unless contradictory information is received later in the reasoning process [McD82]. Truth maintenance also requires support for non-monotonic reasoning, so that facts that become false can be retracted.

In terms of the inference engine, CLIPS fares better overall. CLIPS uses a forward chaining inference strategy based on the Rete-pattern-matching algorithm, but does not support backward chaining. Some researchers, however, have developed an extension to CLIPS for backward chaining using forward chaining. They achieved this by creating a data structure of linked lists that captures relationships among ground facts, premises and conclusions, and by traversing this network using a depth-first strategy with recursion [AAK+90]. The other strong inferencing feature of CLIPS is the ability to partition the knowledge base using control facts and defmodules which alleviates some problems of program execution and control. Truth maintenance is also supported by the *logical* conditional element. When a fact with dependent facts is retracted, the dependent facts are retracted as well. Otherwise, unless specifically retracted, facts remain on the fact-list in working memory. There are also several versions of CLIPS that handle fuzzy logic that may prove useful for temporal applications [Ril96].

Two other requirements are desirable for a temporal expert system. The first requirement is for a design that shows a "time-line" view of events. Some researchers contend that "Allen-style" graphs of temporal relations obscure the "flow of time" because all intervals can be connected to each other and because the graphs may not show the duration of the intervals visually [Dor92]. The second requirement is to be able to keep an historical repository of events for analysis. Issues here include how the shell stores time-varying data, how it searches through that data from a time-line perspective, the number of data values to retain, and the size of working memory.

Although our workflow application requires neither a time-ordered view of events nor access to historical data, CLIPS was found to be deficient in these requirements. Support for a time-line view of knowledge is not provided unless the user application defines one. A time-line view implies a *linear* method of linking events. Historical information is also viewed as a changing linear sequence over time. One way to store and represent past information could be to maintain an archive of historical facts that are related to the current facts. CLIPS, however, stores facts as hashed entries which makes a sequential search impossible without designing some alternate data structure for tracking values over time. Although the ability to read incoming data files is supported by several I/O functions, CLIPS provides no support for database access in the standard versions. The Washington University School of Medicine, however, has developed some database library functions linking CLIPS to Sybase; and KQML (Knowledge Query and Manipulation Language), developed by Foss Friedman-Hill can be used by CLIPS [Ril96]. Historical archiving of facts is not supported, nor is access to historical or temporal databases.

Figure 1. Workflow process for requests for engineering design changes.

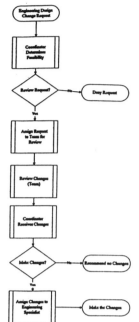

USING A TEMPORAL EXTENSION TO CLIPS IN A WORKFLOW APPLICATION

Our workflow expert system application uses temporal knowledge to assist a group of engineers who are evaluating requests for design changes. Figure 1 shows the workflow process for these requests. A Coordinator determines initial request feasibility. If the request is feasible, the proposal is routed to the review team. Once the team has completed its evaluation, the Coordinator makes the final decision to recommend changes. If a positive recommendation is made, the changes will be implemented. Time plays several roles in this workflow process. The Coordinator may choose from among several design teams, and thus may select the team with the earliest expected completion time. The request itself may be subject to deadlines. During the review process, the team members may run into delays from other members and this condition may need to be brought to the Team Leader's or Coordinator's attention. Two points in the workflow call for a yes/no decision on the part of the Coordinator, so this feature called for providing a way to interact with the shell. The first decision is to determine the feasibility of the review, with a "no" answer halting the program. Similarly, at the end of the program, the Coordinator must again decide whether to recommend that the changes be implemented. If the changes are accepted, then an Engineering Specialist implements them.

The temporal relationships in the program are generated by using our user-defined function *timerels*,

which takes three arguments: the name of the engineer, the expected starting time and the expected ending time that he/she works on the change request, as shown by the rule in Figure 2. The function relieves the developer of coding many different rules and creating additional deftemplates to generate the temporal relationships. It first checks for the valid number and type of arguments and then creates a linked list from the values for the engineer that are passed to it. The values themselves originate in a series of deffacts, but they could also be loaded as a data set externally. The *timerels* function returns a set of facts showing the temporal relationships between all the engineers in terms of the intervals during which they are expected to have the design folder. For example, if team member MA is working on the folder from time *t3* to *t5*, and team member MB is working on the folder from *t3.2* to *t4*, then the relation (*during* MB MA) would be generated. These relations can then be used in pattern matching on the left hand side of rules.

One complication in writing the expert system program is the generation of temporal relationships at two separate times in the expert system program. The first call to *timerels* takes place at the beginning of the program to set up relationships between the Coordinator at two phases of the process (CoorA and CoorB), the Team Leader, the chosen Team, and the Engineering Specialist. The second call occurs when control is passed to the Team Leader, who needs to pick the experts for the Team. After the second call, a number of facts that establish relations with the Team Members and all other members of the workflow are established. We retract facts establishing relations between the Team Members, CoorA and the Specialist because there is no interaction among these individuals in the actual process. We leave it to the developer's discretion to retract any unneeded facts instead of coding those retractions in the user-defined function, as these might change with the situation.

Once the time relations are set up, they can be used in writing rules requiring a temporal constraint. For example, the Team Leader must ensure that Team Members' planned starting and ending times stay within the bounds for the whole Team. Such a rule is shown in

Figure 2. Rule calling *timerels* user-defined function.

```
(defrule make-timerels
        (declare (salience 100))
        (phase CoorA)
        (eng (name ?name) (beg_time ?beg_time)
(end_time ?end_time))
=>
(timerels ?name ?beg_time ?end_time))
```

Figure 3. Rule using temporal constraint generated from *timerels*.

```
(defrule check-memb-times
        (phase Team)
        (eng (name ?name1&MA|MB) (folder_name
?fname&~nil) (beg_time ?beg_time) (end_time
?end_time))
        (eng (name ?name2&ATeam|BTeam) (beg_time
?beg_time) (end_time ?end_time))
        (or (overlaps ?name2 ?name1)
            (overlaps ?name1 ?name2))
=>
(printout t "Overlap between Team and Team Members
not allowed" crlf)
(halt))
```

Figure 3. Another rule checks to see if one Team Member has the folder during the same time another one does. In this case, the Member that "finishes before" is in a *during* relation (*during* MB MA) and might need to alert the other individuals in the workflow that it is being held up, or that he/she is free to accept some of the work. Other rules are written using known time points to compute the percentage of the project that was completed, and whether a Team Member is falling behind (actual starting time greater than expected starting time). These would be monitored conditions that need to come to the attention of the Coordinator.

CONCLUSION

We have presented an analysis of an expert system shell, CLIPS, in terms of its existing capabilities to represent and reason about time. We conclude that although CLIPS can be used to represent temporal elements and reason about them, the process is not "automatic." The user must define temporal relationships as deftemplates and deffunctions, and use defrules to generate the relationships. Interval-based representation increases the number of constructs and their complexity. Knowledge base partitioning using defmodules and control facts is needed for complex scenarios so that program execution proceeds as intended. The process of maintaining "older" facts increases in difficulty with the number and types of constructs. CLIPS, however, does provides an entry point for users to add extensions to the source code in C. This ability to both extend the source code and embed CLIPS in other software greatly enhances its attractiveness for developing temporal applications. We wrote such an extension that creates and maintains temporal relationships, which reduces the development burden of writing additional deftemplates, deffunctions, and defrules for our workflow application in engineering. We plan to test our application in terms of performance both with and without using the extension.

REFERENCES

[AAK+90]   M. Aldrobi, S. Anastasiadis, B. Khalife, K. Kontogiannis, and R. D. Mori. CLIPS Enhanced with Objects, Backward Chaining, and Explanation Facilities. In *1st CLIPS Conference Proceedings*, pages 621-641. Houston: NASA, 1990.

[All83]   J. F. Allen. Maintaining Knowledge About Temporal Intervals. *Communications of the ACM*, 26: 832-843, 1983.

[AK83]   J. F. Allen, and J. A. Koomen. Planning Using a Temporal World Model. In *Eighth International Joint Conference on Artificial Intelligence*, pages 741-747. Karlsruhe, West Germany: IJCAI, 1983

[Bar93]   F. A. Barber. A Metric Time-Point and Duration-Based Temporal Model. *SIGART Bulletin*, 4: 30-49, 1993.

[CLI93]   *CLIPS Version 6.0 Basic Programming Guide*. Johnson Space Center, Software Technology Branch, 1993.

[Dor92]   J. Dorn. Temporal Reasoning in Sequence Graphs. In *Tenth National Conference on Artificial Intelligence*, pages 735-740. Menlo Park: AAAI Press, 1992

[GD93]   A. J. Gonzalez and D. D. Dankel. *The Engineering of Knowledge-Based Systems: Theory and Practice*. Englewood Cliffs: Prentice-Hall, 1993

[KG77]   K. Kahn and G. A. Gorry. Mechanizing Temporal Knowledge. *Artificial Intelligence*, 9:87-108, 1977.

[Mal92]   M. A. Maloof. *TPS: Incorporating Temporal Reasoning into a Production System*. Unpublished master's thesis. University of Georgia, 1992.

[MWK93]   J. Mylopoulos, H. Wang, and B. Kramer. Knowbel: A Hybrid Tool for Building

Expert Systems. *IEEE Expert*, 8: 17-24, 1993.

[MH69]   J. McCarthy, and P. J. Hayes. Some Philosophical Problems from the Standpoint of Artificial Intelligence. In B. L. Webber and N. J. Nilsson, editors, *Readings in artificial intelligence*, pages 431-450. Los Altos: Morgan Kaufmann, 1969.

[McD82]   D. McDermott. A Temporal Logic For Reasoning About Processes and Plans. *Cognitive Science*, 6:101-155, 1982.

[PA90]   W. A. Perkins, and A. Austin. Adding temporal reasoning to expert-system-building environments. *IEEE Expert*, 5:23-30, 1990

[PPU+94] B. E. Prasad, T. S. Perraju, G. Uma, and P. Umarani. An Expert System Shell for Aerospace Applications. *IEEE Expert*, 9: 56-64, 1994.

[Ril96]   G. Riley. *CLIPS Frequently Asked Questions*. ftp://hubble.jsc.nasa.gov/pub/clips/clips-faq, 1996.

[Ste91]   K. L. Stephens. T*emporal Reasoning in Real-Time Expert Systems*. Unpublished master's thesis, University of New Brunswick, 1991.

[VK86]   M. Vilain, and H. Kautz. Constraint Propagation Algorithms for Temporal Reasoning. In *Fifth National Conference on Artificial Intelligence*, pages 377-382. Philadelphia: Morgan Kaufmann, 1986.

REASONING ABOUT RELATIONS UNDER UNCERTAINTY

Bingning Dai[1], David A Bell[1] and John G Hughes[2]
[1]School of Information and Software Engineering, [2]Faculty of Informatics
University of Ulster at Jordanstown, Newtownabbey, Co. Antrim, BT37 0QB, N. Ireland, UK
E-mail: B.Dai, DA.Bell, JG.Hughes@ulst.ac.uk

ABSTRACT

Temporal reasoning and reasoning under uncertainty are two important issues in artificial intelligence and expert/knowledge-based systems. The question is raised of how to represent uncertain times and model temporal relationships among them. This in turn gives rise to the issue of how binary relations over a domain can be handled when uncertainty in the domain is modelled according to some relevant theory, such as probability theory or evidential theory. Although there has been work to tackle this information handling problem in some specific cases, no explicit general work exists. This paper identifies a general framework with which the problem can be tackled. A series of algorithms in the setting of evidential theory are presented and some related specific work is analysed accordingly.

KEYWORDS

Dempster-Shafer Theory, Evidential Theory, Reasoning under Uncertainty, Relation, Temporal Reasoning

INTRODUCTION

Temporal reasoning and reasoning under uncertainty are two important issues in artificial intelligence (AI) and expert/knowledge-based systems. Although the research in AI (e.g. [HM94]) has mentioned uncertainty in time, it still focuses on reasoning about uncertain temporal *events* rather than pure temporal *relationships*. For example, consider two events e_1 and e_2 happening at times t_1 and t_2 respectively. AI researchers will focus on deducing the probability of e_2 given the probability of e_1 while there is no uncertainty about the times themselves. On the other hand, in the area of temporal databases, the researchers seem to be more aware of the problem of representing uncertain temporal relationships (e.g.

[DS93]). Take the same example, the events e_1 and e_2 will be thought to have definitely happened while the happening times t_1 and t_2 are uncertain. Thus the uncertainty lies in the values of t_1 and t_2 rather than the happening of e_1 and e_2. Some resources that may cause such uncertainty in temporal information have been identified. These include *scale*, *dating techniques*, *unknown* or *imprecise event times*, and *clock measurements* [Sno95]. Since manipulations in temporal databases require comparisons among these uncertain times, it is necessary to decide the credibility of the alternative possible temporal relationships when no definite relationships can be found. Probability theory is used to model uncertain times in [DS93, Sno95] and hence to evaluate their relationships. Our approach is to considering evidential theory (i.e. Dempster-Shafer theory or D-S theory in short [GB91a, GB91b]) for this purpose.

Since manipulations in databases often require comparing values from particular domains, similar cases appeared in database systems where data values are uncertain [BGP92]. For example, in [Lee92a, Lee92b], the issue of relating two independent variables is addressed, where D-S theory, rather than probability theory, was applied to model the uncertainty and imprecision in data. Applying this approach and some results in the latter area to represent temporal uncertainty may be useful for both AI and database applications. A simple example domain could be in recording temperatures which take values from a bounded set on same scale, and which might have some uncertainty associated with them. Suppose an agent needs to make a decision based on the comparison of two such uncertain temperature values. Now this requires the agent to find out the credibility of the relationship between the two uncertain values.

These considerations give rise to the issue of how binary relations over a domain can be decided when uncertainty in the domain is modelled according to some relevant theory, such as the three most widely applied theories, i.e., traditional probability theory, fuzzy logic, and D-S theory.

One of the most powerful algebraic tools for temporal reasoning, relation algebra, has been used to obtain some very general results about the decidability and completeness of systems of binary relations. It might also be useful for considering questions of complexity [Hir96].

In this paper a general framework for reasoning about relations under uncertainty is set out in connection with relation algebra. According to this framework, first an appropriate theory should be chosen to model uncertainty in the domain and a relation algebra should be defined over the domain. Then two major issues need to be considered. They are how to establish or evaluate the relationship between two uncertain domain values and how to apply the appropriate uncertain theory in relation composition.

The structure of the paper is as follows. The next section gives some basic definitions that will be used in the rest of the paper. Then a general framework for reasoning about relations under uncertainty will be set out. A series of algorithms in the setting of D-S theory will follow. Related work will be considered with respect to the framework. The conclusion section concludes the paper and points out future research directions.

Now some definitions that will be used in the rest of the paper are given.

DOMAINS AND RELATIONS

Definition 1. Domain. *A domain D is a finite set of countable entities denoted by $\{d_1, d_2, ..., d_n\}$.*   ◊

Definition 2. Relation algebra. *A relation algebra $\Re$ is a set of binary relations over some domain D, which are exclusive and complete, i.e., the relations partition the Cartesian product of D, $D \times D$.*
A function rel is: $D \times D \rightarrow \Re$.   ◊

Note that *algebra* is used informally in this definition and the following example since the algebraic properties of a "relation algebra" [Hir96] are not the focus of this paper, though it will be used for our future research.

Definition 3. Composition rules. *A composition rule is a function comp: $\Re \times \Re \rightarrow 2^{\Re}$.*   ◊

The following example illustrates these definitions.

Example 1. *Given a domain D={1, 2, 3, 4}, a relation algebra could be $\Re$={<, =, >}.*
$D \times D$ thus is partitioned into:
$R_< =\{\langle 1,2 \rangle, \langle 1,3 \rangle, \langle 1,4 \rangle, \langle 2,3 \rangle, \langle 2,4 \rangle, \langle 3,4 \rangle\}$,
$R_= =\{\langle 1,1 \rangle, \langle 2,2 \rangle, \langle 3,3 \rangle, \langle 4,4 \rangle\}$, *and*
$R_> =\{\langle 2,1 \rangle, \langle 3,1 \rangle, \langle 3,2 \rangle, \langle 4,1 \rangle, \langle 4,2 \rangle, \langle 4,3 \rangle\}$.
A composition rule could be defined as:
$comp(\langle <,< \rangle)=\{<\}$,
$comp(\langle <, = \rangle)=\{<\}$,
$comp(\langle <, > \rangle)=\varnothing$,
$comp(\langle =,< \rangle)=\{<\}$,
$comp(\langle =, = \rangle)=\{=\}$,
$comp(\langle =,> \rangle)=\{>\}$,
$comp(\langle >,< \rangle)=\varnothing$,
$comp(\langle >, = \rangle)=\{>\}$,
$comp(\langle >,> \rangle)=\{>\}$.   ◊

SOME BASIC DEFINITIONS IN DEMPSTER-SHAFER THEORY

In D-S theory, a *frame of discernment* Θ is a mutually exclusive and exhaustive collection of propositions in a domain, represented as a finite non-empty set.

Definition 4. Mass functions. *A mass function m is:*
m: $2^{\Theta} \rightarrow [0,1]$, *where* $m(\varnothing)=0$, $\Sigma_{x \subseteq \Theta} m(X)=1$.
$X(\subseteq \Theta)$ is called a focal element of m when $m(X)>0$.   ◊

An uncertain variable taking values from the frame of discernment Θ then will be denoted as $\{X_1/m(X_1), X_2/m(X_2), ..., X_n/m(X_n)\}$, where $X_1, X_2, ..., X_n$ are the focal elements of the mass function that represents the evidence supporting the possible values of the variable.

Definition 5. Evidential functions *bel* and *pls*. *Given a mass function m, a belief function bel is defined on 2^{Θ}:*
for all $A \subseteq \Theta$: $bel(A)=\Sigma_{x \subseteq A} m(X)$.
A plausibility function pls is defined as:
$pls(A)=\Sigma_{x \cap A \neq \varnothing} m(X)=1-bel(\bar{A})$, for all $A \subseteq \varnothing$.   ◊

Note that a mass function is a basic probability assignment to all subsets X of Θ. A belief function *bel* gathers all the support a subset A gets from such an assignment. It is not a probability distribution itself.

A GENERAL FRAMEWORK FOR REASONING ABOUT RELATIONS UNDER UNCERTAINTY

This framework highlights three components, i.e. a domain D, a relation algebra $\Re$ and an uncertainty theory UTh, and their connections as explained as follows:

- The relations in $\Re$ are defined based on D;
- UTh is applied to model uncertainty in the domain;
- UTh applies when deriving the relation between two uncertain domain values;
- UTh applies to the composition of two sets of relations.

Using a different UTh results in different actions in the last two connections.

The next two sections give a series of algorithms to be applied in the last two connections. The UTh is chosen to be D-S theory. The algorithms formalise and generalise the algorithms used in our previous work [DBH96] where the particular algebra $\Re$ used is Allen's Interval Algebra [All83]. In the exposition of the algorithms this framework will be explained in further detail.

RELATION CONSTRUCTION

Given the definitions presented early in this paper, D-S theory is employed to model uncertainty in the domain. In this sense, a domain D is equivalent to a frame of discernment Θ_1. Rather than viewing $X \, r \, Y$ as an event (as was done in [BGL96]), where X and Y are two independent random variables in D and r is a binary relation and a member of $\Re$, r is taken to be another independent variable in the relation algebra $\Re$, i.e., $\Re$ becomes another frame of discernment Θ_2. To derive the credibility of $X \, r \, Y$, or say, to reason about the relation r, a constructive method is used. A mass function of r is constructed based on the mass functions of the two D variables. Then the evidential functions bel and pls can be obtained based on the new mass function. This procedure ensures the correctness of bel and pls. The direct formal definition for a belief function bel can be found in [HF92] and the proof for that a belief function built from a mass function does satisfy the criteria in such a definition can be found in [GB91a].

Algorithm 1. Construct uncertain relations.
Input:

$\{\langle D_{1_i}, m_1(D_{1_i})\rangle | \ i_1=1, 2, \ldots, k_1, D_{1_i} \subseteq D, m_1(D_{1_i}) \neq 0\}$,

$\{\langle D_{2_i}, m_2(D_{2_i})\rangle | \ i_2=1, 2, \ldots, k_2, D_{2_i} \subseteq D, m_2(D_{2_i}) \neq 0\}$.

Output:

$Q = \{\langle R_i, m(R_i)\rangle | \ i=1, 2, \ldots, k, R_i \subseteq \Re, m(R_i) \neq 0\}$.

Procedure:

for $i_1:=1$ **to** k_1 **do**
  for $i_2:=1$ **to** k_2 **do**
    $QUEUE(Q, \langle Rel(D_{1_i}, D_{2_i}), m_1(D_{1_i})*m_2(D_{2_i})\rangle)$
  end for
end for.   ◊

Here $QUEUE(Q,Y)$ adds an element $\langle s, m\rangle$ to the queue Q. If s is found in an earlier member, simply add m to that member's mass. Rel is presented in Algorithm 2.

The two inputs are actually two mass functions over D. Each of the functions is represented by a set of pairs consisting of the focal elements and the corresponding masses. This is how UTh is applied to model uncertainty in the domain. The output is similarly represented as a mass function over $\Re$. Thus the credibility of the relation between the two uncertain values can be represented by the bel based on this mass function.

Algorithm 2. Rel.
Input:

$D_1 = \{d_{11}, d_{12}, \ldots, d_{1_{j_1}}\}$,

$D_2 = \{d_{21}, d_{22}, \ldots, d_{2_{j_2}}\}$.

Output:

$S = \{d_1, d_2, \ldots, d_t\}$.

Procedure:

for $t_1:=1$ **to** j_1 **do**
  for $t_2:=1$ **to** j_2 **do**
    $SET(S, rel(\langle d_{1_{t}}, d_{2_{t}}\rangle))$
  end for
end for.   ◊

Here $SET(S, M)$ adds a member M to the set S.

It is worth noting that different $UTh's$ will result in different input parameters for Rel from Algorithm 1. If probability theory is used (as in [DS93, Sno95]), the two inputs of Algorithm 1 will be two probability distributions on the domain. Thus the parameters passed to Rel will be two definite values from the domain. No iteration is needed in this case and Rel will be identical to rel (as defined in Definition 2).

A non-numerical theory can also be used to tackle uncertainty. In that case, the inputs to Algorithm 1 will be two sets of values suggesting the possible values of the two variables. The information is logically disjunctive and no weight is attached to each disjunct. Consequently, no calculation of numerical weights is needed in Algorithm 1 and it will become identical to Rel.

The above two possible uses of different *UTh's* in a way indicate that D-S theory is more expressive because both cases degenerate with respect to the algorithms.

[SHB94] suggests that the *insufficient reasoning principle* be used when no information about the individuals' probabilities is known in a collective. If this is followed, the second case can be transformed into the first case by assigning an equal probability to each individual value. This principle may also be used in *Rel*. Instead of constructing a set of all the $rel(\langle d_{1_i}, d_{2_i}\rangle)$, an equal probability $1/|D_1|*|D_2|$ may be assigned to each $rel(\langle d_{1_i}, d_{2_i}\rangle)$. This requires further combination of evidence in Algorithm 1. It is not considered necessary and therefore not adopted in this paper.

The following example illustrates these discussions.

Example 2. *Given the domain and the relation algebra in Example 1, suppose there are now two uncertain variables represented by D-S theory as:*

$D_1=\{\langle\{1, 2\}, 0.3\rangle, \langle\{2, 3\}, 0.7\rangle\}$,

$D_2=\{\langle\{3\}, 1\rangle\}$.

The intermediate results in Algorithm 1 are:

$\langle Rel(\{1, 2\},\{3\}), 0.3\rangle$
  $=\langle SET(rel(\langle 1, 3\rangle), rel(\langle 2, 3\rangle)), 0.3\rangle$
  $=\langle SET(<, <), 0.3\rangle$ *and*
$\langle Rel(\{2, 3\},\{3\}), 0.7\rangle$
  $=\langle SET(rel(\langle 2, 3\rangle), rel(\langle 3, 3\rangle)), 0.7\rangle$
  $=\langle SET(<, =), 0.7\rangle$.

The output is: $\{\langle\{<\}, 0.3\rangle, \langle\{<, =\}, 0.7\rangle\}$.
Thus:

$bel(\{\varnothing\})=0, pls(\{\varnothing\})=0$;
$bel(\{<\})=0.3, pls(\{<\})=1$;
$bel(\{=\})=0, pls(\{=\})=0.7$;
$bel(\{>\})=0, pls(\{>\})=0$;
$bel(\{<, =\})=1, pls(\{<, =\})=1$;
$bel(\{<, >\})=0.3, pls(\{<, >\})=1$;
$bel(\{=, >\})=0, pls(\{=, >\})=0.7$;
$bel(\{<, =, >\})=1, pls(\{<, =, >\})=1$.

If the insufficient reasoning principle is employed, the two variables would be

$D_1=\{\langle 1, 0.15\rangle, \langle 2, 0.5\rangle, \langle 3, 0.35\rangle\}$ *and*
$D_2=\{\langle 3, 1\rangle\}$.

The intermediate results would be

$\langle Rel(\langle 1, 3\rangle), 0.15\rangle$
  $=\langle SET(rel(\langle 1, 3\rangle)), 0.15\rangle$
  $=\langle SET(<), 0.15\rangle$,
$\langle Rel(\langle 2, 3\rangle), 0.5\rangle$
  $=\langle SET(rel(\langle 2, 3\rangle)), 0.5\rangle$
  $=\langle SET(<), 0.5\rangle$, *and*
$\langle Rel(\langle 3, 3\rangle), 0.35\rangle$
  $=\langle SET(rel(\langle 3, 3\rangle)), 0.35\rangle$
  $=\langle SET(=), 0.35\rangle$.

The output would be $\{\langle <, 0.65\rangle, \langle =, 0.35\rangle\}$.

Note if the insufficient reasoning principle is applied to the previous output $\{\langle\{<\}, 0.3\rangle, \langle\{<, =\}, 0.7\rangle\}$, *the results will be the same.*

If a non-numerical view is adopted, the two inputs would be $D_1=\{1, 2, 3\}$ *and* $D_2=\{3\}$. *The output would be* $Rel(\{1, 2, 3\}, \{3\})=\{<, =\}$.   ◊

RELATION COMPOSITION

Having investigated the first major issue in reasoning about relations under uncertainty, i.e., constructing the uncertain representation for the relationship between two uncertain domain variables, the following algorithms supply solutions to the second major issue — composition of uncertain relations.

Algorithm 3. Composition of uncertain relations.
Input:
  $\{\langle R_{1_i}, m_1(R_{1_i})\rangle|\ i_1=1, 2, ..., k_1, R_{1_i}\subseteq \Re, m_1(R_{1_i})\neq 0\}$,
  $\{\langle R_{2_i}, m_2(R_{2_i})\rangle|\ i_2=1, 2, ..., k_2, R_{2_i}\subseteq \Re, m_2(R_{2_i})\neq 0\}$.
Output:
  $Q=\{\langle R_i, m(R_i)\rangle|\ i=1, 2, ..., k, R_i\subseteq\Re, m(R_i)\neq 0\}$.

Procedure:

for $i_1:=1$ **to** k_1 **do**
  for $i_2:=1$ **to** k_2 **do**
    $QUEUE(Q, \langle Comp(R_{1_i}, R_{2_i}), m_1(R_{1_i})*m_2(R_{2_i})\rangle)$
  end for
end for.   ◊

Here $QUEUE(Q,Y)$ does the same as in Algorithm 1. The *Comp* is now defined.

Algorithm 4. Comp.
Input:
  $R_1=\{r_{11}, r_{12}, ..., r_{1_j}\}$,
  $R_2=\{r_{21}, r_{22}, ..., r_{2_j}\}$.
Output:
  $S=\{r_1, r_2, ..., r_k\}$.

Procedure:

for $t_1:=1$ **to** j_1 **do**
  for $t_2:=1$ **to** j_2 **do**
    $UNION(S, comp(\langle r_{1_t}, r_{2_t}\rangle))$
  end for
end for.   ◊

Here $UNION(S_1, S_2)$ adds the members of S_2 to the set S_1.

The following example shows how these two algorithms work.

Example 3. *Given the domain and the relation algebra in Example 1, suppose the uncertain relationship between the variables d_1 and d_2 and that between d_2 and d_3 are represented by D-S theory as:*

$R_1 = \{\langle\{<\}, 0.3\rangle, \langle\{=, >\}, 0.7\rangle\}$, *and*

$R_2 = \{\langle\{>\}, 1\rangle\}$.

Then the uncertain relationship between d_1 and d_3 is the composition of R_1 and R_2. Using the above two algorithms, the intermediate results in Algorithm 3 are:

$\langle Comp(\{<\}, \{>\}), 0.3\rangle$
$= \langle UNION(comp(\langle <, >\rangle)), 0.3\rangle$
$= \langle UNION(\varnothing), 0.3\rangle$

and

$\langle Comp(\{=, >\}, \{>\}), 0.7\rangle$
$= \langle UNION(comp(\langle =, >\rangle), comp(\langle >, >\rangle)), 0.7\rangle$
$= \langle UNION(\{>\}, \{>\}), 0.7\rangle$.

The output is:

$\{\langle \varnothing, 0.3\rangle, \langle\{>\}, 0.7\rangle\}$.

Note these two algorithms are similar to those constructing uncertain relations. The difference lies in that *comp* returns a set of relations while *rel* returns a single relation. So Algorithm 4 uses *UNION* rather than *SET*. Again the possible relations are combined using a non-numerical method, which is the one adopted in the specific case of Interval Algebra [All83].

In [DBH96] we have proved the correctness of these algorithms when $\Re$ is the Interval Algebra. The proof for the general case here is similar and thus is not dwelt on further. The major point is that the mass functions constructed by them should indeed satisfy Definition 4. Another point needs to be noted is that in the relation construction algorithms, the output mass function is defined on a different frame of discernment from the input mass functions.

RELATED WORK

Based on the framework presented previously, some work that addresses reasoning about relations under uncertainty in specific cases is analysed in this section.

[DS93] is an early piece of work that realised the necessity to reason about the relations among indeterminate times in database systems. The work then developed further into a part of the work in [Sno95]. In their papers, the basic time domain is taken isomorphic to real number. A *probability mass function* is used to model the uncertainty in indeterminate time. Thus their method follows the path of using probability theory. However, as the work is limited to temporal databases, no relation composition was considered.

In [CT91,Che95], fuzzy logic was employed in Allen's Interval Algebra. Fuzzy membership functions π and the necessity degrees of temporal relationships N are deduced in both relation definitions and compositions.

In [Lee92a, Lee92b, BGL96], D-S theory was applied to model the uncertainty in non-temporal domains. *Bel* and *pls* were defined directly from the mass functions of the independent variables. The definitions there differ in that a more restrictive standard was used in *Rel*. In this paper, every pair $\langle d_{1_i}, d_{2_i}\rangle$ in $D_1 \times D_2$ contributes to the result set of relations. While in [Lee92a, Lee92b, BGL96], a special "$r^{\forall}$" is defined for each relation r in the relation algebra in the definition of *bel* while "$r^{\exists}$" is defined for *pls*. This method intuitively excludes the masses of $Rel(D_{1_i}, D_{2_i})$ that are not subsets of R, in the definition of *bel* and only takes into account those $Rel(D_{1_i}, D_{2_i})$ that have non-empty intersections with R, in the definition of *pls*. So far this kind of restrictions are thought to be philosophical issues and left open to specific applications.

CONCLUSIONS

A general framework for reasoning about relations under uncertainty has been presented. It has direct impact on temporal reasoning under uncertainty. More applications in various AI areas will also benefit from its formality and theoretical support. Different approaches to tackling this problem can be compared with respect to this framework thus it helps a specific application to adopt an appropriate method. A prospective research direction is to look into uncertainty in connection with algebraic properties in relations.

REFERENCES

[All83]   J. F. Allen. Maintaining Knowledge about Temporal Intervals. *Communications of the ACM*, 26(11):832–843, November 1983.

[BGP92]   D. Barbara, H. Garcia-Molina and D. Porter. The Management of Probabilistic Data. *IEEE Transactions on Knowledge and Data Engineering*, 4(5):487–502, October 1992.

[BGL96]   D. A. Bell, J. W. Guan, and S. K. Lee. Gerneralized Union and Project Operations for Pooling Uncertain and Imprecise Information.

Data & Knowledge Engineering, 18:89–117, 1996.

[CT91] Z. Chen and F. Terrier. About Temporal Uncertainty. In *Proc. of IEEE/ACM Int'l Conf. on Developing and Managing ES Programs*, pp. 223–230, October 1991.

[Che95] Z. Chen. Fuzzy Temporal Reasoning for Process Supervision. *Expert Systems*, 12(2):123–137, 1995.

[DBH96] B. Dai, D. A. Bell, and J. G. Hughes. Evidential Temporal Representations and Reasoning. In *Proc. of the 6th Pacific Rim Int'l Conf. on Artificial Intelligence*, pp. 411–422, Cairns, Australia, August 1996.

[DS93] C. E. Dyreson and R. T. Snodgrass. Valid-time Indeterminacy. In *Proc. of the 9th IEEE Int. Conf. on Data Engineering*, pp. 335–343, 1993.

[GB91a] J. Guan and D. Bell. *Evidence Theory and its Applications, Vol. 1*, North-Holland, 1991.

[GB91b] J. Guan and D. Bell. *Evidence Theory and its Applications, Vol. 2*, North-Holland, 1991.

[HM94] S. Hanks and D. McDermott. Modelling a Dynamic and Uncertain World I: Symbolic and Probabilistic Reasoning about Change. *Artificial Intelligence*, 66(1):1–55 , 1994.

[HF92] J. Y. Halpern and R. Fagin. Two Views of Belief: Belief as Generalised Probability and Belief as Evidence. *Artificial Intelligence*, 54:275-317, 1992.

[Hir96] R. Hirsch. Relation Algebras of Intervals. *Artificial Intelligence*, 83:267–295, 1996.

[Lee92a] S. K. Lee. Imprecise and Uncertain Information in Databases: An Evidential Approach. In *Proc. of the 8th Int'l Conf. on Data Engineering*, pp. 614–621, 1992.

[Lee92b] S. K. Lee. An Extended Relational Database Model for Uncertain and Imprecise Information. In *Proc. of the 18th VLDB Conf.*, pp. 211–220, Vancouver, British Columbia, Canada, 1992.

[SHB94] S. Shi, M. E. C. Hull, and D. A. Bell. Combining Evidence in Probability Reasoning. In *Proc. of the 9th Int'l Conf. on Applications of Artificial Intelligence in Engineering*, pp. 535–542, July 1994.

[Sno95] R. T. Snodgrass (ed.). *The TSQL2 Temporal Query Language*, Kluwer Academic Publishers, 1995.

A STOCHASTIC APPROACH WITH HIDDEN MARKOV MODEL FOR ACCIDENT DIAGNOSIS IN NUCLEAR POWER PLANTS

Kee-Choon Kwon
Korea Atomic Energy Research Institute
P.O. Box 105, Yusong, Taejon, 305-600, Korea
E-mail:kckwon@nanum.kaeri.re.kr

Jin-Hyung Kim
Dept. of Computer Science, KAIST
373-1 Gusung-Dong, Yusong-Gu, Taejon, 305-701, Korea

ABSTRACT

Classification of the types of accidents at an early stage of the accident in nuclear power plants is crucial for proper action selection. A plant accident can be classified by its time dependent patterns related to the principal variables. The Hidden Markov Model (HMM), a double stochastic process can apply to accident diagnosis which is spatial and temporal pattern problem. The HMM is created for each accident from a set of training data by the maximum-likelihood estimation method which uses a forward-backward algorithm and a Baum-Welch re-estimation algorithm. The accident diagnosis is decided by calculating which model has the highest probability for given test data. The optimal path for each model at the given observation is found by the Viterbi algorithm, the probability of optimal path is then calculated. The system uses a left-to-right model including 6 states and 22 input variables to classify 8 types of accidents and the normal state. The simulation results show that the proposed system identifies the accident types correctly. It is also shown that the diagnosis is performed well for incomplete input observation caused by sensor fault or malfunction of certain equipment.

Key Words : Statistical Pattern Recognition , Diagnosis, Spatial & Temporal Reasoning

I. INTRODUCTION

It is necessary to diagnose the types of accident at an early stage of the accident in nuclear power plants (NPPs) to improve the plant operability and safety when the plant proceed to transient state from normal state. The term diagnosis, as applied to an engineering system or process, means the determination of the cause which brought about an undesirable state or failure of the system. The diagnosis can be done at several different levels, e.g. component, subsystem, function or event[Kim94]. At the proposed accident diagnosis system, diagnosis is made at event level to determine which accident has occurred in NPPs. Diagnosis of the types of accidents in NPPs is crucial for proper action selection. This system is intended to support an operator's decision-making by interpreting the major plant variables and operating status in order to assist the operator in taking action to mitigate the accident consequence. A pattern is defined as a quantitative description of objects, events, or phenomena. Accidents in an NPP are associated with unique time dependent pattern of major variables and equipment status; hence, diagnosis can be treated as a pattern classification problem. The goal of pattern classification is to assign a physical object, event, or phenomenon to one of the prespecified classes. The accident diagnosis systems for NPPs have been developed using such techniques as pretrained artificial neural network (ANN) with rule-based fuzzy logic system[Iko91], backpropagation ANN with dynamic node architecture[Bas92], connectionist expert systems which have ANNs for their knowledge bases[Che93], multiple ANNs which act as individual neuro agents[Koz95], a probabilistic neural network which is based on an estimation of the pattern's probability density[Bar95], the combination of an ANN and knowledge processing [Ohg93], nearest neighbors modeling which is optimized using genetic algorithms[Lin95],double feedforward ANNs with an implicit time measure[Jeo95] and fuzzy reasoning with causes and symptoms[Iij96]. But most of these systems are prototype or under evaluation with a test facility and have not been applied to operating NPPs.

In this accident diagnosis problem, the classification may involve spatial and temporal patterns. Temporal patterns usually involve ordered sequences of data appearing in time. Spatial patterns mean the unique pattern of each accident and the variation of the same accident may occur under different operation conditions. The variant of patterns depends on accident is shown in Figure 1. The ANN and fuzzy logic approach can absorb spatial variation, but can not provide proper solution for temporal variant. So it is reasonable to adopt a double stochastic approach for classification of the patterns. The Hidden Markov Model (HMM), a double stochastic process enables modeling of, not only spatial phenomena, but also time scale distances. HMM can be used to solve the classification problem associated with time series input data such as speech signal or plant process signals. The HMM can provide an appropriate solution by its modeling and learning capability even though it does not have exact knowledge to solve the problems. The HMM has been applied in classifying patterns in dynamic process, e.g. radar target identification[DeW92], human action recognition[Yam92], condition monitoring in electrical machine[Hat93], space network antenna fault monitoring [Smy94], classification of moving light displays[Fie95], and American sign language recognition[Sta95]. It is a novel attempt to apply HMM to accident diagnosis in NPPs.

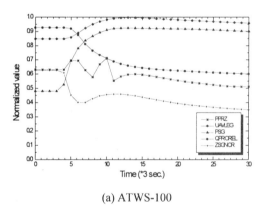

(a) ATWS-100

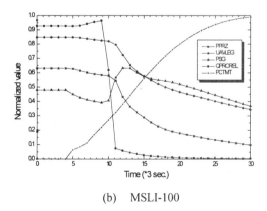

(b) MSLI-100

Figure 1 Variant of Patterns

II. DEVELOPMENT OF AN ACCIDENT DIAGNOSIS SYSTEM

1. HMM for Accident Diagnosis

The problem of accident diagnosis is defined as classifying the types of accident ω_j given sequential input patterns X_t at time t. Input pattern X_t is mathematically defined as an object described by a sequence of features at time t.

$$X_t = \left(x_1, x_2, \ldots, x_i, \ldots x_n\right) \tag{1}$$

The space of input pattern X_t consists of the set of all possible patterns:

$$X_t \subset R^d \; ; \; R^d \text{ is } d\text{-dimensional real vector space.}$$

The k observed data up to time t is defined as Φ_{t-k} :

$$\Phi_{t-k} = \left\{X_t, X_{t-1}, \ldots, X_{t-k+1}\right\} \tag{2}$$

The set of possible accident classes ω_j forms the space of classes Ω :

$$\Omega(t) = \left\{\omega_1, \omega_2, \cdots, \omega_c\right\} ; \text{ c is the number of classes,} \tag{3}$$

Classification task can be considered as finding function f which maps the space of input patterns Φ_{t-k} to the space of classes Ω :

$$f : \quad \Phi_{t-k} \to \Omega(t). \tag{4}$$

There are three basic paradigms of pattern classification; the statistical or decision-theoretic approach, the syntactic or structural approach and neural approach. Statistical approach, as its name implies, a statistical basis for classification of algorithms. A set of characteristic measurements, denoted features, are extracted from the input data and are used to assign each feature vector to one of c classes. Features are assumed generated by a state of nature, and therefore the underlying model is of a state of nature or class-conditional set of probabilities and/or probability density functions. Classifier design attempts to integrate all available problem information, such as measurements and *a priori* probabilities. Decision rules may be formulated in several interrelated ways, for example, by converting an *a priori* class probability into a measurements-conditioned probability or by formulating a measure of expected classification error or risk, and choosing a decision rule that minimizes this measure.

A neural network provides an emerging paradigm for pattern recognition implementation that involves large interconnected networks of relatively simple and typically nonlinear units. It is probably safe to say that the advantages, disadvantages, applications, and relationships to traditional computing are not fully understood. Neural networks are particularly well suited for pattern association applications. The notion that artificial neural networks can solve all problems in automated reasoning, or even all pattern recognition problems, is probably unrealistic.

In a syntactic approach, we must be able to quantify and extract structural information and to access the structural similarity of patterns. One syntactic approach is to relate the structure of patterns with the syntax of a formally defined language, in order to capitalize on the vast body of knowledge related to pattern generation and analysis. Production rules can be used to define the grammar for different classes and can be typically inferred from training examples.

A dynamic process often exhibits sequentially changing behavior. If we defined one short-time period to frame, the probability of the frame transition is different from each accident in NPPs. Therefore, the probability of frame existence and transition between frame can be statistically modeled. The probability of an accident occurring in an NPP is already given and is called *a priori* probability. When the accident has occurred we make a decision only by selecting the types of accident ω with the highest *a priori* probability $P(\omega)$. This decision is obviously unreasonable. It is more reasonable to determine the types of accident after observing the trend of time-series major parameters, namely, to get the conditional probability $P(\omega|\Phi_{t-k})$. This conditional probability is called the *a posteriori* probability. Decision-making based on the *a posteriori* probability is more reliable, because it employs both *a priori* knowledge together with observed time-series data[Hua90]. Classification of unknown pattern X_t corresponds to finding optimal model $\hat{\omega}$ that maximizes the conditional probability $P(\omega|\Phi_{t-k})$ over the types of accident ω. We can apply Bayes rule to calculate *a posteriori* probability,

$$P(\hat{\omega}|\Phi_{t-k}) = \underset{\omega}{max} \frac{P(\Phi_{t-k}|\omega)P(\omega)}{P(\Phi_{t-k})} \qquad (5)$$

The conditional probability $P(\Phi_{t-k}|\omega)$ comes from comparing shapes of the accident models with input observations while *a priori* probability $P(\omega)$ comes from the accident probability which represents how often the accident appears in the NPP. Since $P(\Phi_{t-k})$ is independent of $\hat{\omega}$, we get

$$P(\hat{\omega}|\Phi_{t-k}) \propto P(\Phi_{t-k}|\hat{\omega})P(\hat{\omega})$$

$$= \underset{\omega}{max}\left[P(\Phi_{t-k}|\omega)P(\omega)\right] \qquad (6)$$

In fact, it is difficult to calculate *a priori* probability $P(\omega)$ in NPP, and should satisfies the following equation,

$$\sum_{j=1}^{c} P(\omega_j) = 1 \qquad (7)$$

The accident diagnosis system does not cover all of the accidents occurred in NPPs therefore could not satisfies above equation (7), so we can assume the probabilities of all accidents occurring are equal, therefore, the present observed data controls the decision. $P(\Phi_{t-k}|\omega)$ is called the likelihood of $\hat{\omega}$ with respect to the set of samples. The maximum likelihood estimate of $\hat{\omega}$ is, by definition, that value $\hat{\omega}$ that maximizes $P(\Phi_{t-k}|\omega)$. The HMM can successfully treat an accident classification problem under a probabilistic or statistical framework.

In this classification problem, HMM is used to estimate the conditional probability $P(\Phi_{t-k}|\omega)$. By using HMM, the pattern variability in parameter space and time can be modeled effectively. The HMM uses a Markov chain to model changing statistical characteristics that exist in the actual observations of dynamic process signals. The Markov process is therefore a double stochastic process. Double stochastic processes enable modeling of not only spatial phenomena, but also time scale distances. HMM parameters are estimated from the Baum-Welch algorithm and guarantee a finite improvement on each iteration in the sense of maximization of likelihood. An HMM is trained for each accident from a set of training data, iterative maximum likelihood estimation of model parameters from observed time-series data. Incoming observations are classified by calculating which model has the highest probability for producing that observation.

2. Vector Quantization

Feature extraction is an important task for classification or recognition and is often necessary as a preprocessing stage of data. In feature extraction, data can be transformed from high-dimensional pattern space to low-dimensional feature space. Vector quantization (VQ), the process of approximating a block of continuous amplitude signals by discrete signal is one method of the feature extraction. We are going to choose a sequential VQ algorithm for continuous pattern recognition problem.

The feature mapping algorithm is supposed to convert patterns of arbitrary dimensionality into the response of one-dimensional array of neurons. Suppose that an input pattern has N features and is represented by a vector x in an N-dimensional pattern space. The network maps the input patterns to an output space. The output space in this case is assumed to be 1-dimensional array of output nodes, which possess a certain topological orderness. The question is how to train a network so that the ordered relationship can be preserved. Kohonen proposed to allow the output nodes to interact laterally, leading to the Self-Organizing Map (SOM), in other words Kohonen network[Koh90]. A simple configuration of the SOM is illustrated in Figure 2.

The most prominent feature of the SOM is the concept of excitatory learning within a neighborhood around the winning neuron. The size of the neighborhood slowly decreases with each iteration. A more detailed description of the training phase is provided below:

1) First, a winning neuron is selected as the one with the shortest Euclidean distance between its weight vector and the input vector, where w_i denotes the weight vector corresponding to the ith output neuron.

$$\| x - w_{i^*} \| = \min_i \| x - w_i \| \qquad (8)$$

2) Let i^* denote the index of the winner and let I^* denote a set of indices corresponding to a defined neighborhood of winner i^*. Then the weights associated with the winner and its neighboring neurons are updated by

$$\Delta w_j = \eta (x - w_j) \qquad (9)$$

for all the indices $j \in I^*$, and η is a small positive learning rate. The amount of updating may be weighted according to a preassigned "neighborhood function", $\Lambda(j, i^*)$.

$$\Delta w_j = \eta \, \Lambda(j, i^*)(x - w_j) \qquad (10)$$

for all j. For example, a neighborhood function $\Lambda(j, i^*)$ may be chosen as

$$\Lambda(j, i^*) = \exp(-|r_j - r_{i^*}|^2 / 2\sigma^2) \qquad (11)$$

where r_j represents the position of the neuron j in the output space. The convergence of the feature map depends on a proper choice of η. One plausible choice is that $\eta = 1/t$. The size of neighborhood should decrease gradually.

3) The weight update should be immediately succeeded by the normalization of w_i.

In the retrieving phase, all the output neurons calculate the Euclidean distance between the weights and the input vector, and the winning neuron is the one with the shortest distance.

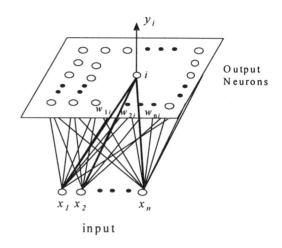

Figure 2   A Network for Self-Organizing Map

3. Application of the Hidden Markov Model

3.1 The Hidden Markov Model

The basic theory of HMM was introduced in late 1960s and implemented for continuous speech recognition in mid-1970s. After this application, HMM has been successfully applied to real problems which can not be solved by conventional Markov models. HMM has advantages that provide proper solutions by modeling and learning by itself even if it does not have exact knowledge about problem solving.

HMM is represented by a graph structure which consists of N nodes called state and arcs that means a directional transition between nodes. In a graph, the observation symbol probability distribution which models spatial characteristics and initial state probability distribution stored in a node, and state transition probability distribution which models time characteristics stored in an arc. HMM states are not directly observable, and can be observed only through a sequence of observed symbols. To describe the HMM formally, the following model notation for an HMM can be used[Rab89, Hua90].

N : the number of states in the model

L : the number of distinct observation symbols per transition, denoting the set of individual symbols $V = \{v_1, v_2, \cdots, v_L\}$

T : the lengths of the observation sequence, $O_1, O_2, \cdots, O_T$

$S = \{ s_t \}$: a set of states, $S_t \in \{1, 2, \ldots, N\}, \quad t = 1, 2, \ldots, T$

state i at time t may be denoted by $s_t = i$

$A = \{a_{ij}\}$: the state transition probability distribution,

$i, j = 1, 2, \ldots N$,

$a_{ij} = P(s_{t+1} = j | s_t = i)$: the transition probability from state i to state j, the parameter should satisfy the stochastic constraint $\sum_i a_{ij} = 1$

$B = \{b_j(k)\}$: the observation symbol probability distribution,

$b_j(k) = P(v_k | s_t = j)$, $\quad k = 1, 2, \cdots, L$

observation probability of kth symbol v_k in state j, the parameter should satisfy the stochastic constraint $\sum_k b_j(k) = 1$

$\Pi = \{\pi_i\}$: the initial state distribution, where

$\pi_i = P(s_1 = i)$, $i = 1, 2, \cdots, N$ satisfying

$\sum_i \pi_i = 1$

S_I : a set of initial states,

S_F : a set of final states,

N_I : the number of initial states,

N_F : the number of final states.

An HMM can be represented by the compact notation $\lambda = (A, B, \Pi)$. Specification of an HMM involves the choice of the number of states, N, the number of discrete symbols, L, and the specification of three probability densities with matrix form, A, B and Π.

3.2 Training and Classification

Training means that the characteristics of input patterns to be modeled by the parameter of $\lambda = (A, B, \Pi)$. An HMM is applied to a classification problem under the assumption that we can precisely determine the model parameters for given observations. But it is difficult that this assumption is exactly realized because of the complexity of the problem and having a local optimal not a global optimal. At present, we are satisfied to find local optimal in parameter optimization methods. In this paper, we use the maximum likelihood estimation for training. This method maximizes the following equation given input observations O.

$$P(O|\lambda) = \sum_S \pi_{s_0} \prod_{t=1}^{T} a_{s_{t-1} s_t} b_{s_t}(O_t) \qquad (12)$$

It calculates the probability of given observation symbols at all paths from initial to final state. The model parameter which maximizes the above equation is efficiently computed by forward-backward algorithm and Baum-Welch re-estimation algorithm. The forward variable $\alpha_t(i)$ can be defined :

$$\alpha_t(i) = P(O_1, O_2, \cdots, O_t, s_t = i | \lambda) \qquad (13)$$

This is actually the probability of the partial observation sequence to time t and state i which is reached at time t for a given λ. We can calculate $\alpha_t(i)$ by forward algorithm as follows:

Step 1: $\alpha_1(i) = \pi_i b_i(O_1)$, for all states i (if $i \in S_I$

$\pi_i = \dfrac{1}{N_I}$; otherwise $\pi_i = 0$) $\qquad (14)$

Step 2: Calculating $\alpha()$ along the time axis, for $t = 2, \cdots, T$, and all states j, compute:

$$\alpha_t(j) = [\sum_i \alpha_{t-1}(i) a_{ij}] b_j(O_t) \qquad (15)$$

Step 3: Final probability is given by:

$$P(O|\lambda) = \sum_{i \in S_F} \alpha_T(i) \qquad (16)$$

The backward variable $\beta_t(i)$ which is used to optimize the model parameter with forward variable, can be defined as :

$$\beta_t(i) = P(O_{t+1}, O_{t+2}, \cdots, O_T | s_t = i, \lambda) \quad (17)$$

i.e. the probability of the partial observation sequence from $t+1$ to the final observation T, given state i at time t and the model λ and $\beta_t(i)$ can be calculated by backward algorithm as follows:

Step 1: $\beta_T(i) = \dfrac{1}{N_F}$, for all states $i \in S_F$, otherwise

$\beta_T(i) = 0 \qquad (18)$

Step 2: Calculating $\beta()$ along the time axis, for $t = T-1, T-2, \cdots, 1$ and all states j, compute:

$$\beta_t(j) = [\sum_i a_{ji} b_i(O_{t+1}) \beta_{t+1}(i)] \qquad (19)$$

Step 3: Final probability is given by: $P(O|\lambda) = \sum_{i \in S_I} \pi_i b_i(O_1) \beta_1(i) \qquad (20)$

The most difficult problem in HMM is how to adjust the model parameters (A, B, Π) to maximize the probability of the observation sequence given a model. The iterative algorithm used in HMM-based recognition is known as the Baum-Welch algorithm. The *a posteriori* probability of transitions, $\gamma_t(i, j)$, will be defined as the probability of a path being in state i at time t and making a transition to state j at time $t+1$, given the observation sequence and the particular model. It can be computed as :

$$\begin{aligned}\gamma_t(i,j) &= P(s_t = i, s_{t+1} = j | O, \lambda) \\ &= \frac{\alpha_t(i)a_{ij}b_j(O_{t+1})\beta_{t+1}(j)}{P(O|\lambda)} \\ &= \frac{\alpha_t(i)a_{ij}b_j(O_{t+1})\beta_{t+1}(j)}{\sum_{k \in S_F}\alpha_T(k)}\end{aligned} \quad (21)$$

Similarly, *a posteriori* probability of being in state i at time t, $\gamma_t(i)$, given the observation sequence and model is

$$\gamma_t(i) = P(s_t = i | O, \lambda) = \frac{\alpha_t(i)\beta_t(i)}{\sum_{k \in S_F}\alpha_T(k)} \quad (22)$$

At this point, $\overline{a_{ij}}$, $\overline{b_j}$, $\overline{\pi_i}$ of re-estimated new model $\overline{\lambda}$ can be computed as:

$$\overline{a}_{ij} = \frac{\sum_{t=1}^{T-1}\gamma_t(i,j)}{\sum_{t=1}^{T-1}\sum_j \gamma_t(i,j)} = \frac{\sum_{t=1}^{T-1}\gamma_t(i,j)}{\sum_{t=1}^{T-1}\gamma_t(i)} \quad (23)$$

$$\overline{b}_j(k) = \frac{\sum_{t \in O_t = v_k}\gamma_t(j)}{\sum_{t=1}^{T}\gamma_t(j)} \quad (24)$$

$$\overline{\pi}_i = \gamma_1(i) \quad (25)$$

Thus, if $\overline{\lambda}$ is iteratively to replace λ and repeat the above re-estimation calculation, it can be guaranteed

that $P(O|\lambda)$ can be improved until some limiting point is reached.

Classification or recognition means to find the best path in each trained model and select the one which maximizes the path probability for a given input observation. Therefore given the observation $O = O_1, O_2, \ldots, O_T$ and the model $\lambda_i = (A_i, B_i, \Pi_i)$ $i = 1, 2, \ldots C$, the classified model λ^* satisfies the following equation:

$$\lambda^* = \frac{max}{i} P(O|\lambda_i) \qquad 1 \le i \le C \quad (26)$$

The Baum's Forward procedure can be used in efficiently obtaining the probability $P(O|\lambda)$ given input observation in a model. This probability is the summation of $P(O, S|\lambda)$ over all possible state sequences S. The joint probability of O and S, i.e. the probability that O and S occur simultaneously is simply the product of $P(O|S, \lambda)$ and $P(S|\lambda)$.

But we can use optimal state sequence instead of all possible state sequences for more efficient classification. This means find the best path with highest probability, i.e. with maximum $P(O, S|\lambda)$ rather than $P(O|\lambda)$. Therefore the classified model λ^* satisfies the following equation

$$\lambda^* = \frac{max}{i}\frac{max}{S} P(O, S|\lambda_i) \quad 1 \le i \le C \quad (27)$$

A formal technique for finding this single best state sequence is called the Viterbi algorithm, which works as follows:

Step 1: Initialization. For all states i,
$$\delta_1(i) = \pi_i b_i(O_1) \quad (28)$$
$$\psi_1(i) = 0 \quad (29)$$

Step 2: Recursion. From time $t=2$ to T, for all states j,
$$\delta_t(j) = \max_i[\delta_{t-1}(i)a_{ij}]b_j(O_t) \quad (30)$$
$$\psi_t(j) = arg\,max_i[\delta_{t-1}(i)a_{ij}] \quad (31)$$

Step 3: Termination. (* indicate the optimized results)
$$P^* = \max_{s \in S_F}[\delta_T(s)] \quad (32)$$
$$s_T^* = arg\,max_{s \in S_F}[\delta_T(s)] \quad (33)$$

Step 4: Path (state sequence) backtracking. From time T-1 to 1
$$s_t^* = \psi_{t+1}(s_{t+1}^*) \quad (34)$$

4. Design of the On-line Accident Diagnosis System

The training data are provided off-line with the test simulator. The simulator refers to 3-loop 993MWe Westinghouse pressurized water reactor. Major variables and equipment status are combined for input symptom vector in each accident when the simulator emulates an accident situation. The collected input symptoms are vector quantized for feature extraction, which means N-dimensional measurement space transmitted to 1-dimensional feature space. A vector quantized code book is the training input of the HMM classifier. The block diagram of the accident diagnosis system is shown in Figure 3.

The accidents are simulated in the test simulator by activating malfunctions during normal operation. We then get parameters, such as temperature, pressure, flow, pump status, or valve open/close. The accidents that can be diagnosed by the accident diagnosis system are:

Accident 1 ; ATWS (Anticipated Transient Without Scram)
Accident 2 ; FWLB (FeedWater Line Break inside containment)
Accident 3 ; LSLC (Loss of Steam generator Level Controller signal)
Accident 4 ; MSIV (Main Steam Isolation Valve closure)
Accident 5 ; MSLI (Main Steam Line break Inside containment)
Accident 6 ; MSLO (Main Steam Line break Outside containment)
Accident 7 ; PORV (Power Operated Relief Valve stuck open)
Accident 8 ; SGTR (Steam Generator Tube Rupture)

The input symptom vectors are a collection of principal variables and status of major equipment from accident simulation in the test simulator. The following 22 major variables and equipment status are used to classify the eight different types of accidents and one normal state.

(1) pressurizer pressure
(2) pressurizer level
(3) reactor coolant average temperature
(4) steam generator pressure
(5) steam generator level
(6) net charging flow
(7) reactor power
(8) reactivity
(9) average fuel temperature
(10) feedwater line flow
(11) main steam line flow
(12) steam flow from steam generator
(13) steam pressure from steam generator
(14) secondary radiation monitoring
(15) containment pressure
(16) containment temperature
(17) containment humidity
(18) pressurizer relief tank pressure
(19) pressurizer relief tank temperature
(20) net power
(21) position of main steam isolation valve.
(22) reactor trip signal

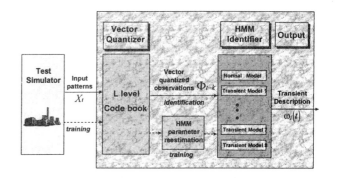

Figure 3 Block Diagram of the Accident Diagnosis System

The test data are collected from the real-time test simulator using the shared memory which is the most efficient method of inter process communication. Shared memory provides data communication between more than two processes share the certain logical memory space. The test simulator is executed every 0.2 second and the calculated simulation variables are stored in the shared memory. The accident diagnosis process get the data from shared memory every 3 seconds. First of all, the collected test data are normalized based on minimum and maximum value and then, vector quantized by the SOM clustering method. The SOM, unsupervised artificial neural network, clustered input vector into L disjoint sets. In our implementation, we chose 150 clusters for an optimal solution after several attempts. It means every input vector is assigned to one of 150 clusters. The code book size is 40, this means the system receives 40 time interval input vectors during 2 minutes to classify the types of accidents. In initial 2 minutes, test results are incorrect because of the code book size is less than 40. The system should wait until get 2 minutes input vectors. We called this initial stage a "Ready mode" and next time step, the system get another 40 time interval input vectors as sliding window method. This data processing method is shown in Figure 4.

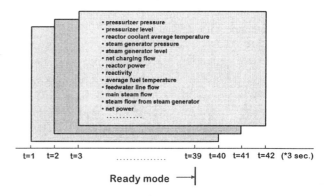

Figure 4 Data Processing Method

A left-to-right HMM has been considered appropriate for processing those signals whose properties change over time. The underlying state sequence associated with the model has the property that as time increases the state index increases or stays the same state, i.e. the state always proceeds from left to right. This model consists of 6 states which have less than 2 direct transitions to the right state; the state transition probability a_{ij} satisfies the following condition :

$$a_{ij} = 0 \quad \text{for} \quad j < i \quad \text{or} \quad j \geq i+3 \quad (35)$$

Few initial conditions are given to this model, and these initial condition are equivalent to all accident models. When the transition is applied to 3 ways, the initial value of a_{ij} are 0.333, to 2 ways like, just before the last state, the initial values of a_{ij} are 0.5, to 1 way, like last state, the initial value of a_{ij} is 1.0. By assuming that the observation symbol probability is equivalent to each state, observation symbol probability $b_j(v_k) = 0.006667$ to satisfy the equation :

$$b_j(k) = \frac{1}{L}, \quad \sum_{k=1}^{L} b_j(k) = 1 \text{ and } L = 150. \quad (36)$$

The initialized HMM state is illustrated in Figure 5. In this system, we use 9 models for 8 types of accidents and 1 normal state.

The training is performed by calculating α value from forward algorithm, β value from backward algorithm, and re-estimate from Baum-Welch algorithm in each model given multiple input observations. The re-estimation is done until the convergence condition, $P(O|\bar{\lambda}) \geq P(O|\lambda)$, i.e. new model estimates are more likely to produce the given observation sequence O, is satisfied in each model.

The probability, $P(O|\lambda)$, is calculated by the optimal path which is obtained by Viterbi algorithm for given input observations in each model. We classify accidents by examining which model has the highest probability for given input observations. The prototype system was implemented in HP747i industrial workstation and the programming done with "C" language.

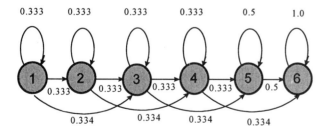

Figure 5 Initialized HMM with 6 states

5. Experiment

We collected 9 training data set as 10%, 25%, 42%, 50%, 68%, 75%, 85%, 92% and 100% of reactor power per accidents and normal state from the test simulator. Each training data consist of around 40 time interval input vectors. The SOM and HMM training is performed with off-line. The experiments were carried out off-line and on-line. The off-line test is performed with 4 test data set, 60%, 80%, 88%, and 95% of reactor power per accidents. Table 1 shows the best path probability of each model which is the result of Viterbi algorithm when the test data are from 95% of reactor power. In this case, model #2 accident, ATWS, has the highest probability, i.e., the ATWS accident has occurred. In the case of one sensor fault or equipment malfunction, the accident diagnosis system exactly classified the accident type as shown in Table 2, for model #4, LSLC. However, it could not correctly classified the accident when more than 2 sensor faults or equipment malfunctions occurred. The recognition rate of off-line test is 96.8% and more detailed test results are shown in Table 3. The SGTR accident at 60% of reactor power was misdiagnosed as an unknown accident.

The on-line test was performed using 5 test data as 78%, 86%, 90%, 96%, and 100% of reactor power per accidents. One hundred % of reactor power data are trained data and the other data are not trained data. The recognition rate of on-line test is 100% and more detailed test results are shown in Table 4. In this case, recognition time is different from each accident because a distinctive feature is appeared different time intervals after the accident occurred.

III. CONCLUSION

We proposed an accident diagnosis system based on the HMM stochastic modeling approach. We can classify accidents by recognizing the patterns of accidents which is a novel attempt to expand the application area of HMM to accident diagnosis. After proper training using the train input vectors, the prototype system exactly recognized the accidents from input vectors. At present, this prototype system is implemented by on-line to accept continuous time series data in real time. But, there are a lot of problems that should be solved before its actual application to an operating nuclear power plant. However, this system have advantages such as an easy expansion of accident types and observation symbol sequences, and a relatively short time needed for training than neural network applications. Further effort is being made to improve the system performance and to overcome the noisy input data.

Table 1 Likelihood Probability (Case I)

Accident Model	Likelihood Probability
Normal	0.0
ATWS	2.782e-62
FWLB	1.251e-76
LSLC	9.594e-87
MSIV	1.889e-94
MSLI	7.204e-101
MSLO	8.267e-106
PORV	9.918e-111
SGTR	5.534e-115

Table 2 Likelihood Probability (Case II)

Accident Model	Likelihood Probability
Normal	0.0
ATWS	0.0
FWLB	0.0
LSLC	2.163e-90
MSIV	1.535e-99
MSLI	1.327e-106
MSLO	1.140e-112
PORV	7.370e-118
SGTR	2.917e-122

Table 3 Off-line Test Result

NPP Operating Mode \ Types of Accident	ATWS	FWLB	LSLC	MSIV	MSLI	MSLO	PORV	SGTR
60%	Y	Y	Y	Y	Y	Y	Y	X
80%	Y	Y	Y	Y	Y	Y	Y	Y
88%	Y	Y	Y	Y	Y	Y	Y	Y
95%	Y	Y	Y	Y	Y	Y	Y	Y

Y : correct, X : incorrect

Table 4 On-line Test Result

Types of Accident \ NPP Operating Mode	78%	86%	90%	96%	100%
ATWS	Immediately	Immediately	Immediately	Immediately	Immediately
FWLB	Immediately	Immediately	Immediately	Immediately	Immediately
LSLC	Immediately	Immediately	Immediately	Immediately	Immediately
MSIV	Immediately	Immediately	Immediately	Immediately	Immediately
MSLI	after ~40 sec.	after ~40 sec.	after ~40 sec.	after ~40 sec.	after ~40 sec.
MSLO	after ~30 sec.	after ~3 sec.	after ~3 sec.	after ~20 sec.	after ~15sec.
PORV	after ~3 min.	after ~3 min.	after ~170 sec.	after ~140 sec.	after ~90 sec.
SGTR	after ~48 sec.	after ~20 sec.	after ~27 sec.	after ~24 sec.	after ~24 sec.

REFERENCES

[Bar95]   Y. Bartal, J. Lin, R. Uhrig, "Transients Identification in Nuclear Power Plants Using Probabilistic Neural Networks and the Problem of Knowledge Extrapolation," 9th Power Plant Dynamics, Control & Testing Symposium, pp.49.01~49.08, Knoxville, TN, USA, May 24~26, 1995.

[Bas92]   A. Basu, *Nuclear Power Plant Status Diagnostics Using a Neural Network with Dynamic Node Architecture*, MS Thesis, Iowa State University, 1992.

[Che93]   S. W. Cheon, "Application of Neural Networks to a Connectionist Expert System for Transient Identification in Nuclear Power Plants," Nuclear Technology, Vol.102, pp.177~191, May 1993.

[DeW92] M. R. DeWitt, *High Range Resolution Radar Target Identification Using the PRONY Model and Hidden MARKOV Models*, MS Thesis, Air Force Institute of technology, WPAFB OH, USA, Dec. 1992.

[Fie95] K. H. Fielding, D. W. Ruck, "Recognition of Moving Light Displays Using Hidden Markov Models," Pattern Recognition, Vol.28, No.9, pp.1415~1421, 1995.

[Hat93] E. Hatzipantelis, J. Penman, "The Use of Hidden Markov Models for Condition Monitoring Electrical Machines," Sixth International Conference on Electrical Machines and Drives, Sep. 8~10, 1993.

[Hua90] X. D. Huang, Y. Ariki, M. A. Jack, Hidden Markov Models for Speech Recognition, Edinburgh University Press, Edinburgh, 1990.

[Iij96] Takashi Iijima, Application Study of Fuzzy Logic Methods for Plant-State Identification, HWR-432, Jan. 1996.

[Iko91] A. Ikonomopoulos, R. E. Uhrig, L. Tsoukalas, "A hybrid Neural Network - Fuzzy Logic Approach to Nuclear Power Plant Transient Identification," AI91 Frontiers in innovative Computing for the nuclear Industry, Jackson, Wyoming, pp.217~226, Sep. 15~18, 1991.

[Jeo95] E. Jeong, K. Furuta, S. Kondo, "Identification of Transient in Nuclear Power Plant Using Neural Network with Implicit Time Measure," Proceedings of the Topical Meeting on Computer-Based Human Support Systems:Technology, Methods, and Future, Philadelphia, PA, USA, pp.467~474, June 25~29, 1995.

[Kim94] I. S. Kim, "Computerized Systems for On-line Management of Failures:A State-Of-The-Art Discussion of Alarm and Diagnostic Systems Applied in the Nuclear Industry," Reliability Engineering and System Safety, Vol.44, pp.279~295, 1994.

[Koh90] T. Kohonen, "The Self-Organizing Map," Proceedings of the IEEE, Vol.78, No.9, pp.1464-1480, Sep. 1990.

[Koz95] R. Kozma, Y. Yokoyama, M. Kitamura, "Intelligent Monitoring of NPP Anomalies by Adaptive Neuro-Fuzzy Signal Processing System," Proceedings of the Topical Meeting on Computer-Based Human Support Systems:Technology, Methods, and Future, Philadelphia, PA, USA, pp.449~456, June 25~29, 1995.

[Lin95] J. Lin, Y. Bartal, R. Uhrig, "Using Similarity Based Formulas and Genetic Algorithms to Predict the Severity of Nuclear Power Plant Transients," 9th Power Plant Dynamics, Control & Testing Symposium, pp.53.01~53.09, Knoxville, TN, USA, May 24~26, 1995.

[Ohg93] Y. Ohga, H. Seki, "Abnormal Event Identification in Nuclear Power Plants Using a Neural Network and Knowledge Processing," Nuclear Technology, Vol.101, pp.159~167, Feb. 1993.

[Rab89] L. R. Rabiner, "A Tutorial on Hidden Markov Models and Selected Application in Speech Recognition," Proceedings of the IEEE, Vol.77, No.2, pp.257~285, Feb. 1989.

[Smy94] P. Smyth, "Hidden Markov Models for Fault Detection in Dynamic Systems," Pattern Recognition, Vol.27, No.1, pp.149~164, 1994.

[Sta95] T. E. Starner, *Visual Recognition of American Sign Language Using Hidden Markov Models*, MS Thesis, MIT, Jan. 1995.

[Yam92] J. Yamato, J. Ohya, K. Ishii, "Recognizing Human Action in Time-Sequential Images Using Hidden Markov Model," Proceeding of Computer Vision and Pattern Recognition, pp.379~385, Urbana-Champaign, IL, USA, 1992.

Two Orthogonal Sub-algebras of the Interval Algebra

Debasis Mitra and Willie G. Brown

Department of Computer Science, Jackson State University

P.O. Box 18839, 1400 J.R. Lynch St

Jackson, MS 39217

E-mail: dmitra@stallion.jsums.edu

Abstract

Allen's interval algebra is the foundation of interval-based qualitative temporal reasoning. In this paper we have exposed two orthogonal sub-algebras of the former. Existence of these two sub-algebras lets us derive clusters of consistent components of a temporal constraint network. This allows (1) efficient query answering for some types of queries, (2) isolating regions of inconsistency in a temporal constraint network, and (3) deriving possible *causal* and *containment* information between temporal entities in the database.

Key words: Temporal constraint propagation, interval-based qualitative reasoning, interval sub-algebras, causal and containment networks.

1   Introduction

From amongst different qualitative temporal representation schemes the interval-based one is the most expressive. But reasoning in this scheme is an NP-hard problem. Hence the problem has been attacked from different angles. One of these angles is to discern some sub-problems which are tractable. However, physical significance of such sub-problems are rarely investigated. Also, different types of information could be derived from a set of temporal constraints. For example, possible cause and effect relations could be inferred by looking at temporal relations between a set of pairs of temporal entities, like events and fluents. The problem of temporal reasoning has never been seriously looked into from these angles, although exploring how to specify causal relations from a temporal point of view has been done before (see [VG94] for a review). In this work we have exposed that the composition operation, which is the foundation of reasoning in this context, actually supports one to derive such semantically meaningful information. But our main contribution in this work is to show the significance of two orthogonal sub-algebras of the interval algebra in dividing a temporal constraint problem into a smaller set of manageable chunks. In this paper we have presented some theoretical results in this direction along with the sketch of an algorithm for this purpose.

Section two of this article introduces the interval algebra and section three introduces the two sub-algebras of it. In the following section we have discussed how these two sub-algebras could be utilized for deriving semantically meaningful temporal constraint networks (SMTCN). The next two sections discuss the issues of constraint propagation. Significance of this identification of semantically meaningful sub-algebra is discussed next. The last section is the conclusion of the paper.

2   Interval-based temporal reasoning

A qualitative constraint between a pair of

temporal intervals could be one of the following relations: *before, meet, overlap, during, start, finish*, inverse of these six relations, and *equal*, depending on the relative position of the endpoints of the relevant intervals. A pair of such relations could be *composed* to derive new information between a pair of intervals through a third one. For example, suppose the constraint between the interval A and the interval B is 'overlap,' and the constraint between intervals B and C is 'meet.' Then the composed relation between intervals A and C is 'before.' One could develop a composition table (thirteen by thirteen) for composing these temporal relations. A composed constraint may be a disjunction of some so called *primitive* relations, as described above. For example, the result of composition between 'overlap' and 'overlap' could be one of {before, meet, overlap}. This raises the question about what happens when the temporal relation between a pair of intervals are not accurately known in the beginning. An incomplete information with a disjunctive set of a few primitive relations could be provided between a pair of time-intervals by the user. Such a set could be a disjunction of all thirteen primitive relations (tautology) if no constraint was present before.

A disjunctive composition scheme has been devised by Allen[All83], which is used as a standard in the area of interval-based reasoning. It involves taking union of all individual composition results between each pair of primitive relations within the two disjunctive sets which are to be composed. Finally, the result of the union operation is intersected with the existing constraint between the final pair of intervals (A to C, in the above example). This step makes sure that the final result is consistent with the derived information as well as the *old* constraints. A number of temporal constraint propagation algorithms are developed[All83, vB90, Lad88, LR92, LMN91] based on this composition technique.

3   Two   orthogonal   sub-algebras

Primitive relation-composition table harbors some intriguing information (for example, see

	b	m	o
b	b	b	b
m	b	b	b
o	b	b	b,m,o

	s	d	f
s	s	d	d
d	d	d	d
f	d	d	f

<u>Rows</u>: Relation between temporal entities A to B. <u>Columns</u>: Relation between temporal entities B to C. <u>Table entries</u>: Relation(s) between temporal entities A to C.
Table 1.

Rodriguez[RA95]). Careful observation reveals two sub-algebras hidden in that table. These sub-algebras are over two subsets of the set of primitive relations. These two sets are {before, meet, overlap}, and {start, during, finish, equal}. Each of these two sets is closed under the composition operation, meaning that the relations resulting from the composition operation between any two members of any of the above sets are always within the same set (Table 1). They are also trivially closed under the union and intersection operations, but they are not closed under the inverse operation because the inverse relations are not included in any of these sets, although the corresponding sets of inverse relations also form similar (isomorphic) sub-algebras. Another point to note is that each of these two sets lie within the conceptual neighborhoods on the partial order of primitive temporal relations, first noticed by Freksa[Fre93].

There are some physical significance of these two sets. If an interval A is either *before*, or *meets* or *overlaps* another interval B, then A could *cause* B. We call this set the **causal set**, and the corresponding sub-algebra as the **causal sub-algebra**. The reason behind this nomenclature is that if A starts before B starts, *and* the former interval finishes before the latter does, then the former could potentially cause the latter temporal entity. It is difficult to perceive how A could be considered as the cause of B under any other circumstances, if a delay between a cause and its effect is presumed to be necessary (as required under the post-Newtonian physics). One should be careful here that A is a *possible* cause for B, as known from the temporal information, but it is not necessary that that is indeed the case. Inferring real causal relation from such temporal relation is a tricky issue. For example, Haddaway[Had96] considers 'starts' as the only valid *causing* relation. One could argue that causing-entity could also be *finished-by* the affected-entity. However, in this work our main objective is not really to extract actual causal entities from temporal relations, but to generate two orthogonal sub-algebras of the interval algebra. Inferring that {before, meet, overlap} form a possible candidate for causal relations is an observation. A point to note here is that incorporating 'starts' in this set (as needed in Haddaway's formalism) still keeps it closed under the composition operation.

The significance of the second set has been observed before[All83], although the algebra has not been noticed so far. If A is either *starts*, or *finishes*, or *during* or *equal* B, then A is said to be *contained* in B. That is why the set of these relations is called the **containment set**, and we call the corresponding sub-algebra as the **containment algebra**. Such containment relations could be utilized to derive *reference intervals*, to which other temporal entities would be contained for isolating the reasoning tasks[Koo89].

The union of the two subsets is the full set of all 'forward' (or non-inverse) primitive relations, and the union of those two subsets and the corresponding inverse subsets is the full set of all primitive relations. Hence, the causal set and the containment set are orthogonal to each other. A point to note here is that these *forward* relations also form an algebra under composition. We call it as the **forward set**, and the corresponding algebra as the **forward algebra**.

4 Derived networks

In a given temporal constraint network (TCN), temporal relations between each pair of intervals are represented as labels on directed arcs in the network. Each label is a disjunctive set of the primitive temporal relations, with cardinality between one (in which case the relation is known accurately) and thirteen (in which

case nothing is known between the related time-intervals). Given a TCN, it is possible to create other networks where each label is a subset of the corresponding original label on the given TCN, in such a way that the label on the created network belongs to only one of the above subsets, causal or containment. Thus, a given TCN could be split into multiple number of such derived networks. We call these derived networks *semantically meaningful temporal constraint networks* (SMTCN). A simple example with a two node-TCN is given below.

```
A --------------------------> B
           (b,a,m,f-inv,d-inv)
```

This TCN will split into three SMTCNs.

```
A --------------------------> B
              (b,m)

A <-------------------------- B
               (b)

A <-------------------------- B
              (f,d)
```

A nice property of the SMTCNs is that they form directed acyclic graphs (DAG) as would be shown in the next section. Actually it is easier to extract directed graphs with only 'forward' labeling from a given TCN, before one generates the SMTCNs. Such a graph which have arcs with labels either within the 'forward' set or being tautologies (the full set of all 13 primitive relations), is called a *forward TCN* (FTCN). Arcs with tautology as label are considered to be **'absent'** in the TCNs.

5 Conditions for Consistency

While deriving the SMTCNs one could check their consistency, thus checking the overall consistency of the original TCN. SMTCNs could be of two types: *causal-SMTCNs* and *containment-SMTCNs*. Actually it is easier to first derive

DAG FTCNs, and then extract SMTCNs from them. The following theorem shows the usefulness of this scheme.

Theorem 1: Existence of a cycle in an FTCN implies inconsistency, except for the trivial case of 'equality' relation.

Proof sketch: It is a trivial observation (over the primitive composition table) that in a consistent temporal constraint network such a situation could not occur. For example, if $A - [l] -> B$, and $B - [m] -> C$, then $C - [p] -> A$ is not possible, where l, m, and p are disjunctive non-empty subsets of the *forward*-set of temporal relations. □

Note that the theorem is valid for any type of SMTCN also, i.e., identifying a cycle while generating SMTCNs from a given TCN indicates inconsistency in the latter.

This theorem ensures that an FTCN (or a SMTCN), which is not a DAG is inconsistent. This condition is checked while deriving FTCNs from a given input TCN in our proposed algorithm.

For the other theorems the following assumptions will be used.
In a DAG SMTCN, $A_1 -> A_2 -> A_3 -> ... -> A_n$ is a chain of arcs with labels in either the causal-set (if it is a causal-SMTCN), or the containment-set (if it is a containment-SMTCN). There also exists a given initial relation between $A_1 -> A_n$.

Theorem 2: Relation between A_1 and A_n could be *overlap* for a causal-SMTCN, if and only if all labels between A_i's and A_{i+1}'s, for $i = 1, ... n-1$, have *overlap* as an element. In all other cases A_1 to A_n can only have *before*.

Proof: Observing the composition table (Table 1) for primitive relations, one could find that *before* and *meet* is closed under composition, producing only *before* as a result. This, in turn indicates that *overlap* could never be produced as a result of composition unless it is present as one of the two operands. Hence, a sequence of composition operations never produces *overlap* unless it exists on all the labels in the chain. □

Theorem 3: Relation between A_1 and A_n could

be *starts* for a containment-SMTCN, if and only if all labels between A_i's and A_{i+1}'s ($\forall i \in [1, n-1]$) have *starts* as an element; and relation between A_1 and A_n could be *finishes* for a containment-SMTCN, if and only if all labels between A_i's and A_{i+1}'s have *finishes* as an element.

Proof: A similar observation as in the proof of theorem 2, on the composition table (Table 1) proves the validity of this theorem. □

FTCNs and SMTCNs are extracted from a given TCN, and while extracting them the above theorems could be utilized to keep an eye on the consistency. One has to note here that all the generated SMTCNs together still comprises only a different representation of the given TCN. No regular consistency checking algorithm has run on them yet. Each SMTCN is likely to be a set of connected components of the graph, since it is unlikely that all the given nodes of the TCN will form a single 'causal' chain or a single 'containment' chain. Detection of inconsistency on a SMTCN (or on an FTCN) also localizes the sources of inconsistency of the given TCN, which is an important issue in many situations (for example, in gene-sequencing research[GR93]) . In the next section we have provided the outline of an algorithm for extracting SMTCNs from a given TCN, which also propagates some amount of constraints.

6 Consistency checking

Suppose a given TCN is a graph $T = (V, E)$, where V is a set of p number of nodes (time-intervals $n_1, ... n_p$), and E is a set of q number of labeled and directed arcs $(e_1, ...e_q)$ constrained initially, where each e_i is an *initially constrained* arc between a pair of nodes (n_j to n_k) with a label l_{jk}, the label being a non-null proper subset of the thirteen-element set of primitive interval relations. The following is a sketch of the algorithm for extracting SMTCNs from T.

Algorithm *generate-SMTCNs.*
Step 1: Let us rewrite this graph (T) in such a way that all labels have only 'forward' relations. For this purpose we may have to create a pair

of arcs with directions reverse to each other, in case the label on the original arc in T has relations in both the 'forward' and the 'inverse' sets. The corresponding pair of arcs will have relations only from the forward set. Those arcs, which have labels having relations only in the 'inverse' set, will be replaced by the corresponding opposite arcs (with 'forward' labels, inverse of the original 'inverse' labels).

Step 2: Create an acyclic partial graph N, with each of the unidirectional arcs of the redrawn T from step 1. Then for each of the bi-directional arcs *try* to add (if it does not form a cycle) each of those pairs to two copies of N, since both of the pairs are not supposed to be on the same FTCN. Detection of a cycle while adding any arc to a partial graph N will eliminate N from the set of partially developed FTCNs. This step will create a set of all DAG FTCNs corresponding to T.

Step 3: For each of the DAG FTCNs generated from the previous step, create a pair of SMTCNs - one causal and the other containment, with appropriate labels (causal or containment) picked up from the original labels in the original FTCN. Employ Theorems 2 and 3 to propagate constraints (to eliminate some primitive relations from labels).[1] Any null label would indicate inconsistency.

End algorithm *generate-SMTCNs.*

Generating SMTCNs involve creating a new arc for each arc of the TCN, where labels on any SMTCN belong to only one of the two sets, the causal-set or the containment-set. In order to handle labels belonging to the inverse-sets (of the causal or the containment sets) on the given TCN, arcs with reverse direction are created, with the corresponding forward relations on their labels. If any arc has both forward and reverse relations, then two separate SMTCNs have to be created corresponding to the reverse directions for the same arc.

For example,

---

[1] For example, if all the labels on the intermediate chain of arcs (within a SMTCN) from A_1 to A_n have either *before* or *meet*, and a direct arc from A_1 to A_n has {before, overlap, meet} as their labels, then the result of constraint propagation would cause the elimination of *overlap* and *meet* from the latter label, because they are inconsistent with the former labels.

```
A  ----------------------------->  B
             (a,mi,si)
```

will create two new arcs in two different SMTCNs,

```
A  <----------------------------  B
             (b,m)
```

```
A  <----------------------------  B
               (s)
```

Actually our algorithm takes care of inverse relations in the step 2, so that all SMTCNs created by it will be DAGs only.

If no DAG FTCN is found by Step 2 of the algorithm then obviously the input TCN is inconsistent. But if the Step 2 generates an FTCN which does not cover all the arcs of the input FTCN then one would be able to isolate a part of the input TCN which is responsible for inconsistency. Similarly, while creating SMTCNs also such zones of inconsistency could be detected. Of course, the above algorithm will have to be modified to isolate such zones of inconsistency.

Note that the step 3 of the algorithm would create a cluster of isolated connected component-SMTCNs (of each of the two types), for each of the generated FTCNs. There are three types of arcs in any SMTCN: 'present' arcs with the corresponding causal or containment labels, 'absent' arcs with the tautology as their labels, and what we call as 'null' arcs with the *Null* set as their labels. A third type of arc appears when in the original FTCN, an arc contains only relations of the different type than the type of the derived SMTCN, e.g., causal label for containment-SMTCN or vice versa . The leads one to the following theorem.

Theorem 4: Any pair of node in a connected component belonging to a SMTCN must have either an arc with the corresponding type of labels (causal or containment) or an 'absent' arc with the tautology as its label.

Proof: Suppose a pair of nodes A and B belongs to one of the connected components of a causal-SMTCN, and originally that arc had only containment label in the corresponding FTCN from which the SMTCN is generated. Since A and B are connected, there has to be at least one path between these two nodes, where all the arcs in the path have causal label on each of them. If the path is of direction $A \to \ldots \to B$, then there has to be either an arc $A \to B$, with causal label on it, or the arc should be 'absent' with tautology as its label, as demanded by the causal algebra. This implies that the arc $A \to B$ or $B \to A$ cannot have 'null' relation between them.

The same can be proved for the containment-SMTCN also. □

Detection of such a 'null' arc within a connected component of a SMTCN would let us declare the corresponding FTCN to be *inconsistent*, in the step 3 of the algorithm generate-SMTCNs.

7   Significance of SMTCNs

If a problem domain is such that the interval relations are restricted to any of the two sub-algebras discussed here, then by using our theorems one could check for consistency of the given TCN (which is naturally a SMTCN) quite efficiently. From this point of view we have identified two 'easy' special cases of interval-based reasoning. These cases are easy because they constitute sub-problems of constraint propagation on pointizable sets[vB90]. These cases also bear semantically more meaningful information (like possible causal-chains or containment-chains) which are useful in some applications. For instance, deriving this type of probable causal information from a set of temporal information is an important issue in the field of distributed computing[RS96]. Containment-chains could also be used for deriving the reference intervals for the given temporal information[Koo89]. Many of the important queries in temporal databases[MP93] could be efficiently answered from the availability of such chains (SMTCNs).

In case one has to derive SMTCNs from a TCN where the problem domain is not restricted as discussed in the previous paragraph, one may argue that the number of derived SMTCNs would be exponentially large. But it should be noted that since SMTCNs are derived from un-constrained initially given network, the actual worst case number of FTCNs are bounded by $O(2^e)$, where e is the number of only *initially constrained arcs*, which typically is much less than the total number of arcs in a TCN, and out of which some FTCNs would be eliminated for having cycles. Furthermore, since the chance of having a primitive relation and its inverse relation in the same label is much less in any real-life situation, this actual number of FTCNs would be quite less. Creation of SMTCNs would divide the problem of consistency checking over isolated smaller networks. Gaining efficiency through such divide and conquer strategy has been noted long back[All83].

Temporal reasoning involves answering some types of questions over a given TCN. Most of these questions are of the type: (1) whether the TCN is consistent or not, (2) what is the relation between a pair of queried intervals (nodes) which is consistent with respect to the other constraints in the TCN. (3) A third type of query could be: whether a temporal entity (represented by an interval) could *cause* or *contain* another temporal entity (represented by a second interval). Such queries could be handled when consistent SMTCNs are generated. The first type of query could also be approximately answered in our scheme, as is done after running many other approximate algorithms. The answer to the second type of questions could be given very efficiently from the SMTCNs, in case the involved pair of nodes lie on the same SMTCN. If they lie across two different SMTCNs, the answer has to be derived by running special algorithms for constraint propagation across SMTCNs.

Our reasoning scheme could be also extended towards quantitative representation[Mei91]. A particular case of interest is where uncertainties exist over the end points of intervals. In that case primitive relations within any label gets restricted to the so called 'adjacent' relations in a conceptual neighborhood[Fre93]. Causal or containment sets are also adjacent according to this partial order of primitive relations. This also suggests that the number of derived SMTCNs might be smaller, in case the incompleteness of relations between intervals are created explicitly because of uncertainty over the end-points of those intervals.

One dimensional temporal reasoning is applicable to gene-sequencing research work in molecular biology[GR93]. For instance, in human genome project different labs produce gene sequences from clones of human chromosomes. Their results are sometimes contradictory, often creating disjunctive interval-relations between identified 'units,' which could be molecular sequences of some biological significance. A temporal reasoning algorithm could help resolving those problems, and creating 'consistent' sequences. In this application domain, identification of clusters (SMTCNs) where inconsistency is localized, could be of great help, rather than informing that the whole network is inconsistent. Here, each causal-SMTCNs would be a possible cluster of properly sequenced units, and each containment-SMTCN would be a possible spatial hierarchy of some identified units.

8 Conclusion

In this paper we have reported some results of our ongoing research on deriving sub-networks of a given TCN where the labels of the former are within one of the two specific sub-sets of the set of thirteen primitive interval relations. These two subsets relate to some semantically meaningful information. Similar direction have been alluded to before in the research works within the point-based reasoning[vB90, GC95]. In this article we have provided some theoretical results of this research. We have also proposed here an algorithm which utilizes these results to derive sub-networks of a given unconstrained temporal constraint network. Significance of the two interval sub-algebras are also being explored here.

Some advantages of our representation are the following.
(1) From our representation scheme one could derive feasible causal information (or containment information) through the semantically meaningful temporal constraint networks (SMTCN). This has a bearing in applying temporal reasoning towards many fields like diagnosis, distributed computing, query answering from temporal databases, planning, and gene-sequencing.

(2) Generation of SMTCNs isolate clusters of temporal intervals related to each other in a partial order within each SMTCN. This also helps in improving the efficiency of reasoning in a real-life situation, as was suggested before in this line of work[All83].
(3) Temporal reasoning with each of the sub-algebras identified here are tractable. In case a given temporal constraint satisfaction problem belongs to one of these two sub-algebras our techniques/algorithms are useful (and efficient) there.
(4) One can isolate some inconsistent portions of a TCN from the rest of the consistent parts of the TCN. Normally temporal reasoning algorithms do not address this problem. They do not identify any specific region of inconsistency in the graph.

References

[All83]   J. F. Allen. Maintaining knowledge about temporal intervals. *Communications of the ACM*, 26(11):510–521, 1983.

[Fre93]   Christian Freksa. Temporal reasoning based on semi-intervals. *Artificial Inteligence*, 54:199–227, 1993.

[GC95]   Alfonso Gerivini and Matteo Christiani. Reasoning with inequations in temporal constraint networks. In *Workshop note on Spatial and Temporal Reasoning, IJCAI-95*, 1995.

[GR93]   John N. Guidi and Thomas H. Roderick. Inference of order in genetic systems. In *Proceedings of the International Conference on Intelligent Systems for Molecular Biology, Bethesda*, 1993.

[Had96]   Peter Haddaway. A logic of time, chance, and action for representing plans. *Artificial Intelligence*, 80:243–308, 1996.

[Koo89]   Johannes A. G. M. Koomen. *The TIMELOGIC Temporal Reasoning Syustem*. Technical Report, TR231, Computer Science

Department, University of Rochester, 1989.

[Lad88] Peter B. Ladkin. Satisfying first-order constraints about time intervals. In *Proceedings of the Seventh National Conference on Artificial Intelligence*, 1988.

[LMN91] Rasiah Loganantharaj, Debasis Mitra, and Gudivada Naidu. Efficent exact algorithm for finding all consistent singleton labelled models. In *Proceedings of IEEE Robotics and Automation conference, Nice, France*, 1991.

[LR92] Peter B. Ladkin and Alexander Reinefield. Effective solution of qualitative interval constraint problems. *Artficial Intelligence*, 57:105–124, 1992.

[Mei91] Itay Meiri. Combining qualitative and quantitative constraints in temporal reasoning. In *Proceedings of the National Conference of Am. Asso. of Artificial Intelligence*, 1991.

[MP93] Angelo Montanari and Barbara Pernici. Temporal reasoning (chapter 21). In *Temporal Databases: theory, design and implementation, eds. Abdullah Tansel et al*, pages 534–562, The Benjamin/Cummings Publishing Co., Inc., 1993.

[RA95] Rita V. Rodriguez and Frank D. Anger. Reasoning with unsynchronized clocks. In *Proc. of workshop on Spatial and Temporal Reasoning, IJCAI-95*, 1995.

[RS96] Michel Ranynal and Mukesh Singhal. Capturing causality in distributed systems. *IEEE Computer*, 49–56, February 1996.

[vB90] Peter van Beek. *Ph.D. dissertation: Exact and Approximate Reasoning about Qualitative Temporal Relations*. University of Alberta, Edmonton, Canada, 1990.

[VG94] Lluis Vila and Lluis Godo. On fuzzy temporal constraint networks. *Mathware and Soft Computing*, 3:315–333, 1994.

A NEW MODEL FOR PROJECTING TEMPORAL DISTANCE USING FUZZY TEMPORAL CONSTRAINTS

Rasiah Loganantharaj and Stanislav Kurkovsky

Automated Reasoning Laboratory, Center for Advanced Computer Studies
University of Southwestern Louisiana. Lafayette, LA 70504
Email: raja@usl.edu, stan@usl.edu

ABSTRACT

In this paper, we focus on the representation of, and the reasoning with, metric bounds with possibilistic distributions. Metric bounds have been used to represent the prepositional quantitative relations between events. For example, the ACM student chapter meeting will take at least 45 minutes and at most 65 minutes is a metric bound on the event meeting. Metric bounds use two extreme bounds for duration of an event even though it rarely occurs. In this paper we show how to associate possibilistic distribution with the metric bounds, and thus makes the model more accurate. When we associate the possibilistic distribution with metric bounds, the composition of metric intervals becomes quite complex as we have shown in this paper. For computational convenience, we provide a method to approximate the distribution and show how to compute the composition of complex composite distributions. The representation can be further enriched by incorporating temporal transition rules. We discuss about the representation and the propagation of temporal constraints using the rule. We illustrate our representation and reasoning mechanism with an example.

1. INTRODUCTION

Temporal projection is understood in the literature as projecting an occurrence of an event in the future using the current domain information and a current event. Applying default logic and other kinds of formal logic for default temporal projection has miserably failed as has been demonstrated by the Yale Shooting problem [Hank86] [Hank86][Hank86]and its variations [Shoh86]. In this paper, we are not interested in the default temporal projection problem, but we will focus on projecting the time duration from a current event to a future event using the domain knowledge and the current event.

Let us consider a mundane task of going to work. John left home at 7am to his friend's house from where they car-pool to office. He reached his friend's house after 10 minutes. He waited for 5 minutes till all other passengers arrived. The trip from there to office took 30 minutes. From this episode we can conclude that it took 45 minutes for John to reach office from home. The information presented here are facts about some past event. Suppose John repeats the same activities at the same time every working day, we can predict that John will be at office at 7.45am every working day. This is, however, an unrealistic modeling of activities since it is highly unlikely that an event of this sort repeats exactly at the same time every day. A more realistic model accommodates a window of time frame, an interval, for each event which we call a metric bound. For example, John waits at least 5 minutes and at most 10 minutes for the passengers is a metric bound. A version of John going to work can be stated with metric bounds as following: John leaves home as early as 6.45am and as late as 7.15am. It takes at least 8 minutes and at most 12 minutes to reach the place to car-pool. The waiting time to gather other passengers takes at least 5 minutes and at most 10 minutes. The trip to office will takes at least 21 minute and at most 39 minutes after picking up the passengers. From this information we can predict that John will be in his office as early as 7:19 am and as late as 8:16 am.

The modeling of events using the metric bounds can be further improved. A metric bound accommodates two extreme boundaries and does not describe a typical or a highly probable duration for an event. To capture the activities accurately we associate possibilistic distribution to the metric bound. Thus, the prediction of the temporal distance may be accurate than otherwise. The model with metric bounds with possibilistic distribution can be further improved by incorporating first order temporal constraints, or a set of rules. Typical examples of first order temporal constraints are: When it rains the trip takes at least 40 minutes and at most 60 minutes. The same trip may take 10 minutes more during the morning rush hours, from 8am to 8.15am.

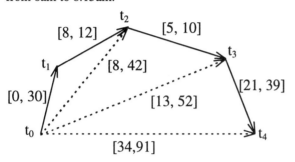

Figure 1

In this paper we address the basic issues involved in representing and reasoning with metric bounds with possibilistic distribution. We then enrich the representational language with first order temporal constraints in the form of rules.

2. PRELIMINARIES

There are two time elements: points and intervals. An interval is a continuous span of time from one end point to the other, thus, an interval can be presented as a pair of end points. An event can occur instantaneously or over a period. Here, we follow the notations of the temporal constraint network of Dechter et al. [Dech89].

Suppose ti and tk are a pair of end points of an event, say E. The unary constraint of the temporal variables t_i and t_k cab be specified as $t_{iLow} < t_i < t_{iHigh}$ where t_{iLow} and t_{iHigh} are respectively the lowest and the highest allowable vales for t_i.

Suppose the event E takes at least t_{ikLow} and at most t_{ikHigh}, the binary constraint is specified as

$t_{ikLow} < t_k - t_i < t_{ikHigh}$,.

In a temporal constraint network, each node models a time point, and a label on a directed arc represents the binary constraint or a metric bound between the respective nodes or points. Suppose, L_{ik} is the label of an arc from node i to node k, representing a pair of time points t_i and t_k respectively. L_{ik} is $[t_{ikLow} , t_{ikHigh}]$.

Composition operation: Suppose L_{ik} is $[t_{ikLow} , t_{ikHigh}]$ and L_{kn} is $[t_{knLow} , t_{knHigh}]$. The composition of the labels L_{ik} and L_{kn}, having the notation L_{ik} o L_{kn}, is given by $[t_{ikLow} + t_{knLow} , t_{ikHigh} + t_{knHigh}]$.

Let us illustrate how to use metric bounds using the example, John going to office. The temporal constraint network for the example is given in Figure 1 where t_0 is the reference time that is set to 6.45am, t_1 represents the

time John leaves home, t_2 and t_3 respectively represent the time John reaches his friend's house, and the time he leaves to office after collecting all the passengers. t_4 represents the time when John reaches his office.

Figure 1 shows the original metric temporal constraints and the propagated metric values along the arcs. Metric bounds have been used for representing and reasoning with the prepositional quantitative temporal constraints.

3. REPRESENTATION OF, AND REASONING WITH, METRIC BOUNDS

A metric bound simply states the boundary values of a binary constraints between a pair of temporal variables. Suppose, John reaches his office within 28 to 32 minutes every day after picking up all the passengers. Assume that there are possibilities for him to reach the office as early as 21 minutes and as late as 39 minutes after picking up all the passengers. In representing this information in metric bounds, we provide the two extreme boundaries for the duration of the event even though these boundary conditions rarely occur. This information can be presented with trapezoidal distribution of possibility proposed in [Dubo89] as shown in Figure 2. Here we assume that the possibility of reaching the office in 21 to 28 minutes linearly varies from 0 to 1. Similarly, we assume that the

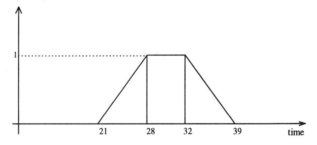

Figure 2

possibility of reaching office in 32 to 39 minutes varies linearly from 1 to 0. This distribution can be succinctly represented as [21(0), 28(1), 32(1), 39(0)]. In general, a possibilistic distribution can be represented as a sequence of adjacent time points and their corresponding values and it takes the form $[t_1 (v_1), \ldots t_k (v_k)]$ where v_j is the posibilistic value at t_j, and it is assumed that v_i linearly varies to v_{i+1} from t_i to t_{i+1}.

A possibility distribution can take an arbitrary form, but for the computational convenience, it is usually assumed that either the possibility value is linearly varying, or remains constant between a pair of adjacent points.

Now, we will discuss how to propagate metric bounds with possibilistic distribution. Suppose the distribution of the label L_{ik} is $f_{ik}(x)$ and L_{kn} is $f_{kn}(x)$. The composition of L_{ik} and L_{kn}, denoted by $L_{ik} \, o \, L_{kn}$, is given by $f_{in}(z)$ where,

$$f_{in}(z) \quad = L_{ik} \, o \, L_{kn}$$
$$= f_{ik}(x) \, o \, f_{kn}(y)$$
$$= max \, (min \, (f_{ik}(x) \, , \, f_{kn}(y)) \, | \, x+y = z)$$

When we assume the possibility distribution is varying linearly or remains constant between adjacent points or nodes, there exist only a few primitive distributions from which other distributions can be constructed. We provide composition of posibilistic distribution for many possible combinations of the primitive distributions. The composition of composite distributions will then be a

For example, the composition of $[t_1 \, (v_1), \, t_2 \, (v_2), \, ..., \, t_m \, (v_m)]$ and $[t'_1 \, (v'_1), \, t'_2 \, (v'_2), \, ..., t'_n \, (v'_n)]$ is given by
$$[t_1 \, (v_1), \, ..., \, t_m \, (v_m)] \, o \, [t'_1 \, (v'_1), \, ..., \, t'_n \, (v'_n)] =$$
$$\cup \, [t_i \, (v_i), \, t_{i+1} \, (v_{i+1})] \, o \, [t'_k \, (v'_k), \, t'_{k+1} \, (v'_{k+1})]$$
$$\text{for} \ i = 1 \text{ to } m\text{-}1, \text{ and } k = 1 \text{ to } n\text{-}1$$

If we can compute the composition of primitive relations, then we should be able to compute compositions of any complex distributions. Let us now consider the composition of a pair of primitive distributions as shown in Figure 3. Here, it is assumed that $h_1 < h_3 < h_2$, and $h_4 < h_3$. C' is the point at which $f_{ik}(x) = h_3$. Thus, $C' = a + (b-a)(h_3 - h_1)/(h_2 - h_1)$.

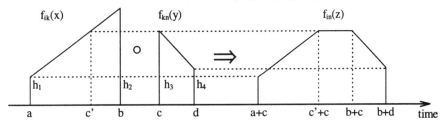

Figure 3

union of all the compositions of their components. Suppose the labels L_{ik} and L_{kn} have composite distributions. The composition of the composite distributions are given by

$$L_{ik} \, o \, L_{kn} \quad = \cup \, f_{ik}(x) \, o \, f_{kn}(y)$$
$$\text{for every } f_{ik}(x) \in L_{ik} \text{ and } f_{kn}(y) \in L_{kn}$$

Suppose, $f_{ik}(x)$ and $f_{kn}(y)$ has the similar distribution as shown in figure 3 except for $min(h_1, h_2) > max(h_3, h_4)$ The composition of the distribution is given as shown in Figure 4

The composition of other primitive distributions are shown in the following Figures 5 through 11. In the distribution shown in Figure 5, $h_4 < h_2 < h_3$.

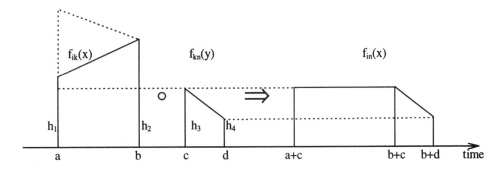

Figure 4

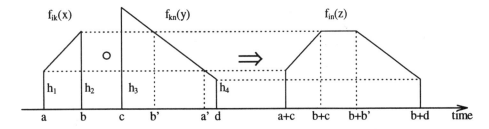

Figure 5

When max $(h_1, h_2) < $ min (h_3, h_4), the composition of $f_{ik}(x)$ and $f_{kn}(y)$ is given as shown in Figure 6.

When $h_3 < h_2 < h_4$ in the distribution shown in Figure 7, the composition of $f_{ik}(x)$ and $f_{kn}(y)$ is given by $f_{in}(z)$ of Figure 7.

When $h_1 < h_4 < h_2$ in the distribution shown in Figure 8, the composition of $f_{ik}(x)$ and $f_{kn}(y)$ is given by $f_{in}(z)$ of Figure 8.

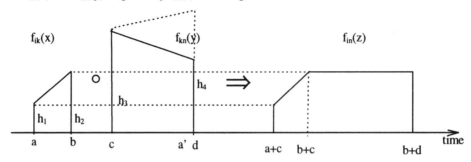

Figure 6

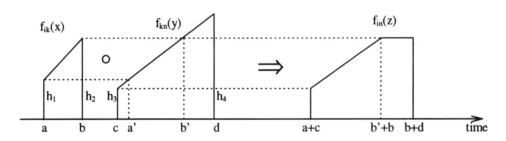

Figure 7

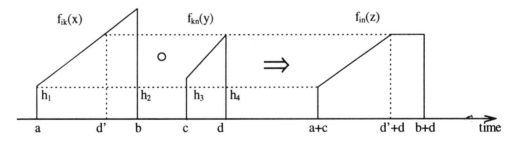

Figure 8

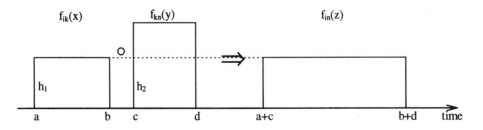

Figure 9

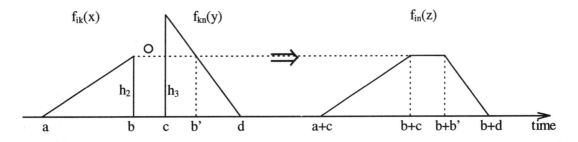

Figure 10

The special cases of the above distributions reduce to the following distributions and their compositions are given in the following figures.

Also assume that heavy traffic starts at 7:15am and continues through 8am. With these first order temporal constraints, the problem is represented more realistically.

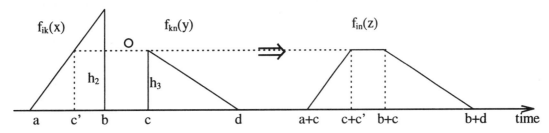

Figure 11

When h_1 and h_4 in the distribution of Figure 5 become zero, $f_{in}(z)$ reduces to the one shown in Figure 10.

When h_1 and h_4 in the distribution of Figure 3 become zero, $f_{in}(z)$ reduces to the one shown in Figure 11.

4. REPRESENTING AND REASONING WITH FIRST ORDER TEMPORAL CONSTRAINTS

We have described how to represent prepositional fuzzy temporal constraints and reason with them by propagating their constraints in the temporal network. By including first order temporal constraints in the form of temporal rules we can enhance the richness of the representational language and the reasoning capabilities of the resulting system. We will introduce this with an example. Consider the statement that John reaches his office within 28 to 32 minutes while there is a possibility of reaching the office as early as 21 and as late as 39. The corresponding fuzzy distribution is given as [21 (0), 28 (1), 32 (1), 39 (0)]. Suppose, the previous distribution is given for a normal day and when it rains, the time taken to reach his office is given by another distribution, say [26 (0), 33 (1), 37 (1), 44 (0)]. Further, let us assume that the distribution is given for a light-traffic hours. When the traffic is heavy, we may have a different distribution and it is given by [25 (0), 32 (1), 36 (1), 43 (0)] for a normal day, and [29(0), 36 (1), 40 (1), 47 (0)] for a rainy day.

A first order temporal constraints of this form can be represented as a temporal transition rules from the current state to the next state. Therefore, our temporal constraint rule will take the following form:

    if <current_state> & [<other_conditions>] & <time_interval>

    then <next_state> <possib_distribution>.

The conditions within [..] are optional. Using the syntax of the temporal transition rule, let us represent the example we described above. Here t_3 and t_4 respectively represent the state of leaving to office after collecting all passengers, and reaching the office.

    if t_3 & [6am, 7:15am]
      then t_4 [21 (0), 28 (1), 32 (1), 39 (0)].
    if t_3 & [7:16am, 8am]
      then t_4 [25 (0), 32 (1), 36 (1), 43 (0)].
    if t_3 & rains & [6am, 7:15am]
      then t_4 [26 (0), 33 (1), 37 (1), 44 (0)].
    if t_3 & rains & [7:16am, 8am]
      then t_4 [29 (0), 36 (1), 40 (1), 47 (0)].

5. APPLICATIONS

In this section we look at one of the applications to illustrate the usefulness of our approach to represent fuzzy metric bounds and their propagation.

Table 1 Numeric data for the travel routing problem

From (t_j)	To (t_k)	Most Likely Bounds (t_{ik})	Distribution
S	C	[20min, 25min]	[20min (1), 25min (1)]
S	D	[45min, 50min]	[45min (1), 50min (1)]
S	E	[15min, 20min]	[15min (1), 20min (1)]
C	O	[15min, 20min]	[15min (1), 20min (1)]
D	O	[25min, 30min]	[25min (1), 30min (1)]
E	O	[30min, 35min]	[30min (1), 35min (1)]

Example. Travel routing problem

John lives in a town where there is a major traffic jam during an annual international festival that lasts for a week. John goes to office after leaving his child at school. There are three possible routes to his office from the school : via locations C, D and E. Let the nodes H, S, C, D, E, and O respectively represent Home, School, locations C, D, E, and the Office. The possible routes are given in Figure 12.

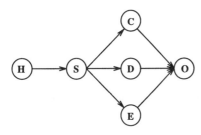

Figure 12 Possible routes of John

The earliest and the latest possible time John leaves home is respectively 7:15am and 7:50am, but most likely he leaves between 7:25am and 7:35am. The time taken for him to drive to school is likely to be between 10 and 12 minutes. To reduce the traffic jam during the festival week, the traffic officers have implemented a schedule: the traffic reaching S before 7:30am is directed to location C. Traffic reaching S between 7:30am and 7:40am is directed to location D, while the traffic reaching S after 7:40 goes through location E. John's estimates of reaching various locations during the festival week is summarized in Table 1.

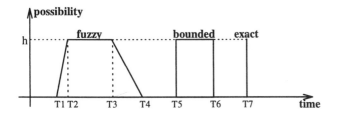

Figure 13 Representation of primitive temporal intervals

For representational convenience, some primitive distributions can be combined together. For example, a triangular distribution followed by a rectangular and another triangular distribution with the same height can be represented as a trapezoidal distribution. More specifically, the primitive distributions [T_1 (0), T_2 (h)], [T_2 (h), T_3 (h)], and [T_3 (h), T_4 (0)] are combined as a trapezoidal distribution, or trapezoidal fuzzy interval [T_1, T_2, T_3, T_4] (h) according to Dubois and Prade [Dubo88][Dubo89]. Figure 13 shows trapezoidal fuzzy interval [T_1, T_2, T_3, T_4] (h) regular (bounded) interval [T_6, T_6] (h) and temporal point [T7] (H).

Syntactical representation in scripts is slightly different (Figure 14). The same intervals will be expressed

```
Hierarchy(minute,m) begin
        hour, h=minute*60;
end;

Script begin
node        H(" John's home -initial point"),
        S("John drops off his child at the school -
          forking point"),
        O("John's office --target point"),
        C("Intermediate point C"),
        D("Intermediate point D"),
        E("Intermediate point E");

fact
        Start(H) at
        fuzzy(7h:15m, 7h:25m, 7h:35m, 7h:50m);

rule  R1( if H after 0m then S
        within bounded(10m, 12m),1),
        R2( if S before 7h:30m then C
        within bounded(20m, 25m),1),
        R3( if C after 0m then O
        within bounded(15m, 20m), 1),
        R4( if S between 7h:30m and 7h:40m then D
        within bounded(45m, 50m), 1),
        R5( if D after 0m then O
        within bounded(25m, 30m),1),
        R6( if S after 7h:40m then E within
        bounded(15m, 20m),1),
        R7( if E after 0m then O
        within bounded(30m, 35m),1);
end;
```

Figure 14 Script for the proposed system to solve traveling route problem

as follows:

fuzzy interval	fuzzy(T1, T2, T3, T4), h
regular interval	bounded(T5, T6), h
point	exact(T7), h

6. SUMMARY AND DISCUSSION

The representation of, and the reasoning with, incomplete, imprecise and uncertain temporal constraints

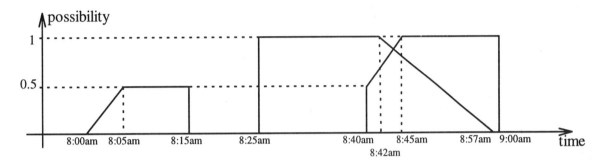

Figure 15 Solution to the travel routing problem

In this example, the duration for each segment from home (H) to office (O) is represented as a bounded intervals. The starting point (Home) will be represented as a fuzzy interval [7h:15m, 7h:25m, 7h:35m, 7h:50m] (1). The rules of Figure 14 do not have any explicit branching, thus, we have only unconditional transitions from one point to another. In the end of the propagation we will have a set of 3 results. This happens because of the implicit branching in rules R2, R4 and R6 (Figure 14). They provide different routes with mutually exclusive temporal constraints. For example, rule R2 has constraint "before 7h:30m" and rule R4 has constraint "between 7h:30m and 7h:40m".

Numerically these three results may be represented as follows (Figure 15):

Route 1 (H-S-C-O):
[8h:00m (0), 8h:05m (0.5), 8h:05m (0.5), 8h:15m (0.5)]

Route 2 (H-S-D-O):
[8h:40m (0.5), 8h:45m (1), 8h:45m (1), 9h:00m (1)]

Route 3 (H-S-E-O):
[8h:25m (1), 8h:42m (1), 8h:42m (1), 8h:57m (0)]

We can see that the earliest and the latest time when John reaches his office is respectively 8:00am and 9:00am, and there is no chances for John to be there between 8:15am and 8:25am. The first route via C results with the distribution with the maximum possibility of 0.5, which means that this route may be taken only when John is uncertain about leaving home. The route through D takes longer time to reach office. In fact, he may reach the office later than he would if he had left home later to use route through E. If John takes the third route, then his chances of arriving earlier are much higher than with the second route. This means that John should leave home only after 7:30 am.

are very challenging. An event usually takes place over an interval, thus, it is represented by a pair of end points: one representing the beginning, and the other representing the finishing point. If we represent some past events with their exact duration, then propagating the constraints and finding the relations between a pair of events become a trivial task. Without exact information about the duration of an event, quantitative metric bounds are used to represent the range of duration of an event. Suppose, a meeting will last at least 30 minutes and at most 45 minutes, the duration of the meeting will be represented as a metric bound, [30 45]. Dechter et al. [Dech89] have described how to represent and to propagate metric bounds over a temporal constraint network. Since the metric bounds represent pair of extreme boundaries for a duration of an event instead of capturing highly probable duration for the event, propagation of these bounds over a temporal constraint network will yield weaker result. When we associate possibilistic distributions with metric bounds and propagate these constraints, we get a stronger result. In section 3 of this paper, we describe the representation of possibilistic distribution, and their compositions for propagating constraints. A possibilistic distribution can be an arbitrary complex function. For the computational convenience, we limit the distribution to a function that either varies linearly, or remains constant over a pair of adjacent points. We have also shown how to compute composition of complex distributions that consists of primitive distributions.

The representation can be further enriched by incorporating temporal transition rules. We discuss about the temporal transition rules in section 4. In section 5, we illustrate our representation and reasoning mechanism using an example. Our implementation of the temporal reasoning system includes possibilistic distribution on metric bounds, and temporal transition rules. Our system accepts the facts and rules for specifying temporal projection problem as has shown in Figure 14. The results of the example in section 5 are obtained by running our system. Our approach can be applied to solve many problems including scheduling, project planning, and prediction of temporal relations between the current and the future events.

ACKNOWLEDGMENT

This research was supported by a grant from Louisiana Education Quality Support Fund (LEQSF 1993-96RD-A-36).

REFERENCES

[All83]   J. F. Allen, Maintaining Knowledge about Temporal Intervals, in *Communications of ACM*, Vol. 26, No. 11, pp 832-843, 1983

[All84]   J. F. Allen, Towards a General Theory of Action and Time, in *Artificial intelligence*, Vol. 23, No. 1, pp 123-154, 1984

[Dech89]   R. Dechter, I. Meiri and J. Pearl, Temporal Constraint Network, in the *Proceedings of Principles of Knowledge Representation and Reasoning*, May 1989.

[Dubo88].   D. Dubois and H. Prade, *Possibility Theory*, Published by Plenum Press, New York, 1988.

[Bubo89]   D. Dubois and H. Prade, Processing Fuzzy Temporal Knowledge, *IEEE Transactions on System, Man, and Cybernetics*, Vol. 19, No. 4, August 89, pp 729-744.

[Hank86]   S. Hanks and D. McDermott, Default Reasoning, Nonmonotonic Logics, and the Frame Problem, in *the Proceedings of American Association for Artificial Intelligence Conference*, 1986, pp 328 - 333.

[McD82]   D. McDermott, A temporal Logic for Reasoning About Process and Plans, *Cognitive Science* 6, pp 101-155, 1982

[Shoh86]   Y. Shoham, Chronological Ignorance, Time, Nonmonotocity, Necessity and Causal Theories, in the *Proceedings of American Association for Artificial Intelligence Conference*, 1986, pp 389-393

[VanB89]   P. van Beek, Approximation Algorithms for Temporal Reasoning, in *the Proceedings of IJCAI*, 1989, pp 1291-1296.

AN EXPERIMENTAL LANGUAGE TRANSLATION SYSTEM FOR ATIS

Bin Wu

Department of Computer Science
University of Pennsylvania
Philadelphia, PA 19104
bwu2@eniac.seas.upenn.edu

Abstract

The drive to automate information services has required the rapid advance of computer-aided language translation in many applications such as the translating phone. We present an Experimental Language Translation System for the Air Traffic Information Service(ATIS) in this paper. It can be used to translate English into Mandarin Chinese or vice versa in the ATIS domain. The headed phrase model and associated probabilistic finite state machines are used in our system. The dependency trees are introduced for representing phrases in both English and Chinese. We developed lexical transfer rules required for mapping analyzed English phrases into corresponding expressions in Mandarin Chinese. By evaluating the effectiveness of our approach against a random sample of test data, we found that the fluent and correct translation rate reaches 94%.

Keyword : Natural Language Processing, Machine Translation, Translating Phone.

1 Introduction

Since the invention of computers, computer-aided translation from one language to another has been one of the ambitious goals of computer scientists and linguists. In past decades, there have been many advances in automatic translation between different languages such as English, German, French, Chinese and Russian. After an initial phase of enthusiastic work, research in Machine Translation reached a "dark age" with the release of the highly critical Automatic Language Processing Advisory Committee (ALPAC) report in 1964 [1].

The work was revitalized after introduction of the AI-based techniques in early 1970's. Whereas the earlier research primarily based on the syntactic knowledge, the present research is directed towards the use of semantic and pragmatic knowledge [1, 2, 3].

The recent survey by Bates and Weischedel asserts that the time is ripe for great leaps forward in the generality and utility of natural language processing(NLP) systems [2]. However, domains must be narrow enough so that constraints on the relevant semantic concepts and relations can be expressed using current knowledge representation techniques.

AI-based techniques require understanding the text in certain extent before translating it. The program modules implementing this technique are for understanding the meanings and/or structures, and storing them in an intermediate form, called an inter-form for brevity. Examples of inter-forms include the semantic net, frames, the headed phrase models, ..etc. This intermediate forms are used in further processing of sentences or phrases for automatic translation, query-answering, and retrieval of natural language database items. Thus, there are two stages of processing in this kind of implementation. The first one is analysis, which transforms the text into an inter-form, and the second one is translation, which translates the inter-form into a phrase or sentence in another language. These software modules that implement these processes are called analyzer and generator(translator).

The recent approaches to NLP are mainly based on semantics in order to overcome the deficiencies of syntactic methods that resulted in some meaningless translations. There are a number of implementations that use semantic representations [1, 3, 6]. These approaches assume that understanding the meaning of a phrase or a sentence is essential to represent it, preferably using a language-independent form, before it can be translated.

There are different kinds of representations, such as the Predicate Calculus, the Conceptual Dependency(CD) proposed by Dr. Schank, Frames introduced by Dr. Minsky, the Semantic Net and others [1, 3]. Each modeling approach can be suited for some cases, but may have some drawbacks in other cases. The choice depends on the application. Recently, Dr. Hiyan Alshawi of AT&T Bell Lab introduced the headed phrase model for the machine translation [4]. In this model, much more attentions are paid to word relations, semantic and syntactic, in the analysis stage. The advantage of this model is its capability to combine semantic and syntactic relations into one representation. In the translation stage, the dependency tree in one language is transformed into the corresponding one in another language. This simplifies the translation process.

2 The Machine Translation System for ATIS

This paper presents our work in developing a natural language translation system especially designed to pave the way for translating phone which has been listed as one of the grand challenges in AI [5]. With this kind of phone, a Chinese speaker can converse with an English speaker in real time. This requires solutions to a number of currently unsolved problems.

We have to restrict our research in a narrow domain because this ambitious goal can not be reached in one step. At the Information Principles Laboratory of AT&T Bell Labs, a speech-to-speech translation system from English into Chinese has been developed for the Air Traffic Information Service(ATIS). This speech translation system has three modules as shown in Figure 1. First, a speech recognizer is used to recognize English speech and convert it to its English text version. Then, a language translator translates the English text into its corresponding Chinese text. Finally, a speech synthesizer generates the synthetic voice in Mandarin. In this paper, we focus our attention on the language translation subsystem, the second module, which is designed and developed by our group. Our ATIS corpus contains 1900 words in its vocabulary. We will show some progress of our translation system in dealing with ambiguity, and incomplete phrases ...etc.

2.1 The Language Translation System

The language translator itself contains an analyzer, a translator, and a generator. The analyzer understands the meaning of an input word string while preserving its structure, and stores the information in our inter-form. A language dependent form called a dependency tree is used as our inter-form in the translation system. The translator takes the English dependency tree and creates its corresponding Chinese dependency tree. Then, the generator produces Chinese texts based on the Chinese dependency tree. We focus our attention on the translation of phrases with numbers, dates, times and codes. Following are some typical phrases and sentences that we need to handle in the ATIS.

1. Show me the ground transportation in Las Vegas on Thursday June seventeenth nineteen ninety three.

2. What type of aircraft is US Air flight one two one six?

3. I will take Eastern flight number six four five leaving Boston at two twenty one P M.

4. I would like the two hundred and sixty six dollar fare.

5. I'd like to arrange for two friends to fly into Los Angeles next Saturday evening.

We have a collection of about 20,000 spoken English sentences for ATIS. We use a language and translation model based on cascading a certain type of probabilistic finite state machines(FSA). Our major tasks were to construct an inventory of automata for both spoken Mandarin and English, such as dates, times, and codes (e.g. flight numbers), a lexicon associating particular words with these automata, and the lexical transfer rules required for mapping analyzed English dates, times, and codes into corresponding expressions in Mandarin. Finally we evaluate and test the correctness and effectiveness of our machine translation system against a set of random samples of our test data. There are five software modules stored in five files. They are:

1. Chimacb.cl contains all the automata for spoken Mandarin dates, times, numbers and flight codes.

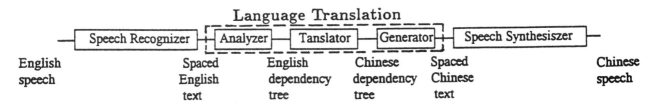

Figure 1. The block diagram of the speech to speech translation system from English to Mandarin.

2. Chilexb.cl contains all the lexical rules associated with each word and its corresponding machine.

3. Ectmacb.cl contains all the macro functions for translating dates, times, numbers, and flight codes.

4. Ectlexb.cl contains all the translation rules to associate each word with the corresponding translation macro function.

5. Chidepb.cl contains all the data needed fro adjusting the cost so that the correct translation will have the lowest cost.

This system is developed using Common Lisp running under the UNIX operating system on Sun workstations.

2.2 Problems and Difficulties in the Language Translation from English to Chinese

Chinese words incorporate shape and complex hierglyphic meanings into an ideographic language which is different from alphabetic languages used in most countries of the world. In addition, there is no explicit tense for verbs in Chinese. Therefore, there are significant problems and challenges in developing our Language Translation System, as follows:

1. Differences in word order. The word orders are quite different between English and Chinese.

$$August\, 12, 1994 \rightarrow 1994(year)8(month)12(day)$$

2. The introduction of new words and the deletion of existing words. In the date example, the words 'year', 'month' and 'day' should be added in the translated Chinese phrase.

$$May \rightarrow 5(month)$$

There is no 'May' standing for the 5th month of a year in Mandarin. The words '5' and 'month' should be combined for 'May' in Mandarin.

3. Differences in word abbreviations.

$$AM \rightarrow (morning)$$

4. "1 to N" or "N to 1" ambiguous matching in the translation.

"Northwest Airline", "Northwest Airlines", "Northwest", or "Northwest Orient Airlines"
    $\rightarrow$ "XiBei Hangkong KongSi"
"flight no. one twenty three" or
"flight no. one two three"
    $\rightarrow$ "HangBan 123"

We are going to discuss how these problems can be solved in the later sections.

3   The Analyzer

The function of an analyzer is to extract and preserve the meanings and structures of the phrases or sentences in the chosen representational form. In our machine translation system, a sentence is usually split into phrases, and the analysis function of our analyzer is restricted to individual phrases in a sentence. The headed phrase model introduced by Dr. Alshawi is used in our analysis system[4]. Each sentence is analyzed independently. There are two steps in the analysis. The first step extracts the head word from the phrase or sentence. The second one is concerned with building the representation of the phrase, which leads to a dependency tree, a structure describing the relational content of the input phrases.

Before we can actually implement the translation, a completed analysis for all the input strings should be finished first. The result of the analysis is an ordered dependency tree which is generated based on our analysis model. An ordered dependency tree is a dependency tree in which the dependent subtrees of each node are grouped into two sets, its left and right dependents, and an ordering relation is imposed on these dependents. To do the actual analysis, each pattern of a certain type of phrase is presented as an ordered dependency tree which is built in terms of the corresponding finite state machine. These machines are driven by a set of input lexicons. Each lexicon is associated with a machine or a macro function, and thus associated with a corresponding finite state machine. Based on Dr. Alshawi's model, a lexicon data base is developed in Chinese for ATIS, and each lexicon entry has the following pattern:

$$(lx < word >< machine - macro >< probability >)$$

The $< probability >$ field tells how likely a phrase headed by $< word >$ will be accepted by the $< machine - macro >$. The $< word >$ field on Chinese side might be a word or a group of words that matches an English word in terms of meaning because unlike English, the space between Chinese words is not always the separation of lexicons. Our ordered dependency tree produced by the analyzer will be also used later for generation, so we want to maintain the consistency between each English word and its corresponding Chinese word or words.

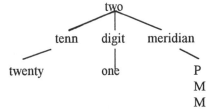

Figure 2. The dependency tree of a phrase.

Each machine-macro is defined as a macro function which is formatted as a nested structure. It has a starting state called rank, some stopping states and a set of "actions". The probability field tells how likely a transition is from state q to state q'. The rank field defines the key token of a input phrase. The key token is the word which invokes the machine-macro, and will be the head of the dependency tree. In picking up the head word, our analyzer will test all possible choices. Therefore, each word in a given phrase has the possibility to be the head word, but our analyzer will only select the one with highest possibility. For example in parsing the phrase "TWO TWENTY ONE P M", we will have a dependency tree headed by "TWO" as shown in Figure 2.

The relation on each arc of the tree links the head node and its dependent. Its corresponding machine-macro $hour - c!$ will be defined as:

```
(defmacro hour - e!
(phon cost)
(make - phrase
  : phon '(, phon)
  : head 1
  : rank 'hour1
  : base '((m 1 , phon hour)
  : ldep '(,(lf tenn      hour1 hour2)
         ,(lf digit     hour2 hour3)
         ,(lf meridian  hour3 hour4)
         ,(lf M         hour4 hour5))
  : abov '(,(up hour5 1 hour unk loc loc2 att lres))
  : cost cost
))
```

In this machine-macro, the "abov" field defines a stopping state with a set of up movements which allow an individual dependency tree of a single phrase to be connected into a huge dependency tree for a whole sentence. The goal of our analyzer is always to get a full analysis for each input string. Unfortunately, it is often possible to get several completed analysises for a single phrase because of the ambiguity. By carefully arranging the above relations, this can be controlled in some degree. On the other hand, we can also describe the machine-macro in terms of a finite state machine as follows:

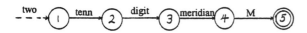

In this diagram, the dashed arrow shows the starting state, the solid arrows show the parsing paths, and the double circle means the stopping state. The relation on each arc of the tree links up the head node and its dependent.

4 Translation

In the implementation of the translation, we want to match each single English word with its corresponding Chinese translation after a completed analysis of each phrase. A transfer entry can be defined as the following format:

$$(tr < english >< chinese >< transfer - macro >$$
$$< probability >< class1 >< class2 >)$$

The $< probability >$ field gives the likelihood of the correct translation from $< english >$ into $< chinese >$. The $< class1 >$ and $< class2 >$ fields are optional depending on the actual field in that node of the dependency tree. Basically these two classes are used to avoid ambiguity. For instance, the phrase "five twenty one" can be translated either as a time or a flight code. We might want to add different classes to the head node "five" to make a distinction so that the generator will produce the appropriate translation for the situation. This will be explained in more detail in the later section. To make a correspondence between the translation stage and the analysis stage, each word appears as a lexicon entry and must have a corresponding transfer entry.

For each pair of Chinese and English words, there is a transfer macro function describing the related arcs and relations attached to that word. In writing the transfer macro, all the nodes of the dependency tree on English side are numbered by the same way in which its corresponding Chinese dependency tree is numbered. It is not necessary that the transfer macro describes the whole dependency tree headed by that word. Actually it only describes a small portion of the analysis tree, usually only one level up or down from that node. A boundary arc which connects two subtrees should only appears once in the whole file. Macro functions are defined as follows:

```
: indx   < start node number >
: rind   < start node number >
: srce   '((m < start node number >,
              < english - form >)
           (a < from node number n >   < relation >
              < to node number m >)
           ...... )
: targ   '((m < start node number >,
              < chinese - form >)
           (a < from node number n >   < relation >
              < to node number m >)
           ...... )
: cost   cost  : rcst cost
```

[1] The *indx* and *rind* fields contain the head node number of a sub-dependency tree headed by this particular word which evokes this transfer macro. The *srce* field and *targ* field contain a description for the head word and a set of edges going out from that node on both English side and Chinese side. The head words on both English and Chinese side are passed through parameter $< english - form >$ and $< chinese - form >$.

In this stage, we only make the graph transfer, and the ordering relation is not taken into consideration. Now let's see how the trees are transferred for the phrase "TWO TWENTY ONE P M".

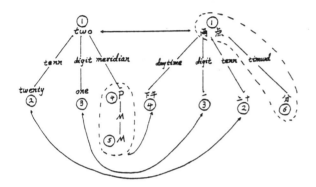

In the graph, the node number is written within each node, and the dashed arrows show how each node of the two trees is matched. And the dashed circle indicates that there are some new words introduced either on the English side or the Chinese side. We might want to translate them as one unit. The actual transfer macro function might look like:

```
(defmacro hour − ec!
  (english − form chinese − form cost)
  (make − tlex
    : indx 1
    : rind 1
    : srce '((m 1 , english − form)
    (a 1 tenn 2)
    (a 1 digit 3)
    (a 1 meridian 4)
    (m 5 M)
    (a 4 M 5))
    : targ '((m 1 , chinese − form)
    (a 1 daytime 4)
    (a 1 digitt 3)
    (a 1 tenn 2)
    (m 6 |  |)
    (a 1 timwd 6))
  : cost cost : rest cost
  ))
```

---

[1] The lower case "english" and "chinese" are used as symbols.

5 Experimental Results

5.1 Testing and Improvement

As mentioned before, we always try to eliminate ambiguity. We currently set the user defined parameter – "the generation-select" to ten to allow at most ten possible translation outcomes since this is an experimental translation system. The output with highest likelihood usually comes out first. In testing over one hundred samples of days, times, numbers, and flight codes from ATIS corpus, one hundred percent of them can be translated correctly. The problem is that over ninety five percentage of them are ambiguous. The correct output often has a lower likelihood and comes out later than the incorrect ones. We can control this by adjusting the cost for each possible translation, so that the correct output will have the lowest cost and come out first. To do so, another file called chidepb.cl is built to adjust the cost and force the correct translation to come out first on the Chinese side. The entry of this file has the following format:

$$(dep \ < head - class1 >< relation \ r >$$
$$< dependent - class2 >< number >)$$

The number field stands for $P(< dependent > | < head >< relation >)$ which tells how likely it is for a dependent with *class2* to occur given relation r and a head with *class1*. For instance, in translating "*flight number one twenty*" into a corresponding Chinese phrase "*flight number one two zero*", we want to reduce the cost for twenty being translated into "*two zero*". So we can write an entry as

$$(dp \ \_FCODE \ tentwo1 \ \_TENTWO1 \ 1.0)$$

In this case the head node of the flight number is the one which always has a class field *_FCODE*, and its dependent twenty has a class field *_TENTWO1* when twenty is translated into "two zero". These two nodes are linked by an arc under the relation called *tentwo1*. By setting the number field to one, the cost for translating twenty into two zero can be reduced, because the cost is computed by taking negative log of that number. So the higher number field will result in a lower cost. With the help of the dependency adjustment, the ambiguity is avoided.

I have evaluated the correctness and effectiveness of our system against a set of randomly chosen data from our ATIS corpus. With the execution time parameter setting : generation-select = 20, translation-nondet = 100, translation-select = 10, analysis-select = 2, the results of testing are shown in the following table:

Table 1. The correct translation rate of our MT system.

TestParameters	*Rate*
fluent and correct translation rate	94%
nongrammatical but understandable rate	3%
incorrect translation rate	3%
correct and understandable rate	97%

As much as 94% of the testing data can be translated fluently and correctly. Three percent translations are nongrammatical but understandable. Only 3% translation are incorrect. One hundred percent of them have at least one completed analysis. This means that we get a complete analysis for each input phrase. The correct translations have been obtain even for some uncommon cases of usages such as : "April twenty one" and "Flight one four twenty one". The incorrect translations occur for some uncommon usages such as: "Wednesday the next week" and "Flight eighteen hundred".

5.2 Some Solutions

Because of the complexity of the natural language itself, the problems in machine translation are usually caused by differences in word orders, the introduction of new words, differences in word abbreviations, and ambiguous matching in the translation. The disambiguation is the most difficult and important one. Usually, a transfer macro should include all the possible arcs that attach to the word evocating this macro. In order to get the correct translation, we apply the contextual disambiguation in our system. When there are too many possible relations attaching to that node and the same word could be translated into different Chinese words under different situations, we break the transfer macro into several smaller ones. This can force the generator giving the correct translation under a particular context. The obvious example for this are the numbers. A number without the context from zero up to nine can either start a time automaton, a number automaton, or a flight code automaton. But we don't want to write all those possible arcs from time automata, number automata, and flight code automata into one huge transfer macro. On the contrast, we use the class field mentioned earlier which can apply some restrictions to a certain context to break the huge macro into four small machines. So each transfer macro take care of only one type of translation. For instance, the number "five" with class field "*_TIME*" will evocate the transfer macro "hour-ec!" and will be translated into "WuDian" which means five o'clock in Chinese, and number "five" with class field "*_NUMBER*" will evocate transfer macro "*digit − ec!*" and translated as a general number.

Another specially case that we need to take care of is the flight code. For example, "flight number one twenty three" is equivalent to "flight number one two three", and "flight one twenty" is equivalent to "flight one two zero". In Chinese, we always say those digits of a flight number one by one. For instance, this means that twenty can be translated as "twenty", "two ten", or "two zero". We can force the correct translation coming out under a particular context by naming a distinct relation for each possible translation and giving a different class field on the Chinese side. So the basic idea is to have different macros for translating twenty into "twenty", "two ten", and "two zero".

6 Conclusion

This machine translation system has been developed for a specific domain of the Air Travel Information Service. Based on our experiments, the fluent and correct translation rate is 94%, and the correct and understandable translation rate is as high as 97%. I believe that the techniques developed here are quite suitable for this specific domain. After some improvements in the performance, this system can be used in the natural language translation for more complex text.

Acknowledgements

I am grateful to Dr. Alshawi for his distinguished guidance and his valuable suggestions. I would also like to thank Dr. Adam Buchsbaum and Dr. Jisheng He for their suggestions and interesting discussions.

References

[1] Hutchins, W. G., "Machine Translation–Past, Present and Future", Ellis Horwood, 1992.

[2] Bates, M and Weischedel, R. D. ed., "challenges in our Language Processing", Cambridge University Press, 1993.

[3] Grishman, R., "Computational Linguistics", Cambridge University Press, 1986.

[4] Alshawi, H., " A Headed Phrase Model for Natural Language Translation", (to be published soon).

[5] Reddy, R., " Grand Challenges in AI", ACM Computing Surveys, Vol. 27, No.3, 1995, 301-303

[6] Charniak, E., " Natural Language Learning", ACM Computing Surveys, Vol. 27, No.3, 1995, 301-303

Extracting Design Models from Natural Language Descriptions

Walling R. Cyre

Associate Professor of Electrical Engineering

The Bradley Department of Electrical Engineering

Virginia Polytechnic Institute and State University

Blacksburg, VA 24061-0111

E-mail: cyre@vt.edu

Abstract

Valuable information that is useful or necessary to certain design activities is often available only in natural language text. To be useful, the information must be extracted manually and formal design models must be generated. These extracted models are . used for simulation studies, product evaluation and patent analysis. The approach considered here to automating model extraction from natural language documents includes parsing, semantic analysis, coreference detection and model generation. This paper focuses on the semantic analysis step and the semantic domain of digital computer systems. Conceptual graphs, a form of semantic networks, have been selected as the knowledge representation. With word meanings represented by conceptual graphs, semantic analysis proceeds by unification of these graphs, as guided by the parse trees developed in the parsing step.

Keywords: Knowledge Acquisition, Natural Language Processing, Knowledge Representation

Introduction

Valuable information that is useful or necessary to certain design activities is often available only in natural language text, and must be translated manually to formal notations. For example, simulation models of components must be derived from product descriptions in manufacturer's data sheets, before the components can be used in the design of new systems. To determine if an existing patent should be licensed for a new product design, or if a proposed design infringes on it, the patent must be analyzed, and possibly a model of the artifact developed. While modeling requires considerable expertise, the work is tedious and repetitive, and does not attract the skills needed. This paper describes the ASPIN system being developed to automatically extract modeling information from English text.

Information extraction [CowJ96] is the activity of scanning natural language texts and filling instances of templates (semantic frames) from a predetermined library of templates. Typically the library of templates is relatively small. A text usually has multiple instances of templates filled to various degrees, and an instance of a template may be filled from multiple expressions in the text. The goal of model extraction is to construct formal models of systems from natural language descriptions of digital computer systems and components. Model extraction here uses a very large number of quite simple templates, and requires that these simple templates be integrated when filled. From the integrated template instances, formal design models in conventional modeling notations are generated.

Natural language processing is often decomposed into the following steps: lexical analysis, parsing, semantic analysis, coreference detection and pragmatics (model generation, here). This paper focuses on the semantic analysis step, but does describe the other tasks as they are being implemented in the ASPIN system. The technical documents of interest here are actually written in a sublanguage of English, which simplifies the problem considerably. This is especially true when the semantic domain is restricted. The documents of interest here describe digital computers and their major components. This 'sublanguage' uses fewer words and grammatical constructs than standard English. Also, words tend to have fewer senses (meanings). However, technical English uses some unusual constructs and word senses. For example in the phrase 'a memory write is performed', the infinitive verb form is used as a noun. In standard English this would be expressed 'writing the memory is performed.'

The natural language source documents are actually long strings of characters with frequent embedded blanks. Lexical analysis breaks the document string into lexical units or tokens. These tokens are words, acronyms, symbolic names, and punctuation. Lexical analysis may use morphological analysis to determine forms of verbs and number for nouns. The ASPIN system uses a simple scanner and dictionary for lexical analysis. Currently, morphological analysis is avoided by storing all interesting forms of words in the dictionary. The lexical analyzer is built into the parser, so during dictionary look-up, the grammatical categories of words are retrieved. Words not found in the dictionary are classified as identifiers: acronyms or symbolic names, which are the equivalent of proper nouns in standard English.

Syntactic analysis or parsing determines if a string of tokens is a sentence and if so, determines the grammatical structure of the sentence. The ASPIN system uses a context-free grammar and chart parser for this purpose. Although English is not a context-free language, the parser successfully analyzes quite sophisticated sentences, and often proposes multiple possible structures. Semantic analysis eliminates meaningless syntactic analyses and attempts to generate a knowledge representation of the meaning of the sentence. Ambiguous sentences may, of course, have multiple valid meanings, one of which is adopted. To support semantic analysis, the knowledge representation used in the ASPIN system is a form of semantic network called conceptual graphs [SowJ84]. Formal modeling languages, such as flowcharts or dataflow diagrams are not useful here because they are semantically incomplete. Even more current and comprehensive design notations such as STATEMATE [i-Lo91] and the VHDL hardware description language [IEEE88] are inadequate.

Knowledge from individual sentences must be integrated. This is accomplished by detecting references (expressions) having the same referent and merging these references. This coreference detection employs a system of rules for deciding when two references have the same referent. The author of a document usually assumes a reader has a certain amount of knowledge in the subject area, so coreference detection may require additional knowledge to support unification. Adding this knowledge amounts to abductive reasoning. During and after integration, certain consistency checks can be made.

At the start of the data transfer from the DLM, the DLM interface logic circuits provide a GP DLM REGISTER LOAD signal to the DLM register file of the geometry processor to enable that register file to input the data packet.

Figure 1. An Example Sentence

The final step is the generation of a formal model in the desired target notation from the unified conceptual graph. This involves tracing though the conceptual graph and translating subgraphs to constructs in the target language using a library of subgraph/model pairs. The model fragments must then be merged based on connectivity in the conceptual graph. In the current ASPIN project, two target languages have been of interest: VHDL [IEEE88] and a graphical formalism developed here for visual feedback of the interpretation of sentences in an interactive application.

To illustrate the model extraction process, the sentence of Figure 1 from a U.S. Patent [PelA90] is used as a running example.

Conceptual Graphs

The knowledge representation selected for the ASPIN project is conceptual graphs [SowJ84], a form of semantic networks. A conceptual graph is a directed hypergraph. Its nodes represent concepts and its hyperarcs represent conceptual relations among concept nodes. (Traditionally, conceptual graphs are described as bipartite graphs.) In conceptual graphs, a hyperarc is directed out of its source node, and is directed into all of its target nodes. A concept node is labeled by a concept type and a referent marker. The set of concept type labels form a lattice based on supertypes and subtypes. In the present domain of interest, digital systems, the upper level types are object, behavior and attribute. Objects may be devices or values, and behaviors may be actions, states or events Carrier (wire or bus) is a subtype of device. Program and data are subtypes of value. The referent marker of a concept may indicate a generic (unspecified) concept by a *, a specific individual concept by a #, or, in the case of a state concept a conceptual graph describing the state. Additional referent markers are used, but need not be mentioned here. Conceptual graphs can be denoted in a textual form, where concepts are indicated by their labels enclosed in square brackets, and relations by arrows and their labels enclosed in parentheses. The first label or a concept is the concept type and the second the referent marker. A conceptual graph representing

```
[enable : *] -
        (inst) -> [signal : *]
        (patient) -> [counter : #]
        (purpose) -> [increment] -
                        (opnd) -> [value : #1]
                        (result) -> [value : #1],,.
```

Figure 2. An Example Conceptual Graph

'a signal enables the counter to increment the value' is
illustrated in Figure 2.

This graph shows five relations and five concept nodes:
enable, signal, counter, increment and value. The enable
concept is related to signal, counter and increment
concepts by the relations instrument, patient and
purpose, respectively. Signal is the agent of the enable
action. Value is both the operand and the result of the
increment action. The number 1 in the two value
concepts indicates they are just multiple appearances of
a single node. Most of the 60 relation types used in this
domain of interest relate one, two or three nodes.

Four elementary operations are defined on conceptual
graphs: copy, restrict, join and simplify. The copy
operation just reproduces a conceptual graph. The
restrict operation replaces a concept type label by a more
specific label or replaces a generic referent marker by an
individual one. The type and referent must remain
compatible. For example, a node [memory:#]
representing a specific individual register can be
restricted to [register:#], but not to [counter:#]. The third
operation, join, merges two identical concepts into one
concept by deleting one and reattaching all its incident
relations to the remaining node. (Any two nodes can be
joined if they can be restricted to become identical, as
[computer:*] and [microprocessor:#] can be. When
nodes are joined, it is possible for a pair of nodes to be
linked by two appearances of the same relation. The
simplify operation elides one copy of the relation. These
operations, with additional constraints on join, are
adequate to generate conceptual graphs from text
[SowJ86].

```
(sentence
        (advl   (prep at)
                (np
                        (np (det the) (head (noun start)))
                        (of of)
                        (np
                                (np (det the) (noun data) (head (noun transfer)))
                                (pp (prep from) (np (det the) (head (id dlm)) (# 16)))))
        (, ,)
        (np (det the) (id dlm) (noun interface) (noun logic) (head (noun circuits)) (# 318))
        (pred
                (avs (verb provide))
                (np (det a) (id gp) (id dlm) (noun register) (noun load) (head (noun signal )))
                (advl
                        (prep to)
                        (np
                                (np (det the) (id dlm)(noun register) (head (noun file)) (# 115))
                                (of of)
                                (np (det the) (noun geometry) (head (noun processor)) (# 36))))
                (advl
                        (to to)
                        (vinf enable)
                        (np (det that) (noun register) (head (noun file))))
                        (advl
                                (to to)
                                (vinf input)
                                (np (det the) (noun data) (head (noun packet))))))))
        (. .))
```

Figure 3. Parse Tree for Example Sentence.

Parsing

The parsing strategy chosen for the ASPIN project is a bottom-up, parallel chart parser [WinT83]. This method finds all valid parse trees of a sentence for the given grammar, so the syntactic ambiguity is known. The parser scans input, delimiting words by spaces or punctuation. As a token is encountered, it is looked up in the dictionary to retrieve its possible grammatical categories (parts-of-speech). Tokens not found in the dictionary are classified as identifiers and are assumed to be the names of objects (values or devices). The parser uses a set of about 100 context-free productions. When the end of a sentence (period) is encountered, the chart is scanned for complete parse trees of the sentence. These trees are retained for semantic analysis.

The example sentence of Figure 1 has 47 words and yields three valid parse trees using the ASPIN grammar. The chart for this grammar has 1241 constituents. The preferred parse tree for this sentence is shown (in simplified form) in Figure 3. Many intermediate (renaming) nodes have been left out for clarity and the internal structures of noun phrases has not been shown. To explain the LISP-like notation, lines 2 through 8 describe the structure of the adverbial (advl) phrase 'at the start of the data transfer from the dlm 318.' The entire sentence constituent (sentence) consists of an adverbial, a comma, a noun phrase (the subject) and a predicate whose main verb is 'provides'. The multiplicity of parse trees in this example is due to different attachments of prepositional and infinitive phrases, as discussed later.

Semantic Analysis with Conceptual Graphs

Given a parse tree for a sentence, semantic analysis attempts to construct a conceptual graph from the sense graphs in the dictionary which define the meanings of the words of the sentence. A sense graph is based on a conceptual graph which is stipulated to be canonical, and for a verb is similar to a case frame [FilC68] . It has one distinguished node called its root, and which is the concept associated with the word being defined. The relations of a sense graph have both a relation type and a list of possible syntactic markers which identify the structural clues to the relation in English sentences. For example sense graphs for the words 'at', 'provide' and 'transfer' are shown in Figure 4. Note that the token 'provide' actually maps onto the concept type 'send'.

In the 'transfer' sense graph, the transfer concept is the root. The agent of the transfer is either the subject (subj) of an active sentence or the object of the prepositional

```
at:        [action] -> (begin : at) -> [event].

provide:   [send] -
                (agent : subj, by) -> [device1]
                (result : obj) -> [value]
                (dest : to) -> [device2],.

transfer:  [transfer] -
                (agent : subj, by) -> [device]
                (opnd : obj) -> [data]
                (src : from) -> [device]
                (dest : to, into) -> [object],.
```

Figure 4. Sample Sense Graphs

phrase marked with the preposition 'by' in a passive sentence. If a prepositional phrase introduced by either 'to' or 'into' is nearby and has a device concept as its object, it can be the destination device to which the transfer is directed. The operand value of the transfer action is found in the grammatical direct object. In general, the sense graphs for verbs are the most complex, usually having many pendant concepts. Rather than having all possible relations represented in each verbal sense graph, only the most frequent ones are represented there. The less frequent relations are handled by sense graphs defining the prepositions or subordinating conjunctions that mark the relations. Prepositions have sense graphs with a single relation joining the object of the prepositional phrase with the modified noun phrase concept. (See the 'at' sense graph.) Nouns representing objects (values and devices) have only single concepts as their sense graphs. Adjectives typically have simple sense graphs similar to prepositions with a single 'attribute' relation.

Processing a parse tree to construct a conceptual graph begins by replacing each leaf node with a sense graph for that word from the dictionary. When all children of any node have been replaced by sense graphs, a sense graph for that node may be formed. The sense graph for an internal node of the parse tree is formed by joining the sense graphs of all its children nodes. If an internal node of the parse tree has only one child, its sense graph becomes the sense graph for that node. If a node has multiple sense graphs, one of the child sense graphs is selected as the basis for that node's sense graph. This approach is founded on X-bar theory [ChoN70], which posits that a grammatical construct has a head element which is qualified or restricted by its premodifiers and postmodifiers. The selection of the head child node is dictated by the grammatical rule that formed the node's constituent. For example, the rule '*pred => avs* np advl advl*' used to detect the predicate of the example sentence also asserts that the first element (avs) is the head. In predicates, the first element is usually the head. In noun and verb phrases, the last element is often the head. Once the head is selected, each child is joined to it by joining the root concept node of the child with an uninstantiated, pendant concept of

the head sense graph which has a suitable marker. For example, the [reg_file:that] concept in Figure 5 will join with the [device:2] concept of the 'enable' sense graph. Similarly, the root [input] of the 'input' sense graph will join with the [action] concept of the 'enable' graph. In many cases suitable pendant concepts with appropriate marker choices are not available. In these instances, it is

```
(sentence
    (advl
            (prep at)
            (np
                    ( np   [start: #] <- (begin:obj) <- [action]   )
                    (of of)
                    (np    [transfer] -
                                    (agent:subj,by) -> [device]
                                    (opnd:obj) -> [data: *]
                                    (src:from) -> [device]
                                    (dest:to,into) -> [object: #] -
                                                        (name) -> ['dlm']
                                                        (name) -> ['16']   )
        (, ,)
        (np    [circuits: #] -
                    (name) -> ['dlm interface logic circuits']
                    (name) -> ['318']    )
        (pred
            (avs (verb    [send: *] -
                            (agent:subj,by) -> [device]
                            (result:obj) -> [value]
                            (dest:to) -> [device]     ))
        (np    [signal: *] -> (name) -> ['gp dlm register load signal']   )
        (advl
                (prep to)
                (np    [reg_file: #] -
                                (name) -> ['dlm register file']
                                (name) -> ['115']   )
                    (of [device] -> (part) -> [device]  )
                    (np    [processor: #] -
                                        (name) -> ['geometry processor']
                                        (name) -> ['36']   )
            (advl
                (to to)
                (vinf  [enable: *] -
                            (agnt:subj,by) -> [device:1]
                            (inst:subj,by) -> [value:2]
                            (patient:obj) -> [device:3]
                            (purpose:to) -> [action] -
                                            (agent) -> [device:3]
                                            (enabler) -> [value:2]   )
                (np    [reg_file: that] -> (name) -> ['register file']   )
                (advl  [input] -
                                (opnd:obj) -> [value:#] -> (name) -> ['data packet']
                                (agent:subj,by) -> [device:1]
                                (dest) -> [device:1]
                                (inst:on,using) -> [carrier]   )
        (. .))
```

Figure 5. Partial Semantic Analysis of a Parse Tree.

necessary to find a sense graph for the marker preposition or conjunction that has concepts that join with the respective roots of the head and child graphs. This is how the adverbial 'at the start of the data transfer from the dlm' is attached to the sense graph for 'provides'.

To illustrate semantic analysis, Figure 5 shows a partially analyzed parse tree for the example sentence. At this stage in the analysis, most of the leaves of the tree have been replaced by sense graphs, and some internal nodes have been processed. Notice that when a token is an identifier (id), it's sense graph is an object concept with the identifier as its name. Now, it may turn out that unification of sense graphs fails at some internal node. In this case, backtracking is performed to see if alternative sense graphs were available for leaf nodes. If so, analysis resumes with the alternatives until either a sense graph is formed for the root (sentence) node, or

until all choices fail and semantic analysis fails. Continuing the semantic analysis of the example parse tree yields the conceptual graph of Figure 6.

Returning to the multiplicity of parse trees for the example sentence, multiplicity is due to alternative constituents that prepositional and infinitive phrases can be attached to. Of the three parse trees generated from syntactic analysis, one variant attaches the prepositional phrase 'to the dlm register file of the geometry processor' to the noun phrase 'a gp dlm register load signal.' This is semantically unacceptable because a 'value' cannot be 'to' anything. Semantic analysis of that tree will fail because there is no sense graph for 'to' which accepts a value as its source concept. Another of the trees attaches the infinitive clause 'to input the data packet' to the main verb 'provides'. This could be an acceptable 'purpose' for the verb, but if this attachment is made, then there is no suitable site for attaching the

```
[send: *] -
        (agent:subj,by) -> [circuits: #] -
                            (name) -> ['dlm interface logic circuits']
                            (name) -> ['318']
        (result:obj) -> [signal: *2] -> (name) -> ['gp dlm register load signal']
        (dest:to) -> [reg_file: #1] -
                            (name) -> ['dlm register file']
                            (name) -> ['register file']
                            (name) -> ['115']  )
                            (part) -> [processor: #] -
                                            (name) -> ['geometry processor']
                                            (name) -> ['36']
        (purpose) -> [enable: *] -
                    (agnt:subj,by) -> [device:1]
                    (inst:subj,by) -> [signal: *2]
                    (patient:obj) -> [reg_file: #3]
                    (purpose) -> [input] -
                                (opnd:obj) -> [value:#]-
                                                (name) -> ['data packet']
                    (agent:subj,by) -> [reg_file: #3]
                    (inst:on,using) -> [carrier]
                    (enabler:enable) -> [signal: *2]
        (begin:at) -> [start: #] <- (begin:obj) <- [transfer] -
                                    (agent:subj,by) -> [device]
                                    (opnd:obj) -> [data: *]
                                    (src:from) -> [device]
                                    (dest:to,into) -> [device: #] -
                                                    (name) -> ['dlm']
        (name) -> ['16']
```

Figure 6. Conceptual Graph for the Example Sentence

other infinitive phrase 'to enable that register file,' since a verb cannot have two purposes (unless they are coordinated to a single purpose with a conjunction).

In some sentences, more than one parse trees may generate a well-formed conceptual graph. That is the sentence is semantically ambiguous. The present approach is to accept the first conceptual graph assembled for the sentence node. Future plans include finding all graphs and selecting the 'best' one based on how well it integrates with the graphs generated for previously analyzed sentences. That is, semantic analysis should be context sensitive.

Coreference Detection

The next task in model extraction is to unify conceptual graphs generated for the different sentences of the document. This requires finding concepts in the graphs which should be joined because they have the same referent, they are coreferences. The ASPIN approach is to unify each new sentence with the graph accumulated from previous sentences in the text. The unification task is characterized by having many concept nodes that could be joined because they have compatible types, but having very few (about zero to two) that should be joined because they actually have the same referents.

The problem of avoiding erroneous joins is handled by looking at the relations incident with the candidate nodes. These are called restrictive relations. As an example, consider the noun phrase 'the DLM interface logic circuits in the tree traverser controller', and the noun phrase 'the interface circuits.' If the second phrase is encountered shortly after the first phrase, the second is probably a coreference to the same device (circuits). On the other hand, if the second phrase occurs first, most

readers would not consider the phrases to be coreferences because the more specific reference (highly restricted by modifiers) should occur first to define the concept. In the ASPIN system, conceptual subgraphs headed object type concepts are compared using a system of rules to decide if they are coreferent with any objects encountered previously [ShaS94]. To a more limited extent, coreference detection has been investigated for action type concepts.

The second problem of having very few coreferences to glue the sentence graphs together is addressed by adding design knowledge: common knowledge that a designer might be expected to know. A bit of knowledge such as 'a register file can simultaneously read and write data in registers selected by addresses' in conceptual graph form might be useful in unifying the example sentence with other sentences from the document. This approach has been studied [KamR95], but not implemented yet.

Model Generation

Once a unified conceptual graph has been obtained from the given document, the desired type of formal model can be generated from it. Some work has been done on generating VHDL models from conceptual graphs [CyrW94], but the discussion here will only describe generation of graphical models because of space limitations. The graphical modeling notation used in the ASPIN system is a combination of component-connectivity diagrams [GajD94], Petri nets [PetJ77] and dataflow graphs [DemT79]. This combination was chosen for graphical feedback in an interactive situation where the user enters a sentence and the system displays the interpretation of that sentence in graphical form [ThaA94] for confirmation of the analysis. The user can

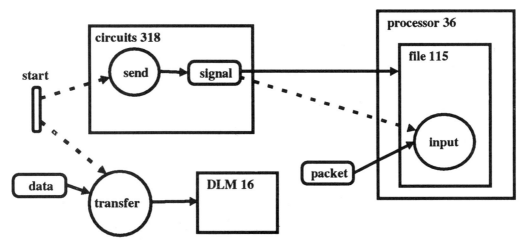

Figure 7. A Graphical Interpretation of Example Sentence.

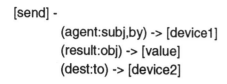

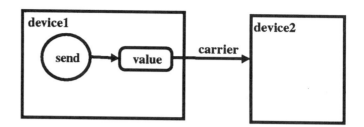

Figure 8. Sense Graph Model

accept or edit conceptual graph indirectly by editing the displayed graphical interpretation [BalS94]. The graphical interpretation for the example sentence appears as Figure 7. For convenience, the example sentence is "At the start of the data transfer from the DLM, the DLM interface logic circuits provide a GP DLM REGISTER LOAD signal to the DLM register file of the geometry processor to enable that register file to input the data packet."

This graphical interpretation is formed by unifying merging interpretations of the sense graphs used in analysis of the original sentences. The interpretation or model fragment for the sense graph of 'provide' (send) is shown in Figure 8. In forming the model, only the sense graph concepts having references in the original sentence are actually displayed, this avoids cluttering the model with extraneous information.

Conclusions and Future Work

Design models, or at least fragments of design models can be extracted from the wealth of knowledge available in natural language text form. This knowledge can play a useful engineering role in new product development. Currently, the ASPIN system is a collection of prototype programs in various languages (Pascal, 'C', C++ and Prolog) which demonstrates the critical steps in the model extraction process. Currently, the ASPIN software components are being up-graded and integrated in C++ for application to a requirements extraction problem. The dictionary is also being expanded. Future plans include more comprehensive testing to develop model extraction performance data.

Acknowledgments

This research was supported in part by Semiconductor Research Corporation (Contract 92-DJ-230), the National Science Foundation (Grant MIP-9120620) , and Virginia's Center For Innovative Technology (Grant INF-92-005).

References

[BalS94] S. Balachandar and W. R. Cyre, Knowledge Acquisition Tool using Visualization and Back Annotation," Proc. Workshop on Knowledge Acquisition using Conceptual Graphs, College Park, MD, 95-115, August 20, 1994.

[ChoN70] N. Chomsky, "Remarks on Nominalization," in R. Jacobs and P. Rosenbaum, Readings in English Transformational Grammar, Ginn and Co., Waltham, MA, 1970.

[CowJ96] J. Cowie and W. Lehnert, "Information Extraction," CACM, 39(1), 80-91,January, 1996.

[CyrW94] W. R. Cyre, J. R. Armstrong, M. Manek-Honcharik, and A. Honcharik, "Generating VHDL Models from Natural Language Descriptions," Proc. Euro-VHDL, Grenoble, France, 474-479, September 19-23, 1994.

[DemT79] T. DeMarco. Structured Analysis and System Sepcification, Yourdon Press, NY, 1979.

[FilC68] C. J. Fillmore, "The Case for Case," in Universals in Linguistic Theory, Bach and Harms, eds., Holt, Reinhart and Winston, Chicago, 1-90, 1968

[GajD94] D. Gajski, F. Vahid, S. Naranay and J. Gong, Specification and Design of Embedded Systems, Prentice, Hall, NJ, 1994.

[i-Lo91] i-Logix, The Languages of Statemate, I-Logix Inc., Burlington, MA, January 1991.

[IEEE88] IEEE Standard VHDL Reference Manual, IEEE, New York, 1988.

[KamR95] R. Kamath and W. Cyre, "Automatic Integration of Digital Systems Requirements using Schemata," in G. Ellis, R. Levinson,

W. Rich and J. Sowa (eds), Conceptual Structures: Applications, Implementation and Theory, Springer, Berlin, 1995.

[PelA90] A. Pelham, W. Steiner and W. Didden, Fast Architecture for Graphics Processor, U.S. Patent 4,967,375, United States Patent Office, Washington D.C., October 30, 1990.

[PetJ77] J. L. Peterson, "Petri Nets," Computing Surveys, Sept. 1977.

[ShaS94] S. Shankaranarayanan and W. R. Cyre, "Identification of Coreferences with Conceptual Graphs," Proc. 1994 International Conference on Conceptual Structures, College Park, MD, 45-60, August 16-20, 1994.

[SowJ84] J. Sowa, Conceptual Structures, Addison-Wesley, Reading, MA, 1984.

[SowJ86] J. F. Sowa and E. C. Way, "Implementing a Semantic Interpreter Using Conceptual Graphs," IBM J. Research and Development, Vol. 30, 57-69, January, 1986.

[SteW74] W. P. Stevens, G. J. Meyers and L. L. Constantine, "Structured Design," IBM Systems Journal, 21 (2), 115-139, 1974.

[ThaA94] A. Thakar and Walling Cyre, "Visual Feedback for Validation of Informal Specifications," MASCOTS'94, Durham, NC, 411-412, January 31-February 2, 1994.

[WinT83] T. Winograd, Language as a Cognitive Process, Vol. 1: Syntax, Addison-Wesley, Reading, MA, 1983.

OFFICEASK – A NATURAL LANGUAGE INTERFACE TO STRUCTURED DATA ARCHIVES

Matthias Kaiser and **Andreas Dengel**
German Research Research Center for Artificial Center for Artificial Intelligence (DFKI)
P.O. Box 20 80, 67608 Kaiserslautern, Germany
email:{kaiser, dengel}@dfki.uni-kl.de

ABSTRACT

In this paper we propose a system allowing natural language queries for data in a structured archive. The system not only reacts appropriately to spoken remarks, but also is minimally restrictive on the users language customs in order to meet real world requirements. Thus, it accepts specifics of human communication on several levels, such as indirect speech acts, non-grammatical morphological and syntactical phenomena. Two very important features of our system are its flexibility to be modified to work on another natural language and to be ported to different data archives with a minimum of effort. The system currently runs on a lap top in two languages, English and German.

INTRODUCTION

Modern computers – especially when connected to data networks, data archives or knowledge bases in a broader sense – can provide us with huge amounts of data within a very short time. Consequently we easily get overflown with information so that it becomes difficult, if not impossible, to deal with it efficiently, i.e., filter relevant data and incorporate it into our knowledge while ignoring others. The challenge for researchers in document analysis and natural language processing is to develop tools which can be used just to perform the filtering and extraction of data from the large amounts of information offered according to our needs and wishes in a form most suitable for us, i.e., performing most effective transfer of desired information while interacting with the human user in a most natural/convenient way [Jac92, LR82, Ing94].

In this paper we present a brief outline of *OfficeAsk*, a natural language interface that is capable of providing answers to questions on the basis of a data archive. In particular, it provides a tool for the filtering of relevant information from a data archive about business documents on the basis of spontaneously presented natural language questions which are translated into formal queries for data search in the archive. Here the focus is on the robustness and flexibility of the NLP interface to react appropriately even to nearly unrestricted and possibly slightly non-grammatical language input as it occurs in real world communication.

Most classical NLP systems are just not really able to work under real life communication conditions. Because natural language is considered to be a formal language of a more complex kind, every item in an utterance must be described in a formal way in order to build a provable best solution [And91, LR82, Ull88].

The only "systems" that do understand natural language are human beings not tending to treat language as a complex formal provable system [Tal78, Lak87, Lan91, RK89]. Thus, the strategy realized in *OfficeAsk* is based more on modern research results in cognitive linguistics than in computational linguistics. The sceptisism towards cognitive linguistics uttered by computer scientists stems mostly from their deductive way of solving problems which have problems at certain levels of quantity (and quality) of problem structures. Instead a "generate & test" approach has been choosen instead of applying deductive methods.

STARTING POINT

Since natural language is the most natural means of human information exchange we are keen to build an information system capable of reacting to natural language queries. Features we find desirable to hold for natural language interfaces for computer systems in general and hence motivational for the design of *OfficeAsk* are:

- robust interpretation of spontaneous spoken language that may contain errors of various kinds
- easiness to port to different domains
- easiness to port to work in other natural languages

These requirements demand a new approach in order to achieve robustness and intelligence of a natural language interface like *OfficeAsk*. The approach must cope particularly with the following problems:

- free use of language in questions makes a complete parsing intractable, the dictionary would be too expensive, the syntax too complex, and real world language performance is often non-grammatical
- the archive contains structured data on a general functional level, items contained in the structured objects may not be in a dictionary, however large it may be – such as proper names – subject to user questions

In order to give an idea of the intuitive background which founded the NLP strategy realized in *OfficeAsk* by considering a simple example: Imagine you want to buy a train ticket. First you have to communicate with the person behind the counter selling tickets. You may talk with the ticket seller about many other things but in order to get the ticket you want, you have to give him just the information he needs. So in prototypical communication situations (such as standing in front of a ticket counter) you are expected to know what your communication partner can do for you, the communication partner is expecting you to act as he/she expects you should and, if you meet the expectations, you are allowed to communicate freely and with flexibility.

OfficeAsk performs as the ticket seller characterized above: it is able to communicate in a limited domain, interpreting user questions that may be freely spoken and which may not always use standard grammar. If the goal of a NLP system is to provide a natural way of communication with a computer, this challenge is not met if the user has to learn/remember exactly which words to use (because they are contained in the system's dictionary) and which syntactic constructions to obey. If this does not hold, the communication fails. It does not make much sense for the user to behave according to restrictions on natural language that indeed make it a formal one instead of learning a mnemonic way of interaction with a computer right from the start.

Besides this, natural language interaction with a computer system has also economic advantages by lowering the amount of training a person needs to handle the system successfully. Moreover, systems that do not require special user training for their handling can work in areas they could not be used, forcing a training necessity on their human communication partners such as in public services (call centers, ticket services, city information systems etc).

Today, the most common form of intuitively motivated interfaces are graphical user interfaces. Items which are the subject of interest are selected and activated by a keyboard or – more frequently – a mouse click. These interfaces work well as long as data can be depicted in an intuitive way and does not exceed a certain amount of complexity – within one level as well as hierarchies. When the complex-

ity of quantitative and/or qualitative data features grows, it becomes more and more difficult and time-consuming to click the relevant entities on the screen explicitly to represent queries to the computer system this way. In natural language the selection of relevant items is much more natural, non-relevant question possibilities do not have to be ignored explicitly – we say just what we want, without having to tell (or go around with the mouse).

The linguistic basis for the approach is that language two subsystems which can be designated as the grammatical and the lexical. The grammatical elements of an utterance determine the majority of the structure, while the lexical elements contribute the majority of its content [Tal78, Lak87, Lan91]. The grammatical specifications in a sentence, thus, provide a conceptual framework for the conceptual material that is lexically specified. Since generally, across the spectrum of languages, the grammatical elements that are encountered, taken together, specify a crucial set of concepts, this set is highly restricted: only certain concepts appear in it, and not others – a phenomenon which is one key to the high degree of portability from one natural language to another in the *OfficeAsk* architecture. We use an anchor point parsing mechanism to determine a conceptual skeleton that is capable of letting us reliably predict lexical fillers.

In order to give an illustration of the performance of *OfficeAsk* we will use as an example the domain in which *OfficeAsk* was originally designed to work, namely that of a structured document message archive.

EXEMPLARY APPLICATION

In the *Document Analysis* department at DFKI a system for the analysis of business letters – called *OfficeMAID* – has been developed [DBH95]. This system is capable of identifying logical objects in printed business letter [Den94], such as *sender* of the letter, *recipient*, *date*, *subject*, *name* of the writer. Moreover, letters can be classified according to the message they contain [HD93, WH95], e.g., whether they are intended to an *offer*, an *order*, a *confirmation* etc. The results of the logical object identification as well as the document classification process are kept in a relational database (see Figure 1).

With the help of this document message archive we can answer questions about analyzed documents like:

```
Who received letters from John Smith?
(asking for the sender and/or writer of
letters to John Smith) ??
Who sent offers to Miller conc. workstations?
(asking for letters whose recipient is Miller,
whose subject is workstations and  whose mes-
sage type is offer)
Give the last request on database software in
july 1996 from Italy.
```

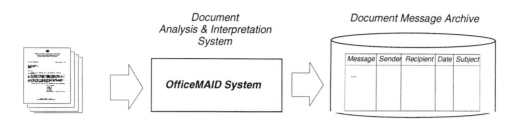

FIGURE 1   Document analysis and interpretation as an exemplary application for the *OfficeAsk* system

In order to react appropriately to the questions and requests shown above, *OfficeAsk* has been designed as an interface for the retrieval of the archive data. Strictly speaking we can say that it works as a double interface: it is an interface to the user and to the data archive at the same time. Natural language questions are translated into formal representation. This representation is used as a database query to retrieve the desired data. Finally, the retrieved data is presented to the user in natural language.

SYSTEM OVERVIEW

THE CONCEPT OF THE APPROACH

In order to perform as desired, the natural language interpretation strategy relies on two fundamental assumptions:

- *Extralinguistic assumption*
  In order to determine interpretative context features, we have to provide the expectations which the interpreter may have to work on when evaluating language entities. This means that the user knows these expectations about the context which discourse is based on and must comply with them to get the desired results.

- *Linguistic assumption*
  A fundamental feature of language is that it has two subsystems which can be designated as the grammatical and the lexical. The grammatical elements of an utterance determine the majority of the structure, while the lexical elements contribute the majority of its content [Tal78, Lak87, Lan91]. The grammatical specifications in a sentence, thus, provide a conceptual framework, a kind of skeletal structure for the conceptual material that is lexically specified.

Note that more generally, across the spectrum of languages, the grammatical elements that are encountered, taken together, specify a crucial set of concepts. This set is highly restricted: only certain concepts appear in it, and not others, a phenomenon which is one key to the high degree of portability from one natural language to another in the *OfficeAsk* architecture.

The set of grammatically specified notions constitutes the fundamental conceptual structuring system of language. This cross-linguistically selected set of grammatically specified concepts provides the basic schematic framework for conceptual organization within the cognitive domain of language. For this reason knowledge about it is of great relevance for the interpretation of utterances in humans and also in our system.

The terms "grammatical" and "lexical" as employed here require clarification. The distinction between the two is made formally, i.e., without reference to meaning on the basis of the traditional linguistic distinction between "open-class" and "closed-class". A class of morphemes is considered open if it is large and easily augmentable in relation to other classes, as is in our case the material contained in a dynamic dictionary. A class is considered closed if it is relatively small and fixed in membership.

We can identify the particular classes belonging to these two types. The open classes of elements, i.e., the lexical classes are, quite simply, the nouns, verbs, and adjectives. If open-class forms comprise the roots of nouns, verbs, and adjectives then everything else is closed-class and is here considered to be, quite generally "grammatical". Closed-class forms are of three broad kinds: overt, abstract, and complex. Among the overt kind are such bound forms as inflections and derivations, such free forms as determiners, prepositions, conjunctions, and particles. Within abstract, or implicit, closed-class forms are grammatical categories. Here this term is intended to encompass not only such major grammatical categories as noun and verb, but also any morpheme category that is structurally distinguished in a language, grammatical relations, word order, and perhaps also paradigms and "zero" forms. And thirdly, there are regular combinations of simpler closed-class forms, tending to have a unified or integrated semantic function here called "grammatical complexes" that include grammatical constructions and syntactic structures. The static dictionary now is a collection of such closed-class items, that helps to find out about the conceptual structuring of an utterance. Up to this point it contains mainly grammatical entities of the first kind mentioned, but will be extended in future work. We use this notion of conceptualization power of closed-class entities to build heuristic templates which provide more specific information with respect to a certain data archive schema.

In particular, prepositions such as *to, from, about* etc. and question words such as *who, what, when* etc. as closed-class elements play a major role in our approach. Thus in our domain we introduce closed-class words together with an annotation saying which information type, or in a sense, semantic type such as *recipient, sender* etc. is pointed at by a preposition or question word.

The preposition *to* for example, seen in a question asked within the domain of the data of documents mentioned above points to the term following this preposition (being the prepositional object) as having the information type *recipient*. So the entry for *to* in the static dictionary looks something like:

```
(to recipient).
```

The question word *who* interpreted in this domain can ask for a *recipient*, a *sender* or a *name* but not e.g. for a date. The corresponding dictionary entry would be:

```
(who recipient, sender, name)
```

THE GENERAL STRATEGY

OfficeAsk acts on a small amount of specific statically provided linguistic knowledge, acquires the main amount of linguistic data by generation of its own annotated dictionary and interprets user utterances specifically in the context of the domain provided by the used data archive about document messages [Wei90].

As it currently works as a natural language data retrieval system *OfficeAsk* is made up of the following components:

- data archive or knowledge base: containing structured knowledge items as e.g. business documents explicitly structurally analyzed by a document analysis system
- a dictionary being subdivided into partitions according to the specific function of words in the particular knowledge domain and not in any traditional syntactic or broader semantic respect
- a dictionary generation component which automatically generates a dictionary on the basis of a given data base and its corresponding data base schema

- a small dictionary containing grammatically essential entries specifying the conceptual structure of user utterances
- the heuristic parsing module generating hypotheses for instantiation of variables being given in or inferred from a question/request provided by the human user
- a hypothesis evaluation and selection module for the selection of preferable interpretations of natural language items

The working principle of this interface is bottom-up/top-down. On one hand, the natural language parsing strategy employs fix expectations introduced by the general domain-specific context as top-down-inputs for the parsing process. On the other, many hypotheses are in turn results of a bottom-up mechanism detecting natural language items as elements of questions/requests providing evidence for certain interpretation.

The whole parsing strategy can be seen as an abduction process concluding question interpretation hypotheses from mostly uncompletely instantiated language items and pre-stored domain-dependent knowledge forming an expectation frame. Thus an interpretation is sequentially extended by hypotheses. They help to explain the use of certain language items in a question as effects of the cause. This cause is a knowledge gap at the users side to be instantiated by the answer of the question. In the following we will present our view on questions as the basic form of language utterances the interface has to interpret in a little more detail.

THE TREATMENT OF QUESTIONS

If we ask a question, the communication partner is able to answer in three different ways:

- a question variable such as *who, what, when, where* etc which is to be instantiated by information of a certain kind, i.e. the name of a person, an inanimate object name (common noun for example), a point in time, an expression denoting a location etc.
- a (possibly complex) predicate, which forms a true proposition by taking an instantiation of the question variable; such a predicate may be a property such as *red* or a verb-noun-construction such as *has written a letter in May 1990*
- information restricting the search space of instantiations of the question variable by providing a constraint. A potential variable instantiation must meet such as *who of the people in this room or when in May 1993*. In the first case, of all the people a con-

straint is made to those in a certain location, namely in this room. In the second case, of all points in time an interval is given, that of May 1 to May 31, 1993 [Kai92].

Furthermore, the interpretatin of a question depends on the knowledge domain that is in focus. If we use the data archive described above, every question aims at a set of potential answers, namely those which can be generated using knowledge from this archive and no other source of knowledge. If we ask for example:

```
Who wrote love letters to my secretary from
Italy in August 1994?
```

The answer found using the data archive could be negative, which may be true in respect to this knowledge source since private letters are not included here. We call a certain knowledge base forming the search space for answers to questions in the answer domain. So in general, a question consists of the following parts:

```
Q = (variable,constraint,predicate,domain)
```

After having promoted the basic principles behind our approach we now turn to a little more explicit and illustrated outline of the system's components which are shown in Figure 2.

COMPONENTS OF *OFFICEASK*

DOCUMENT MESSAGE ARCHIVE

The knowledge forming the basis of *OfficeAsk*'s answering capabilities is a relational database which can be thought of as a table of varying complexity [Cha93].

The column headers name logical parts of documents such as *recipient, sender, date, subject, name of person* having signed the document and *type of message* the document provides as well as the text body carrying the specific information for the person the letter was written to (see also Figure 1).

The rows of such a table are just continuous numbers as document identifiers. For example, the column *recipient* and *No. 5* will point to the entry (or table cell) containing the recipient of document No. 5. Note that in general, table cells may contain very complex data as e.g. a text body, containing perhaps tables itself [Ull88, WW93].

Basically we can distinguish between two kinds of information in an archive as described here:

- meta information as headers and
- labels determining the schema of the archive information which is structured and labeled according to the archive schema [Kai95].

While – in our case – row labels (as belonging to the meta information making up the archive schema) name whole documents, column headers (also belonging to the meta information) name logical objects and message type of documents. We can think now of a document as a tuple:

```
document = [sender, recipient, date, subject,
name, message type, text body]
```

Having such a structurea data archive at hand, we can ask questions or utter indirect speech acts which are to be interpreted as questions of the following ki [Sea75]:

```
Give me document No. 3!
```

The answer would be just to list all columns named above the row labeled No. 3. To the question:

```
I would like the recipient of the letter writ-
ten by Mr. Jones on April 13., 1994.
```

The answer would be the table cell containing the recipient of the letter whose date cell contains April 13. 1994 and whose sender contains Mr. Jones. To the question:

```
What was the subject of the letters from Ar-
gentina in July 1990 to Mr. Mueller?
```

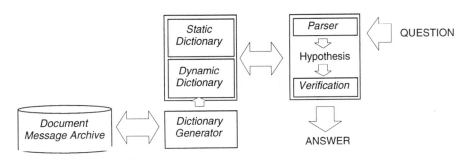

FIGURE 2   The architecture of the *OfficeAsk* system.

The answer is just the subject cells of all letters with the following specifications in the respective table cells:

- *sender contains the word Argentina*
- *recipient is marked to be Mr. Mueller*
- *letters were written at a date greater than or equal to July 1 at 1990 and less than or equal to July 31, 1990*

On the basis of this data archive the dynamic part of the dictionary is automatically generated to provide language material which is necessary to communicate with the archive data.

THE DYNAMIC ARCHIVE-SPECIFIC DICTIONARY

It is a trivial fact that in order to be able to talk about some entities we must have words to denote them. Thus we must have a certain amount of vocabulary corresponding to a collection of images of certain ontological items. In *OfficeAsk* the ontological entities are documents and their logical parts as provided by the data archive. The data archive has a dynamic property in that it grows according to the appearance of new results provided by the document analysis system feeding it.

Also it may shrink when certain items are removed from the archive which are found to be not relevant any more. This fact must be reflected by the dictionary as having an entry for all relevant lexical expressions in all documents. This can practically not be achieved by a conventional dictionary because proper names do not appear in conventional collection. Moreover, a dictionary should not only contain entries in form of just strings that match those in – say a document – but should also give some linguistic information about them. In our approach entries in the dictionary need only to contain information important in the given domain. Therefore it suffices to specialize the dictionary to this domain. Entries have the following form:

```
(wordform (document part
    (number of document ...) ...))
```

We can see that the information given in the dictionary entries has two very important properties:

- it can be generated automatically from the data archive contents
- dictionary entries are characterized by a kind of semantic information

An example of a lexical entry generated would be:

```
(Mueller (recipient (2 5 18)) (sender (1 3
15)))
```

This means that the wordform *Mueller* occured in the recipient part of documents 2, 5 and 18 and in the sender part of documents 1, 3 and 15.

The information accompanying the wordforms in the dynamic dictionary can be thought of as a kind of semantic type information, e.g., the word *Mueller* is of type recipient in document 2, 5 and 18 and of type sender in document 1, 3 and 15. These semantic types are of course strictly dependent on the domain reflected in the data archive.

The advantage of this information is to focus on certain documents and document parts according to the mentioning of a wordform which refers to those respective parts. Having the dictionary entry exemplified above at hand we can say that all we know about *Mueller* is that it (we do not know whether he or she from this entry) is recipient of documents 2, 5 an 18 and sender of documents 1, 3 and 15. This is important knowledge for the answering of questions in the domain as will be shown below.

In the dictionary stop words like articles are not permitted to become entries. We have shown that the result of the availibility of such a dictionary is to think of the annotations made to the word form entries as a kind of semantic description. This description is valid only in the domain of the data archive and not of general purpose type. For every entry we know that the word in question refers to certain parts of documents, denoting them as it were, although this denotation is very ambiguous, as a wordform often refers to many documents and within these can refer to several parts. However, ambiguity is something we always come across when dealing with natural language. The ambiguities are more and more reduced the more language material we have available – say – in an utterance. If we find in an utterance not only the word *Mueller* but the combination *Peter Mueller*, the denotation of this phrase is figured out according to the set theoretical operation of intersection as follows:

```
(Mueller (recipient (2d 5, 18)) (sender (1,
3, 15)) (name (11 15)))
(Peter (recipient (2 9)) (sender (1 18)))
denote(Peter Mueller) = intersect ((2 5 18) (1
3 15) (11 15)) with (((2 9)
(1 18)) = denote((Peter Mueller),(recipient of
document No. 2))
```

Note that the intersection above is performed on the component lists of the word entry list since we cannot match lists of different types, such as recipient with sender for instance. By a type we understand here the information of document parts annotated to a wordform, so the wordform *Mueller* can be recipient, sender, or name, *Peter* can be recipient or sender. That means, that the phrase *Peter Mueller* can, according to the types of its components be of type recipient or sender which is the intersection of the type annotations of both components. Now the respective information about the occurance of these two wordforms in documents is matched accordingly i.e. type for type, and we come up with the result shown above.

The generation process of such a dictionary is very simple. Every wordform in a logical component of an ontological entity – here a document – except stopwords becomes a dictionary entry, annotated with the name of the logical object in which it occured and the identification of the ontological entity – in our case the document number containing the logical object where the wordform was found. The process can be described concisely as follows:

(1) take wordform *n* in document part *p* of document d

(2) if *n* has no entry in the dictionary, generate an entry of the form *(n (p (d)))*

(3) if an entry for the wordform exists, insert the document part *p* and the number of the document into it

Thus we get dictionary entries of the form:

```
(wordform (logical object name (numbers of
documents containing the wordform in the
specified logical object)))
```

So the same wordform can be entered into the dictionary several times if it is found in different logical objects. There are more clever ways to build such a dictionary but we want to show the principle here.

Thus we created a dictionary containing domain-specific wordforms and helpful quasi-semantic annotations automatically. The dictionary can be easily extended parallel to an extention of the data archive – the user is totally freed from the burden of any dictionary maintainance.

A problem which can occur often with such a kind of dictionary is that we can only find wordforms of the same grammatical inflection. We cannot use an accurate morphological device to lemmatize wordforms to wordstems because then we would have to characterize each word in the dictionary by adding certain morphological information which cannot be done automatically.

In order to avoid this trouble, we use a heuristic cutting of presumable endings to match wordforms up to three of the right most letters of a word.

The search process tries first to find the wordform which is a component of an utterance. If this cannot be found the search process tries to find a dictionary entry that is equal to the search string up to – say two letters before the end (which might in fact be a morphological suffix). So in the search suffixes are heuristically stripped off the

wordforms which is not always accurate according to grammatical rules but rather stable at work.

THE STATIC DICTIONARY

The static dictionary is a collection of words annotated with specific information which is also domain-specifically designed but cannot be generated automatically. It is very small in size and need not be maintained by the user when the data archive should be extended.

Before looking closer at the contents of this kind of dictionary we give a motivation for its presence in the system since this is a major feature of our system which makes the design worth being called a new approach. A fundamental feature of language is that it has two subsystems which can be designated as the grammatical and the lexical. The grammatical elements of an utterance determine the majority of the structure, while the lexical elements contribute the majority of its content [Tal78, Lak87, Lan91].

Since generally, across the spectrum of languages, all gramamatical elements in this static dictionary specify a crucial set of concepts. This grants the high degree of portability of the *OfficeAsk* architecture from one natural language to another.

In the following section we will illustrate how these two kinds of dictionary suffice as linguistic database for the interpretation of natural language utterances about a restricted domain determined by the data archive.

THE PARSING MODULE

The parsing module uses several kinds of knowledge on different levels of abstraction:

- knowledge about domain characteristics given by the data archive schema
- lexical knowledge about wordforms and their domain-specific semantic features
- knowledge about disambiguation principles and interpretation methods from the static dictionary
- general heuristics of decomposition and meaning construction

The general strategy here is a buttom-up/top-down abduction strategy (as opposed to deductive methods used in most other parsing paradigms). Abduction is the name given to a kind of inference leading from descriptions of data to the best explanation of these data. We can characterize abductive reasoning as the process of finding the most plausible cause for the presence of a certain manifest effect [Kai93]: Given

$$e_1, e_2, ..., e_n \rightarrow c_1, c_2, ..., c_m \rightarrow c!$$

where $e_1, e_2, ..., e_n$ is a set of manifest effects, $c_1, c_2, ...,$ c_m is a set of causes that could have led to the effects and $c!$ is the most plausible cause explaining the effects present.

We can see from this schema that the course of abductive reasoning is a multi-layered process consisting of the following steps:

- selection of relevant data
- generation of a set of alternative hypotheses explaining the data
- evaluation of each hypothesis according to criteria of typicality and plausibility as well as selection of the best one
- verification of the certainty of the best hypothesis by evaluating its distance to the alternatives and by evaluating how well each effect given as a premisse for the abduction process is accounted for by the hypothesis in terms of its explanatory power

For the different steps in this process different kinds of knowledge are utilized. In the following, we will outline the function of the parsing module by considering an example.

```
I wish I knew who wrote offers to Mueller con-
cerning databases.
```

This is – obviously – a valid request as if asked to a human communication partner. It is, however, not formulated as a question, rather as a wish that can be interpreted as a question:

```
Who wrote offers concerning databases to Muel-
ler?
```

Since the only thing *OfficeAsk* can do is answer questions which it supposes the user to know, it tries to interpret this utterance as a question, otherwise it could not act appropriately anyway.

The first step of the parsing module is to filter out relevant data, i.e., data that make sense to the module in the way that it can access further knowledge about them from its knowledge bases. In this step all wordforms which are not in either of the dictionaries are discarded since no further information about them is available. Every wordform which can be found in any knowledge base is provided together with its annotated information. So we get something like:

```
(who question word (recipient, sender,
name)) [from static dictionary] (offer
(message type 2 5 11)) [from dynamic dic-
tionary] (to preposition (recipient)) [from
static dictionary] (Mueller (recipient 2)
(sender 11 15) (name 15)) [from dynamic dic-
tionary (concerning preposition (subject))
[from static dictionary] (databases (subject
1 2 8 10 15)) [from dynamic dictionary]
```

Now the documents that might be relevant can be selected from the data archive simply by looking at all document numbers within the annotations of known wordforms, so we see that the documents (1, 2, 3, 8, 10, 11, 15) are chosen to be potentially relevant at this stage of parsing. The process continues taking into consideration the information from the static dictionary:

- *who* is a variable which can be instantiated by a recipient, a sender or a name, if none of these is explicitly given in the utterance later on, all three are to be given as instantiations of *who*.
- *to* is a preposition binding a recipient in this domain of discourse. So the word *Mueller* immediatley following *to* is bound and taken to be an explicitly provided recipient. Consequently *who* in this utterance does obviously not ask for a recipient, since it is provided, but only for sender and name so far.
- *concerning* is a preposition (we treat it as one here for simplicity) binding the following wordform as a subject, in this case the word databases.

Now we know that only those documents containing the word *Mueller* in their recipient part and the word databases in their subject part at the same time are still of interest. We find that this is the case with document No. 2 only. An answer now must contain the document parts to be instantiations of the question word *who*, namely sender and name. Thus, finally a valid answer to the question, which was presented as the wish:

```
I wish I knew who wrote offers to Mueller con-
cerning databaseses.
```

would be *sender* and *name* of those having written/sent offers to Mr./Mrs. *Mueller* concerning databases.

Without being able to go into detail we hope to have given an idea of how our parsing module works using several kinds of explicitly given knowledge stored in the data archive and the dictionary components as well as implicit knowledge about the domain.

Normally in this process, either hypotheses found are clear or there is no way to rank them without discarding all but the best ones. There is also some more knowledge involved to solve some tricky cases of hypothesis generation based on heuristics which is not necessary to know for understanding the general approach.

Note that since the system works obviously on a data archive containing German letters, the dynamic dictionary also contains German wordforms, their English equivalents cannot be handled since this would mean *OfficeAsk* to act like a translation system which it is not. So we must use the word *Reisebericht* instead of *business trip report* or something similar.

CONCLUSION

The motivation for the system *OfficeAsk* was to find a sound basis suitable for the convenient design, development, adaption and modification of NLP tools to meet specific requirement imposing a minimum of effort to tailor solutions. It is to state that successful and commercially valid NLP software is not necessarily in proportion to the amount of theoretical input, but mainly is a consequence of a clear understanding of the use of natural language as a means of communication in general, and the most natural way to communicate and exchange information between two knowledge systems of which one is a human being.

OfficeAsk accepts spontaneously uttered natural language input and reacts robustly even towards non-grammatical construction. It is easily portable to other natural languages as its computional requirements are so small that it can be used on a laptop. The system is capable of adapting automatically to new knowledge domains by generating a dynamic dictionary on the basis of a database schema. We think that *OfficeAsk* can be counted as a successful approach towards the design and implementation for real life communication systems imposing minimal specific requirements on the user to work with. It can be used in a very wide range of application areas such as call centers, tourist information services, scientific information and communication systems, etc.

OfficeAsk not only runs on document message types but is currently also applied to a real world tax advisory data base and further tested on the basis of spoken questions via a speech recognition system.

REFERENCES

[And91] R. Anderson, *Cognitive psychology and its implications*, New York (1991).

[CK93] S. Chandan, R. Kasturi, *Structural Recognition of Tabulated Data*, Proceedings ICDAR, Tsukuba City, Japan (1993), 516-519.

[DBH95] A. Dengel, R. Bleisinger, R. Hoch, F. Hönes, M. Malburg and F. Fein, *OfficeMAID — A System for Automatic Mail Analysis, Interpretation and Delivery*, Proc. DAS94, Kaiserslautern (1994).

[Den94] A. Dengel, *About the Logical Partitioning of Document Images*, Proc. SDAIR94, Las Vegas, NV (April 1994), 209-218.

[HD93] R. Hoch and A. Dengel, *INFOCLAS — Classifying the Message in Printed Business Letters*, Proc. SDAIR93, Las Vegas, NV (April 1993), 443-456.

[Ing94] P. Ingwersen, *Information Science as a Cognitive Science*, in: H. Best (ed.), Informations- und Wissensverarbeitung in den Sozialwissenschaften: Beiträge zur Umsetzung neuer Informationstechnologien, Opladen (1994), 23-56.

[Jac92] P.S. Jacobs, *Text-Based Intelligent Systems*, Lawrence Erlbaum Assoc. Inc., Hillsdale (1992).

[Kai92] M. Kaiser, *Präsuppositionen möglicher, sinnvoller, angemessener und relevanter Fragen im Diskurs*, doctoral dissertation, University of Leipzig (1992).

[Kai93] M. Kaiser, *An application of a neural net for fuzzy abductive reasoning*, TR-93-044, ICSI, International Computer Science Institute Berkeley (1993).

[Kai95] M Kaiser, *TABLEX: A system for the generation of natural language descriptions of tabularly arranged information*, Proc. CSUN95, Los Angeles (1995).

[Lak87] G. Lakoff, *Women, Fire, and Dangerous Things: What Categories Reveal about the Mind*, University of Chicago Press (1987).

[Lan91] R.W. Langacker, *Concept, Image, and Symbol – The Cognitive Basis of Grammar*, in: R. Dirven, R.W. Langacker (eds.) Cognitive Linguistic Research, Vol. 1, Mouton–de Gruyter , N.Y. (1991).

[LR82] W.G. Lehnert and M.H. Ringle, *Strategies for Natural Language Processing*, Lawrence Erlbaum Associates Inc., Hillsdale (1982).

[RK89] H. Ritter and T. Kohonen, *Self-organizing semantic maps*, Biological Cybernetics, 61 (1989), 241-254.

[Sea75] J.R. Searle, *Indirect Speech Acts*, in: P.Cole, J.L. Morgan (eds.) Syntax and Semantics. Vol.3. Speech Acts. New York/San Francisco (1975).

[Tal78] L. Talmy, *The Relation of Grammar to Cognition A Synopsis*, in: David Waltz (ed.), Proceedings of TINLAP-2, Association for Computing Machinery, New York (1978).

[Ull88] J.D. Ullman, *Principles of Database and Knowledge-Base Systems*, Computer, Vol. 1 (1988).

[Wei90] H. Weigant, *Linguistically motivated principles of knowledge-base systems*, in: P.S. Jacobs, Dortrecht Domain-Specific Techniques. In Paul S. Jacobs, Dortrecht et a., Foris publications (1990).

[WH95] C. Wenzel, R. Hoch, *Text Categorization in Scanned Documents Applying a Rule-based Approach Classifiers*, Proc. SDAIR95, Int'l Symposium on Document Analysis and Information Retrieval, Las Vegas, NV (April 1995).

[WW93] X. Wang and D. Wood, *An Abstract Model for Tables*, University of Waterloo, Dept. of Computer Science, Research Paper (1993).

ADAMCO — AN AGENT ARCHITECTURE WITH DOMAIN INDEPENDENT, ADAPTIVE, MULTIPLE COORDINATION BEHAVIOR

Markus Lohmann, Andreas Schmalz, and Christof Weinhardt*

University of Giessen, Computer Science in Business,
Licher Strasse 70, D-35394 Giessen, Germany
Email: {markus.lohmann, andreas.schmalz, christof.weinhardt}@wirtschaft.uni-giessen.de

ABSTRACT

The agent architecture ADAMCO for multi–agent–systems that support the allocation of bounded resources within a distributed problem solving process is presented in this paper. This architecture can be applied to different domains using different coordination mechanisms for the allocation. In order to assign an adequate coordination mechanism to a specific problem class an adaptive coordination behavior is modelled. Starting from the requirements for such an architecture the modules of an ADAMCO agent are derived and explained using two real world scenarios.

INTRODUCTION

In this paper an **A**gent Architecture with **D**omain independent **A**daptive **M**ultiple **CO**ordination Behavior, briefly ADAMCO, is introduced as an architecture for a multi–agent–planning–system based on various coordination principles, namely that of hierarchy, that of markets, and others in between them [MYB87]. The agents involved in ADAMCO should be enabled to solve complex planning problems in business administration. Essentially, these problems concern the allocation of bounded resources within the problem solving process.

There exists a lot of literature on planning problems from the operational research point of view, where centralized methods and algorithms are used to manage problems of high complexity. While centralized exact methods often require an exhaustive data set and have a poor performance, heuristics sometimes run with less data and much faster but may converge far from the optimum solution. Pure parallelization of parts of the considered solving procedure does not bring the effects expected by the applicants, since usually the same data set is needed and communication effort cannot be compensated by better performance of these partial processes.

Thus, it is obvious to go one step further trying to exploit synergetic effects by delegating a complex real world problem to a group of (software) agents each having its own competence in a more or less specific domain. Many real world problems have an inherently distributed structure. Hence, it is natural to look for an adequate, analogously decentralized IT–infrastructure for problem solving, e.g., production planning, transportation planning/logistics or CPU allocation in a distributed computing environment.

From a more abstract point of view, in each domain the problem to be solved is considered as a (customer's) order to be fulfilled by one or more organizational units that are represented by (more or less intelligent) software agents. Members of such agent teams should cooperate (i) in order to solve common tasks or to achieve a common objective, i.e. an optimum solution, and (ii) in order to get synergy generated by this kind of cooperation.

With ADAMCO a distributed infrastructure is developed that enables planners to apply different problem solving techniques on different problem classes in various domains in order to improve the quality of their solutions.

The main goal of this work is to analyze the agent architecture for such a multi–agent–system. Therefore, after explaining a scenario for decentralized planning in business the basic requirements for the architecture are introduced. Starting from these requirements, the agent architecture is derived thereafter and then explained in more detail using two real world scenarios. Some concluding remarks in the last section wind up the paper.

*****Acknowledgement:** This work is part of the program 'Distributed Information Systems in Business' supported by the DFG under contract We 1436/3-1.

A SCENARIO FOR DECENTRALIZED PLANNING IN BUSINESS

For the domain of transportation decentralized planning gains increasing importance. This is based on the effect of international mergers and cooperations that can be observed, recently. The different forwarding agencies involved in a merger are organized as autonomous profit centers. But local decisions usually converge far from the global optimum solution for the affiliated group. Therefore, it is necessary to enforce a coherent behavior of the group. This can only be achieved by using adequate coordination mechanisms.

Consider, as an example, an affiliated group consisting of four forwarding agencies $\{A, B, C, D\}$ and six destinations $\{a, b, c, d, e, f\}$ as shown in figure 1.

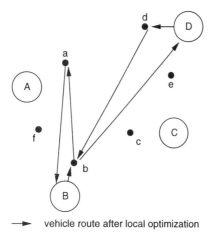

— vehicle route after local optimization

Figure 1: A scenario in the transportation domain

Agencies D and B receive orders to carry a cargo from d to b (D) and from b to a (B), respectively. Each agency decides to fulfil its order based on a locally optimum decision. But usually, this is not the global optimum of the affiliated group. It would be a better solution if agency D handels both orders (first from d to b and then from b to a) assuming no violation of time restrictions. Thus, there is a need of using a coordination mechanism to reallocate the distributed resources and to enable cooperative work between the profit centers. One possibility is to allocate the orders via an auction. Hence, the agent with the highest bid will get the order(s). Another possibility is to use a operational research method to reallocate the resources.

But there exists no coordination mechanism that guarantees a global optimum solution for every allocation problem. Which coordination mechanism is used crucially depends on the problem structure and on the

given individual situation of the members of the affiliated group.

Similar structured allocation problems can be found in other domains, so that the question of finding adequate coordination mechanims is a key problem in many domains. This leads to the conclusion that coordination can be studied in a domain independent way.

Considering a multi–agent–system a profit center is represented by an agent. If the multi–agent–system has to solve the problem of allocating distributed resources it has to use an adequate coordination mechanism as well. Jennings points out that "coordination ... is perhaps the key problem of the discipline of distributed artificial intelligence" [Jen96].

As shown in the pharagraph above coordination can be handled domain independently, which suggests that a corresponding multi–agent–system architecture should consist of domain dependent as well as of domain independent features. Deatailed requirements for an agent architecture that holds for different domains will be presented in the next section.

REQUIREMENTS FOR A DOMAIN INDEPENDENT ARCHITECTURE

A single agent in a multi–agent–system allocating resources for different kinds of orders has two main tasks:

- to solve the (sub-) problem on its own and/or

- to initiate a coordination mechanism by forwarding the whole order or a certain part of it.

Thus, each agent can act in different roles within the problem solving process:

    (i) starting a coordination mechanism

    (ii) participating in the coordination procedure initiated by another agent.

If the agents of a multi–agent–system should be able to react and to initiate multiple coordination mechanisms, they need a specific architecture. In our context the agent architecture must have the flexibility to interact within different domains using multiple coordination mechanisms. Based on the main objective mentioned in the first section, the following requirements for such an agent architecture are derived:

COSTO: An agent must be able first to **communicate** with other agents and second to **sto**re information received when communicating with others or reflecting the states of the local resources. This requirement must be fulfilled by all agent architectures.

DAPROC: To solve problems or parts of problems the agent furthermore needs domain dependent local **da**ta and **pro**blem **c**ompetence.

COMEC: The agent needs the ability to participate in different **c**oordination **mec**hanisms. Therefore, these mechanisms must be integrated in the agent architecture in order to react correctly corresponding to the actual coordination procedure. The implementation of the mechanisms should be domain independent. Thus, the mechanisms can be used in different domains without being readapted.

ASSIG: On the other side an agent has to know when to initiate a certain coordination mechanism. Which mechnism is chosen depends on the problem class. As a result an **assig**nment of problem classes to coordination mechanisms is necessary.

DOMDI: To interact and cooperate with other agents there is a need of protocols. To separate the domain specific information there should exist a **dom**ain **d**ependent and a domain **i**ndependent protocol. The moduls should be separated in domain dependent and domain independent ones. Hence, only the domain dependent modules have to be changed for a new application domain.

TRADIF: For the assignment of problem classes to coordination mechanisms in a domain independent way the agent has to **tra**nsform a real world problem into a **d**omain **i**ndependent **f**orm. As already mentioned in [BG88] different problem representations should be considered when designing multi–agent–systems. According to the relevant characteristics the problem can be classified and assigned to an adequate coordination mechanism.

ADCOB: As the relation between problem classes and coordination mechanisms can not be completely specified in advance, it is necessary to fill these gaps of unknown relations. Furthermore, new unpredictable problem classes appear, that require the **ad**aption of the agents regarding the **co**ordination **b**ehavior. This can only be achieved if the agent that initiated the coordination process receives a feedback about the efficiency of the coordination mechanism that describes the quality of the solution in order to improve the coordination behavior. This adaptive behavior belongs to the desirable agent characteristics [Mae94, SPW+96].

In the next section the components and protocols of the agent architecture ADAMCO will be presented.

ADAMCO AGENTS

To meet all requirements that are postulated in the last section ADAMCO based agents must be structured as shown in figure 2.

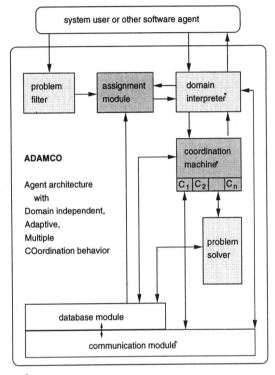

*communicating modules

Figure 2: Structure of an ADAMCO agent

The non shaded modules represent the basic features of an agent whereas the shaded ones show the extensions according to the requirements of a domain independent agent architecture. While the light shaded modules are domain dependent the dark shaded ones are domain independent.

The modules of an ADAMCO agent and their interfaces will be discussed in the next paragraphs of this section with respect to the above mentioned requirements.

The Communication Module

The first ADAMCO module to be described is the COMMUNICATION MODULE. According to requirement COSTO the ability of communication is one of the basic features of an autonomous agent. The COMMUNICATION MODULE is modelled using object oriented techniques and is implemented in C++. ADAMCO agents communicate by sending and receiving KQML messages [FLM95]. Its implementation is realized using the KAPI library [1].

The corresponding C++-class provides the following methods:

- register the agent at a given agent name server

- unregister the agent at the name server

- query the agent name server for other registered agents using the ping–pong protocol for the detection of registered but dead agents

- send a KQML message to a single agent (unicast), to multiple agents (multicast) or to all registered and living agents (broadcast)

- receive KQML messages from other agents

Filtering of incoming messages is done by a facilitator object. ADAMCO modules that have to receive messages register at its facilitator by sending a list of subjects/ontologies they want to react on. This is the same principle as applied to the registration of event driven callback functions. The modules that exchange messages with modules of other agents are called *communicating modules*. Messages containing information concerning the environment of the agent will be stored in the DATABASE MODULE explained in the next paragraph. The COMMUNICATION MODULE, therefore, provides an interface to all other communicating modules and puts received information into the database.

The Database Module

The DATABASE MODULE is the local information base of each agent. It contains an entry for any resource controlled by the agent. Furthermore, technical information like adresses are stored here. Using the DATABASE MODULE the following items can be accessed:

- id and access point of each controlled resource

- utilization of each controlled resource

- names of the implemented coordination mechanisms

- name and URL of the agent name server

- name, URL and resource information of other agents

---

[1] KAPI is a KQML Application Programmer's Interface developed by Enterprise Integration Technologies Corporation and Lockheed Martin Missiles and Space Company, Inc.

While the entries for each controlled resource are updated periodically, the agent's environment information is updated only if one of the communicating modules receives a corresponding message. The DATABASE MODULE belongs to the basic modules of an autonomous agent as described in requirement COSTO.

The Problem Solver

The PROBLEM SOLVER represents the agent's isolated capabilities as described in requirement DAPROC. It is a domain dependent module that contains specific algorithms or just the knowledge of how to solve a given problem of this domain. Considering the transportation domain this might be the capability to solve a Dial–a–Ride Problem using a meta–heuristic (e.g., genetic algorithms or tabu search) as well as it might be the evaluation of allocating an incoming transportation order to a certain vehicle fleet. The PROBLEM SOLVER allocates the agent's local resources (e.g., the vehicle fleet in the above mentioned example) and, therefore, is responsible for updating the local resource information in the DATABASE MODULE (see last paragraph). It receives the orders from the COORDINATION MACHINE.

The Coordination Machine

The COORDINATION MACHINE is modelled as a domain independent module that controls a set of coordination mechanisms (requirement COMEC). This set is used to implement collaboration in a multi–agent–system. Coordination is necessary if more than one agent is involved in solving a given problem and if synergy should be exploited. As already mentioned in the introduction and the scenario from the transportation domain, we assume a relation between the kind of problem that has to be solved and an adequate coordination mechanism.

Coordination mechanisms may be realized as hierarchical as well as market like allocation or as any mechanism between these two extremes [MYB87]. These mechanisms can be adopted from the disciplines of computer science (e.g., load–balancing algorithms from [SK90]), operations research (e.g, Genetic Algorithms from [Hol75, Gol89, TNJ91]) and/or economics (e.g., auctions [Vic61, GSW96]). This adoption is possible because the mechanisms do not depend on the kind of resource that has to be allocated. For example Genetic Algorithms can be applied to job–shop–scheduling, traveling salesman problems and other combinatorical optimization tasks.

During the coordination process agents may act in different roles. Considering auctions as market like coordination mechanisms there exist usually one auctioneer and some bidding agents. Because every agent can accept orders and because there is no central coordination agent, each ADAMCO agent should be able to act in both roles. Agents that are able to act in multiple roles also are considered in [DWS96].

Because the COORDINATION MACHINE belongs to the communicating modules there is an interface to the COMMUNICATION MODULE. Furthermore, it accesses the DATABASE MODULE to read the adresses of other agents in order to communicate with them. Each mechanism registers at the local database (see database module). It passes a given (sub-) problem to the PROBLEM SOLVER and receives the corresponding results. The COORDINATION MACHINE gets the problem description with all necessary details from the DOMAIN INTERPRETER (see domain interpreter).

The Problem Filter

The PROBLEM FILTER is a domain specific module. It provides an interface between incoming orders (from system users or from other agents) and the ASSIGNMENT MODULE. The main task of this module is to extract the domain specific parts from the order description. The residual domain independent information is passed to the ASSIGNMENT MODULE which is thereby enabled to identify the main characteristics of the order. These characteristics may be:

- Is the problem decomposable or not?

- If it is decomposable, is there a predefined decomposition or is it possible to decompose the problem in arbitrary parts?

- Which parameter is used for the decomposition (time, object, etc.)?

- What kind of problem arises after the decomposition: single instruction multiple data (SIMD), multiple instruction multiple data (MIMD) etc.?

- What time constraints must be fulfilled while processing the problem?

- Are there any priorities that must be considered?

Answers to these questions outline the characteristics of the problem in a domain independent way as postulated in requirement TRADIF. This is necessary for the assignment of the problem to a suitable coordination mechanism.

The Assignment Module

According to the requirements ASSIG and TRADIF the main task of the ASSIGNMENT MODULE is to assign a domain independent problem description to a domain independent coordination mechanism. Therefore, the ASSIGNMENT MODULE is a domain independent module. Applying ADAMCO agents to the field of planning problems, it performs a classification of planning problems. This can be achieved in the following way: Answering the questions mentioned in the last paragraph the problem can be represented as a binary string by switching on the corresponding detectors (see figure 3).

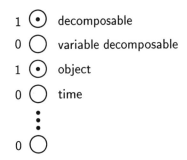

Figure 3: Output of the PROBLEM FILTER used as detectors

The ASSIGNMENT MODULE uses rules to find an adequate coordination mechanism for the given problem, i.e. a mechanism that generates an efficient problem solution meeting all constraints given by the problem description. The coordination mechanisms implemented in the COORDINATION MACHINE are read from the local database. In addition, it is important to know which agents actually are registered in the multi–agent–system. This is necessary because, e.g., an auction based mechanism only makes sense if there are some bidding agents. If only a few agents are registered a hierarchical allocation might outperform the auction because there may be more time spent for the auction than gained by the better solution produced by the auction. Considering all available information the rules of the ASSIGNMENT MODULE propose a coordination mechanism that should be used to solve the problem.

Using a fixed set of rules the module would have to know all possible problem characteristics for all possible system states, which is impossible. That is why the rule set must be adaptive. To fulfill this feature (requirement ADCOB) the ASSIGNMENT MODULE is implemented as follows:

After the problem has been solved by the multi–

agent–system the ASSIGNMENT MODULE gets a feedback from the DOMAIN INTERPRETER about the quality of the solution. According to this feedback the rules having proposed the coordination mechanism will be rated. This rating is taken into account if more than one rule matches a given situation. The rule with the highest rating will be chosen to select a coordination mechanism. In order to be able to react on new situations, i.e. problem characteristics and system states that are not modelled in the rule set, new rules must be created. This can be done by applying *Genetic Algorithms* [Hol75] to the rule set. Using the *Simple Classifier System* introduced by Goldberg [Gol89] an adaptive ASSIGNMENT MODULE as described above can be realized.

The Domain Interpreter

The last module to discuss is the DOMAIN INTERPRETER. This is a domain dependent communicating module. One of its main tasks is to decompose – if necessary – the problem under consideration and to make the synthesis when all parts have been solved. Another important task is to transfer details of the problem to the proposed coordination mechanism. Considering a decomposable problem these details are: the problem part to be solved, time constraints, quality of service etc. As already mentioned in the last paragraph the DOMAIN INTERPRETER has to rate the solution of the problem. This is done after delivering the solution to the system user or software agent that submitted the job. It is a communicating module because it is responsible for distributing domain specific information concerning the problem under consideration if the problem is solved by a set of agents.

Whenever more than one agent is involved in the problem solving process there must be protocols controlling the communication between the agents. These protocols are briefly described in the next section.

The Protocols in ADAMCO

Figure 4 (next page) shows the connection between two ADAMCO agents via the corresponding protocols.

The COMMUNICATION PROTOCOL defines the technical details of the communication between two or more agents. This includes how to establish a connection to other agents, how to send a message, etc.

The coordination process is controlled using the CO-ORDINATION PROTOCOL, i.e. COORDINATION MACHINES are linked together via this protocol (requirement COMEC). Considering an auction this means that after an invitation for bids the deadline for submitting bids must be supervised. Finally the contract is awarded to the winner.

The last protocol is the DOMAIN PROTOCOL. As mentioned in paragraph *The Domain Interpreter*, this protocol controls the submission of domain specific information concerning the problem under consideration (requirement DOMDI).

The last two protocols define the rule set for the communicating modules. They define a *virtual* communication between agents – virtual, since the agents are physically connected only via the COMMUNICATION PROTOCOL.

In the next section the application of the architecture to two different domains will be explained in order to show how domain specific problems are solved with ADAMCO.

APPLYING THE ADAMCO ARCHITECTURE TO "REAL WORLD" DOMAINS

In this section two scenarios are introduced to explain how to use the agent architecture for real world applications. The first scenario considers again the domain of transportation. In this domain different forwarding agencies are organized as profit centers, say depots, within an affiliated group. Each single profit center disposes of several trucks and is represented by an ADAMCO agent. A central planning or coordination unit does not exist, the aim is to find a decentralized solution of a Dynamic Vehicle Routing Problem where every single incoming order is assigned to depots immediately. Within this scenario each depot can receive orders and it has to decide whether to accept the order or not. Accepting the order, (i) the agent uses its own knowledge of vehicle routing to compute the local solution integrating this order into its actual tour set or (ii) it tries to exploit synergy within its agency by collaborating with other depots.

Keeping in mind that in the domain of transportation various kinds of problems exist the PROBLEM FILTER transforms the problem in a domain independent form. The relevant characteristics for the coordination process are used by the ASSIGNMENT MODULE in order to form problem classes and to assign them to coordination mechanisms. An order can be decomposable or not. The decomposability may concern the cargo or the shipment and can be predefined or variable. Other problem characteristics are the computation time needed for the problem, the priority of a problem and the maximal number of agents involved

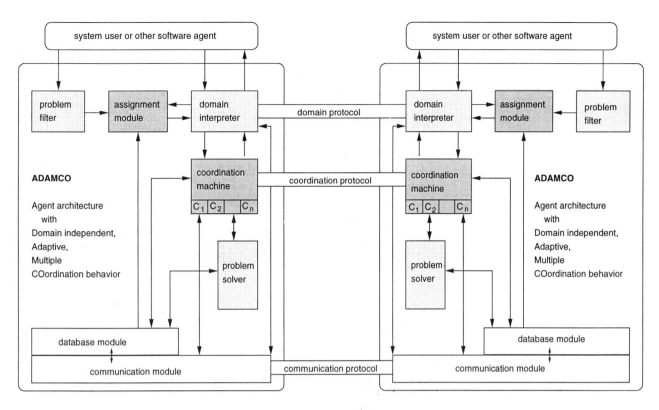

Figure 4: Two ADAMCO agents and the corresponding protocols

in the problem solving process. After the coordination mechanism has been chosen the DOMAIN INTERPRETER delivers the information to execute this coordination mechanism.

In order to illustrate this process, take the example of an order from Berlin to Munich. The cargo and the shipment can not be decomposed. The assignment module takes these characteristics and the information from the DATABASE MODULE and assigns the Vickrey auction coordination mechanism to this problem class. For a problem class where a decomposition is not possible this mechanism will lead to a pareto–optimum allocation [Vic61, Wei94]. The agent has to send a broadcast message to the other agents. The PROBLEM SOLVER of each agent works on the problem and sends a message containing the agents evaluation of that transportation job back to the agent that initiated the auction. When all agents have replied or a timeout occurs, the agent with the highest bid gets the order to the price of the second highest bid. Finishing the coordination process the domain interpreter returns a feedback to the ASSIGNMENT MODULE including information about the quality of the coordination mechanism. With this feedback the strength of

the rules in the ASSIGNMENT MODULE can be changed [SW89, Mae94].

In a second scenario CPU allocation in a distributed computing environment is considered. The problem to be solved is a two dimensional parameter analysis of a dynamical system, e.g., the logistic function $x_{t+1} = \alpha x_t(1 - x_t)$ (for further references see [Dev85]). The long term behavior of the system will be analyzed for $\alpha \in [0; 4)$ and $x_0 \in [0; 1)$.

The PROBLEM FILTER has a mathematical knowledge base that identifies the problem as a decomposable SIMD problem where all subproblems have the same priority. There are no other constraints. The goal is to solve this problem as fast as possible.

The computing environment consists of several heterogenous nodes (regarding CPU speed and system architecture). Each node has the same local PROBLEM SOLVER that is able to do the parameter analysis. The ASSIGNMENT MODULE reads from the DATABASE MODULE that there exist heavy loaded and some light loaded nodes. It decides to use a sender–initiated allocation policy [SK90]. The DOMAIN INTERPRETER decomposes the problem into several equal sized parts that are passed to a sender–initiated coordination mechanism. The parts are sent to the light loaded nodes (hierarchical assignment), one per node using the COORDINATION PROTOCOL. After its subproblem is solved another part is sent to that node until the whole problem is solved. The solution of each subproblem is returned to the DOMAIN INTERPRETER that performs a synthesis of the partial solutions. Finally, the solution of the whole problem is given to the system user.

The performance of the chosen coordination mechanism can be evaluated as follows: The system user spends money for the problem solving. The faster the system solves his problem the more money he has to pay. The agent that initiates the coordination has to buy CPU seconds from the other nodes as it is done in computing centers. The faster the node the more money the initiating agent has to spend. In that way the surplus can be measured.

Another method to evaluate the mechanism is to compare the performance with the one achieved when solving reference problems. But then the actual situation (the jobs in the system) must be taken into account.

CONCLUSION

Using the agent architecture ADAMCO it is possible to develop multi–agent–systems for different domains without changing the main features COMMUNICATION MODULE, DATABASE MODULE, COORDINATION MACHINE and ASSIGNMENT MODULE. The core of ADAMCO is the adaptability concerning the coordination behavior. Adaption is necessary (i) for an assignment of a specific problem class to an *adequate* coordination mechanism, (ii) to be able to react to unpredictable problem classes and situations and (iii) to draw conclusions from the coordination process in order to improve the assignment.

In our project the agent architecture ADAMCO is applied to different decentralized operational planning domains. The main focus is set on the domain of transportation. In the corresponding multi–agent–system different methods for problem solving – based on AI and OR methods – are integrated in different agents. This implies that problems must be represented in various forms according to the available problem solving methods.

The next step is to implement and test the COORDINATION MACHINE with different simple auctions and multi–staged auctions [GSW96]. The results will be used to extend the rule set of the ASSIGNMENT MODULE.

REFERENCES

[BG88]   A.H. Bond and L. Gasser. An Analysis of Problems and Research in DAI. In A.H. Bond and L. Gasser (eds.), *Readings in Distributed Artificial Intelligence*, pp. 3-35. Morgan Kaufmann. 1988.

[DWS96]   K. Decker, M. Williamson and K. Sycara. Matchmaking and Brokering. To appear in *Proceedings of the Second International Conference on Multi–Agent–Systems*. 1996.

[Dev85]   R.L. Devaney. *An Introduction to Chaotic Dynamical Systems*. Addison–Wesley. 1985.

[FLM95]   T. Finin, Y. Labrou and J. Mayfield. KQML as an agent communication language. In Bradshaw, J.M. (eds.), *Software Agents*. MIT Press. 1995.

[Gol89]   D.E. Goldberg. *Genetic algorithms in search, optimization, and machine learning*. Addison–Wesley. 1989.

[GSW96]   P. Gomber, C. Schmidt and Ch. Weinhardt. Efficency and Incentives in MAS-Coordination. To appear in *Proceedings of the 5th European Conference on Information Systems*. 1997.

[Hòl75]   J. Holland. *Adaption in Natural and Artificial Systems*. The University of Michigan Press. 1975.

[Jen96]   N. R. Jennings. Coordination Techniques for Distributed Artificial Intelligence. In G.M.P. O'Hare, N.R. Jennings (ed.), *Foundations of Distributed Artificial Intelligence*, pp. 187-210. Wiley. 1996.

[Mae94]   P. Maes. Modeling Adaptive Autonomous Agents. In C. Langton (ed.), *Artificial Life Journal*, Vol. 1, No. 1&2, pp. 135-162. MIT Press. 1994.

[MYB87]   T. Malone, J. Yates and R. Benjamin. Electronic markets and electronic hierarchies. *Communications of the ACM*, pp. 484-496. June 1987.

[SW89]   M.J. Shaw and A.B. Whinston. Learning and Adaptation in Distributed Artificial Intelligence Systems. In Gasser, L. and

Huhns, M.N. (eds.), *Distributed Artificial Intelligence*, Vol. II, pp. 413-429. Morgan Kaufmann. 1989.

[SK90]   N.G. Shivaratri and P. Krüger. Two Adaptive Location Policies for Global Scheduling Algorithms. In *Proceedings of the 10th International Conference on Distributed Computing Systems*, pp. 502–509. IEEE Comp. Soc. Press. 1990.

[SPW+96]   K. Sycara, A. Pannu, M. Williamson, D. Zeng, and K. Decker. Distributed Intelligent Agents. IEEE Expert, Vol. 11, No. 6, pp. 36-46. December 1996.

[TNJ91]   S.R. Thangiah, K.E. Nygard, and P.L. Juell. GIDEON: A Genetic Algorithm System for Vehicle Routing Problems with Time Windows. In PROCEEDINGS OF THE 7TH IEEE CONFERENCE ON ARTIFICIAL INTELLIGENCE APPLICATIONS, pp. 322-328. Miami. 1991.

[Vic61]   W. Vickrey. Counterspeculation, Auctions, and Competitive Sealed Tenders. *Journal of Finance*, Vol. 16, pp. 8-37. 1961.

[Wei94]   Ch. Weinhardt. Auctions and Negotiations for Business and Financial Problem Solving with Distributed Artificial Intelligence. In H.J. Müller und S. Kirn (eds.): *Proceedings to the 11th European Conference on Artificial Intelligence*, pp. 59-64. 1994.

CONSTRAINT-BASED RESOURCE ALLOCATION FOR AIR CARGO TRANSFER PLANNING

Hon Wai Chun

Department of Electronic Engineering, City University of Hong Kong,
Tat Chee Avenue, Kowloon, Hong Kong
Email: eehwchun@cityu.edu.hk

ABSTRACT

In this paper, we documented results produced by a prototype resource allocation system that performs transfer planning for an air cargo handling facility using a constraint-based approach. The key problem is to decide, for each export and import air cargo container, which transfer deck should it be placed at before or after transfer from the aircraft, through which path within the building facility should it be transferred and at which time, and how the empty container should be transferred out to or from the storage area. This resource allocation problem is modelled as a constraint-satisfaction problem (CSP) in this research. The key objective is to optimise the use of the air cargo handling facility, to minimise any delay to aircraft departure, and to fulfil customer service commitments. Another objective is to be able to perform reactive scheduling if flight arrival or departure time has changed, and if the number and type of containers have changed. This paper describes how this problem can be modelled, the constraints that are involved, and the scheduling algorithm used.

INTRODUCTION

This paper describes a constraint-based model and algorithm for an air cargo transfer planning prototype system. This system determines how each export or import air cargo container, or *unit loading device* (ULD), is handled before or after it is taken to or from the aircraft. Several resources must be scheduled for each ULD – the conveyor belt, the hoist or crane, the transfer deck on the ground floor, the transfer times, and the transfer path for the empty ULD. Since resources are valuable and limited within the air cargo handling facility, they must be allocated efficiently taking into consideration all the necessary constraints. Constraints such as the type and priority of a ULD, the arrival or departure time of the flight, the availability and capacity of the resources, and the service commitments to the clients of the air cargo handling company. This research was performed using flight data from the Hong Kong International Airport; one of the busiest international airport in the world.

Our research focuses on designing and developing a constraint-based resource allocation algorithm [PUGE94b] that can assist the human planner by generating a detailed transfer plan for each export and import ULD. This plan will allow the human planner to estimate resource requirements such as the number of workers needed to build up or break down containers at different times of the day. Currently, due to the complexity of the problem, a human planner is not capable of producing a long-term transfer plan that is detailed down to the ULD level.

Figure 1 is a diagram of the simplified air cargo handling facility that was used as the test case for this research. Although the building structure is fictitious, it was modelled after and contains all the essential characteristics of a real air cargo handling facility. For simplicity, the building has three main floors – the workstation floor used for building up and breaking down of containers, the ground floor to take or received the container to or from the aircraft, and the empty container floor to store empty ULDs. The building contains three main transfer paths – the stacker crane which is used for long-term storage, the outer hoist for normal traffic, and the inner hoist for high priority or late cargo. Not shown in Figure 1 is the floor that is used to store empty ULDs. For export, an empty ULD must first be transferred to the workstation floor for container build-up. For import, after ULD breakdown, the empty ULD must be transferred to the empty ULD storage floor. The transfer planning system will also schedule the transfer of these empty ULD.

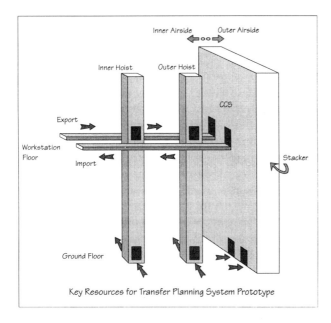

Figure. 1 The simplified building structure used by the transfer planning system.

Figure 2 is a floor plan of the simplified workstation level used by our transfer planning system. The key resources are the two conveyor belts, the two hoists, and the stacker crane. In our simplified building, there is one conveyor belt for import and one for export. Each conveyor belt can access the two hoists and stacker crane. The transfer rate for each resource may be different.

Figure 3 is a diagram of the simplified ground floor layout used by the transfer planning system. There are seven transfer decks that can be used for export containers and three transfer decks for import containers. However, two of the three import transfer decks feeds to both the outer hoist and the stacker crane. A key constraint is that each transfer deck can only be used by containers of the same flight at any one time-period. This is to avoid

potential human error in transferring a container to a wrong flight.

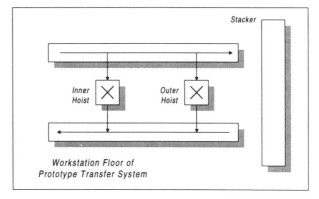

Figure. 2 The workstation level floor plan used by the transfer planning system.

The input to the transfer planning system is a flight schedule that contains the airline code, flight number, flow (either export or import); the scheduled time of arrival or departure, and number and types of ULDs in the aircraft.

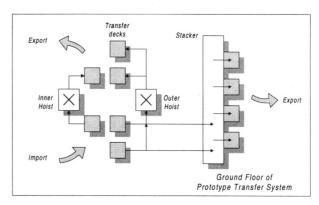

Figure. 3 The ground level floor plan used by the transfer planning system.

Our constraint-based scheduling algorithm allocates five main resources to each ULD - (1) a transfer path within the air cargo handling facility, (2) the transfer time, (3) the transfer deck on the ground level, (4) the empty ULD transfer path, and (5) the empty ULD transfer time.

THE OBJECT-ORIENTED CONSTRAINT-BASED MODEL

To solve air cargo transfer planning, we represented the resource allocation task as an object-oriented [LEPA93] constraint-satisfaction problem (CSP) [KUMA92]. Although CSP or constraint-programming has a relatively long history [STEE80], with constraint language extensions found in Prolog [COLM90, VANH89], and Lisp [SISK93], it is only recently that constraint-programming became more popular with the availability of the ILOG's C++ class libraries [PUGE94a]. This library provided a very efficient and clean implementation of constraint-based programming features in a conventional language. It also integrated constraint programming with object-orientation. Our scheduling algorithm was implemented using the ILOG C++ class libraries.

In general, any scheduling and resource allocation problems can be formulated as a constraint-satisfaction problem (CSP) which involves the assignment of values to variables subjected to a set of constraints. CSP can be defined as consisting of a finite set of n variables $v_1, v_2, ..., v_n$, a set of domains $d_1, d_2, ..., d_n$, and a set of constraint relations $c_1, c_2, ..., c_m$. Each d_i defines a finite set of values (or solutions) that variable v_i may be assigned. A constraint c_j specifies the consistent or inconsistent choices among variables and is defined as a subset of the Cartesian product: $c_j \subseteq d_1 \, x \, d_2 \, x \, ... \, x \, d_n$. The goal of a CSP algorithm is to find one tuple from $d_1 \, x \, d_2 \, x \, ... \, x \, d_n$ such that n assignments of values to variables satisfy all constraints simultaneously.

The constrained variables used by the transfer planning scheduling algorithm are represented as attributes of a *ULD Assignment* class. Figure 4 is a Unified Modelling Language (UML) class diagram with the key domain classes used by the transfer planning system. There are seven main domain classes:

- **Flight** - Each instance of the *Flight* class represents an individual flight. Attributes of this class store information loaded from the flight schedule. In particular, there are two types of flight -- export and import.

- **ULD** - Each *ULD* object represents one single ULD in a flight. Attributes of this

class include the type of ULD, the priority, and the ULD identity number.

- **Flight Assignment** - This class contains resource assignment information for one particular flight. It contains a reference to a flight object and lists of ULD assignment objects. There are three types of ULD assignments, depending on the priority and type of ULD.

- **ULD Assignment** - This class contains the resource assignment information for one ULD. All the constrained variables used by the CSP scheduling algorithm is stored within objects of this class. This includes a constrained variable to represent the transfer path the ULD should take. The domain of this variable includes all the possible transfer paths that are available to this type of ULD. There is a constrained variable to represent the transfer start time. The domain of the start time depends on whether the ULD is an export or import ULD and will range between an earliest and latest time. There is a constrained variable to represent the transfer deck to place the ULD. The domain of the transfer deck variable depends on whether the ULD is an export or import ULD. There is also a constrained variable to represent the transfer path for the empty ULD, and a constrained variable to represent the empty ULD transfer time.

- **Resource** - Each *Resource* subclass has a throughput attribute. The subclass includes the *Transfer Path* and the *Transfer Deck*. Objects of this class keep track of resource availability.

- **Transfer Deck** - Each instance of the *Transfer Deck* class represents one transfer deck on the ground level.

- **Transfer Path** - Each instance of the *Transfer Path* class represents one transfer path within the air cargo handling facility.

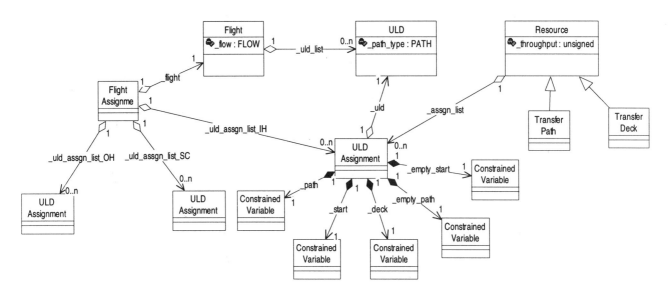

Figure. 4 The UML class diagram of some of the classes used in the transfer planning system.

TRANSFER PLANNING CONSTRAINTS

Some of the constraints that are considered by the transfer planning system were mentioned in the previous sections. The following are other constraints that are also considered by the scheduling algorithm:

- **No resource conflict** - This is of course the most basic constraint. Resources assigned to different ULDs cannot conflict with each other. For example, the same hoist cannot be used to transfer two different ULDs at the same time.

- **Temporal ordering** - The sequence which ULDs are dispatched are sequentially ordered for two reasons. First, for export ULDs, dispatch from the stacker crane is performed before normal ULDs and urgent ULDs. The second reason to order the ULDs is to reduce the search time by eliminating equivalent redundant solutions.

- **Empty ULDs transfer times** - For export flights, an empty ULD must be transferred to the workstation floor before the earliest transfer start time. Empty ULDs from

import flights can be transferred after the latest transfer completion time.

- **Number of allocated transfer decks** - The number of transfer decks to be allocated to a flight will depend on the total number of ULDs for that flight.

- **Transfer deck rule** - This is a constraint to encode the transfer deck rule which forces the transfer deck to process only one flight's ULD at a time.

The following are constraint diagrams developed by the author to illustrate temporal constraints. The horizontal line ranges from the earliest transfer start time (ETST) to the latest transfer completion time (LTCT). This line also represents the domain of the temporal constrained variables, shown as upward arrows. Arrows sliding along this line is analogous with variables taking different values from a domain. Each temporal constrained variable represents the transfer start time of one ULD. The rectangular box represents the duration of the transfer that depends on the selected transfer path. The "less than" symbol within the circle represents the temporal constraints.

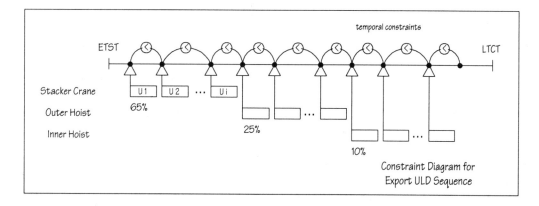

Figure. 5 Constraint diagram for export ULDs.

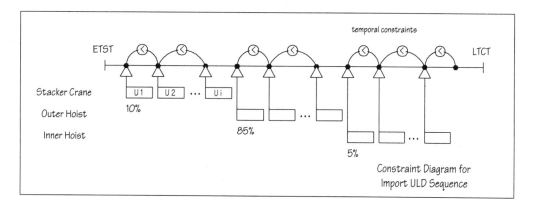

Figure. 6 Constraint diagram for import ULDs.

For export ULDs within the same flight, all ULDs are temporally sequenced in a chain as illustrated in Figure 5. The temporal constraints for ULDs of the same type are used as "redundant constraints" to reduce search time. However, for import flights, ULDs are only temporally sequenced within the same transfer path, as illustrated by Figure 6.

The percentage in the diagram indicates the statistical percentage of ULDs of different type. For example, for export flights, around 65% of the total ULDs for any one flight are taken from long term storage, around 25% are normal ULDs, and around 10% are late arrival ULDs.

THE TRANSFER PLANNING SCHEDULING ALGORITHM

The scheduling algorithm developed for this research consists of a collection of goals. The CSP search algorithm tries to solve each goal sequentially. The main goal is defined as follows:

```
void main() {
  ...
  IlcSolve(TransferPlanSearch());
  ...
};
```

IlcSolve is the built-in CSP search algorithm provided by the ILOG Solver C++ class library. TransferPlanSearch is the main goal and is defined as follows:

```
ILCGOAL0(TransferPlanSearch) {
  // Export flights have higher priority.
  IlcAnd(AllocateExportULD(),
         // then import flights
         AllocateImportULD(),
         // store empty ULDs away
         AllocateImpEmptyULD(),
```

```
// retrieve empty ULDs for export
AllocateExpEmptyULD()
    ) ;
}
```

TransferPlanSearch itself consists of four other goals that are ordered according to priority – export has highest priority since flight departure time should not be delayed, next is import flights, and then the empty ULDs. ILCGOAL0 is a macro to define goals using ILOG Solver. As an example, the first goal of allocating resources to export ULDs can be defined as follows:

```
ILCGOAL0(AllocateExportULD) {
    …
    ULDAssignment* pU ;
    while (pU=(ULDAssignment*)iter()) {
      // assign transfer path for ULD,
      IlcAnd(AssignTransferPath(pU),
             // then the transfer time,
             AssignTransferTime(pU),
             // the transfer deck,
             AssignTransferDeck(pU),
             // and then fire demon to check
             // additional constraints
             CheckExpConstraints(pU)
             );
    }
}
```

Besides the transfer plan generation, our prototype also performs reactive scheduling to handle to following two types of dynamic changes:

- **Change in number of ULD** - The number of ULDs in a flight might not be accurate and may change. Changing this number will cause the transfer planning system to recompute the number of ULDs in each category, and reschedule all affected ULDs.

- **Equipment breakdown** - If a resource breaks down, the transfer planning system reschedules all the ULDs that were previously assigned to that resource according to constraints.

The transfer planning prototype displays the results of the schedule as a Gantt chart. Each row represents one resource. A resource is either a transfer path, transfer deck, or unassigned transfer path. Each rectangle in the Gantt chart represents the allocation of a resource to a ULD for a period of time.

SYSTEM IMPLEMENTATION

The object-oriented constraint-based [LEPA93] allocation system was implemented in C++ using the ILOG Solver class library [PUGE94a] and the RTL Scheduling Framework developed by Resource Technologies Limited [CHUN96a, CHUN96b]. The graphic user interface that displays the generated schedule as a Gantt chart was developed using C++ graphic components provided by ILOG Views. The system was developed using platform independent coding and can execute within Windows 95/NT or Unix environment. ILOG was chosen since it is the fastest implementation of CSP in the market and, with a C++ environment, the system can interface easily to external systems.

The performance of the scheduling algorithm was an important consideration for this research and a considerable amount of time was spent in improving the system performance. The performance improved from the initial 5 to 6 minutes to half a minute required to schedule one whole day's flight on a Pentium-based 133MHz personal computer. Most of the performance gains are obtained through the addition of redundant constraints.

Our test cases had on the average around 300 ULDs daily and since each ULD requires three resources to be allocated - the transfer path, the empty transfer path, and the transfer deck, the system had to allocate around 900 items total in half a minute - while considering all the constraints defined. Reactive scheduling only takes a few seconds to perform as well. Due to the large number of items to be scheduled, it is currently not feasible for a human to accurately produce a daily transfer plan down to the ULD level.

Reactive scheduling is the ability to react to changes such as changes in aircraft loading, additional flights, flight cancellations, or flight delays. In our system, reactive scheduling only takes a fraction of the time needed by the initial scheduling. Traditionally, reactive scheduling involves regenerating a new schedule all over again. In our approach, the reactive scheduling algorithm reuses the current schedule as much as possible and only makes changes to the original schedule, if absolutely necessary, in order to solve problems caused by unexpected changes.

Although the users would like an "optimal" solution, in reality an "optimised" solution will be sufficient. The reason is that an "optimal" solution is guaranteed to be non-optimal as soon as it is put into use – flights are not always punctual, import and export cargo loading will change, flights might be added, etc. Our approach is to use a scheduling algorithm that generates a heuristically optimised solution and couple that with a quick reactive scheduling component, also using the same heuristics, to deal with last minute changes.

In certain cases, depending on the flight schedule and aircraft loading, the scheduling problem might be over-constrained. In those cases, the system first relaxes constraints related to import ULDs since they are less time critical. In cases where the system is still over-constrained after import constraints are relaxed, the system must then violate export ULD constraints and warn the user. For these cases, the physical cargo handling facility is over its maximum capacity and there is not much a scheduling algorithm can do to improve the situation.

CONCLUSIONS

This paper documents our research in modelling air cargo transfer planning as a constraint-satisfaction problem. The constraint-based scheduling algorithm was tested using air cargo data from the Hong Kong International Airport. The results showed that the scheduling algorithm was able to generate a more efficient schedule in much less time than a human scheduler. Currently, the scheduling task is too complex and time consuming to be produce manually down to ULD precision. Long-term planning is, of course, out of the question.

ACKNOWLEDGEMENTS

Research described in this paper was performed in cooperation with Mr. Albert Pang of Resource Technologies Limited (http://www.rtl.com.hk/~rtl) in Hong Kong and with assistance from ILOG (http://www.ilog.com) in Singapore.

REFERENCES

[CHUN96a] H.W. Chun, K.H. Pang, and N. Lam, "Container Vessel Berth Allocation with ILOG SOLVER," The Second International ILOG SOLVER User Conference, Paris, July, 1996.

[CHUN96b] H.W. Chun, M.P. Ng, and N. Lam, "Rostering of Equipment Operators in a Container Yard," The Second International ILOG SOLVER User Conference, Paris, July, 1996.

[COLM90] A. Colmerauer, *An Introduction to Prolog III*, Communications of the ACM, 33(7), pp.69-90, 1990.

[KUMA92] V. Kumar, "Algorithms for Constraint Satisfaction Problems: A Survey," In *AI Magazine*, 13(1), pp.32-44, 1992.

[LEPA93] C. Le Pape, "Using Object-Oriented Constraint Programming Tools to Implement Flexible "Easy-to-use" Scheduling Systems," In *Proceedings of the NSF Workshop on Intelligent, Dynamic Scheduling for Manufacturing*, Cocoa Beach, Florida, 1993.

[MACK77] A.K. Mackworth, "Consistency in Networks of Relations," In *Artificial Intelligence*, 8, pp.99-118, 1977.

[PUGE94a] J.-F. Puget, "A C++ Implementation of CLP," In *ILOG Solver Collected Papers*, ILOG SA, France, 1994.

[PUGE94b] J.-F. Puget, "Object-Oriented Constraint Programming for Transportation Problems," In *ILOG Solver Collected Papers*, ILOG SA, France, 1994.

[SISK93] J.M. Siskind and D.A. McAllester, "Nondeterministic Lisp as a Substrate for Constraint Logic Programming," In *Proceedings of the Eleventh National Conference on Artificial Intelligence*, Washington, DC, pp.133-138, July, 1993.

[STEE80] G.L. Steele Jr., *The Definition and Implementation of a Computer Programming Language Based on Constraints*, Ph.D. Thesis, MIT, 1980.

[VANH89]   P. Van Hentenryck, *Constraint Satisfaction in Logic Programming*, MIT Press, 1989.

[WALT72]   D.L. Waltz, "Understanding Line Drawings of Scenes with Shadows," In *The Psychology of Computer Vision*, McGraw-Hill, pp.19-91, 1975.

ROBOT PATH PLANNING BY A MULTIRESOLUTION A^* SEARCH AND SOME ADMISSIBLE HEURISTICS FOR IT

Antti Autere and Johannes Lehtinen

Department of Computer Science, Helsinki University of Technology,
Otakaari 1 M, SF-02150 Espoo, FINLAND
email: aau@cs.hut.fi, jle@cs.hut.fi

ABSTRACT

In this paper, a point to point robot path planning problem is studied. It occurs in industry for example in spot welding, riveting and pick and place tasks.

A new A^* -based method is presented. The algorithm searches the robot's configuration space with many different resolutions at the same time. When a path candidate goes far from the obstacles, a coarser resolution is used. When it goes near the obstacle surfaces, a finer resolution is used.

A^* is applied because of the possibility to generate better guiding heuristics. A better admissible heuristic roughly means that A^* using it expands fewer configuration space nodes, which is known a priori.

A known AI-method utilizing *relaxed models* is applied to generate admissible heuristics. Constructing relaxed models involves removing details from the base level problem to get simplified ones. The heuristics are then obtained by solving these simplified problems.

A simulated robot workcell is provided for demonstrations. The path planning of a 5-degrees-of-freedom industrial robot appears to be reasonably fast.

Keywords: Robotics, Heuristic Searching, A^*, Planning, Path Planning, Geometric Path Planning

INTRODUCTION

Robot path planning in known environments refers to finding a collision free path from an initial robot configuration to a desired goal configuration. This problem occurs in industry in spot welding, riveting and pick and place tasks, to mention some examples. Figure 6, at the end of the paper, illustrates an example path for a 2-joint robot manipulator. In Fig. 6, "S" is a start configuration and "E" is the goal.

When an automated production line is operating, it is often very expensive to stop it to re-program robots, e.g. to deal with new products. A more economical way is to generate the movements of the robots off-line, in a simulator, and then download the programs into the robot controllers. Minor modifications may still be needed but the time the production line needs to be "off duty" is small compared to the manual teaching of the robots.

Path planning has received much attention over the years and many different approaches have been presented. They can be roughly categorized into *potential field*, *cell decompositions* and *roadmap* or *skeleton* methods [Lat91] and [HA92].

Potential field methods generally employ repulsive potential fields around obstacles and an attractive field around the goal. Path planning is then done by following the negative gradient of the combined potentials. The major drawback is that the potential function will often lead the path to some local minimum, from which it cannot escape without additional help. Many techniques for escaping from local minima have been presented, e.g. in [Lat91] and [BL91]. It is possible to construct potential functions that are free from local minima but they must be calculated numerically, see e.g. [Lat91], [BL91] and [Con94]. However, present computers cannot keep, for example, a 6-dimensional numerical potential function in memory.

Cell decomposition approaches are based on decomposing (either exactly or approximately) the set of free configurations into simple non-overlapping regions called cells, e.g. [ea86] and [Lat91]. The adjacency of these cells is then represented in a connectivity graph that is searched for a path. The drawback is that all the cells must be constructed before a path can be found. However, recently a hierarchical method to do this has been reported in [BH93] and [BH95].

Roadmap or skeleton approaches attempt to retract or map the set of feasible motions onto a network of one-dimensional lines, called the roadmap, skeleton, visibility graph or subgoal network. The path planner tries to connect the start and goal configurations to this network and then search it to find the (optimum) path. The roadmap is usually generated by preprocessing (sampling) the robot's configuration space. There are many ways to do this. Examples of this approach are found in [Cla90], [Lat91], [KL94b] and [KL94a] to mention a few.

The nodes of the roadmap can also be constructed by the A^* algorithm. Reference [War93] is an example of this. In the reference, A^* first searches with a coarse resolution to produce a set of collision free nodes. Then, it testes if there exists a collision free path from the start to the goal going through some of these nodes. If not, then A^* uses a finer resolution etc.

In the roadmap methods, including the last one, it is usually not known that a path exists from the start to the goal via the roadmap nodes. The path will be searched after the roadmap is generated. This may cause unnecessary overhead.

In this paper, we discuss the principles of the new A^*-based method that searches with both coarser and finer resolutions *at the same time*. The coarser resolutions correspond with bigger step sizes and the finer resolutions smaller ones. Bigger step sizes are used when the search proceeds far away from obstacles and smaller ones when a path candidate is near the obstacle surfaces. Further, a path from the start configuration to a node *has already been found* before the node is included in the roadmap.

ROBOT'S WORK AND CONFIGURATION SPACE

This section shortly introduces some commonly used notions related to robot path planning.

Let A denote the robot, W its *work space*, and C its *configuration* space. The left picture of Figure 6 shows a 2-joint robot manipulator in its work space, and the right picture of Fig. 6 shows the corresponding configuration space. A point $q \in C$ specifies the position of every point in A with respect to a coordinate system attached to W [ea83]. This point, q, is called a *configuration* of A. All the positions of the points of A specified by $q \in C$ belong to the robot's work space W.

The subset of C consisting of all the configurations where the robot, i.e. its every point, has no contact or does not intersect the obstacles in W is called the *free space* and denoted C_{free}. The complement of C_{free}

consists of obstacles and is denoted C_{obst}. C_{obst} is composed of some distinct subsets. These are called *configuration space obstacles* or, in short, C-obstacles. To determine wether a configuration q is in C_{free} or in C_{obst}, in a robot simulator, some geometric calculations are needed. This is called *collision testing*.

Let d be the dimension of C, i.e., the number of the joints or the degrees-of-freedom of A. The configuration $q \in C$ is represented as a d-dimensional vector $(q_1, q_2, ..., q_d)$. Every q_i has m_i distinct values, so C is a graph of $m_1 * m_2 * ... * m_d$ vectors. Let us call it a *configuration graph*.

The graph has edges that connect each node q with all its neighbors. For example, a neighbor of $(q_1, q_2, ..., q_d)$ is $(q_1, q_2 + \frac{q_{max}}{m_2}, ..., q_d)$ with only one component allowed to be changed. So, if C is d-dimensional, then every node has $2 * d$ neighbors.

All searching for path planning is done in the robot's configuration space. We map every configuration q onto a corresponding search node n. The nodes are vectors $n \in Z^d$, where Z^d is the set of vectors whose components are integers, and d is the dimension of C. For example, the neighboring configurations $q = (q_1, q_2, ..., q_d)$ and $q' = (q_1, q_2 + \frac{q_{max}}{m_2}, ..., q_d)$ are mapped onto $n = (n_1, n_2, ..., n_d)$ and $n' = (n_1, n_2 + 1, ..., n_d)$, respectively. This is called a *basic search graph*. The neighboring relation of this graph is similar to the one that the configuration graph has.

THE A^* ALGORITHM

In brief, A^* is an ordered best-first search algorithm that always examines the successors of the "most promising" node based on the function: $f(n) = g(n) + h(n)$, where $g(n)$ is the length of the shortest path from the starting node to n. The *heuristic*, $h(n)$, is an estimate of the length of the shortest path from n to any goal node.

The A^* algorithm maintains two sets of nodes: a tree of paths already obtained, called CLOSED, and a priority queue containing a subset of the leaves of CLOSED, called OPEN. After removing a node n from OPEN, it is tested if the robot, located at n, contacts or intersects with the obstacles, i.e, the collision test is done. If n belongs to C_{free}, then it is placed on CLOSED.

ON A^* THEORY. Using A^* enables us to get shorter path planning times by constructing "better" heuristics for the search. If we have a better heuristic, then we know that the A^* using it is also "better".

Definition 1. [Pea84] p. 77. A heuristic function h is *admissible* if it underestimates the cost of the optimal solution from a node n to the goal, $h(n)^*$, i.e.

$$h(n) \leq h(n)^* \quad \forall n. \tag{1}$$

Definition 2. [Pea84] p. 83. A heuristic function $h(n)$ is *monotone* if it satisfies the triangle inequality:

$$h(n) \leq c(n, n') + h(n') \quad \forall (n, n') \mid n' \in succ(n), \tag{2}$$

where $c(n, n')$ is the cost of the edge from n to n'.

Theorem 1. [Pea84] pp. 82-83. Monotonicity implies admissibility.

Definition 3. [Pea84] p. 85. An algorithm A_2^* *largely dominates* A_1^* if every node expanded by A_2^* is also expanded by A_1^* except, perhaps, some nodes for which $f(n) = C^*$, where C^* is the cost of the optimal solution.

Theorem 2. [Pea84] p. 85. If $h_2 \geq h_1$ and both are monotone, then A_2^* (using h_2) largely dominates A_1^* (using h_1).

RELAXED MODELS

This section describes one method of generating admissible heuristics, namely utilizing *relaxed models*.

Relaxed models are a well-known source of admissible heuristics [Pea84], [Val84], [MP89], [HMV92] and [PD95]. They are abstract problem descriptions generated by ignoring constraints that are present in base-level problems [HMV92]. The intuitive reason that abstractions generate admissible heuristics is because they add short-cut solution paths by simplifying the original problem [PD95]. Several authors emphasize, however, that relaxed problems should be easily solvable compared with their original counterparts in order to speed up the overall computation time, [Val84], [Pea84], [MP89], [HMV92].

Let a search problem be a three-tuple $\langle S, c, G \rangle$, where S is a set of states; $c : S \times S \mapsto \Re$ is a positive cost function; and $G \subseteq S$ is a set of goal states. A problem *instance* is a problem together with an initial state [PD95]. Formal representation of a relaxed model is given in the next definition.

Definition 4. [PD95]. An *abstracting transformation* $\phi : S \mapsto S'$ removes certain details (e.g. constraints)

from the original problem $\langle S, c, G \rangle$ and produces a *relaxed problem* or a *relaxed model* $\langle S', c', G' \rangle$ iff ϕ reduces all costs and expands all the goals:

$$(\forall s, t \in S) c'(\phi(s), \phi(t)) \leq c(s, t) \tag{3}$$

$$(\forall g \in G) \ \phi(g) \in G', \tag{4}$$

where $c(s, t)$ and $c'(\phi(s), \phi(t))$ are the costs of the shortest paths between the corresponding nodes. Note that an abstracting transformation does not have to be a *function*, in contrast to [PD95]. We will utilize this in the Section HEURISTICS below.

The next theorem shows that heuristics generated by optimizations over relaxed models are monotone and thus admissible.

Theorem 3., [Val84], [Pea84]. Assign every node $n \in S$ of the original problem a heuristic $h(n)$ that is the cost of the optimal solution of a relaxed problem instance with starting node $\phi(n) \in S'$. Then $h(n)$ is monotone $\forall n \in S$.

Proof [Pea84] p. 116: Suppose $h(n)$ and $h(n')$ are the heuristics assigned to nodes $n, n' \in S$, respectively. These heuristics are minimum costs from the corresponding nodes $\phi(n), \phi(n') \in S'$ to a goal $\phi(g) \in G'$. Thus $h(n) \leq c'(\phi(n), \phi(n')) + h(n')$, where $c'(\phi(n), \phi(n'))$ is the relaxed cost, since otherwise $c'(\phi(n), \phi(n')) + h(n')$ would constitute the optimal cost from $\phi(n)$. Monotonicity follows from equation (4): $c(n, n') \leq c'(\phi(n), \phi(n'))$ $\square$

To satisfy the conditions of definition 4, it suffices that s and t are **neighboring nodes** in the original problem S. Finally, it is easy to prove the following theorem:

Theorem 4. If $\langle S'', c'', G'' \rangle$ is a relaxed model of $\langle S', c', G' \rangle$ and $\langle S', c', G' \rangle$ is a relaxed model of $\langle S, c, G \rangle$, then $\langle S'', c'', G'' \rangle$ is a relaxed model of $\langle S, c, G \rangle$.

THE MULTIRESOLUTION IDEA

"BASIC" A^* -VERSION. In the following, $n \in Z^d$ refers to the node whose successors, $succ(n) \in Z^d$, A^* is expanding. In short, let $n' = succ(n)$. In the "basic" A^* -application, all n's form a subset of the immediate neighbors of n (there is an edge connecting n and n'). The costs between the n and n' are: $c(n, n') = 1 \ \forall n, n'$. Figure 1 illustrates a 2-dimensional search graph ($d = 2$) without any obstacles. We will call it a *basic search graph*.

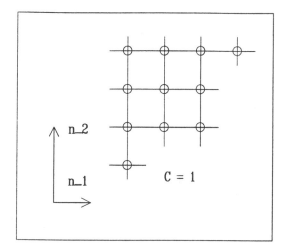

Figure 1: Basic search graph

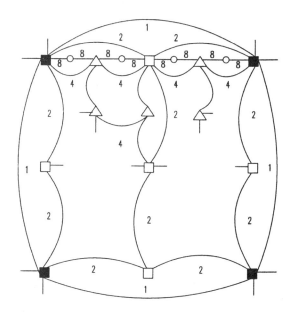

Figure 2: Modified search graph

Figure 3 shows a 2 DOF path planning example produced by the above basic A^* algorithm. In Fig. 3, grey areas are C-obstacles. Thin black lines illustrate the tree of the path candidates expanded by A^*. The path from start to goal is the bold black line. The goal is on the right end of the path. Small black dots are the expanded nodes (that are in CLOSED). In Fig. 3, there are 2716 nodes in CLOSED and 2864 collision tests completed. The collision test is the most time consuming single operation during the search.

This method have been found to produce long searching times for realistic path planning tasks, hours or so, see e.g. [ea96].

MODIFIED SEARCH GRAPH. Figure 2 shows another 2-dimensional graph. We will call it a *modified search graph*. It is similar to the one in Fig. 1 except it has many additional edges between the nodes. In Fig. 2, all the edges are present only in the upmost line of the nodes, for clarity.

In Fig. 2, the nodes $n = (n_1, n_2) \in Z^2$ marked with black squares have all their components, n_1 and n_2, divisible by 8. We say that their m-*value* is 8. The m-value is synonymous with the *resolution* of the node. The nodes with white squares have all their components divisible by 4 and *not* by 8. Their resolution is thus 4, i.e. $m = 4$. Further, all the components of the nodes with white triangles are divisible by 2 and not by 8 and 4 ($m = 2$). Finally, all the components of the white sphere -nodes are divisible by 1 and not by 8, 4, and 2 ($m = 1$).

The costs $c(n, n')$, are *not* measured by $m = | n - n' |$. We calculated them as follows. Let the dimension of the search space be d and the maximum allowed m

-value or resolution be max. In the multiresolution algorithm, m will always be 2^k ($k = 0, 1, ..., max_k$). For example, in Fig. 2, $d = 2$ and $max = 8$. If $m = 8$ for both n and n', then $c(n, n') = d^0 = 1$, see the black square -nodes. If $m = 4$ for both n and n', then $c(n, n') = d^1 = 2$, see the white square -nodes. If $m = 2$ for both n and n', then $c(n, n') = d^2 = 4$, see the white triangle -nodes, etc. The formula for the costs is: $c(n, n') = d^{log_2(max/m)}$.

On the other hand, if n has the resolution m and n' has m' so that $m > m'$, then $c(n, n')$ is determined as if both the nodes had the same m-value: m'. See, for example, the cost of the edges between the black squares and the white triangles, in Fig. 2.

Actually, the modified search graph is a relaxed model of the graph that the multiresolution A^* expands. This is because all the edges between the neighboring nodes are not present, or the corresponding costs are infinite, when the multiresolution algorithm is running. We will discuss this next.

MULTIRESOLUTION A^*. In the multiresolution search, the decision whether a configuration $q_j \in C_{free}$ *will or will not* be mapped onto the search node $n_j \in Z^d$ depends on how close that configuration is to the C-obstacles. See Figure 4 and compare it to **Fig. 3**.

When a configuration q corresponding to the node n is far away from C-obstacles, then its successors n' also are far away from n. These distances are measured by the number of free configurations between n and C-

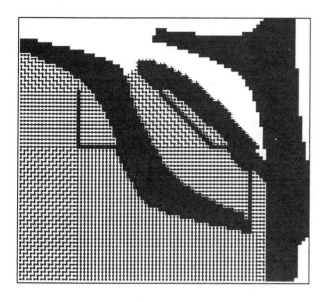

Figure 3: A 2-D search of the basic A^*

Figure 4: A 2-D multiresolution search

obstacle surfaces, or n and n'. This means that only low cost edges of the graph shown in Fig. 2 are present.

Instead, if the node n is near C-obstacle surfaces, then its successors are also near n. Now, also high cost edges shown in Fig. 2 are present. So, the modified search graph, discussed earlier, is a relaxed model of the graph the multiresolution search expands. This is because some edges present in the modified search graph are now missing.

Figure 4 has been produced by the algorithm based on these principles. In Fig. 4, there are 208 nodes in CLOSED and 1197 collision tests completed cf. Fig. 3 (2716 and 2864).

Why are we doing the multiresolution search ? On average, there are some large obstacles in C_{obst} and large connected areas in C_{free}, see e.g. Figures 3, 4 and 6. We want to "delay" the expansion of the nodes near the C-obstacle surfaces. This is done by assigning the corresponding edges higher costs. It follows that if a path exists that goes further away from the C-obstacle surfaces, then it may be found sooner. The second reason is that the collision testing algorithms usually are slower when a node (the configuration) is near the C-obstacle surfaces.

However, the problem that the basic A^* solves is different from the one the multiresolution A^* solves. Thus, these algorithms can only be compared with each other empirically.

A SUCCESSOR GENERATING EXAMPLE.

There are a few quite simple rules to implement the multiresolution search. Their purpose is to produce the successors for the current node n, i.e. to expand n. Then A^* inserts n into CLOSED and the successors into OPEN. All the simulation results have been produced by A^* based on the rules.

The rules guarantee that all the surface nodes of any big C-obstacle will be expanded eventually. On the other hand, the algorithm finds routes that go around small C-obstacles, so their surfaces are not of interest.

Unfortunately, there is no space to fully describe all the rules. More details will be found in [AL97]. We will discuss here a somewhat simplified example of the successor generation. The next pseudo code generates a subset of the successors n' of n. Note the following comments. $n = (n_1, n_2, ..., n_d)$ as before. Let $k = 1, ..., d$ represent the normalized direction vector $(0, .., 1, .., 0)$, where the index of the number "1" is k. The subroutine "PROBE(k, t, n)" returns t that is the number of the free configurations $(q \in C_{free})$ between n and a C-obstacle along the direction k. $t = 0, ..., n.m$, where $n.m$ is the m-value or the resolution of n. In this example, $n.m = 8$. The first subroutine call is: "RULE-1(1, n, n'-set)".

RULE-1(k, n, n'-set):

```
(1) PROBE(k, t, n)
(2) IF (t = 0)  THEN
          n' = (n_1,..,n_j+1,..,n_d)
          for all j <> k such that n'
          is in C_free, determine n'.m;
```

```
(3) IF (t = 1)  THEN
       n' = (n_1,..,n_k+1,..,n_d),
       determine n'.m;
(4) IF (2 <= t < 4) THEN
       n' = (n_1,..,n_k+2,..,n_d),
       determine n'.m;
(5) IF (4 <= t < 8) THEN
       n' = (n_1,..,n_k+4,..,n_d),
       determine n'.m;
(6) RULE-1(k+1, n, n'-set)
(7) include all n' into the n'-set
       if they are not already there;
```

END OF RULE-1

While k increases from 1 to d, there is a variable t_{min} that records the minimum of the t-values. Let the maximum allowed m-value or resolution be max. Then:

```
(8) IF (t_min = max) THEN
(9)     n' = (n_1,..,n_j+max,..,n_d)
          for all j, n'.m = max;
(10)ELSE
(11)  IF (t_min = n.m) THEN
(12)      n' = (n_1,..,n_j+n.m,..,n_d)
            for all j, check n'.m,
            accept n' if n'.m >= n.m;
(13)  IF (t_min < n.m) THEN
(14)      n' = (n_1,..,n_j+n.m,..,n_d)
            for all j, determine n'.m;
(15)include all n' into the n'-set
       if they are not already there;
```

The successors n' that "RULE-1" generates may have m-values that are lower than the one n has. Lines (11) and (12) can be seen as a kind of inverse to "RULE-1": they generate n' such that $n'.m \geq n.m$. The resolution cannot be higher than max (lines (8) and (9)). Finally, the costs $c(n', n)$ are assigned similarly as was done in MODIFIED SEARCH GRAPH.

HEURISTICS

In this section, we will present admissible heuristics for the multiresolution algorithm and use them in the experiments of the next section. First, we will introduce a method of decomposing a d-dimensional problem into several lower dimensional problems that together form a relaxed model of the former one.

CONSTRUCTING ϕ USING PROJECTIONS.
Assume a problem $\langle N, c, G \rangle$ whose nodes $n \in N$ and $g \in G$ are vectors in Z^d. We will construct a relaxed problem $\langle N', c', G' \rangle$ as follows.

We define an abstracting transformation $\phi : N \mapsto N'$:

$$\phi(\pi_j(n)) = \pi_j(n), \qquad (5)$$

where $n \in N$ and $\pi_j(\cdot) \in N'$ ($j = 0, 1, 2, ..., d$). $\pi_j(\cdot)$ is a *projection*. For example, if $d = 2$ then $\pi_1((n_1, n_2)) = (n_1, s_2)$ and $\pi_2((n_1, n_2)) = (s_1, n_2)$ ($s \in Z^d$). Note that ϕ is a function of π_j but *not* necessarily a function of n. If we accept the latter, then we can do the following.

Let us consider a 2-dimensional example. Let a node $n = (n_1, n_2) \in N$, the start node $s = (s_1, s_2) \in N$, and the goal node $g = (g_1, g_2) \in G$. We define $\pi_1((n_1, n_2)) = (n_1, s_2) \in N'$ and $\pi_2((n_1, n_2)) = (s_1, n_2) \in N'$ for all nodes in N. It follows that the start node s remains the same. In particular, $\pi_1(g) = (g_1, s_2) \in N'$ and $\pi_2(g) = (s_1, g_2) \in N'$. The goal node stays where it was, in other words, we define an additional mapping $G' \ni \pi_0(g) = g$, where the argument $g \in G$.

Next, we will define the costs c' between the nodes $\pi_j(n_1)$ and $\pi_j(n_2)$, where n_1 and n_2 are now vectors in N (Z^2). First, if the first coordinate of n_1 and n_2 is the same, say r, then $c'(\pi_1(n_1), \pi_1(n_2)) = 0$, since they have been mapped onto the same node (r, s_2). Second, if $n_1 = (s, v)$ and $n_2 = (t, w)$ so that $t \neq s$, and the nodes are neighbors in N, then $c'(\pi_1(n_1), \pi_1(n_2)) = min\{c(n_1, n_2)\}$. In this paper, also $v = w$.

Similarly, if the second coordinate of n_1 and n_2 is the same, say r, then $c'(\pi_2(n_1), \pi_2(n_2)) = 0$, since they have been mapped onto the same node (s_1, r). If $n_1 = (v, s)$ and $n_2 = (w, t)$ so that $t \neq s$, and the nodes are neighbors in N, then $c'(\pi_2(n_1), \pi_2(n_2)) = min\{c(n_1, n_2)\}$.

Finally, there remains a question how to define the costs $c'_1 = c'(\pi_0(g), (g_1, s_2))$ and $c'_2 = c'(\pi_0(g), (s_1, g_2))$. Let us find the minimum lengths of the routes from s to (g_1, s_2) and (s_1, g_2) by searching and call them t_2 and t_1, respectively. Then we assign $c'_1 = t_1$ and $c'_2 = t_2$.

We have now constructed a transformation ϕ that satisfies definition 4 and, as a result, we have a relaxed model $\langle N', c', G' \rangle$ of $\langle N, c, G \rangle$. The interpretation is that we have decomposed a two-dimensional problem $\langle N, c, G \rangle$ into two one-dimensional subproblems called together $\langle N', c', G' \rangle$. Solving them for a particular node $n' \in N'$ essentially needs two one-dimensional searches and is computationally effective.

The above argument that we call here the **projection principle** can easily be extended to higher dimensional search spaces.

We calculate the heuristic for $\langle N, c, G \rangle$ by by summing the pathlengths obtained by solving the subproblems. This is the same as solving the relaxed problem $\langle N', c', G' \rangle$. It follows from theorem 3 that the heuristic is monotone and thus admissible.

However, we must be sure that the optimum paths for the subproblems *can be found*. This usually means that the problem $\langle N, c, G \rangle$ *already is* a relaxed model of the original problem, constructed in a proper way. We will discuss this next.

MANHATTAN-MAXRES.

This heuristic is like the Manhattan distance from a node n to the goal that is most often used in conjunction with the basic A^*.

Suppose the robot manipulator has d degrees-of-freedom (DOF). The **Manhattan distance** between a node n and the goal node g is:

$$\|n - g\|_{L^1} = \sum_{i=1}^{d} | n_i - g_i |, \qquad (6)$$

where $n, g \in Z^d$, i.e., n_is and g_is are integers. We cannot use the Manhattan distance directly because the distance between a node and its successor depends on on the node's resolution (the step size). Therefore we must use the maximum resolution value, max, see MODIFIED SEARCH GRAPH.

This is done as follows. First, every $| n_i - g_i |$ -term is divided by max. Second, if there is a non-zero reminder, then 1 is added to the result. For example, if $max = 4$ and $| q_i - g_i | = 10$, then $h_i = 2 + 1 = 3$. The **Manhattan-maxres** -heuristic is the sum of these h_i-values: $h(n) = \sum_{i=1}^{d} h_i$. One can argue that when max is high then $h(n)$ may not be very effective because the actual path lengths may be very long compared to it.

Using the Manhattan distance as a heuristics corresponds to solving a relaxed problem of the base level problem $\langle S, c, G \rangle$: all the C-obstacles have been ignored. Let us call it $\langle S', c', G' \rangle$ that now equals $\langle N, c, G \rangle$ discussed above.

Calculating the Manhattan distance corresponds to solving d one-dimensional subproblems called together $\langle N', c', G' \rangle$. They can be constructed from $\langle N, c, G \rangle$ by using the projection principle (in d dimensions). It follows from theorems 3 and 4 that the Manhattan distance heuristic is admissible.

MANHATTAN-MULTIRES.

The Manhattan-maxres heuristic may be very uninformative if the maximum resolution max is high. To utilize more

our multiresolution algorithm, we will construct a better heuristic. Consider a d-dimensional version of the *modified search graph* in Fig. 2. Remember, it is a relaxed model of the graph that the multiresolution algorithm expands during the search. This is since all the obstacles have been ignored. So, we call it $\langle S', c', G' \rangle$ that equals $\langle N, c, G \rangle$.

Next, we apply the projection principle to it and obtain d one-dimensional subproblems called together $\langle N', c', G' \rangle$. We get the **Manhattan-multires** heuristic by solving these separately and summing the optimum path lengths. Again, we conclude from theorems 3 and 4 that the heuristic is admissible.

However, solving each one-dimensional subproblem needs one-dimensional searching, in contrast to the Manhattan-maxres heuristic. This searching is done by the multiresolution algorithm. The algorithm always finds the optimum path because there are no obstacles involved. The Manhattan-multires heuristic is always bigger or as big as the Manhattan-maxres for every node and thus is better, see ON A^* THEORY.

K-DOF-MULTIRES.

Consider again the d-dimensional version of the *modified search graph* in Fig. 2. We decompose it into two subspaces: the first one is k-dimensional ($k \leq d$) and the second one is $d - k$-dimensional. The first subspace corresponds to the robot's k first degrees-of-freedom (DOFs) and the second one the $d - k$ rest DOFs.

First, we ignore the obstacles in the second subspace and calculate the Manhattan-multires heuristic in it, as was done previously. Let us call this subspace Z_2^{d-k}. Suppose $Z^d \ni n = (n^k, n^{d-k})$, where $n^{d-k} \in Z_2^{d-k}$. Then, let $h(n^{d-k})$ denote the Manhattan multires-heuristic for n^{d-k}.

Second, we preserve the obstacles in the first subspace but ignore the $d - k$ last degrees of freedom of the robot operating there. If we have, e.g. a usual 5- or 6-DOF industrial robot, then we ignore the 2 or 3 last DOFs, respectively. As a result, we have a robot manipulator whose wrist is "amputated".

Third, we simplify the "amputated" robot by modeling it only by a couple of polygons that are inside the volume covered by the original robot. The original robot model usually consists of tens of polygons. This makes the collision tests more effective. Let us call this simplified robot model a *reduced robot*.

Fourth, we do collision tests by using the reduced robot. Some configurations that originally are in C_{obst}, by using the original robot, may now be in C_{free}. This is because the volume covered by the reduced robot is a subset of the volume covered by the original robot. It means that some costs between the neighboring nodes decrease from infinity to a small number. Also, since we only have a k-DOF robot, the optimum path can be found by searching in the first subspace. Let us call this subspace Z_1^k. Suppose $Z^d \ni n = (n^k, n^{d-k})$, where $n^k \in Z_1^k$. Then, let $h(n^k)$ denote the heuristic for n^k.

Obviously, the space $Z_1^k \times Z_2^{d-k}$ is a relaxed model of the modified search graph. The heuristic, called the **k-DOF-multires** is: $h(n) = h(n^k) + h(n^{d-k})$. We can now apply the projection principle to show that this heuristic is admissible.

Let $Z_1^k \times Z_2^{d-k}$ correspond to $\langle N, c, G \rangle$. To construct a relaxed model $\langle N', c', G' \rangle$, we define the projections $\pi_j(\cdot)$ $j = 0, 1, ..., d - k + 1$ as follows. For example, let $k = 3$ and $d = 5$. Let a node $n = (n_1, n_2, ..., n_5) \in N$, the start node $s = (s_1, s_2, ..., s_5) \in N$ and the goal node $g = (g_1, g_2, ..., g_5) \in G$. First, $G' \ni \pi_0(g) = g$, as before. Second, $\pi_1(n) = (n_1, n_2, n_3, s_4, s_5) \in N'$. Third, $\pi_2(n) = (s_1, s_2, s_3, n_4, s_5) \in N'$ and $\pi_3(n) = (s_1, s_2, s_3, s_4, n_5) \in N'$. $\pi_1(n)$ projects every n onto the 3-dimensional subspace Z_1^3. $\pi_2(n)$ and $\pi_3(n)$ project every n onto one-dimensional subspaces that together form Z_2^{5-3}. The costs are defined analogously to the projection principle example, see HEURISTICS.

So, we have constructed $\langle N', c', G' \rangle$, a relaxed model of $\langle N, c, G \rangle$ that further is a relaxed model of the modified search graph. It follows from theorems 3 and 4 that the k-DOF-multires -heuristic is admissible. The k-DOF-multires heuristic is always bigger or as big as the Manhattan-multires for every node and thus is better than it.

Actually, we calculate the $h(n^3)$ -part of the heuristic for every n by the breadth-first search in Z_1^3 starting from the "projected goal" $(g_1, g_2, g_3, -, -)$. The search is done by our multiresolution algorithm in 3 dimensions. The calculations were done off-line i.e. before starting the multiresolution search of the original problem. The overhead time for doing this is about 1.5 minutes ($k = 3$).

Utilizing the k first DOFs of the robot to guide the search process is not new in robotics, see e.g. [Lat91]. However, we have applied it in conjunction of our multiresolution A^* algorithm and shown that the resulting heuristic is admissible.

EXPERIMENTAL RESULTS

ROBOT TEST CELL. Figure 5 shows the robot test cell. It has been adapted from [HA92]. The actual dimensions may differ from those in the reference.

There is a 5-DOF Adept industrial robot with a L-shaped object attached to its gripper. The left picture shows the starting configuration and the right picture the goal.

The robot has to bend its wrist to get the L-shaped object out of the wicket. Then it has to lift its linear axis and to rotate its first and second link to avoid the middle block, see Figure 7.

SIMULATIONS. We record the program execution times, the number of the collision tests and the number of the nodes in CLOSED and OPEN. The times reported are CPU times in seconds on a SGI Indigo 2 computer with 150 Mhz R4400 processor and 96 megabytes of main memory. The collision test was done using a software package called *RAPID* (Rapid and Accurate Polygon Interference Detection) [et.96]. The collision test is the most expensive single operation in the search process. The discretizations for the Adept robot are 150x100x40x100x40 corresponding about 2 cm maximum movement of the robot (in W).

We have used four algorithm versions in the tests. The first one is the basic A^*. It has a heuristic that is similar to the 3-DOF-multires heuristic. The basic A^* was very slow: we had to stop the searching since the main memory was full. This happened after about 1.5 hours, 970000 collision tests, 590000 nodes in CLOSED, and 270000 nodes in OPEN.

The second version is the new multiresolution A^* algorithm. It has three variants corresponding to the three heuristics presented. The first one uses the Manhattan-maxres heuristic (**algorithm 2.1**). The next one uses the Manhattan-multires heuristic (**algorithm 2.2**). The third one uses the 3-DOF-multires heuristic (**algorithm 2.3**). The off-line calculation time for the 3-DOF-multires heuristics was about 1.5 minutes. This is not included in the results since these calculations can be utilized for many different path planning problems in the same robot cell.

Tables 1-4 show the simulation results. The first column of the tables shows the algorithm version, the second one the planning times in seconds, the third one the number of the collision tests, the fourth one the nodes in CLOSED, and the fifth one the nodes in OPEN.

CONCLUSIONS

Tables 1-4 show that the new multiresolution idea, embedded in the structure of A^* algorithm, clearly outperformed the basic A^* search that had to be stopped during the search.

	TIME	COLL.	CLOSED	OPEN
2.1	368	108404	3631	16824
2.2	296	88126	2231	8998
2.3	236	71586	1781	7204

Table 1: Results (max=8)

	TIME	COLL.	CLOSED	OPEN
2.1	135	37343	1536	8824
2.2	56	16288	400	2817
2.3	32	9355	283	2165

Table 2: Results (max=16)

	TIME	COLL.	CLOSED	OPEN
2.1	485	125459	2927	14963
2.2	189	55752	906	4688
2.3	186	55192	876	4460

Table 3: Results (max=32)

	TIME	COLL.	CLOSED	OPEN
2.1	560	150938	3333	16847
2.2	232	58028	948	4865
2.3	197	57468	918	4637

Table 4: Results (max=64)

The searching times are comparable to the measured computing times, and the number of the collision tests. The amount of the used memory, on the other hand, is comparable to the sum of the nodes in OPEN and CLOSED.

In this simulation example, the best searching times did not vary very much as a function of the *max* -value, the maximum allowed resolution, see the last lines of the Tables 1, 3, and 4. An exception is the Table 2 ($max = 16$) that shows the shortest planning times of all.

The Manhattan-maxres -heuristic produced searching times that were about 2.5 times the ones produced by the Manhattan-multires. Table 1 is an exception ($max = 8$). The amounts of the used memory with the Manhattan-maxres were about 3 times the amounts got by using the Manhattan-multires. Again, Table 1 is an exception (1.8 times).

The 3-DOF -multires -heuristic was not much better than the Manhattan-multires. The latter produced, on average, 1.2 times longer searching times than the former. Table 2 is an exception (1.7 times).

On the whole, the searching times of the multiresolution algorithm seem to be reasonably short for on-line path planning in low dimensional robot configuration spaces, e.g. $d = 5, 6$.

FUTURE WORK. We are now studying if the heuristics obtained by sampling the search space with a coarse resolution can be shown admissible.

ACKNOWLEDGEMENTS:

This work was supported by the Academy of Finland. We also thank the anonymous referees for their comments.

References

[AL97]   Antti Autere and Johannes Lehtinen. Robot motion planning by a^* search on a modified discretized configuration space. In *IEEE/RSJ Int. Conf. on Intelligent Robots and Systems (IROS'97) (submitted to)*, September 8-12 1997.

[BH93]   Michael Barbehenn and Seth Hutchinson. Efficient search and hierarchical motion planning using dynamic single-source shortest paths trees. In *IEEE International Conference on Robotics and Automation*, May 1993.

[BH95]   Michael Barbehenn and Seth Hutchinson. Efficient search and hierarchical motion planning by dynamically maintaining single-source shortest paths trees. *IEEE Trans. Rob. Autom.*, 11(2):198–214, 1995.

[BL91]   Jerome Barraquad and Jean-Claude Latombe. Robot motion planning: Distributed representation ap-

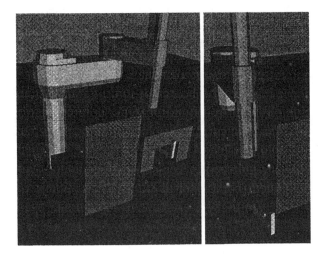

Figure 5: Robot test cell

proach. *The International Journal of Robotic Research*, 10(6):628–649, 1991.

[Cla90]   Bernhard Clavina. Solving findpath by combination of goal-directed and randomized search. In *IEEE International Conference on Robotics and Automation*, 1990.

[Con94]   Christopher I. Conolly. Harmonic functions and collision probabilities. In *IEEE International Conference on Robotics and Automation*, Vol.4 May 1994.

[ea83]   Tomas Lozano-Peres et. al. Spatial planning: A configuration space approach. *IEEE Trans. Computers C-32(2):108-120*, 1983.

[ea86]   Tomas Lozano-Peres et. al. A simple motion planning algorithm for general robot manipulators. In *5th AAAI, Philadelphia*, 1986.

[ea96]   Antti Autere et. al. Robot motion planning by enhanced *a** algorithm. In *27th International Symposium on Industrial Robots (ISIR)*, October 6-8 1996.

[MP89]   Jack Mostow and Armand E. Prieditis. Discovering admissible heuristics by abstracting and optimizing: A transformational approach. In *IJCAI-89*, pages 701–707, 1989.

[PD95]   Armand Prieditis and Robert Davis. Quantitatively relating abstractness to the accuracy of admissible heuristics. *Artificial Intelligence*, 74:165–175, 1995.

[Pea84]   Judea Pearl. *Heuristics: Intelligent Search Strategies for Computer Problem Solving*. Addison-Wesley, 1984.

[Val84]   Marco Valtorta. A result on the computational complexity of heuristic estimates for the *a** algorithm. *Information Sciences*, 34:47–59, 1984.

[War93]   Charles W. Warren. Fast path planning using modified *a** method. In *IEEE International Conference on Robotics and Automation*, May 1993.

[et.96]   S. Gottschalk et.al. Obbtree: A hierarchical structure for rapid interference detection. In *Computer Graphics Proceedings, SIGGRAPH 96*, August 4-9 1996.

[HA92]   Y. K. Hwang and N. Ahuja. Gross motion planning - a survey. *ACM Computing Surveys*, 24(3):219–284, 1992.

[HMV92]   Othar Hansson, Andrew Mayer, and Marco Valtorta. A new result on the complexity of heuristic estimates for the *a** algorithm. *Artificial Intelligence*, 55:129–143, 1992.

[KL94a]   Lydia Kavraki and Jean-Claude Latombe. Randomized preprocessing of configuration space for path planning: articulated robots. In *IEEE/RSJ/GI International Conference on Intelligent Robotics and Systems (IROS'94)*, Vol.3 September 1994.

[KL94b]   Lydia Kavraki and Jean-Claude Latombe. Randomized preprosessing of configuration space for fast path planning. In *IEEE International Conference on Robotics and Automation*, Vol.3 May 1994.

[Lat91]   Jean-Claude Latombe. *Robot Motion Planning*. Kluwer Academic Publishers, 1991.

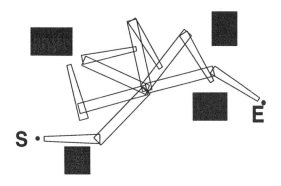

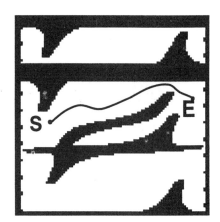

Figure 6: A 2-D path planning example

Figure 7: The resulting path

ENFORCING SOCIAL LAWS
IN
GOAL DIRECTED COMMUNICATION

Manas Ranjan Patra and Hrushikesha Mohanty
Artificial Intelligence Laboratory,
University of Hyderabad, Hyderabad, India, 500 046
Email: hmcs@uohyd.ernet.in

Abstract

Multi-Agent systems are virtual representations of social systems consisting of coarse grained intelligent entities, called *agents*. The motivation behind considering such paradigms is to emulate social behaviour among agents. Since an agent may not have the capabilities to accomplish all its tasks, it seeks cooperation of other agents in its society. Such goal directed cooperative efforts require communication among agents. In this paper, we present a scheme to manage communication in a large society of agents where there could be heavy communication traffic load due to all possible interactions among agents. The purpose of our communication scheme is two fold; (1) to manage communication traffic load by employing an agent partitioning strategy, (2) to maintain a behavioral discipline among agents by enforcing certain social laws during communication. Enforcement of social laws also ensures cooperative attitude among agents. The applicability of the proposed communication scheme is shown by taking an example from corporate sector.

Key words : Agent society, communication, agent partitioning, social laws

1. Introduction

Of late, intrinsic limitations of an intelligent agent have been overcome by considering a class of Decentralized Artificial Intelligence (DzAI) systems, called Multi-agent Systems (MAS) [DEMAZEAU 90]. Simulation of agent societies have attracted DzAI research community to understand key concepts underlying solution of complex problems that can not be solved in isolation [ELLIOTT 93, GASSER 93, MATSUBAYASHI 93]. Issues like multi-agent problem solving [DURFEE 88], resource allocation [FONSECA 96], and planning [GEORGEFF 88, MARTIAL 90, TADEPALLI 92] have been studied by drawing experience from human societies. Communication is a natural means to realize agent interactions. A high-level communication structure that is necessary for coordination of complex social activities has been investigated in [WERNER 88]. Contract net [SMITH 80] also provides a high-level communication structure for solving one's problem through a set of communications namely, task announcement, bidding, award of task and result communication. The problem solving strategy in Contract net takes the participation of an agent for granted, in the sense that an agent bids for a task if it fulfills the requirements for accomplishing a task. Here, there is no mention of whether an agent can exercise its free will or not in deciding its participation in the problem solving process. Among other factors, participation of an agent in any group activity would largely depend on one's free will and inter-agent relationship akin to a social paradigm. But allowing agents to act totally according to their free will may lead to chaos. Thus in order to control such chaos in agent societies one should enforce certain social laws. The notion of social law has been used in [SHOHAM 92] to avoid conflicts in case of positive collaboration but it does not provide a method to deal with agents who do not abide by social laws.

In this paper, we present a scheme to manage communication in a society consisting of a large number of agents with varied service providing capabilities. We propose to enforce certain social laws in the agent society in order to ensure behavioral discipline among communicating agents. The agent society is partitioned into different groups in order to manage massive communication traffic load.

The rest of the paper is organized as follows: We begin by providing a scenario that motivated us to develop such an agent communication scheme. In section 3, we explain a macro structure of an agent society and provide the motivation for using social laws in agent communication. In section 4 we outline the agent structure required for the purpose. The proposed communication scheme, taking into account social laws has been algorithmically dealt in section 5. In section 6, we provide an analysis of our communication scheme. Section 7 summarizes our work.

2. A scenario

Procurement of items is one of the key activities of any organization, which is accomplished according to some organizational

procedures. This activity requires communication within the organization as well as with vendors. While making quotation calls many allied factors other than the normal office procedures also contribute to the activity, viz., relationship of customer (the department /organization) with vendors, past performance of vendors etc. In a typical situation when a department intends to make certain purchases; it seeks for a list of approved vendors, from relevant authority in the organization, who may be contacted for supply of required items. The department then calls for quotations from individual vendors to send their offers. Next the department does the vendor selection for placing the final purchase order based on some predefined criteria. The entire procurement procedure generally involves communications within the organization (among relevant departments like purchase, finance, stores etc.) as well as with entities external to the organization (i.e. vendors). The present work is motivated towards managing such a complex communication scenario.

3. Agent Society : A Macro Structure

The conglomeration of several agents can be viewed as an agent society whose functionalities can be studied in the light of activities that take place in a human society. Agents try to achieve goals which may be self-centered or social through cooperation, negotiation and collaboration. Below, we discuss elements of an agent society.

3.1 Agent partitioning

Management of communication traffic in a large society of agents has been a major concern for DzAI researchers. This is because with the increase in the number of agents, the communication overhead also increases dramatically resulting in a poor scalability. In order to overcome this problem one may adopt an agent partitioning strategy so as to partition a large agent society into smaller groups of agents on the following basis :

Fixing the group size

The size of each group can be predefined or user can have the flexibility to choose the size of each group depending on the requirement.

Geographical location

Agents can be grouped depending on their geographical locations i.e., agents within a geographical distance limit are grouped together.

Heterogeneous grouping

It is basically a pool of different kinds of resources necessary for solving specific problems. In this grouping technique, agents with varied capabilities required to deal with a specific task are grouped together. Thus, each group is self-contained as far as the solution of a particular type of problem is concerned.

Homogenous grouping

It refers to a pool of similar resources. Here, agents with similar capabilities are grouped together and are identified by the tasks they can perform. Such a grouping technique eliminates the need to store capability information of similar agents redundantly. Thus any agent seeking help for a particular task can contact the group possessing the desired capability.

In our work, we consider assemblies of homogenous agents. After dividing an agent society into smaller groups of agents, we designate one of the group members in each group as *facilitator*. In the example scenario, we have two broad categories of agents, viz., service requesting agents (representing departments) and service providing agents (representing vendors). Further, we group vendors into different categories depending on the items they are capable of supplying e.g., stationery, equipment, raw materials etc. We employ a *facilitator* agent with each group of

vendors such that a customer contacts only the facilitator agent which in turn broadcasts the purchase requirements within its group invoking individual offers. This information is later passed onto the requesting department in the form of quotations. Next the department does the vendor selection for placing the final purchase order based on some predefined criteria. We also associate a *facilitator* agent with service requesting agents. This facilitator represents the authority in the organization which maintains up-to-date information of vendors approved by the organization for departmental purchases. The motivation behind introducing such a facilitator is to eliminate redundant storage of same vendor information with all departments, and thus ensuring consistency.

3.2 The Communication pattern

Multi-agent systems are essentially based on concurrent object-based systems where the key attributes are objects (or agents) and message passings [FISHER 94]. Though many of the concurrent object-oriented systems employ point-to-point message communication but it restricts the power of multi-agent systems due to the following reasons.
- an agent must know specifically the address of the receiving agent.
- lack of naturalness because an agent may not always know in advance the address of an agent who could possibly provide necessary service.

A more natural way of requesting for service is that an agent may broadcast its requirements and in response to that potential and desirous servicing agents may express their willingness to serve. In spite of the fact that broadcast strategy is quite a natural way of communication but its use is not encouraged in multi-agent systems due to perceived difficulties in its implementation. We overcome such a problem by adopting group communication strategy which handles intra-group communication through broadcast and inter-group communication through multicast technique.

In our scheme, an agent requesting for service on a goal context sends a request message to its own group facilitator for providing information regarding agents who can provide services that helps an agent to achieve its goal. The facilitator in turn broadcasts the request message to its group members in order to find agent(s) who may provide the desired service. Failing this, the facilitator may multicast its request to facilitators of other service providing groups with a view to find agents capable and willing to provide desired service(s). This information is fed back to the service requesting agent which selects a set of agents to whom it can make service request. Selection of participating agents on a goal context is guided by two criteria: approachability and serviceability. An agent A is approachable for service by agent B if the later knows that the former agent can provide the required service; and in addition, agent B knows that it has not defaulted to a service from agent A which is a consequential service to a service provided by agent A. Similarly, an agent on receiving a service request, evaluates whether a service should be provided or not based on the serviceability of an agent. An agent A is considered to be serviceable for a service S by agent B, if A has not defaulted earlier to the service requirement S' of agent B where S' is consequent to service S. The concepts can be explained from the scenario described in section 2. An agent A asks for quotation on a particular item (s) to a vending agent B if agent A knows that B supplies items required by A and also the case that A has made the payments for earlier supplies. Here we say that agent B is approachable to agent A in the context of service S. On the other hand B provides service to A if A has not defaulted in payment for earlier supplies made by B. The pattern of communication has been shown in figure 1 (minute details are omitted for clarity of the figure).

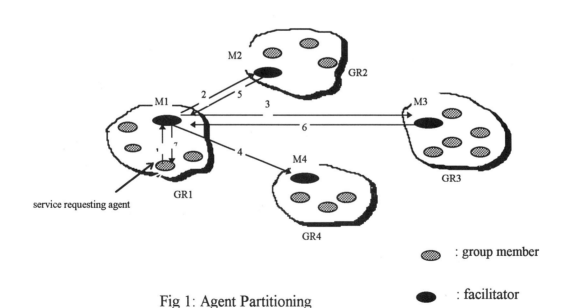

Fig 1: Agent Partitioning

⬚ : group member

⬤ : facilitator

3.3 Enforcement of Social law

As evident from above discussions, communication is central to multi-agent systems without which a group of agents constituting a society would degenerate into a set of individual agents with chaotic behaviour. Thus imposition of certain social laws during agent communications is essential for ensuring productive interactions among agents either for achieving a single global goal or separate but interdependent individual goals. Society has a law and ideally agents in a society are law abiding and when an agent does not abide by the law it has to be punished in the sense that it may not receive expected behaviour from other agents.

In our scheme, social law embodies certain rules which are necessary to ensure cooperative behavior among agents during problem solving. When agents abide by certain agreed upon laws, according to one's own capability an agent can provide service to other agents in its society and can also benefit from others whenever necessary. We take care of both the aspects, viz., when agents abide by social laws and when they do not. Communication necessary during cooperative problem solving is influenced by social laws. In our work, an agent's communication faculty depends on its relationship with its intended recipient. Our mechanism uses the law to constrain the strategy of sharing so as to ensure cooperative attitude among agents.

Below, we provide certain social laws to be followed in an agent society during problem solving which in turn effect an agent's communication strategy.

Law of collaboration : An agent seeking a service normally collaborates with another agent endowed with the desired service providing capabilities.

e.g. A customer trying to procure items contacts a vendor who is capable of supplying the desired items.

Law of correspondence : Normally, an agent provides service to one from which it receives service in a consequential transaction.

e.g. If a customer has received service from a vendor then it takes care that payment is made for the purchased items, and it includes the vendor's name in the list of vendors from whom quotation may be asked in future.

Law of relation preservation : If an agent can not provide service to one from which it has received service earlier, due to some genuine reasons (like overload, nonavailability of resource etc.) then it must express its inability by communicating appropriate message, thereby taking care that the relationship is not stained.

e.g., If a vendor cannot satisfy the requirements of a customer then it expresses its genuine reasons (say, the time constraint is not acceptable/inability to meet desired specification for the item) for not being able to quote for the intended item.

Law of commitment honouring : Once an agent commits to provide certain service to another agent, the service commitment should be honoured. Dishonoring of a service commitment may lead to negative relationship among agents.

e.g., Once a vendor submits its offer it must adhere to the terms and conditions of the offer by all means.

Law of reactive behaviour : An agent may excommunicate another agent by removing it from its servicing list, if the latter repeatedly dishonors service commitments.

e.g., If a vendor fails to supply ordered items within a stipulated time or does not adhere to the committed specification of items or violates the terms and conditions of the deal in any manner then a customer organization has every right to blacklist the vendor for all future deals.

Law of relation revival : Either of the agents, those having stained relationship between them, may decide to revive positive relationship for collaboration.

e.g., A vendor once blacklisted by a customer may try to revive its relationship in some manner. Similarly, a customer may also try to revive relationship with a blacklisted vendor from its side either deliberately or if the situation demands (say, when customer knows that the required item(s) can only be supplied by the vendor once blacklisted).

4. Agent : A Micro Structure

An agent constitute the fundamental entity of an agent society. In this section, we formalize the agent knowledge required to participate in problem solving. An agent's world consists of self knowledge and neighborhood knowledge i.e. knowledge of other agents in the society, as described below:

```
Agent = {
          self identity          : agent_no.
          group identity         : group_no.
          group facilitator      : facilitator_no.
          Own capabilities       : < skills >
          Commitments            : < services being
                        provided, pending services >
          Memory                 : < past events >
          Social laws            : < rules >
          Society                : < relations >
          }
```

Similarly, the knowledge base of a facilitator can be described as below :

```
Facilitator  = {
          self identity          : facilitator_no.
          group identity         : group_no.
```

```
          group members  : < list of agents
                        belonging to its group >
          facilitator list   : < list of other group-
                        facilitators in society >
          group capabilities : < capabilities of
                        other groups in society >
          }
```

5. Agent Communication Scheme

In an effort to achieve a goal, an agent may solicit service(s) from other agents. From the point of time when an agent requests for service till it actually receives the desired service, the entire process involves several activities like,

- participant selection
- service commitment
- commitment evaluation
- societal relation maintenance

Further, when the agents engage themselves in cooperative efforts they communicate with each other at various stages. Below, we algorithmically explain the associated communications :

Algorithm :

Activities at the Service Requesting Agent **(SRA):**

A1. **SRA** needing some service sends a message to facilitator to provide a list of servicing agent (called, **ServList**) who may be contacted for the required service

A2. On receiving the **ServList**, **SRA** evaluates its *approachability* to agents listed in **ServList** based on some criteria evolved from past interactions with the servicing agents

A3. **SRA** sends messages to those agents which succeeded the approachability test in step A2 (this constitutes an **AppchList**), asking for their service offers (e.g., quotation calls)

A4. **SRA** selects its servicing agent (**SA**) from among the list of agents in **AppchList**

A5. Send formal message to **SA** to provide service (e.g., to make the supply of intended items)

A6. Waits till requested service is received
If service is received within the stipulated time then goto step A7
else add **SA** to the commitment dishonoured service list, termed as **ComDisList**

A7. On receiving desired service
provide consequential service to **SA** (e.g., make payment on receipt of items)

Activities at the facilitator of **SRA** :

B1. Facilitator refers to a directory of servicing agents, called **DSA**

B2. Sends a **ServList** (prepared out of **DSA**) to **SRA**

Activities at facilitator of servicing agents (**SAs**) :

C1. Receive service request from **SRA**

C2. Send messages to servicing agents of own group to provide offers

C3. Send servicing agent ids to **SRA**

Activities at Servicing Agents (**SA**) :

D1. On receipt of a service request (e.g., on receiving a quotation call)
SA evaluates *serviceability* of agent **SRA** based on criteria evolved from past interaction with **SRA** and ability to satisfy the service requirements of **SRA** (e.g., required specification of item(s), time constraint, availability of items etc.).

D2. If the *serviceability* test of step D1 succeeds then **SA** sends its service offer to **SRA** and

enters this offer in its service commitment list (**ComList**)
else sends a message to **SRA** expressing its inability to provide the desired service

D3. On receipt of a formal request to actually provide service (e.g., on receipt of a supply order)
SA Checks whether it is a committed service, if yes then **SA** provides service to **SRA** else **SA** may not provide service

D4. After providing a service, **SA** records whether consequential service to its service is received from **SRA** or not (e.g., payment for supplied items is received or not)

As explained in the above procedure, the communication faculty of an agent is equipped with necessary functionalities to act according to the laid down laws (as described in sec. 3.3). To start with each agent considers every other agent as approachable and serviceable. As agents interact with each other on different context they follow the social laws and accordingly communicate messages. An agent may commit to provide service considering its load and availability of resources. But once service commitment is made, the committing agent should honour it by executing the task; failure of which may eventually lead to excommunication. However, an agent may revive positive relationship with an agent through appropriate message communication. We have simulated our communication management strategy in an agent society using Borland C++ under MS-Windows [PETZOLD 92, BOOCH 92]. The implementation details can be seen in [RAVI 96].

6. Communication Analysis

The proposed communication scheme is analyzed on the basis of the following lemmas and theorems.

Theorem 1 : *An agent A collaborates with another agent B, if both of them satisfy each others service requirements.*

Proof : At the service requesting agent,

In step A2 agent SRA takes a decision whether to collaborate with agent SA or not.

If the test at step A2 succeeds then SRA communicates its intention to collaborate with SA

Similarly, at the servicing agent,

In step D1, SA decides whether to collaborate with SRA or not, in providing the required service.

If the test at D1 succeeds then in step D2, SA communicates its service offer.

Hence theorem 1.

Lemma 1: *Agent A reciprocates service to agent B, if A has received service from B in a consequential transaction.*

Proof : The proof of this lemma is evident from step A7 where agent SRA provides service to SA as a consequence to the service received from SA.

Lemma 2 : *Agent A on failing to provide a requested service explains its inability explicitly.*

Proof : The proof of this lemma is evident from step D2.

Theorem 2 follows from the above lemmas.

Theorem 2: *An agent who receives service from another agent has to reciprocate service or should explain its inability on failing to provide service.*

Proof : In the else ... part of step D2, agent SA expresses its inability to provide service and in step D3 SA provides requested service to SRA. This proves theorem 2.

Lemma 3 : *By all means an agent A should provide service P, committed to an agent B.*

Proof : The proof of this is evident from steps D2 and D3.

Lemma 4 : *An agent A can excommunicate agent B, if the latter dishonors a committed service.*

Proof : The proof of this is evident from the else part of step A6.

Lemma 3 and lemma 4 leads to theorem 3.

Theorem 3 : *An agent that makes a service commitment must provide the committed service, failure of which eventually leads to excommunication.*

Proof : In step D2, agent SA makes the service commitment and in step D3, agent SA provides the committed service. If step D3 is not executed by agent SA then in step A6, agent SRA adds SA into the ComDisList showing its reactive behaviour.

Theorem 4 : *An agent A may decide to revive positive relationship with agent B, if A expects some benefit from B or A's request for help is persistently dishonoured.*

This is a deliberative action of any agent where an agent may offer itself to collaborate with another agent inspite of earlier stained relationships. For example, if a service requesting agent (SRA) has blacklisted an agent (SA) because the latter has once dishonoured a service commitment, even in this case the former may choose to ask for service from the latter. Similarly, a vendor (SA) may choose to provide service (say, supply ordered items) even if it has not received its consequential service from the service requesting agent (say, payment is not made for earlier supplies).

7. Conclusion

The motivation behind the work reported here was to study the communication involved in goal directed problem solving in an agent society where agents are potentially preexisting with their respective capabilities. We have proposed a communication scheme that works towards the management of heavy communication traffic load prevalent in a large society of agents. An agent partitioning strategy has been adopted in order to manage communication effectively. We have enforced certain social laws in order to ensure behaviour discipline among communicating agents.

Acknowledgment

We thankfully acknowledge the programming support rendered by M.Ravi and P.S.Viswaprasad for testing of our communication scheme.

References

[BOOCH 92] Booch, G., Object-Oriented Design withApplications,Benjamin/Cummings Pub.,1992.

[DEMAZEAU 90] Demazeau Y., Muller J.P., Decentralized A.I., Proc. of the First European workshop on Modeling Autonomous Agents in a Multi-Agent World, Elsevier Sc. Pub, 1990.

[DURFEE 88] Durfee, E.H., Lesser, V.R., Corkill, D.D., Coherent Cooperation Among Communicating Problem Solvers, In Readings in Distributed Artificial Intelligence, Bond, A.H., Gasser, L., (eds), Morgan Kaufmann Pubs, Inc., 1988, pp 268-284.

[ELLIOTT 93] Elliott C., Using the Affective Reasoner to Support Social Simulation, Proc. Int. Joint Conf. on AI. 1993, pp 194-200.

[FISHER 94] Fisher M., Representing and Executing Agent-Based Systems, Lecture notes on AI, LNAI-890, Springer Verlag , 1994, pp 307 - 323.

[FONSECA 96] Fonseca, J.M., Oliveira, E., Garcao, A.S., A DAI Based Resource Management System, Proc. 1st Int. Conf. on the Practical Application of Intelligent Agents and Multi-Agent Technology, 1996, pp 263-277.

[GASSER 93] Gasser L., Social Knowledge and Social Action : Heterogeneity in Practice, Proc. Int. Joint Conf. on AI, 1993, pp 751-757.

[GEORGEFF 88] Georgeff, M., Communication and Interaction in Multi-agent Planning, In Readings in Distributed Artificial Intelligence, Bond, A.H., Gasser, L., (eds),

Morgan Kaufmann Pubs, Inc., 1988, pp 200-204.

[MARTIAL 90] Martial, F.V., Interactions Among Autonomous Planning Agents,In Decentralized A.I., proc. of the first European workshop on Modeling Autonomous Agents in a Multi-Agent World Demazeau Y., Muller J.P (eds)., Elsevier Sc. Pub, 1990, pp 105-119.

[MATSUBAYASHI 93] Matsubayashi K., Tokoro M., A Collaboration Mechanism on Positive Interactions in Multi- agent Environments, Proc. Int. Joint Conf. on AI, 1993, pp 346-351.

[PETZOLD 92] Petzold, C., Programming Windows 3.1, Microsoft Press, 3rd edn., 1992.

[RAVI 96] Ravi, M, Viswaprasad, P.S., A Communication Framework for A Multi-Agent System, Master of Computer Applications thesis submitted to Univ. of Hyderabad, 1996.

[SHOHAM 92] Shoham, Y., Tennenholtz, M., On the Synthesis of useful Social Laws for Artificial Agent Societies, Proc. AAAI-92,1992, pp 268-276.

[SMITH 80] Smith R.G., "The Contract net protocol : High level Communication and Control in a Distributed Problem Solver", IEEE Trans. on Computers, C-29, no.12, December 1980, pp 1104-1113.

[TADEPALLI 92] Tadepalli, K. and Parameswaran, N., Reason Maintenance Systems for Multi-agent Planning, in Proc. 10th European Conf. on AI (ECAI-92), edited by Neumann, B., pp 648 - 652.

[WERNER 88] Werner, E., Towards a Theory of Communication and Cooperation for Multi-agent Planning, In Theoretical Aspects of Reasoning About Knowledge: Proc. 2nd Conf., Morgan Kaufman Pubs., 1988, pp 129-142.

MODELING OF UNKNOWN SYSTEMS USING GENETIC ALGORITHMS

Alaa F. Sheta, Ophir Frieder*and Kenneth De Jong
Department of Computer Science, George Mason University, Fairfax, VA 22030
Email: ashetal@site.gmu.edu, ophir@cs.gmu.edu and kdejong@cs.gmu.edu

ABSTRACT

This paper presents a new methodology for modeling and identifying the dynamics of unknown systems. Assuming that an unknown system belongs to a set of models, the methodology attempts to find the model that matches the characteristics of the unknown system and the best set of parameters for this model. To achieve these goals, a Genetic Algorithm (GA) methodology is utilized as a search strategy due to its efficiency and robustness. The proposed methodology consists of three stages: a model selection stage which is concerned with selecting the subset of models that best matches the input to and the output from the unknown system; a best model selection stage which is concerned with selecting the best model from the pre-selected subset of models from stage 1; and a parameter tuning stage which is concerned with the fine tuning of the best model parameters selected from stage 2. Simulation results are presented which show the effectiveness of the proposed method.

INTRODUCTION

Identification is the process of constructing a mathematical model of a dynamical system from observation and *a priori* knowledge [NR86]. Structure identification and parameter estimation of the unknown systems are considered to be important aspects of system identification theory. A mathematical model of a system is required for many purposes (e.g., system analysis and controller design).

In the past few decades, system modeling and identification has attracted a considerable number of researchers; the reason is the wide domain of practical applications. Chemical processes, biomedical systems, transportation, ecology, electric power systems, hydrology, astronautics are but a few examples of such applications. It is often assumed that there is enough prior knowledge about the system structure such that only parameter estimation is required [ED88, EL82, SH96]. However, the determination of a suitable model structure is a critical step in system identification and is equal in importance to parameterization.

The motivation behind this study is to propose a new identification methodology to solve the model identification and parameter estimation problem for unknown systems. The identification procedure consists of three stages of design.

In the first stage, a set of candidate models based on some *a priori* knowledge are evaluated using GAs. The evaluation depends on the minimization of the mean error between the unknown system response and all models in the model set. Only models with small mean error are selected for further investigation.

In the second stage of design a GA is used to select the best model from the pre-selected subset of models and estimate more precisely the parameters of the models selected from previous stage. This information is used to select a single model which best matches the characteristics of the unknown system.

In the third stage, a GA with a modified search space is used to fine tune the parameters of the best model to achieve high performance modeling. Our use of GAs at each stage is based on a number of observations.

GAs have been successfully applied to optimization problems like scheduling [SA94], controller design [VK93], observer design [PR95], parameter estimation [SH96], and system identification [KR92]. The advantages of using GAs in control system design was explored in [CH96]. GAs are capable of searching the multi-model error surfaces effectively. GAs do

---

*O. Frieder is presently on leave from George Mason University and is with the Department of Computer Science at Florida Institute of Technology.

not need any prior knowledge about the characteristics of the noise imposed on the measurements like other search techniques [SH96]. GAs have been used successfully in a number of control optimization problems and provide solutions that are more likely to be global in nature than the traditional ones [PA94].

RELATED WORK

There is already a considerable body of literature regarding the issue of modeling and identification of unknown systems from input-output observations. For example, the work done in [AK71] discusses a canonical realization procedure using an input-output description. In [NU92] an augmented identification algorithm (AUDI) was introduced which simultaneously identifies the model order and parameters of a single-input single-output (SISO) process and which has excellent numerical performance and high computational efficiency. An extension to the same procedure was presented in [NU94]. Recently, a complete diagnostic algorithm for detecting and isolating faults of actuators and sensors of an automotive engine based on describing a model structure for the engine was presented [GR95].

In [SH96] Sheta and De Jong show that the problem of parameter estimation can be solved using GAs. They show that GAs do not need any *a priori* knowledge about the measurements and the system initial conditions. GAs are able to handle noisy measurements and estimate the system parameters efficiently. It was assumed that the system structure is given. In this study, we are trying to solve the problem from a higher level point of view, since we are first selecting the optimal model structure which can represents the unknown system and then estimate its parameters.

Many problems in control systems can be viewed as system identification problems. Often, the object is to find a model that behaves like the actual unknown system. The complexity of nonlinear models lead to a research direction using Artificial Intelligence (AI) techniques. In [NA90] Narendra demonstrates that neural networks (NNs) can be used effectively for the identification and control of nonlinear dynamical systems. One of the drawbacks of using NNs in solving these kind of problems is that NNs usually require a large number of measurements to achieve high performance modeling. In addition NNs do not give any idea about the model structure which is very useful in checking system stability. Also, the evaluation procedure for neural networks is usually based on gradient descent techniques raising the possibility that only locally op-

timal solutions might be reached.

EXPERIMENTS

In this section we provide three examples to show the effectiveness of the proposed idea. Each example consists of an unknown system of increasing complexity from which input-output samples were taken to be used for system identification. For each unknown system a set of candidate models were selected which represent typical *a priori* knowledge an engineer might have concerning the general family of models to be used.

For each example a random sequence of 100 input values was generated and saved, the values of which were uniformly distributed between -2 and 2. This input sequence was then applied to the unknown system to obtain the corresponding system output values. This same input-output sequence is used at each of the 3 stages by the GA to find an optimal set of model parameters.

At each stage, the same GA was used in which the population size, crossover and mutation parameters were selected as 50, 0.85 and 0.01, respectively. The domain of search in stage 1 and 2 of the design was wide enough to catch the values of the model parameters. In stage 3 of the design, a reduced search space was used to get efficient values of the model parameters.

The initial model set for each example was selected based on typical prior knowledge that an engineer can gain from experiments. This knowledge may account for the model order or the number of system states. Typical known models for linear systems are Moving Average (MA) and Auto-Regressive Moving Average (ARMA). A set of nonlinear models that can be used for the identification of nonlinear systems are Wiener, Hammerstein, Wiener-Hammerstein, or the Volterra series models. These models are well known and have been successfully used in system identification and controller design although they have some problems regarding the large number of model parameters and nonlinear relationships between these parameters.

SYSTEM 1

Our first experiment involves a simple example which demonstrates the ability of GAs to identify the degree of an unknown system and to estimate the real system parameters effectively. The unknown system was described by the following equation:

$$y(k) = \phi(\theta, u(k))$$

in which the function ϕ is defined as follows:

$$\phi(\theta, u(k)) = \theta_1 u(k) + \theta_2 u^2(k) + \theta_3 u^3(k)$$

The values of parameters θ^{act} were chosen to be: $\theta_1 = 0.1$, $\theta_2 = 0.2$, and $\theta_3 = 0.4$.

In this case the model set was selected such that it covers all possible model structures from degree one to degree four, corresponding to the notion of an engineer looking for the best low order model of the unknown system. The selected set of models were defined as follows:

- Model 1: $\hat{y}(k) = \theta_1 u(k)$

- Model 2: $\hat{y}(k) = \theta_1 u^2(k)$

- Model 3: $\hat{y}(k) = \theta_1 u(k) + \theta_2 u^2(k)$

- Model 4: $\hat{y}(k) = \theta_1 u(k) + \theta_2 u^2(k) + \theta_3 u^3(k)$

- Model 5: $\hat{y}(k) = \theta_1 u(k) + \theta_2 u^2(k) + \theta_3 u^3(k) + \theta_4 u^4(k)$

DESIGN STAGE 1

In this stage of design, the objective is to use GAs to select a subset of models that can perform like the unknown system. A GA was used to find a set of parameters which minimized the training set error for each of the initial models. In this case the estimated parameters were:
$\hat{\theta}_1 = 0.6542$, $\hat{\theta}_2 = 0.1520$, $\hat{\theta}_3 = 0.2296$, and $\hat{\theta}_4 = 0.0201$.

In figure 1, we represent the error for each model in the model set. From this figure it is clear that models 4 and 5 are achieving the minimum error in this case. Those two models are considered for further investigation, corresponding to the observations that models of degree three or higher are required.

It is important to mention that the values of the estimated parameters in the first stage of design have nothing to do with the actual system parameters. These values just help us to explore the different model characteristics. Thus, we are able to select a subset of models that perform approximately like the unknown system for further investigation in the next stage.

DESIGN STAGE 2

In this stage of design, we repeat the procedure on the reduced set of candidate models. In this case there were two remaining models: model 4 and model 5. The estimated values of the parameters θ^{est} in this stage of design were :
$\hat{\theta}_1 = 0.1166$, $\hat{\theta}_2 = 0.1926$, $\hat{\theta}_3 = 0.3959$, and $\hat{\theta}_4 = 0.0087$.

In figure 2, we show the mean error between the unknown system and the estimated model responses for both model 4 and model 5, clearly indicating that model 4 is achieving better performance in modeling than model 5.

DESIGN STAGE 3

In this stage of design, our goal is to fine tune the parameter of the best model. Model 4 is considered the best model from the previous stage of design. GAs are used with a modified search space based on the values of the parameters from the previous stage. This search space is more constrained than before. The same population size, crossover, and mutation parameters were used in this stage of design. The parameters θ^{est} were calculated as :
$\hat{\theta}_1 = 0.1032$, $\hat{\theta}_2 = 0.2038$, and $\hat{\theta}_3 = 0.3947$.

The mean error in this case using a random test sequence was reduced to a value 0.0097. This results show a good convergence to the nominal system parameters and a high performance characteristics for the system under consideration.

SYSTEM 2

The unknown system in example 2 was described in [NA90]. They have used what are called a series-parallel models to describe unknown nonlinear systems. These models are dynamic models and are a function of the input and the output of the unknown systems.

The unknown system is described by the following difference equation:

$$y(k + 1) = 0.3y(k) + 0.6y(k - 1) + f[u(k)]$$

The function f is described as follows :

$$f(u) = 0.6sin(\pi u) + 0.3sin(3\pi u) + 0.1sin(5\pi u)$$

In [NA90], authors have used neural networks to model the dynamics of the nonlinear system. Neural network have been used to simulate the behavior of the nonlinear function $f[u(k)]$. In our case, we will explore the different structure for this function. We will apply our methodology to identify the dynamics of the unknown function $f[u(k)]$. We have selected four different models for the model set. The nature of these

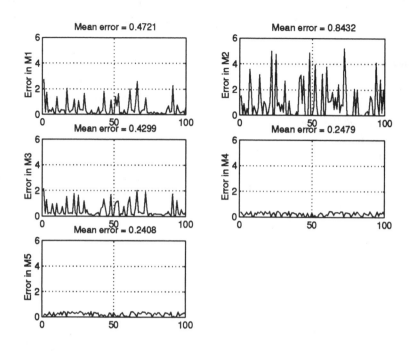

Figure 1: Error difference for all models in the initial model set (stage 1) for system 1

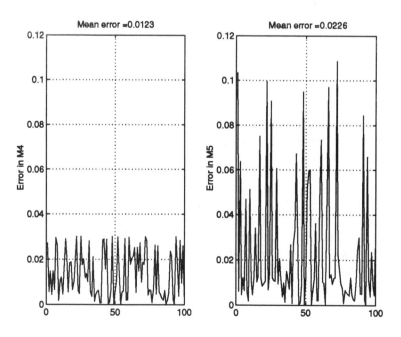

Figure 2: Error difference for M4 and M5 (stage 2) for system 1

models looks like the series-parallel model described in [NA90]. These models have the following characteristics:

$$\hat{y}(k+1) = \hat{\theta}_1 y(k) + \hat{\theta}_2 y(k-1) + N[\theta, u(k)]$$

We have defined the function $N[\theta, u(k)]$ by different ways. A four different models in the model set were introduced:

• Model 1:

$$\begin{aligned} \hat{y}(k+1) &= \theta_1 y(k) + \theta_2 y(k-1) \\ &+ \theta_3 u(k) + \theta_4 u(k)y(k) \end{aligned}$$

• Model 2:

$$\begin{aligned} \hat{y}(k+1) &= \theta_1 y(k) + \theta_2 y(k-1) \\ &+ \theta_3 u(k) + \theta_4 u^2(k) \end{aligned}$$

• Model 3:

$$\begin{aligned} \hat{y}(k+1) &= \theta_1 y(k) + \theta_2 y(k-1) \\ &+ \theta_3 u^2(k) + \theta_4 u^3(k) \end{aligned}$$

• Model 4:

$$\begin{aligned} \hat{y}(k+1) &= \theta_1 y(k) + \theta_2 y(k-1) \\ &+ \theta_3 u^3(k) + \theta_4 \end{aligned}$$

DESIGN STAGE 1

We again used GAs to estimate the parameters of the initial set of models to minimize the total mean error for all models in the model set. The values of the estimated parameters θ^{est} were :
$\theta_1 = 0.1910$, $\theta_2 = 0.5510$, $\theta_3 = 0.0930$, and $\theta_4 = 0.0640$.

In figure 3 we show the error difference for all models in the model set using a random input test sequence. The mean error for the four models in the model set shows that models 1 and 2 are the best models.

DESIGN STAGE 2

In the second stage of design, we have made different changes in the GA parameters in order to be able to select the best model structure. Unfortunately, there was not much improvement in the mean error for either model 1 or 2. The reason is that the dynamics of the unknown function N didn't fit either model 1 or model 2. However, we can use the information gained from the first stage of design to define a new model

structure which contains the $u(t)$ model 1, $u^2(t)$ from model 2 and the $u^3(t)$ from model 3. It is also quite natural to increase the model complexity as long as the error is high for the first and second order models.

Thus the new structure of the function $N[\theta, u(k)]$ will be :

$$N[\theta, u(k)] = \hat{\theta}_3 u(k) + \hat{\theta}_4 u^2(k) + \hat{\theta}_5 u^3(k)$$

We have run GAs to estimate the parameters of the new model. The estimated values of the parameters θ^{est} were: $\hat{\theta}_1 = 0.2790$, $\hat{\theta}_2 = 0.6010$, $\hat{\theta}_3 = 0.7450$, $\hat{\theta}_4 = 0.0220$, and $\hat{\theta}_5 = -0.3610$. The mean error in this case was reduced to a value 0.2329.

DESIGN STAGE 3

We employed the estimation process again to improve the performance of the estimated parameters and to reduce the mean error for the above model. The new estimated set of parameter were :
$\hat{\theta}_1 = 0.3000$, $\hat{\theta}_2 = 0.6010$, $\hat{\theta}_3 = 0.6450$, $\hat{\theta}_4 = 0.0113$, and $\hat{\theta}_5 = -0.3210$.

It is clear that the values for the parameter $\hat{\theta}_1$ and $\hat{\theta}_2$ are equal to the same nominal values of the real plant. The mean error was 0.2202. In [NA90], it was found that the nonlinear function stored in the unknown system was $f(u) = u^3 + 0.3u^2 - 0.4u$. In our case we have found a new nonlinear function $f(u) = -0.321u^3 + 0.0113u^2 + 0.645u$ which also fit to the nonlinearity in the unknown system. In stage 2 and 3 of the design, we have used the same test sequence described in [NA90] to show both the system and the model responses.

In figure 4 (a), we show the error difference between actual unknown system response and the estimated one in stage 2 of the design. In figure 4 (b), we show the system output and estimated response using GAs for stage 3 of the design. The number of measurements used in modeling process is 100.

SYSTEM 3

In example 1 and 2, the initial model set was described in a unified way. That is, the system output was defined as a function of the system input. There exist other interesting model structures which involve the description of the system state variables. These variables represents the actual dynamic factors of a systems. For example, in electronic circuit device, the current across a capacitor and the voltage across an inductor can be represented as states. Thus, trying to achieve certain behavior (i.e, control the values of the current and voltage) implies that we need to know the

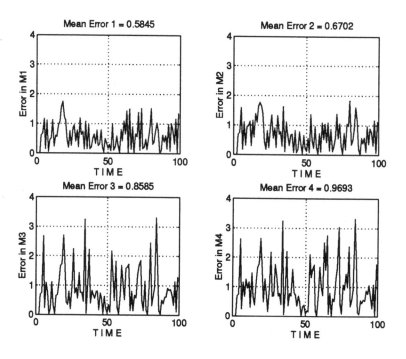

Figure 3: Error for different models in the initial model set (stage 1) for system 2

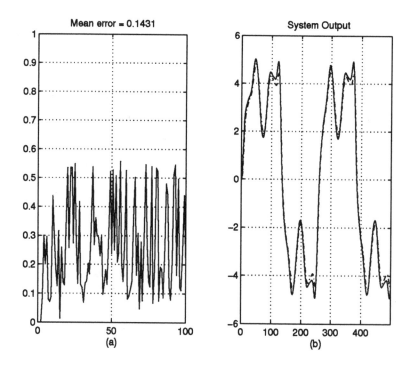

Figure 4: (a) Error difference between the actual and estimated response for the new model (stage 2) for system 2 (b) System output and estimated response (stage 3) for system 2

structure of what is called the state space representation of the circuits.

In example 3 we have tried to achieve a very important goal which is the description of the state space representation of the unknown system. We have assumed that the system output function is known while the inter-relationships between the system input and the system states are unknown. Our model set was constructed by using all possible combinations of the bilinear state and input terms. In addition we have kept the linear state terms fixed in all models.

In general any system can be described by the following form:

$$x(k+1) = \phi(x(k), u(k))$$

$$y(k) = \psi(x(k))$$

The function ϕ and ψ are considered nonlinear functions. In system identification procedure we used to develop models as a function of the input to and the output from the system. This is useful in many cases. But sometimes it is important to know what is the inter-relationships among the system states. This is valuable for system linearization and thus simpler controller design. In this example we will assume that the function ψ is known and the objective is to find the function ϕ. The function $\phi(x(k), u(k))$ is function of three variables the system states $x_1(k)$, $x_2(k)$ and $u(k)$. We have developed the model set in such a way that it usually contains the system states in linear format. Also each model contains some form of bilinear function of the states and the input.

The system under consideration was presented in [WH92]. The nominal system state equation is presented as follows:

$$
\begin{aligned}
x_1(k+1) &= \theta_1 x_2(k) + \theta_2 x_2(k)u(k) \\
x_2(k+1) &= \theta_3 x_1(k) + \theta_4 x_2(k) + \theta_5 u(k) \\
y(k) &= 0.3x_1(k) + 2x_2^2(k)
\end{aligned}
$$

The real values of the above system parameters are: $\theta_1 = 1$, $\theta_2 = 0.2$, $\theta_3 = -0.2$, $\theta_4 = 0.5$ and $\theta_5 = 1$.

The candidate models in the model set were defined as follows:

- Model 1 :

$$
\begin{aligned}
x_1(k+1) &= \theta_{11}x_1(k) + \theta_{12}x_2(k) \\
&+ \theta_{13}x_1(k)u(k) + \theta_{14}u(k) \\
x_2(k+1) &= \theta_{21}x_1(k) + \theta_{22}x_2(k) \\
&+ \theta_{23}x_2(k)u(k) + \theta_{24}u(k)
\end{aligned}
$$

- Model 2 :

$$
\begin{aligned}
x_1(k+1) &= \theta_{11}x_1(k) + \theta_{12}x_2(k) \\
&+ \theta_{13}x_2(k)u(k) + \theta_{14}u(k) \\
x_2(k+1) &= \theta_{21}x_1(k) + \theta_{22}x_2(k) \\
&+ \theta_{23}x_1(k)u(k) + \theta_{24}u(k)
\end{aligned}
$$

- Model 3 :

$$
\begin{aligned}
x_1(k+1) &= \theta_{11}x_1(k) + \theta_{12}x_2(k) \\
&+ \theta_{13}u(k) + \theta_{14} \\
x_2(k+1) &= \theta_{21}x_1(k) + \theta_{22}x_2(k) \\
&+ \theta_{23}u(k) + \theta_{24}
\end{aligned}
$$

- Model 4 :

$$
\begin{aligned}
x_1(k+1) &= \theta_{11}x_1(k) + \theta_{12}x_2(k) \\
&+ \theta_{13}u(k) + \theta_{14}u^2(k) \\
x_2(k+1) &= \theta_{21}x_1(k) + \theta_{22}x_2(k) \\
&+ \theta_{23}u(k) + \theta_{24}u^2(k)
\end{aligned}
$$

DESIGN STAGE 1

We run a GA on the above set of models to find the model which exhibits the best performance characteristics. The estimated values of the parameters θ^{est} were :

$\theta_{11} = -0.288$, $\theta_{12} = 0.915$, $\theta_{13} = 0.161$, $\theta_{14} = -0.293$, $\theta_{21} = -0.842$, $\theta_{22} = 0.991$, $\theta_{23} = 0.012$ and $\theta_{24} = -0.678$.

In figure 5, we show the error difference between the unknown system and estimated model response for all models. The system and the models were tested by a random sequence. Thus, model 1 and 2 will be selected for further investigation.

DESIGN STAGE 2

GA was used to identify the parameters of model 1 and 2. The estimate parameters in this case were :

$\theta_{11} = 0.061$, $\theta_{12} = 0.881$, $\theta_{13} = 0.190$, $\theta_{14} = -0.174$, $\theta_{21} = -0.211$, $\theta_{22} = 0.462$, $\theta_{23} = -0.006$ and $\theta_{24} = 0.995$.

In figure 6, we show the error difference in stage 2 of the design for both model 1 and 2. The mean error for model 2 is much better than that one for model 1 and so it is selected for a fine tuning stage.

DESIGN STAGE 3

In the third stage of design, we fine tune the parameters of model 2. The mean error for model 2 is 0.0181. The estimated values of the parameters were :

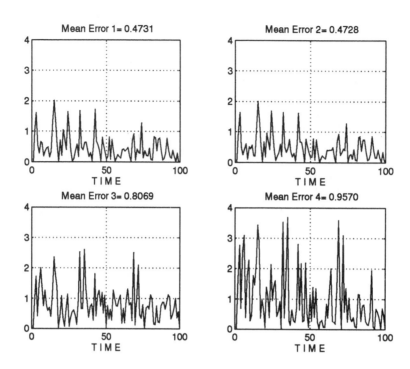

Figure 5: Error different for all models in the initial model set (stage 1) for system 3

$\theta_{11} = -0.037$, $\theta_{12} = 1.031$, $\theta_{13} = 0.195$, $\theta_{14} = 0.052$, $\theta_{21} = -0.195$, $\theta_{22} = 0.512$, $\theta_{23} = 0.004$ and $\theta_{24} = 1.008$.

Interesting results can be derived from the values of the estimated parameters θ. When comparing the values of the parameter θ to the actual model parameters we can see that parameters θ_{11}, θ_{14}, and θ_{23} did not contribute that much in the best model while the rest of the parameters are very close in values to the real one. The reason for this is that θ_{11}, θ_{14}, and θ_{23} do not exist in the real system. This is an indication of the success of our methodology in attempting to find a good model structure.

CONCLUSION AND FUTURE WORK

We presented a new methodology for modeling the dynamics of unknown systems. The methodology utilizes a genetic algorithm: 1) to first search for optimal model or subset of models from a set of candidate models; 2) to estimate the most effective set of parameters for the selected model set, and 3) to improve the quality of the best model by fine tuning the parameter settings.

The method was tested on three systems and showed good performance. The mean error for the models selected in stages 2 and 3 of the design were improved along the way. The results of these experiments indicate that genetic algorithms are a promising tool for these kinds of problems. Genetic algorithms were able to identify the model that best behaves like the unknown system.

This work presents our primary exploration of using GA in system modeling process. Our aim is to extend this idea by using a well known model structure for nonlinear processes like Wiener, Hammerstein, Wiener-Hammerstein, and the Volterra series models. These model structure have some difficulties because of the infinite number of model parameters. We intend to study the different ways procedure of reducing the size of these models and make it suitable for our methodology.

REFERENCES

[GR95]   J. Gertler, M. Costin, X. Fang, Z. Kowalczuk, M. Kunwer and R. Monajemy. Model Based Diagnosis for Automotive Engines-Algorithm Development and Testing on Production Vehicle. In *IEEE*

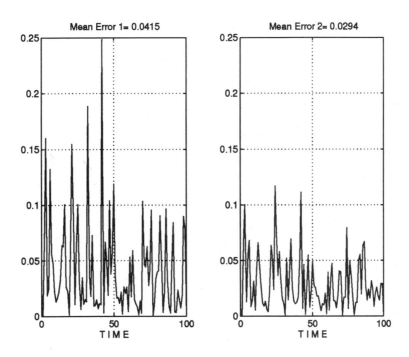

Figure 6: Error difference for model 1 and 2 (stage 2) for system 3

Trans. Contr. Syst. Technology 3, 1, pp. 61-69. 1995

[SA94]   N. Sannomiya, H. Iima and N. Akatsuka. Genetic Algorithm Approach to a Production Ordering Problem in an Assembly Process with Constant Use of Parts. In *Int. J. Systems Sci.*, vol.25, no.9, pp. 1461-1472, 1994.

[VK93]   A. Varsek, T. Urbancic and B. Filipic. Genetic Algorithms in Controller Design and Tuning. In *IEEE Trans. Syst. Man. Cybern.*, vol.23, no.5, 1993

[WH92]   D. A. White and D. A. Sofge. Handbook of Intelligent Control, Neural, Fuzzy, and Adaptive Approaches. Van Nortrand Reinhold, New York, 1992

[SH96]   A. F. Sheta and K. De Jong Parameter Estimation of Nonlinear Systems in Noisy Environment Using Genetic Algorithms. In *Proc. of the IEEE International Symposium on Intelligent Control*, pp. 360-366, 1996

[NU92]   S. Niu and D. G. Fisher. An Augmented UD Identification Algorithm. Int. J. Contr. vol.56, pp.193-211, 1992

[NU94]   S. Niu and D. G. Fisher. Simultaneous Structure Identification and Parameter Estimation of Multi variable Systems. Int. J. Contr. vol.59, no.5, pp.1127-1141, 1994

[AK71]   J. E. Ackermann and R. S. Bucy. Canonical Minimal Realization of a Matrix Impulse Response Sequences. Information and Control, vol.19, pp.224-231, 1971

[ED88]   V. Elden and N. Yildizbayrak. Parameter and Structure Identification of Linear Multi variable System. Automatica, vol.24, pp.365-373, 1988

[NA90]   K. S. Narendra and Pathansarathy, K. Identification and Control of Dynamical Systems Using Neural Networks. In *IEEE Trans. on Neural Networks*, vol.1, IEEE Press, 1990

[EL82]   H. Elliot and W. A. Wolovich A Parameter Adaptive Control Structure for Linear Multi variable Systems. In *IEEE Trans. Aut. Contr.*, vol.27, pp.340-352, 1982

[CH96]   A. J. Chipperfield and P. J. Fleming. Genetic Algorithms in Control Systems Engineering. IASTED Journal of Computers and Control, vol.24, no.1, 1996

[PR95]   L. Porter II and K. M. Passino Genetic Adaptive Observers Engng Applic. Artif. Intell. vol.8, no.3, pp.261-269, 1995

[PA94]   R. J. Patton and G. P. Liu. Robust Controller Design via Eigenstructure Assignment, Genetic Algorithms and Gradient-based Optimization. In *Proc. of the IEE Control Theory Appl.*, 1994

[KR92]   K. Kristinsson and G. A. Dumont. System Identification and Control Using Genetic Algorithms. IEEE Transaction on System, Man and Cybernetics. vol.22, no.5, 1992

[NR86]   J. P. Norton An Introduction to Identification. Harcourt Brace Jovanovitch, 1986

OPTIMIZATION OF PAGING COST IN MOBILE SWITCHING SYSTEM BY GENETIC ALGORITHM

Hee C. Lee
Northern Telecom
Wireless Networks, MS D210
2201 Lakeside Boulevard
Richardson, TX 75082-4399
E-Mail: heelee@nortel.ca

Junping Sun
School of Computer and Info. Sci.
Nova Southeastern University
3100 Southwest 9th Avenue
Fort Lauderdale, FL 33315
E-Mail: jps@scis.nova.edu

ABSTRACT

The maximum bandwidth capacity of the radio frequency channel such as the forward control channel (FOCC) is limited in a mobile switching system. The FOCC has been experiencing severe congestion because the inefficient conventional mobile paging methods cause the bottleneck in the FOCC due to constraints of both the bandwidth and the limited number of radio frequency channels in mobile telecommunication systems. In order to minimize the paging cost and to maximize utilization of the bandwidth in the FOCC, we develop a new paging schema with the optimal partition of paging zones. By using the refined mobile's probability patterns stored in the statistical profile, we employ the genetic algorithm and the derived fitness function to generate the optimal partition of paging zones such that the paging cost to locate a mobile station as well as the bandwidth consumption in the FOCC is minimized.

1 INTRODUCTION

The cellular mobile network system uses a radio frequency channel for the wireless communication between a mobile switching system and a mobile station.

A mobile radio frequency channel is a two-way communication channel. The channel consists of two frequencies with the channel bandwidth of 30 KHz. A number of radio channels are allocated to each cell site of a mobile switching system. These channels are logically divided into many voice channels and a single control channel. These voice channels carry the data and the voice between the mobile switching systems and the mobile stations. Each control channel carries the system control information such as the paging message, mobile station registration, voice channel designation and hand off notification between a mobile switching system and a mobile station (MS).

The control channel is further divided into two different control channel formats: the forward control channel (FOCC) and the reverse control channel (RCC), based on the usage of the control channel. The FOCC is mainly used to deliver paging messages and system orders to the MS. The reverse control channel is used to request an access to the mobile switching system by the MS.

In general, the mobile switching center (MSC) in the system does not have the exact information about the geographical locations where these mobile stations operate in the service area (SA). Most of current mobile switching systems broadcast the same paging messages through all the FOCCs of the cells in the SA whenever the MSC tries to locate a mobile station in order to establish a call connection to the paged mobile station.

A typical mobile switching system operating in large metropolitan areas consists of over 100 cells. In order to locate a mobile station, the mobile switching system has to broadcast the same paging messages through all the FOCCs of over 100 cells in the system, but the mobile switching system receives only one reply message from the paged mobile station. Since the ratio of termination, i.e., responding versus paging ratio, is so low, the broadcasting of paging messages to mobile stations causes significant bottleneck problems in the FOCCs.

In order to alleviate the bottleneck generated by unnecessary paging messages sent through the FOCCs, a number of location tracking strategies have been proposed [YY95] [IV92] [Cim94] [BNK94] [BNK93] [Ste94] [MB94] [Tab95] [HA95]. Since many researchers have shown that the genetic algorithm approach is very promising for various optimization problems in different fields [Koz96], in this paper, we will present a mobile paging strategy that generates paging zones by the genetic algorithm approach, such that the paging cost to locate a mobile station as well as the bandwidth consumption in FOCCs can be substantially reduced or minimized in comparison with the current paging methods in real applications.

1.1   The Topology and Architecture of Mobile Switching Systems

There are three basic elements in the mobile switching system such as the mobile station (MS), the base station, and the mobile switching center (MSC).

The mobile station could be a car phone, hand held, transportable, or any other type of wireless terminals. Each mobile station is identified by a mobile identification number (MIN). The mobile station (MS) transmits the information of network address, data, and voice. It is also tunable on system command to a channel in the radio frequency spectrum allocated to the mobile switching system at the certain pre-programmed power level.

The base stations that cover the service area are located at cell sites (the definition of a cell is given in the next paragraph). Each of base stations receives and processes its radio frequency signals to make them suitable for the transmission between the wired line network and the radio network among all mobile stations interfacing with it.

The cell is a geographical area that the radio signal frequency, which is transmitted from a base station, covers to serve mobile telephone calls. In general, the size of a cell depends on the strength of the radio signal. The stronger the radio signal is, the wider area it can cover, and the larger the size of a cell is. The size of a cell can be adjusted by controlling the strength of radio signal that is transmitted from a base station.

The mobile switching center (MSC) operates as the central coordinator and controller for the mobile switching system. The MSC sends to the MS the different type of information such as the MSC's system identifier, the radio frequency that the MS should use for voice communication, paging response acknowledgment, etc.. The MSC also functions as the interface between the mobile station(s) and the public telephone network (PTN). In addition, the MSC performs these functions such as administration of radio channel allocation, coordination of the grid of cell sites and active mobile stations, and maintenance of the integrity of the MSC as a whole [FP79].

The mobile switching system is organized in the hierarchical manner. The MSC is connected to the top-level public telephone network (PTN). Each MSC controls 100 base stations on an average, and each base station serves a number of mobile stations simultaneously.

In general, a mobile station (the origination) can originate a call from any one of cells to others (the termination) in any geographical location areas. As soon as the MSC receives a request from the originating mobile station, it determines the location of the mobile station and allocates a voice channel available to the mobile station (originator). Once the voice channel is allocated to the originator, the MSC pages the terminating (terminal) mobile, and the MSC connects the call if the paged mobile responds to the paging.

1.2   Forward Control Channel Messages Frame

The forward control channel (FOCC) is a continuous data stream sent at the rate of 10 kbps (kilobits per second) from the base station to the MS. Each FOCC message frame consists of total 463 bits at the transmitting rate of 21.598 message frames per second[ET92]. So the number of paging messages that a FOCC can carry is limited.

1.3   Related Research Work

Since the bandwidth bottleneck of the FOCCs is due to heavy consumption and high rate of occupation, a number of mobile location tracking and paging methods have been studied in order to reduce the FOCCs' consumption in locating a paged mobile station.

An aggregate mobile station tracking strategy and reporting center approach was proposed in [BNK93]. This method designates a subset of cells in a service area as reporting centers. The strategy is described as the following: a mobile station sends the location updating message whenever it enters a new reporting center and a tracking for the mobile station is restricted to the set of adjacent cells of the reporting center. This method is not efficient because of two major reasons. First, if a mobile station does not move far and it moves in and out of the reporting center frequently, the frequency channels will be heavily loaded due to unnecessary paging activities for location update. Second, although it moves a lot around, a mobile station may never report changes of its current location for a long time if it never enters to a reporting center. Also, this strategy can not guarantee that the mobile station responses to the same paging messages sent to the adjacent cells of the last visited reporting center.

The location area approach was presented in [MB94]. The mobile service area is partitioned into several location areas with equal size. Each location area consists of a number of cells. All mobile stations must register whenever they cross over from one location area to another. This strategy keeps track of the exact location area of every mobile station, and it also guarantees that a mobile station can be exactly located whenever the MSC pages the most recent cells in the location area registered by the mobile station, but it is still inefficient for the following reasons. First, although the paging activities are reduced to only these cells in the most recently registered location area, all the registrations must be performed in these cells that are along the borders of the location areas. In addition, this approach makes the reverse control channels for these cells heavily loaded and congested. Second, all cells in the location area are still paged so that the call termination rate is still very low.

The individual mobile station tracking strategy by utilizing the mobility pattern of a mobile station was proposed in [Tab95]. This strategy partitions the service area into several location areas based on each mobile station's mobility pattern. The location areas are arranged in descending order of location probabilities of mobile stations. When it tries to locate a mobile station, the MSC pages the mobile station in these cells of the location area with the highest probability first. If the mobile station is not found in the first paged location area, then the MSC will page the next highest one, and so on. The MSC will not stop paging until it receives a responding message from the paged mobile station or it determines that the paged mobile station is not active. However, this approach only considered the mobile probability pattern for the entire location area but not for each individual cell. It is obvious that the mobile probability distribution of a mobile station is not uniform in the entire location area since the mobile probability pattern always varies from one cell to another.

In the paper, we will present our approach as the follows: First, we will use the cell-based histogram approach to store the mobile usage patterns, so the refined mobile probabilities with better resolution can be derived. Second, we define the fitness or cost function in terms of the mobile probability and paging cost for the genetic algorithm to generate optimal partition of paging zones. Third, we will show how to generate the optimal partition of paging zones by grouping cells together with the help of fitness or cost functions in terms of mobile probabilities. We will also demonstrate that the paging messages are always sent to each of paging zones in the descending order based on the values of probability function whenever a mobile station is paged. So the paging cost can be further reduced since the number of paging messages sent in this approach is less than that in the previous approaches in [MB94][Tab95].

2 OPTIMAL PAGING ZONE PARTITIONING

2.1 The Problem Statement

For the given set of mobile cells, $\{C_1, C_2, ..., C_n\}$ in the problem domain, and the associated probabilities for each mobile station, the set $\{C_1, C_2, ..., C_n\}$ will

be optimally partitioned into a set of paging zones such that paging cost to locate a mobile would be minimized, by genetic algorithm, in terms of the fitness function.

2.2 The Strategy

For each mobile station or mobile user (we will use the mobile station and mobile user interchangeably), we define the multiple location area layers as $\{L_1, L_2, ..., L_k\}$. Our approach defines the location area layers based on the pattern of the mobile users mobility in a day. For example, the location area layer L_1 refers to the working area of a mobile user, L_2 refers to the home area of a mobile user. The location area layer L_i can be used to further partition the mobile service area into at least two location areas: L_{i1} and L_{i2} based on the location probability of a mobile user in each cell C_i where $1 \leq i \leq n$. The location area L_{i1} contains all probable cells, $\{C_1, C_2, ..., C_m\}$ where $m \leq n$, in the area that the mobile might reside in, and the location area L_{i2} contains the remaining cells in the service area.

We define the probability $Prob(C_i)$ where $1 \leq i \leq n$ as the corresponding probability of each cell in which the mobile station or mobile user may reside. Based on the probability $Prob(C_i)$ for $1 \leq i \leq n$, the location area L_{i1} is partitioned into a set of paging zones $P_1, P_2, ..., P_k$ such that

1. $\cup_{i=1}^k P_i = \{C_1, C_2, ..., C_m\}, 1 \leq m \leq n$.

2. $P_i \cap P_j = \emptyset$, for $1 \leq i, j \leq k$, where k is the total number of paging zones to be generated and $\emptyset$ means the empty set.

3. For each $C_i \in L_{i1}$, $pos(C_i) < pos(C_j)$ for $1 \leq i < j \leq m, 1 \leq z \leq k$, and $pos(C_i)$ is the position of C_i in location area L_{i1}.

4. For each paging zone P_z, the probability $Prob(P_z)$ $= \sum_{C_i \in P_z} Prob(C_i)$, where $1 \leq i < j \leq m \leq n$ and $1 \leq z \leq k$.

Later in this paper, we will show how the probabilities of both $Prob(P_z)$ and $Prob(C_i)$ are employed to create the optimal partition of paging zones, $PZ = \{P_1, P_2, ..., P_k\}$, such that $Prob(P_i) > Prob(P_j)$ where $1 \leq i < j \leq k$ and the $Prob(P_i)$ is the probability of each paging zone in which the mobile station or mobile user may reside. The paging cost would be minimized if these paging zones are paged in such order $P_1, P_2, ..., P_k$ when the MSC tries to locate a mobile station. The profile of a mobile station contains the information about the mobile user's multi-layered location areas and paging zones of location area L_{i1}.

The MSC uses this profile for the two following events:

1. Location Area Registration: Whenever an MS moves from one location area to another location area, the MS sends the location area registration message to the closest base station nearby.

2. Paging: When it tries to locate a mobile for establishing a call connection to the mobile station, the MSC obtains the mobiles current location area layer, the location area and the paging zones from the profile of the mobile. If the mobile's current location area is L_{i1}, then the MSC first pages the cells in the paging zone P_1 in which the mobile most probably resides. If the mobile does not respond to the paging message within a fixed short time period, then the MSC pages all the cells in the paging zone P_i ($i = 2, 3, ...$). P_i is paged before P_j where $Prob(P_i) > Prob(P_j)$ and $i < j$. If the mobile station still does not respond to the same paging messages until the last paging zone is paged, it is assumed that the mobile is not active. If the mobiles current location area is L_{i2}, then the MSC pages all cells in the location area L_{i2} at once since all the cells in the location area L_{i2} have the same value of the mobile existence probability 0 and they are treated equally.

The mobility pattern of each mobile station or user is constantly monitored, and the location areas and paging zones are periodically updated and dynamically maintained.

2.3 Histogram for Derivation of Mobile Probability Patterns

Since our approach depends on the knowledge of mobile probability patterns, it is crucial to derive the accurate mobile probability pattern for the partition of paging zones. The derivation process of the mobile probability patterns will be based on the paging/termination patterns of mobile stations in different geographical location areas at different time segments. In practice, we discover that neither the activities of paging/responding in the different time segments of a day nor in the different cells are uniform. In the previous approaches [BNK94] [MB94] [Tab95], neither the non-uniformity of mobile usage patterns in different geographical location areas nor in the different time segments is considered.

We use a cell-based histogram for the purpose of discovering mobile probability patterns based on non-equally divided time segments and the cells in different geographical location areas. So the mobile probability patterns are derived from the data along two dimensions: the dimension of time segments and the dimension of geographical areas. The dimension of geographical areas consists of these cells arranged in the

system-assigned sequential order. For the simplicity, one dimension of array in the histogram is used to store the mobility data for a mobile user from the different cells in different areas. The size of the array is equal to the number of cells in the system and the index of the array represents the cells in a system. Each of elements in the one-dimensional array is sequentially corresponding to the mobile cells. Each array represents a set of collected data for each of time segments.

The cell location of the mobile user is known to the system, and correspondingly the element of the array is increased by one when one of the following events happens:

1. The mobile originates a call.

2. The mobile responds to a paging message.

3. The mobile registers (power on registration, power down registration, and location area registration).

Whenever any one of these events is occurred, the corresponding cells counter is incremented by one.

In order to achieve the accuracy of the mobility pattern data along the dimension of time segments, the data are also collected and segmented based on usage patterns in a day. Suppose, a mobile user's profile shows that: work 8:00 AM - 6:00 PM, social activity 6:00 PM - 10:00 PM, home 10:00 PM - 8:00 AM. The time cycle in a day could be divided into three time segments. However, the time intervals between different activities can not be easily differentiated at an exact time point. For this reason, the time dimension is segmented into six different time segments in this example. In Figure 1, the time segments T_2, T_4, and T_6 cover the boundaries of two different mobile users activity patterns. Table 1, Table 2, and Table 3 give the data of mobile activities in different time segments.

Later on, we will show the mobile probability will be derived from the histogram.

2.4 Creation of Location Area Layer and Location Area

Our approach defines location area layers from histogram tables of the mobility data. The location area layer L_1 is defined by combining the histogram data tables corresponding to time segments T_6, T_1, and T_2. In the same manner, the location area layer L_2 is constructed by combining the mobility data tables corresponding to time segments T_2, T_3, and T_4. The location area layer L_3 is constructed from the histogram data tables corresponding to time segments T_4, T_5, and T_6.

For each location area layer L_i, our method partitions the service area into two location areas, L_{i1} and

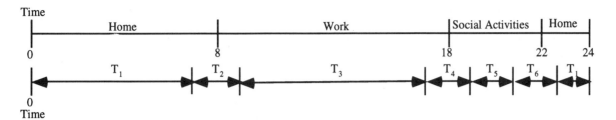

Figure 1: The Segment of Time

Cell Number	C_1	C_2	C_3	C_4	C_5	C_6	C_7	C_8	C_9	C_{10}	$C_{10}...C_{100}$
Count	246	0	836	521	39	657	26	516	769	0	0 ... 0

Table 1: The Histogram Table for Time Segment T_2

L_{i2}. From the combined mobility data collection tables of location area layer, the location area L_{i1} is constructed by including all cells whose count value is non zero and the location area L_{i2} is constructed by consisting all other cells in the service area. As soon as both the location area layer and the location area are defined, the mobile usage data from the histogram can be combined to generate the mobile probability values from different location area layers.

The mobile location probability of each cell $Prob(C_i)$ in the location area L_{i1} is computed from the mobility data tables, and the sample is given as shown in Table 4.

In following sections, we will show that the location area L_{i1} is farther partitioned into the several paging zones in order to obtain minimal paging cost.

2.5 Creation of Paging Zones

The cells in a location area L_{i1} are partitioned to a set of paging zones such that the paging cost over the partitioned paging zones is minimized. With the help of the combined histogram data table of the location area L_{i1}, the probability $Prob(C_i)$ where $1 \leq i \leq n$, is computed as shown in Table 4.

Let $C_1, C_2, ..., C_m$ be m cells in the location area L_{i1}. The m cells are partitioned to a set of paging zones $\{P_i\}$ where $1 \leq i \leq k$. Let $N(P_i)$ be the number of cells in each of the paging zones P_i for $1 \leq i \leq k$ such that

$$m = N(P_1) + N(P_2) + ... + N(P_k) = \sum_{i=1}^{k} N(P_i)$$

The mobile location probability of paging zone P_i is defined as $Prob(P_i)$, which is the probability such that the mobile station resides in the paging zone P_i:

$$Prob(P_1)+Prob(P_2)+...+Prob(P_k) = \sum_{i=1}^{k} Prob(P_i) = 1$$

Since there are extra consumption costs for both the forward control channel and fixed link channel each time when a mobile station is paged, we use the symbol α to denote consumption in the forward control channel per paging message, and β consumption in the fixed link channel consumption in the MSC per paging message. Both consumption costs in terms of α and β are proportional to the size of frames in either the forward control channel or the fixed link channel in terms of bits. For the purpose of simplicity in experiments, we could choose the value 1 for both α and β at this time since the actual values of both α and β might vary from different industry standards.

The paging cost C_p to locate a mobile user is defined as:

$$C_p = N(P_1)\times\alpha\times\beta+(1-Prob(P_1))\times N(P_2)\times\alpha\times\beta+...+$$

$$(1-\sum_{i=1}^{j-1} Prob(P_i)) \times N(P_j) \times \alpha \times \beta +...+$$

$$(1-\sum_{i=1}^{k-1} Prob(P_i)) \times N(P_k) \times \alpha \times \beta \qquad (1)$$

$$C_p = (N(P_1)+\sum_{j=2}^{k}((1-\sum_{i=1}^{j-1} Prob(P_i))\times N(P_j)))\times\alpha\times\beta \qquad (2)$$

The paging cost is incurred if and only if the paging zone is paged. Based on the assumptions and the definition of the mobile location probability, the paging process will stop as soon as the paged mobile responses to the paging message. Since the paging process always starts at the paging zone with the highest probability, the probability of earlier termination is always very high. If this is the case, not every paging zone is necessarily paged. Thus, the paging cost can be substantially reduced to the minimal level, so the consumption of FOCCs is. In the next section, we will

Cell Number	C_1	C_2	C_3	C_4	C_5	C_6	C_7	C_8	C_9	C_{10}	$C_{10}...C_{100}$
Count	4969	2450	0	0	387	8621	7621	4621	3620	2130	0 ... 0

Table 2: The Histogram Table for Time Segment T_3

Cell Number	C_1	C_2	C_3	C_4	C_5	C_6	C_7	C_8	C_9	C_{10}	$C_{10}...C_{100}$
Count	1601	106	22	331	0	1372	21	401	723	0	0 ... 0

Table 3: The Histogram Table for Time Segment T_4

show how these paging zones are configured by applying the genetic algorithm.

2.6 Paging Zone Configuration by Applying the Genetic Algorithm

In this research, we use the individual object (chromosome) which is a string of integers: $1, 2, 3, ..., k$ where k is the number of paging zones in location area L_{i1}. Each string corresponding to a chromosome represents an assignment of paging zones generated from the partition, and each chromosome is associated with its corresponding paging cost. We use the genetic algorithm to transform a set of chromosome (population) into a new population by using genetic operations such as reproduction, mutation, and cross over.

The genetic algorithm continuously produces new populations till it finds the best individual that has the optimal paging cost. When the best individual is found, the string corresponding to the best individual represents the optimal partition that groups the cells in the location area into a number of paging zones, such that $Prob(P_i) > Prob(P_j)$ where $i < j \leq k$.

2.7 Genotype Representation

Let $PZ = \{P_i \mid 1 \leq i \leq k \leq m\}$ be a set of paging zones to be generated by partitioning m cells based on the paging cost function where k is the total number of paging zones and m is the total number of cells.

Let cs be a chromosome string of integers $d_1 d_2 d_3 ... d_x ... d_w$ such that $\forall d_x \ d_x \in \{1, 2, 3, ..., k\}$ where $1 \leq x \leq w \leq m$, x is the cell position number in the chromosome string cs.

The position of each integer digit d_x represents the corresponding cell position number in the sorted cell list of location area L_{i1}. For example, the 5th cell in the sorted cell list of location area L_{i1} is represented by the 5th digit d_5 of the chromosome '112241133...'. The fixed length of a chromosome string cs is equal to the total number of cells in the location area L_{i1}. The character '1' in the xth position of the chromosome represents the cell in the xth position in the sorted cell list of the location area L_{i1} is assigned into the paging zone P_1. The character '2' in the yth position of the

chromosome represents the yth cell in the sorted cell list of the location area L_{i1} is in the paging zone P_2, and so on.

Given 10 cells in the location area L_{i1} and they are sorted as: $\{10, 20, 30, 40, 50, 60, 70, 80, 90, 100\}$. Assume, the cells $10, 40, 70$, and 100 are partitioned into the paging zone P_1, the cells $20, 30$, and 90 are in the paging zone P_2, and the cells $50, 60$, and 80 are in the paging zone P_3. Then the corresponding chromosome string is represented as '1221331321'.

2.8 Genotype to Phenotype Conversion

The mobile location probability of each cell $Prob(C_i)$ in the location area L_{i1} is computed from the mobility data table and is given as in Table 4. In Table 4, the cell C_1 has the mobile location probability 0.16. When the mobile station is paged, the probability, $Prob(C_1)$, for the mobile to respond to the paging message from the C_1 is 0.16.

The type conversion function that is called, $Genotype\_to\_Phenotype\_Conversion$, uses the probability in Table 4. The function takes a chromosome as an input. The function counts the total number of same character i in a chromosome string, i.e., $N(P_i)$, and adds the mobile location probability, i.e., $Prob(P_i)$ of each cell whose corresponding character is i. The function repeats this process for all different integer digits in a chromosome string.

2.9 The Fitness Function and Convergence of the Genetic Algorithm

The paging cost function as well as the fitness function computes the paging cost that the MSC pages the mobile within the given paging zone (chromosome). The fitness function is given as in (1) and (2).

Since the term $\sum_{j=2}^{k} ((1 - \sum_{i=1}^{j-1} Prob(P_i)) \times N(P_j))) \times \alpha \times \beta$ in (1) and (2) is the dominant part of the fitness function, the fitness function is always favor for the paging zone with the highest probability. The higher the value of $Prob(P_1)$ or $Prob(P_i)$ is, the less the value of the term $\sum_{j=2}^{k} ((1 - \sum_{i=1}^{j-1} Prob(P_i)) \times$

$N(P_j)$))$\times \alpha \times \beta$, and the less the paging cost. Since the fitness function is always favor for the paging zone with the highest probability, which implies the lowest paging cost, the best candidate with the lowest paging cost is always selected during the evolution process based on the tournament selection. Our experiments also show that the computation procedure converges within 9-12 generations with optimal partitions for larger number of cells.

2.10 Genetic Operators Used

In this paper, we use three genetic operators (crossover, reproduction, and mutation) to create the new generation [Gol89] as follows:

• *Crossover Operator* :

Two children are generated from the two given parents, parent 1 and 2. Assume the length of a chromosome string is m. The division point x from 1 to $m-1$ is determined randomly by using a uniform probability distribution. The two parents chromosomes are divided at position x into two character string fragments. The fragment A refers to the left hand side of position x and the fragment B refers to the right hand side of position x. The fragment A from parent 1 is copied into child 1 beginning at position 0. The fragment B from parent 2 is copied into child 1 beginning at position x. Likewise, the fragment A from parent 2 is copied into child 2 beginning at position 0. The fragment B from parent 1 is copied into child 2 beginning at position x.

• *Reproduction* :

A child is created by copying the chromosome string of the parent.

• *Mutation* :

First, a child is created by copying the chromosome of the parent. Then one or more digits in the copied chromosome are selected randomly and the selected digits are exchanged by the digits that are randomly selected from 1 to k where k is the maximum number of paging zones in the location area.

2.11 Control Strategy

The tournament selection mechanism is used for both parent selection and old generation replacement selection. The control strategy used in the paging zone configuration genetic algorithm is the steady state control. The details of the steady state control are as follows:

• The population size is fixed over the life of program run.

• The individual is implemented by a single array of characters.

• The parents are selected through the tournament selection method to create new individuals; tournament selection - select n individuals from the population using a uniform probability distribution and select a parent from the n individuals that has the smallest paging cost.

• The individuals are selected through the inverse tournament selection method to be replaced by the new individual; inverse tournament selection - select n individual from the population using a uniform probability distribution and select a individual from n individuals that has the largest paging cost.

• New individuals that are created from the parents by the genetic operators replace the existing individuals in the population.

• The replaced individuals are dead and no longer used as a source of genetic materials in the population.

3 EXAMPLE

The optimal paging strategy proposed in this paper is illustrated in this section with the example of two paging zones.

Step 1) Generating the Histogram

Assume the Mobile Switching System consists of 10 cells, and the data stored in the arrays corresponding to time segments T_2, T_3, and T_4 are the mobile usage data from the time segment T_2:7.00 AM - 9:00 AM, T_3: 9:00 AM - 5:00 PM, and T_4: 5:00 PM - 7:00 PM in a week.

Step 2) Generating the Location Area Layer and the Location Areas

The data corresponding to time segments T_2, T_3, and T_4 in Table 5 are combined to a single table (Table 6) to create the mobile station location area layer for L_2, for example, the 'working area.'

The location area layer L_2 is further partitioned into location areas L_{21} and L_{22}: $L_{21} = \{2, 4, 6, 7, 9\}$ consists of non-zero count cells, and $L_{22} = \{1, 3, 5, 8, 10\}$ consists of zero count cells, where the numbers within the parentheses are the corresponding cell numbers in the location layer.

Step 3) Creating Initial Partition of Paging

a.) Initial Paging Zone Assignment to Cells in the Location Layer

Assume the location area L_{21} is initially partitioned into two paging zones: $P_1 = \{2, 4, 9\}$ and $P_2 = \{6, 7\}$.

b.) Computing the Values of the Fitness Function

At this stage, the fitness function value for each chromosome string is computed. Assume that a chromosome string has the assignment '11221'.

The number of cells in P_1 is $N(P_1) = 3$.

The number of cells in P_2 is $N(P_2) = 2$.

The location probability of P_1:
$Prob(P_1) = 0.17 + 0.35 + 0.03 = 0.55$.

The location probability of P_2:
$Prob(P_2) = 0.24 + 0.21 = 0.45$.

The paging cost for the chromosome string '11221' is calculated as the following:

$C_p = N(P_1) + (1 - Prob(P_1)) \times N(P_2) = 3 + (1 - 0.55) \times 2 = 3.9$

Step 4) Creating New Generation

The tournament selection method is used in order to select individuals (parents) '11221' and '21122'. A genetic operator is applied to these parents to create their children. Assume the crossover operator is used to create the children and the division point 3 is chosen randomly, and two children: '11222' and '21121' are generated.

The inverse tournament selection method selects two replacement individuals, and they are replaced by the children '11222' and '21121'. The Step 4 is repeated 100 times to generate a new generation. At the end, the best individual '21122' is created.

Step 5) Generating Optimal Paging Zones

From the best individual '21122', paging zone $P_1 = \{4, 6\}$ and $P_2 = \{2, 7, 9\}$ are generated from the location layer L_{21}. The expected paging cost is $C_p = 2 + (1 - 0.59) \times 3 = 3.23$ for location layer L_{21}.

In any partition, we always have $Prob(P_i) > Prob(P_j)$ for $1 \leq i < j \leq k$, so here it is also true for $Prob(P_i) > Prob(P_j)$.

4 THE EXPERIMENTAL RESUTLS

The paging zone configuration genetic algorithm ran three times per mobile user. At the first run, we use the genetic algorithm to partition the location area L_{i1} into two paging zones. At the second run, the algorithm partitions the location area into three paging zones and

at the third run, the algorithm partitioned the location area into the maximum number of paging zones such that each paging zone contains only one cell.

The paging costs of each paging zone configuration were compared to the paging cost of system wide paging [MB94] since the system wide paging is widely used in more than 90% of wireless mobile communication systems.

We use the sample data in the service area that consists of 100 cells. The mobile users' location probability in each cell of location area L_{i1} are as shown in Table 8.

As the paging cost analysis shown in Table 9, the paging cost can be further reduced as the number of paging zones is increased. The maximum optimization can also be achieved when the number of paging zones was equal to the number of cells in the location area. However, the number of paging zones in a location area is limited by the constraint in real application. In the wireless mobile communication standard adopted in North America, a cellular phone call connection must be set up and completed, in the worst case, within 10 - 12 seconds. In general, it takes 3 seconds for the system to detect whether a paged mobile responses to the paged message or not. If each of paging zones is paged based on the descending order of the probability value, it implies that the maximum number of paging zones in a location area is constrained by 3 - 4 paging zones in any real applications. For this practical reason, the simulations have been done for only these partitions with the smaller number of paging zones.

The genetic algorithm partitions the location area into the paging zones per mobile user as shown in Table 10:

For the example of Mobile 1:

For the case of two paging zones, the paging zone P_1 consists of cells 6, 7, 8, the paging zone P_2 consists of cells 1, 2, 3, 4, 5, 9, 10, and the optimal paging cost of this configuration is 5.68;

For the case of three paging zones, the paging zone P_1 consists of cells 6, 7, the paging zone P_2 consists of cells 1, 8, 9, the paging zone P_3 consists of cells 2, 3, 4, 5, 10, and the optimal paging cost of this configuration is 4.51;

For the case of ten paging zones, each paging zone contains only one cell in the following order; cell 6, 7, 1, 8, 9, 2, 10, 3, 4, 5 and the optimal paging cost of this configuration is 3.36.

As shown in Table 9 and Table 10, the best optimal paging cost was 1.48 in mobile 6 when the genetic algorithm partitions the location area into 5 paging zones. It saves 98.52compared to the system wide paging strategy whose paging cost is 100 [MB94].

Cell Number	C_1	C_2	C_3	C_4	C_5	C_6	C_7	C_8	C_9	C_{10}	$C_{10}...C_{100}$
Probability	0.16	0.06	0.02	0.02	0.01	0.25	0.18	0.13	0.12	0.05	.....

Table 4: Mobile Location Probability

Cell Number	C_1	C_2	C_3	C_4	C_5	C_6	C_7	C_8	C_9	C_{10}
T_2	0	2	0	6	0	0	5	0	0	0
T_3	0	14	0	25	0	22	16	0	3	0
T_4	0	1	0	4	0	2	0	0	0	0

Table 5: The Location Probabilities for Cell $C_1, C_2, ..., C_{10}$ in Time Segments T_2, T_3, and T_4

Cell Number	C_1	C_2	C_3	C_4	C_5	C_6	C_7	C_8	C_9	C_{10}
Count	0	17	0	35	0	24	21	0	3	0

Table 6: The Location Probability for Cell $C_1, C_2, ..., C_{10}$ at Location Layer L_2

Cell Number	C_2	C_4	C_6	C_7	C_9
Probability	0.17	0.35	0.24	0.21	0.03

Table 7: The Location Probability for Cell C_2, C_4, C_6, C_7, and C_9 in Location Area L_{21}

Cell Number	C_1	C_2	C_3	C_4	C_5	C_6	C_7	C_8	C_9	C_{10}	.....
Mobile 1	0.16	0.06	0.02	0.02	0.01	0.25	0.18	0.13	0.12	0.05	......
Mobile 2	0.16	0.02	0.03	0.63	0.05	0.08	0.01	0.02	0.00	0.00	......
Mobile 3	0.16	0.21	0.09	0.32	0.05	0.17	0.00	0.00	0.00	0.00	......
Mobile 4	0.57	0.02	0.07	0.05	0.01	0.19	0.02	0.01	0.03	0.01	......
Mobile 5	0.05	0.13	0.46	0.01	0.24	0.02	0.07	0.00	0.00	0.00	......
Mobile 6	0.32	0.62	0.02	0.03	0.01	0.00	0.00	0.00	0.00	0.00	......

Table 8: The Probability of the Cells in which the Mobiles Reside

	Mobile 1	Mobile 2	Mobile 3	Mobile 4	Mobile 5	Mobile 6
Sys. Wide Paging	100	100	100	100	100	100
Two Paging Zone	5.68	3.26	3.88	3.92	3.50	2.18
Three Paging Zone	4.51	2.39	3.22	3.05	2.76	1.56
Max. Paging Zone	3.36	1.88	2.64	2.23	2.13	1.48

Table 9: The Paging Costs for Different Paging Methods

Cell Number	Two Paging Zone	Three Paging Zone	Maximum Paging Zone
Mobile 1	{6,7,8}{1,2,3,4,5,9,10}	{6,7}{1,8,9}{2,3,4,5,10}	{6}{7}{1}{8}{9}{2}{10}{3}{4}{5}
Mobile 2	{4}{1,2,3,5,6,7,8}	{4}{1,6,8}{2,3,5,7}	{4}{1}{6}{5}{3}{2}{8}{7}
Mobile 3	{2,4}{1,3,5,6}	{2,4}{6}{1,3,5}	{4}{2}{6}{1}{3}{5}
Mobile 4	{1,6}{2,3,4,5,7,8,9,10}	{1}{3,6}{2,4,5,7,8,9,10}	{1}{6}{3}{4}{9}{2}{7}{5}{8}{10}
Mobile 5	{3,5}{1,2,4,6,7}	{3}{5,2}{1,4,6,7}	{3}{5}{2}{7}{1}{6}{4}
Mobile 6	{1,2,5}{3,4}	{2}{1}{3,4,5}	{2}{1}{4}{3}{5}

Table 10: Partition Results Generated by Genetic Algorithm

5   CONCLUSIONS

The contribution of this research can be summarized as follows:

First, we use cell-based histogram to track mobility pattern. The observation of probability can be made at the cell level instead of location area level[Tab95]. The better resolution of the mobility pattern is achieved.

Second, with the help of the fitness function in terms of mobile probability patterns, we present the optimal partition scheme for paging zones, by applying the genetic algorithm, such that the paging cost is minimized in terms of the defined paging cost function.

Third, our strategy presented in this paper is also dynamic and adaptive. The mobile usage data stored in the histogram can be constantly monitored and dynamically updated.

Fourth, our multi-level partition approach does not only achieve better resolution of mobile probability pattern, but also differentiate the cells with zero probability from non-zero probiblity for each mobile station.

Fifth, our optimal paging algorithm adjusts the size of mobile search domain by observing the individual mobile user's mobility pattern.

We would like to conduct further research with comparisons between our approach and the approach in [Tab95]. One of the key issues is the reasonable design of the simulation model and benchmark for such comparisons. At this time, we believe our approach can be considered as one of the feasible alternatives for the practice in order to improve the performance in existing mobile switching systems.

ACKNOWLEDGEMENTS

The authors would like express many sincere thanks to Dean Edward Lieblein, Barb Edge, and the faculty and staff at School of Computer and Information Sciences, Nova Southeastern University. The authors would also like to thank Dr. Michael J. Laszlo and Dr. S. Rollins Guild for their comments on this research. Many thanks also go to Carol Stern.

References

[BNK94] A. Bar-Noy and I. Kessler. Mobile Users: To Update or Not to Update. In *Proceedings of IEEE Infocom Conference on Computer Communications*, pages 570-576, 1994.

[BNK93] A. Bar-Noy and I. Kessler. Tracking Users in Wireless Communications Networks. In *Proceedings of IEEE Infocom Conference on Computer Communications*, pages 1232-1239, 1993.

[Cim94] I. Cimet. How to Assign Service Areas in a Cellular Mobile Telephone System. In Proceedings of IEEE International Conference on Communications, pages 197-200, 1994.

[ET92] EIA/TIA Interim Standard: Cellular System Dual-Mode Mobile Station - Base Station Compatibility Standard. Telecommunications Industry Association, pages 43-69, 1992.

[Gol89] D. E. Goldberg. *Genetic Algorithms in Search, Optimization, and Machine Learning* Reading Mass.: Addison-Wesley Publishing Company, Inc., 1989.

[FP79] Z. Fluhr and P. Porter. Control Architecture. *The Bell Systems Technical Journal*, AT&T, pages 43-69, 1979.

[HA95] J. Ho and I. Akyildiz. A Dynamic Mobility Tracking Policy for Wireless Personal Communications Networks. In *Proceedings of IEEE Global Telecommunications Conference*, pages 1-5, 1995.

[IV92] T. Imielinski and A.and Virmani. Locating Strategies for Personal Communication Network. In Rutgers Univ. WINLAB Workshop on Networking of Personal Communication Application, pages 1-7, 1992.

[Koz96] J. R. Koza, D. E. Goldberg, D. B. Fogel, and R. L. Riolo. eds., *Proceedings of the First Genetic Programming Conference*, The MIT Press, Cambridge, Massachusetts, 1996.

[MB94] S. Madhavapeddy and K. Basu. Optimal Paging in Cellular Mobile Telephone Systems. In *Proceedings of the 14th International Teletraffic Congress*, pages 493-502, 1994.

[Ste94] K. Steiglitz. Optimization of Wireless Resources for Personal Communication Mobility Tracking. In *Proceedings of IEEE Infocom Conference on Computer Communication*, pages 577-584, 1994.

[Tab95] S. Tabbane. An Alternative Strategy for Location Tracking. *IEEE Journal on Selected Areas in Communications*, 13(5): 880-892, 1995.

[YY95] K. Yeung and T. Yum. A Somparitive Study on Location Tracking Strategies in Cellular Mobile Radio System. In *Proceedings of IEEE Globecom*, pages 22-28, 1995.

DEFINITION OF EXPLORATORY TRAJECTORIES IN ROBOTICS USING GENETIC ALGORITHMS

Maria João Rendas   **Wilfrid Têtenoire**

Laboratoire d'Informatique, Signaux et Systèmes de Sophia Antipolis
CNRS URA 1376, 250 rue Albert Einstein, Les Lucioles 1, 06560 Valbonnne, France.
e-mail: rendas@estrela.unice.fr

ABSTRACT

In this paper we address the problem of defining survey trajectories for robotic platforms equipped of short range perception sensors and using dead-reckoning navigation, explicitly considering the impact of positioning errors in the execution of mapping missions. Assuming that the mapped region is delimited by a set of known objects, each possible exploratory trajectory is evaluated not only with respect to path length and surface coverage but also with respect to the effective utilization of the border of the explored region. We apply stochastic optimization tools (genetic algorithms) to find a set of trajectories which correspond to distinct trade-offs between the individual criteria. New appropriate reproduction and mutation algorithms have been defined, which must deal with the fact that the solution space is the union of several finite dimensional spaces. To cope with the multi-criteria nature of the problem, we use an evaluation/selection strategy based on the notion of Pareto set, which tries to identify the set of non-dominated solutions of the problem. We present results for ideal and real situations, which show that the genetic algorithms are able to find solutions with geometric regularities, learning basic local "filling modes" dependent on the geometry of the region.

INTRODUCTION

This paper addresses the problem of determining a trajectory for exploration of a closed region delimited by known objects, motivated by robotic mapping missions, where an unknown region must be surveyed (in the framework of underwater robotics this can, for example, correspond to the exploration of the inner region of an harbor). This problem is distinct from the traditional path planning problem in that it is fundamentally dependent on the "empty part" of the environment, i.e., in that it must essentially resort to deliberate motions of the robot, being impossible to formulate it as a search for an optimal chain of go-through points, or of a sequence of perception based behaviors.

The mapping problem has been studied by several researchers in the last years. Our work differs from the majority of these studies in that we explicitly address the effect of navigation errors in the execution of survey missions. In fact, the majority of the mapping algorithms that have been proposed consider ideal perfect positioning of the platform along the exploration mission, being based on purely geometric arguments. This kind of robotic task usually has one of two main objectives: either to make sure that if a given kind of object is present it will be detected – one obvious example is mine-hunting – and thus that *all the surface* will be observed; or to acquire an internal representation of the objects present in the explored region, for instance, to be able to study the evolution of the environment between successive observations. In either case, blind application of algorithms that assume perfect platform positioning to a platform using dead-reckoning navigation will fail to achieve the the mission goal: drifting will create "holes" – areas which are not observed, and that increase the probability of not detecting the objects of interest – and deform the internal representation built with respect to reality. With these considerations in mind, our approach is based on giving the robot the possibility of having a better knowledge of the regions that were effectively surveyed, by using available information about the contour of the explored region. Note that the role of perceptual information in this framework is to provide global positioning information at the borders of the explored region, and can be either pre-existing (for instance if the robot started by learning the contour of its workspace) or installed for mission purposes (it can consist, for instance, on a set of landmarks which can be observed using a vision

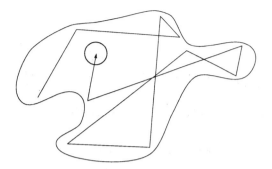

Figure 1: Exploration of a closed region.

system or a set of short-range acoustical positioning systems that cover the border of the surveyed region).

Since for robotic platforms using dead-reckoning positioning uncertainty tends to increase most importantly when the vehicle maneuvers, the exploratory trajectory is assumed to be a sequence of straight line segments, minimizing in this way the accumulated uncertainty with respect to the robot position and orientation.

A second constraint considers that the location of the extreme points of these line segments must coincide with one of the delimiting objects, allowing in this way repositioning of the platform at each of these "contact points".

With these two constraints, the exploratory trajectory t can be defined, for a fixed initial position $p(0)$, by the number of linear segments n_t and by the orientation ϕ of each segment with respect to a global frame, see figure 1,

$$t \leftrightarrow \left(n_t, \{\phi_i(t)\}_{i=1}^{n_t} \right).$$

We further assume that the robotic platform has limited range perception sensors, which define an observed region around each pose of the robot. This region is represented in figure 1 by a gray circle around the robot's position. Our goal is to find a minimal length trajectory which optimizes the use of a priori information, such that the predefined region is completely covered.

PROBLEM FORMULATION

We formalize this problem as a multi-criteria optimization problem, considering the following three criteria :

- the length of the trajectory, $L(t)$;

- the percentage of the complete surface that is effectively observed along trajectory t : $S(t)$;

- the quality of vehicle self-positioning at the vertices of t: $C(t)$.

Our problem can thus be formalized as finding the number n_{t^*}, and the set of orientations $\phi_i(t^*)$, $i = 1, \ldots, n_{t^*}$, such that $S(t^*) \simeq 1$, $L(t^*)$ is minimized, and $C(t^*)$ is maximized. The first two criteria, $L(t)$ and $S(t)$ have an obvious geometric definition, being non-linear functions of the angles that define the geometry of each possible trajectory. They are the two performance indexes considered by the majority of the works on definition of survey trajectories.

The last criterion evaluates the ability of the platform to acquire its position and orientation at the end of each linear segment, and is its explicit consideration is new. We define it using a performance analysis tool based on notions of Statistical Estimation Theory, and which describes the asymptotic properties of Maximum Likelihood parameter estimates, the Cramer-Rao lower bound. Let z be a vector that denotes the available observations, which are probabilistically related to a parameter θ that we whish to determine (estimate), by a known conditionally family of density functions:

$$\{p(z|\theta) \,|\, \theta \in \Theta\}.$$

It is known that the covariance of the estimation error of any unbiased estimator of θ, $\hat{\theta}$, is bounded below by the Cramer-Rao bound $CRB(\theta)$:

$$\mathrm{E}\left\{ \left(\theta\hat{\theta} \right) \left(\theta\hat{\theta} \right)^T \right\} \geq CRB(\theta) = J(\theta)^{-1}$$

where $J(\theta)$ is the Fisher information matrix for parameter θ, defined by

$$J(\theta)_{ij} = \mathrm{E}\left\{ \frac{\partial \ln p(z|\theta)}{\partial \theta_i} \frac{\partial \ln p(z|\theta)}{\partial \theta_j} \right\}.$$

Denote by p_i the pose (position and orientation) of the robot at the i-th "contact point" of the trajectory. We define criterion $C(t)$ as the sum of the volumes of uncertainty (with respect to the vehicle position and orientation) at all the contact points of the trajectory t:

$$C(t) = \sum_{i=1}^{n_t} |\mathrm{CBR}(p_i)| . \qquad (1)$$

Note that the particular definition of this criterion is strongly depend on the available perceptual sensors and on the geometric characteristics of the delimiting objects (shape, texture, etc.).

As it can be easily seen, criteria $S(t)$ and $C(t)$ are highly non-linear functions of the variables of the problem, i.e., this problem is a non-linear multi-variate multi-criteria problem, with the additional difficulty that the dimension of the solution, i.e., the number of linear segments need to cover the complete surface, is not known. Moreover, there is not, in general, *one optimal* solution to the problem, i.e., there may exist several distinct trajectories that lead to the same value of the three criteria. Standard optimization techniques cannot deal with problems with these characteristics. We applied stochastic optimization tools to this problem (genetic algorithms) in order to determine a *set of trajectories* which correspond to distinct trade-offs between the three criteria involved.

DEFINITION OF GENETIC OPERATORS

Considering each possible trajectory as an individual, genetic mutation and reproduction operators have been defined.

For single criteria problem, it is well known that genetic algorithms are guaranteed to find the set of parameter values that yield the extreme values of the optimizing criteria if the set of probabilistic reproduction and mutation operators cover the complete solution space with probability one, i.e, if given any two points t_1 and t_2 in the solution space, there us a sequence of reproduction and mutation operators that apply bt_1 into t_2. Since the solution space of our problem is composed of several disconnected "layers," each one corresponding to the set of feasible trajectories of a given dimension, the usual set of genetic algorithms had to be increased with operators able to change the dimension of the current set of solutions (number of linear segments).

Three different *mutation* operators were designed : elimination of a randomly chosen contact point; introduction of a randomly chosen contact point at a random location; and random modification of a randomly chosen contact point. The first operator enables to decrease by one the dimension of the trajectory, while the second results in an increase of the number of linear segments. The last mutation operator corresponds to the classical genetic mutation operator, and keeps the dimension fixed, enabling the search inside each layer of the solution space. Figure 2 illustrates the three mutation operators.

The probability that the three mutation operators be applied is changed according to the values of the evaluation criteria, in order to speed up convergence process. The choice of the particular mutation operator that is applied is made randomly, favoring, for instance, the application of creation of line segments

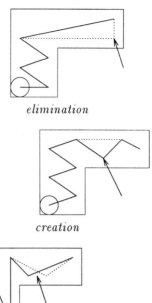

elimination

creation

modification

Figure 2: Mutation operators.

for individuals with low values of the covered surface, and the application of elimination operator for very long trajectories.

The *reproduction* operator (applied to a pair of trajectories) randomly selects two points (one in each parent) and builts a new pair of trajectories by joining the original parent trajectories at the selected points. this operator differs from the traditional genetic operator in that the dimension of the resulting "child trajectories" differs in general from the dimension of the parent trajectories, n_{t_1} and n_{t_2}, being comprised in the interval $[2, n_{t_1} + n_{t_2} - 2]$. Figure 3 illustrates the application of the reproduction operator to two parent trajectories.

The trajectories found by the algorithm must be completely contained in the region to be explored. At each step of the algorithm, a verification is made that the application of the mutation and reproduction operators does not yield invalid solutions. If this happens, another random application of the genetic operators is tried, until an admissible trajectory is generated.

EVALUATION AND SELECTION

Two approaches to the definition of the selection mechanism for this multi-criteria problem have been studied. The first is based on the total ordering of the

parent trajectories

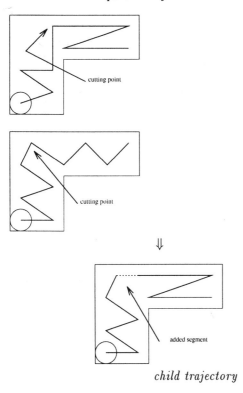

⇓

child trajectory

Figure 3: Reproduction operator.

at an intermediate step, and after a large number of iterations.

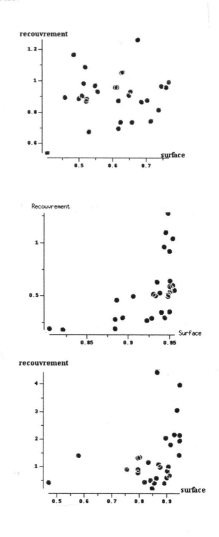

Figure 4: Evolution of re-covered/ observed surface distribution.

population with respect to a single criterion which algebraically combines the values of each separate criterion, and is seen to rapidly lead to strong genetic pressure, leading to strongly homogeneous populations that correspond to one of the possible trade-offs between all criteria.

The second approach uses a selection mechanism based no the notion of Pareto optimality [1], and does not require the total ordering of the population. The individuals in the population are separately evaluated with respect to each of the criteria. For each one, the number of individuals by which it is dominated, i.e., which have better values of *all* the criteria, is counted. The selection process randomly selects the next population, with the probability that a given individual is not selected being proportional to the number of individuals dominating it. This approach has the advantage of identifying a *set* of possible solutions, that correspond to different compromises between the criteria considered. Figure 4 illustrates three snapshots of distribution of the values of percentage of re-covered surface (surface that is covered more than once) and percentage of the total surface that is effectively observed, for three distinct iterations of the algorithm: at the beginning, where a wide spread is observed,

Notice the wide spreading of the initial distribution (which is randomly chosen), and the increasing coverage of the trade-off set by the population of trajectories. This figure shows that the algorithm effectively succeeds in identifying the surface of feasible non-dominated solutions for this problem.

RESULTS

In the first phase of this work, only the requirements of minimal length and complete coverage, which allow an easier and direct geometric interpretation and analysis of the solutions found, have been considered. The interested reader can find the details of the results presented in [2]. Incorporation of the criteria

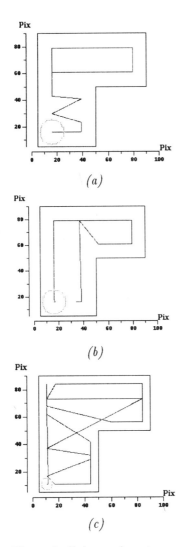

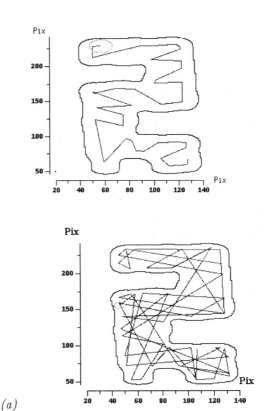

Figure 5: Polygonal contour.

Figure 6: Natural contour.

measuring uncertainty, using (1), presents no conceptual difficulty, implying mainly an increase on computational effort.

Our experiments show that the genetic algorithms are able to "discover" regular geometric solutions, using for instance, sets of parallel lines, diagonals, etc. Figure 5 shows examples of trajectories generated by the algorithm for a simple and geometric contour, for two distinct sizes of the region covered by the perceptual sensors of the robot.

.The two examples in the left of figure 5, which correspond to a large size of sensing region, show that the algorithm is able to find highly regular trajectories, using parallel lines, and zig-zag behaviors. We note that these two clearly distinct solutions to the problem could not be simultaneously obtained using a simple evaluation/selection method which combines the two criteria considered (trajectory length and ob-

served surface), which cold only be obtained using the Pareto approach. The other example in this figure corresponds to a smaller size of the region covered by the perception sensors. Even if the trajectory shown has a more complex geometric structure, it is still possible to recognize the good use of lines parallel to the contour, and of zig/zaging procedures.

In figure 6 we present two trajectories generated by the algorithm (for a large and small sensing region) for a smoother contour, which has been learned by the robot in a previous step. the trajectory shown in the left of this figure corresponds to a large sensing region of the perception sensors of the robot. Een if the contour is not a simple polygon in this case, the solution provided by the algorithm still presents a strong regular geometry, making again use of lines (approximately) parallel to the contour, and of zig-zig segments. This structure is less clear when a more dense recovering of the surface is needed, because the range of the robot's sensors is smaller, as it is the case for the trajectory on the right hand side. However, even if the same basic constructs (parallel lines and zig-zigging behaviors) can still be identified in this case, the nice local coverage of each of the convex components of the surface, which happens in the previous case, is now lost.

CONCLUSIONS

The paper presents a new formulation of the problem of definition of exploratory trajectories, appropriate for platforms having no means of global positioning. We show that genetic algorithms can provide meaningful solutions to the problem. We present examples of exploratory trajectories defined by a conveniently defined genetic algorithm, demonstrating the ability of this tool to exploit the geometry of the problem. Current work integrates the uncertainty criteria in the algorithm, enabling the definition of exploratory trajectories that are "supported" by perceptual knowledge, and whose geometry depends on the local perceptual features of the contour surface, as well as on the global geometrical properties of the surface to be covered.

ACKNOWLEDGMENTS

This work has been partially funded by the European Union through the ESPRIT (Long Term Research) project no. 20185 NARVAL (*Navigation of Autonomous Robots Via Active Environmental Perception*).

References

[1] Carlos Fonseca and Peter Flemming. An Overview of Evolutionary Algorithms in Multiobjective Optimization. In *Evolutionary Computation* , volume 3, number 1, pages 1–16, Spring 1995.

[2] Wilfrid Têtenoire and M. João Rendas. Planification de Trajectoires d'Exploration pour Robots Mobiles avec des Algorithmes génétiques. *CNRS, Laboratoire I3S*, (96-45), September 1996.

PARTIAL EVALUATION OF GENETIC ALGORITHMS

Alberto A. Ochoa Rodriguez and Marta R. Soto Ortiz.

Artificial Intelligence Center.

Institute of Cybernetics, Mathematics and Physics.

Calle 15 # 551 e/ C y D. CP 10400. Vedado. La Habana. Cuba.

ochoa@cidet.icmf.inf.cu

mrosa@cidet.icmf.inf.cu

ABSTRACT

We have introduced the concept of Partial Evaluation (PE) in Genetic Algorithms (GAs) to deal with costly fitness functions. A GAs can be costly for many reasons and the problems are highlighted when a user tries to evaluate the solutions presented by the GAs, or when too much time or resources are required to evaluate the objective functions. Some authors try to solve these problems by using small populations, but often this is not a good solution. PE is intended to evaluate part of the generation or even parts of the individuals. We propose some strategies for solving this problem. Most of the strategies estimate the fitness of individuals who were not completely evaluated.

This contribution discusses a neural network approach to PE. We investigate the relationships between the fitness of an individual, his parents, and the way in which the crossover operation takes place. The results obtained prove to us that the PE is valid.

INTRODUCTION

The goal of this paper is to present an original approach to the problem of dealing with costly fitness functions in GAs [Goldb89][Holl75]. We call this approach Partial Evaluation (PE) in GAs.

Genetic Algorithms are often applied to large problem domains where the domain structure is not very well known and a prior-knowledge is scarce. The aim is for the GAs to yield an efficient exploration of the search space in question. Thus GAs could be considered as a valuable tool for analysis of complex behaviors [Suka91].

Recent interest in a combination of Neural Networks (NNs) and GAs paradigms arises from the wish to make use of the combinatorial power of a GAs to enhance the flexibility of adaptation of the Networks. Themselves behind this motivation lies, of course, the biological roots which both approaches have in common.

A main application of GAs to NNs has been the usage of the GAs as a learning rule to find the weights of Feedforwards-Networks. They have also been used for searching optimal learning rules [Chal90] and efficient net structure and topology [WhiHa89]. It can be said that GAs has been mainly used to improve the performance of the NNs applications. In our research we emphasize the opposite direction of this coupling: *we use NNs to improve the performance of costly GAs.*

The problem of decreasing the cost of GAs is directly related to the amount of evaluations needed to reach near optimal solutions. In the past, this topic had been approached in different ways, one example of them is the steady-state strategy [Sysw89]. Another [Reev93] tries to solve this problem by using small populations. An interesting conclusion of that research was that non-binary alphabets require larger populations than binary ones, and that this size increases with the order of the code. It proposed the use of small population generations, and tries to characterize a minimal population size below which the GAs could not be expected to operate effectively. In contrast to this approach, we prefer to deal with a rather large population size, but keeping the amount of evaluations at a minimum. In occasions, our approach even allows the successful evaluation of a GAs below the theoretical minimum founded in [Reev93].

PE is intended to evaluate part of the generation or even parts of the individuals. We propose some strategies for doing PE; most of them estimate the fitness of the individuals who were not totally evaluated.

WHY DO WE NEED A PARTIAL EVALUATION SCHEME?

When we talk about the cost of a GA, we mean:

- human factor cost,
- resources cost, or
- computational cost (time and storage cost).

Next we give some examples, to illustrate this.

Let us suppose that in an automatic graph drawing system, each chromosome represents a different drawing of one graph [OchRo95]. If we want to obtain highly readable drawings considering the opinion of a person in this task, we must show him every chromosome in each generation. In this example the cost is given by the tedium and the fatigue caused by the task of having to observe each drawing carefully in order to qualify it. Consider that if the GAs only evolves 20 generations with a population of 20 individuals (and these are with security very low parameter values), the user should carry out 400 evaluations. Then, it is more convenient that the person evaluates only part of the drawings and some process evaluates the others, even if this is more time consuming!

Another example is the application of genetic methods in engineering design problems, where the effect of using a given set of parameters has to be found by experiments, or when the evaluation of the design requires a lengthy computer calculation. Such experiments may be costly to set up and run, thus the number of experiments needs to be kept as low as possible.

A final example can be taken from the image processing field. In geometric primitive extraction problems using genetic algorithms, we have to run often the GAs to extract all the primitives of the image. This can be a very costly process and if the objective function is costly, the problem is still harder.

All the previously examples show different forms in which an objective function can be costly, and the where Partial Evaluation could help us.

The next section outlines some strategies for Partial Evaluation implementation of Genetic Algortithms.

SOME STRATEGIES OF PE IN GENETIC ALGORITHMS.

There are different ways to decrease the cost of evaluation of the GAs:

1. Partial Evaluation of the generation: It consists in applying some technique that gives us the possibility of decreasing the number of individuals to be evaluated. This means that only a few individuals of the generation are evaluated and for the rest, its value is estimated.

2. Partial Evaluation of the individuals: It consists in applying some technique that gives us the opportunity of decreasing the cost of evaluation of each individual. Two ways of applying this are:

a. Decomposition of the objective function (OF): The OF breaks down into subfunctions and the value is estimated from the evaluation of some subfunctions.

b. Decomposition of the chromosome: Here, the value of the chromosome is estimated using different sections of the chromosome or subchromosomes.

To achieve these estimations, different techniques can be applied. We can use a small set of heuristic rules (small expert system) or use a neural network trained in learning properties of the relation genotype-phenotype-fitness of individuals and their ancestors.

COMPLEXITY VS COST

The question of what makes a problem hard for GAs has received a good deal of attention as of late. Goldberg [Goldb93] suggests several quasi-separable dimensions of GAs problem difficulty: Isolation, Misleading, Noise, Multimodality and Crosstalk. On the other hand, the question of what makes a problem costly for a GA and how to reduce this cost, has received very little attention from the GA community. We believe that this is a very important problem and there is still much research to be done about how to implement GAs efficiently with respect to the cost of evaluation of its functions.

Among the above mentioned dimensions, Noise is the most significant factor related to both cost and complexity. This relationship is not symmetric. The difficulties in evaluating a fitness function, may lead to noisy results, which increases the complexity of the problem. On the other hand, very often to reduce GA complexity, it is necesary to evaluate a large number of individuals, which increases the cost of the problem.

The complexity and cost of a GA are very weakly correlated: an easy GA problem, can be very expensive, whereas a very hard GA problem can be very cheap. This fact is one thing that makes PE possible and useful. Let us to explain a little more.

PE has an associated computational cost, independent of its implementation. Therefore, in the Neural Networks approach, if we are interested in the reduction of the computational cost of the GA, then the training and construction of the networks cannot be more expensive than computing the objective functions for the estimated individuals.

Note that the class of all problems for a fixed GA complexity, have a range of costs associated with it: from very low to very high computational costs. Assume we can measure complexity by the number N of functions evaluations the GA needs to converge to a solution. Now if PE estimates M individuals, it will be successful for all problems where the time cost to evaluate the M individuals, is more than the time cost of PE implementation.

It is very easy to construct a problem, where we can use partial evaluation with success. Take any fitness function, run GA for K generations, train the networks, and then run the remainder generations until convergence is reached, using the networks to estimate some function values. Now, if the resulting scheme is more time consuming than the GA without PE, try again, but make your functions sleep for a while.

It is worth noting, that occasionally, we are not interested in the reduction of the problem's computational cost, but in some other type of cost, even if the first increases.

A STRATEGY FOR PARTIAL EVALUATION BASED ON NEURAL NETWORKS

The NN approach to PE is based in the following assumptions or necessary conditions:

1. Independence of cost and complexity.

2. Existence of regularities in the evolution process.

3. Minimal negative incidence of the estimation error in GA performance.

4. Reusability.

In this section we will explore a strategy for PE, which uses knowledge of the fitness of two individuals and the place where crossover occurs, in order to estimate the fitness of its offsprings. We analyse the case of a Simple Genetic Algorithm (SGA) [OchSo96] with standard crossover, mutation and roulette wheel selection. The aim of the experiments is only to present some interesting aspects of PE, and outline some of its possibilities.

Three functions were used in our initial experiments. Two of De John's functions[Goldb89]:

$$f(x, y)=100* (y - x^2)^2 + (1 - x)^2, \quad x, y \in R, \quad -1 \leq x, y \leq 1$$

$$f(x, y, z)= x^2 + y^2 + z^2, \quad x, y, z \in R, \quad -5.12 \leq x, y \leq 5.12$$

and one obtained by translating and scaling Gaussian distributions:

$$f(x,y)= abs(3* (1- x)^2* exp((-x^2)- (y+ 1)^2) - 10* (x/5-x^3-y^5)*exp(-x^2 - y^2) - 1/3*exp(-(x + 1)^2 -y^2)), \quad -2 \leq x, y \leq 2, \quad x, y \in R$$

The results we will show in tables 1, 2, and 3 are for the last function.

More complex functions were also studied but presented similar behavioural pattern result with respect to the simple 3_variable PE scheme: <fitness, fitness, cross-point>.

The first set of experiments aimed to explore the satisfaction of the second PE assumption, namely that there exists some regularities in the 3 variable patterns and that their use allows the estimation of the individual fitness.

We train a classification NN with 3_variable patterns taken out from a GA run. Then we used the network to estimate the fitness values of the individuals in a subsequent GA run. We repeat this second step 20 times for each NN trained. Using the actual evaluation of the individuals, we can calculate the percentage of success that is reached in several types of 3_variable patterns. Table 1 presents the average results over 20 runs.

The % of success means the average percentage of times that the neural network returns the correct value for the functions. For example, when crossover break down occurs in the chromosomes at the element X, the network was able to estimate correctly the fitness of an individual 66.06 % of times (independently of the fitness of his parents).

The second set of experiments aimed to explore the satisfaction of third PE assumption.

A GA with Partial Evaluation architecture (GAPE) sometimes calls the objective function and sometimes calls the trained neural network to estimate the objective function value. In order to determine how often the network can be called, we use the percentage of success obtained during learning.

Table 1. Successful estimations for different types of patterns. X and Y are codified in 10 bits.

Pattern type	% of success
Any pattern	61.49
Pattern with good parents	87.88
Pattern with bad parents	40.00
Pattern with CP in the last bit	60.37
Pattern with CP in the first bit	67.45
Pattern with CP in X	66.06
Pattern with CP in Y	55.70

CP- Crossover Point.

It is obvious that the estimation error of the NN evaluation of an individual can introduce variations in the number of offsprings it will have in the next generation. These variations can seriously damage the selection pressure distribution in a generation, and therefore make a GA unable to converge.

In [Reev93] the author characterizes a minimal population size below which the GAs could not be expected to operate effectively. We think that at this minimal limit, a GA running for a few generations must have a maximal affectation from the estimation error of NN evaluation. We decided to study the performance of a GAPE at this limit and compare the results with the performance of its SGA equivalent.

Our experimental results showed us, that provided the second assumption holds, PE is valuable tool to operate a GA effectively at its operation limit. This is very important for real world applications, when we are interested in human factor's or resource's cost.

Table 2 presents a comparison between SGA and GAPE with respect to convergence in a maximization task with the selected function of this paper. We ran both algorithms 350 times. The SGA had to do 20 function evaluations in 5 generations, whereas the GAPE made an average of 17 function evaluations, in 7 generations of 7 individuals. This means that a previously trained NN made an average of 32 estimations. The table shows an evident superiority of the GAPE scheme.

In all the experiments of this paper we used a Fuzzy Artmap Neural Network [CarpGros91]. The confidence level parameters of this network allow us to control both, the size of the clusters in the 3-variable pattern space and the estimation variance of the fitness functions.

We repeated the experiments in table 2 for different values of confidence level (ranging from 0.7 to 0.95) in order to change the estimation variance. The results were similar.

Although we are not addressing this specific problem here, we can give a theoretical explanation as to this result. The selection method we used was the noisiest - roulette wheel which has a high variance. Therefore, if the PE estimation variance is of the same order as that of the variance of the choosen selection method, then the PE will be sufficient. Moreover, if a problem with a given GA-complexity can be solved with a noisy selection method like roulette wheel, we can select a better selection method and let PE work in a less restricted scenario.

At this point arises the problem of when and how to train the NN for the cases of large and small populations. In small populations, seems unrealistic that we can get the sufficient amount of training samples in one run of the GA. In large GA populations the limiting factor is time resource.

The solution to the above problem is not simple and is related to the fourth PE assumption: reusability.

Let us give a very trivial example, to illustrate what we mean by reusability.

Suppose that we are maximizing the function X^2 and codifying the variable X with a 15-bits chromosome. If both parents are "good", then if the crossover occurs at the least significant section of the chromosome then the resulting offspring will also be "good". This PE rule will always work, moreover it will work for many other problems (for instance X^k k>2). Since this is a PE rule, we can simply add it to the implemented evaluation scheme.

For nontrivial relations NN is required. The PE knowledge codified in the NN has to be reusable across several dimensions or spaces:

1. space of variable ranges,

2. space of resolutions,

3. space of parameters,

4. space of chromosome's interpretations.

Let us illustrate with an example, what we mean by interpretation. In many applications, the evaluation of a chromosome depends on a certain set of parameters. For example in an edge detection problem, each chromosome represents an edge configuration of a given image. We consider, that the image defines a chromosome's interpretation.

Experiments carried out in [OchSo96] consisted in training a neural network for a definite range of values of the variables and the use of this training in GAPE afterwards, but in a different range The capability to estimate success across ranges is an. interesting proper-

Table 2. Comparison between SGA and GAPE respect convergence.

SGA vs. GAPE.		
Parameters	SGA	GAPE.
Number of generations	5	7
Population size	4	7
Amount of individuals to evaluate	20	49
How often is found a value < 0.005 ?	0 in 350 runs	72 in 350 runs
How frequent the optimum is found ? (%)	0	20
Amount of evaluations	20	17 (average)

ty for niching problems [Mafh95]. Here, we study the capability of GAPE to estimate success across resolutions.In order to train the neural network, we ran GAs using m bits to codify each variable in to cases: for m=4 and m=10. Then, we calculated the % of success for n bits, where   n > m. n=6, 8, 10 for m=4 and n=12 for m=10. The experiment totaled 20 runs for each pair <m,n>. These results are shown in Table 3.

The capability of a GAPE scheme to estimate success across   resolutions, have far reaching influence in the performance of a GAs. We can think of a GAPE evolving during the first K generations at a given resolution, and then switching on to a greater resolution. The cost of many operations, which occurs at each chromosome bit for example mutation, can be reduced to a large extent. Moreover, some problems are more easily solved at low resolutions, because the search space is smaller.

In this context reusability means that the estimation success across the resolution dimension must be kept at reasonable values. It is worth noting, that not all patterns will be preserved across ranges, resolutions, or space of parameters. However, here is the important issue, that if some patterns are preserved, and we are able to find and use them correctly,   then Partial Evaluation is still possible!

CONCLUSIONS

In this paper we developed a partial evaluation scheme that reduces the overhead of evaluating costly GAs. This is specially important for situations in which the evaluation function is expensive in terms of time, resources or when a human is in the loop. We investigate the relationships between the fitness of an individual, his parents, and the way in which the crossover operation takes place.

We investigate the relationships between the fitness of an individual, his parents, and the way in which the crossover operation takes place.

For the sake of simplicity, we showed the fundamental ideas in simple examples. Nevertheless, we have obtained similar results in far more complex cases.

It is worth noting, that although we had presented a 3-variable PE scheme for SGA, many others variants are possible. For example, we can consider using uniform crossover or other genetic operator and/or parameter.

At the present, we are actively developing this approach

Table 3. Percentage of estimation success across resolutions.

	% of success in estimation			
Patterns	m= 10 bits Mean	n= 12 bits Best	n= 12 bits Mean	n= 12 bits Worst
Any pattern	61.49	63.7	57.6	43.3
Pattern with good parents	87.88	88.7	84.6	81.6
Pattern with bad parents	40.00	70.0	50.7	37.4
Pattern with CP in the last bit	60.37	73.3	65.5	52.6
Pattern with CP in the first bit	67.45	68.9	54.3	50.0
Pattern with CP in X	66.06	69.1	62.8	51.7
Pattern with CP in Y	55.70	57.7	51.4	41.8

in several directions, and are studying its impact in different complex problems.

REFERENCES

[CarpGros91] Carpenter G. A, Grossberg S, Rosen D. B. Fuzzy ART: Fast Stable Learning and Categorization of Analog Patterns by an Adaptive Resonance System. Neural Networks, v. 4, pp. 759-771, 1991.

[Chal90] D. Chalmers.The evolution of learning: an experiment in genetic connectionism. D.S. Touretzky, J.L. Elman, T.J. Sejnowski, and G.E. Hinton, editors, Proceedings of the 1990 Connectionst Models Summer School,San Mateo, California, Morgan Kaufmann, 1990.

[Goldb89] D.E. Golberg. Genetic algorithms in search, optimization and machine learning. Addison Wesley, 1989.

[Goldb93] D. E. Golberg. Making genetic algorithms fly: a lesson from the Wright brothers. Advanced Technology for Developers, 2, 1-8.

[Holl75] J. H. Holland. Adaptation in Natural and Artificial Intelligence Systems. The University of Michigan Press, Ann Arbor, 1975.

[Mafh95] Mahfoud, Samir W, D.E..Golberg Niching Methods for GA. The IlliGAl Report 95001. 1995.

[OchRo95] Ochoa R. A., Rosete S. A. Automatic graph drawing by genetic search. Proceedings of the 11th ISPE/IEE/ IFAC International Confere-nce on CAD/CAM, Robotics & and Factories of the Future. pp. 982-987. 1995.

[OchSo96] A. Ochoa, M. R. Soto and M. García. Partial evaluation in genetic algorithms: an approach based in neural networks.

Technical Report. Institute of Cybernetics, Mathematics and Physics. Havana, Cuba, December 1996.

[Reev93] C. R. Reeves. Using genetic algorithms with small populations. David Shaffer, editor, Proceedings of the Fith International Conference on Genetic Algorithm, p.p. 92-99. 1993.

[SuKa91] K. Susuki and Y. Kakasu. An approach to the analysis of the basins of the associative memory model using genetic algorithms. Richard K. Belew and Lashon B. Booker, editors, Proceedings of the Fourth International Conference on GAs, pp. 539-546, Morgan Kaufmann, July 1991.

[Sysw89] G. Syswerda. Uniform crossover in genetic algorithms. David Schaffer, editor, Proceedings of the Third International Conference on Genetic Algorithm, pp. 2-9, San Mateo California, 1989.

[WhiHa89] D. Whitley and T. Hanson. Optimizing neural networks using faster, more accurate genetic search. David Shaffer, editor, Proceedings of the Third International Conference on Genetic Algorithm, pp. 391-396, San Mateo California, 1989.

KNOWLEDGE DISCOVERY IN
HIGH-VOLTAGE INSULATORS DATA

Manuel Mejía-Lavalle, Guillermo Rodríguez-Ortiz and Gerardo Montoya-Tena[+]
Unidad de Sistemas Informáticos   [+]Unidad de Materiales Eléctricos
Instituto de Investigaciones Eléctricas (IIE), Temixco, Morelos, México, 62490
Email: mmlavalle@iie.org.mx

ABSTRACT

The integration of four artificial intelligence tools that combine the ID3 algorithm and the case-based reasoning method of the nearest neighbor are proposed to solve the problem of characterizing the flashover on high-voltage insulators. The first tool uses data mining to build a classification or decision tree from historic data, the second, generates production rules, the third, operates the decision tree as an expert system, and the last, makes tests with known cases to evaluate classification accuracy. These tools are applied to the high-voltage insulators flashover problem, and the results are compared against other machine learning tools: C4.5, FOIL, CN2 and OC1.

Key words: Data Mining Tools, Knowledge Discovery, Machine Learning, Electric Power Application.

INTRODUCTION

Data mining has been employed with success in various fields and in many real world problems, as in large-scale telecommunications networks [SS96], in nuclear fuel plant production process [Lee86], or in electric energy generation [Her95]. Data mining is applied to huge volumes of historical data mainly with the expectation of finding knowledge, or in other words, when it is sought to determine trends or behavior patterns that permit improve the current procedures of marketing, production, operation, maintenance, or others. In summary, data mining or knowledge discovery, is the nontrivial extraction of implicit, previously unknown, and potentially useful information from data [Sha+91].

Some of the traditionally used computer techniques to accomplish data mining are: neural nets, induction of decision trees, decision rules and case-based methods.

In particular, the advantages of the induction of the decision tree technique, based on the ID3 algorithm [Qui79, Qui86], are:

● The knowledge is extracted directly from the recorded historical data.

● Is not necessary to accomplish a knowledge acquisition stage with a human expert, which is considered a bottle neck when building expert systems [MSR+91].

● Updating knowledge is a simple process, which is appropriate when knowledge and data are changing dynamically with time.

● Opposed to classic neural nets, decision trees describe the knowledge discovered in explicit way and the trees can be easily transformed to production rules.

● ID3 is an algorithm that does not need much computational time in the learning stage when knowledge is extracted. This is important for on-line applications.

In this paper four artificial intelligence tools developed at the Electrical Research Institute (IIE) are described. These tools are based on the integration of the ID3 induction algorithm and the nearest neighbor case-based method. The first tool builds the decision tree from historic data, these records are called the learning examples and this tool performs the learning stage or extraction of knowledge; the other three tools are components of the exploitation stage and operate using the decision tree previously created by the induction tool to:

a) Transform the decision tree into production rules so that the user can easily interpret the extracted knowledge.

b) Operate as an expert system, in which the user can present hypothetical cases, and the system makes questions until a conclusion is reached.

c) Measure the classification accuracy or efficiency of the decision tree using known cases that are different from the learning examples, and to obtain the proportion of these cases correctly classified by the tree. This measure is in itself an estimate of how reliable is the extracted knowledge.

In this paper, the high-voltage insulator flashover problem is described together with its characteristics that are appropriate for the employment of the developed tools. After, the results are presented and compared with more complex machine learning tools: C4.5, FOIL, CN2 and OC1. The paper concludes with a summary of the results. Although this paper focuses on the high-voltage insulator flashover problem, our work can be applied to other types of similar problems as well.

HIGH-VOLTAGE INSULATOR FLASHOVER PROBLEM

The high voltage lines that transmit electricity are normally supported using posts or towers. Non-conducting material pieces or external insulators are placed between the line and the post to prevent the establishment of electric conduction between the energized line and the post. The external insulators are subject to environmental action. With time the isolating surface is covered with conducting pollutant particles such as industrial or saline deposits, to such a degree that an electric flow arc, or flashover, between the line and the post can be established which damages the unit and causes interruption of power supply.

The traditional solution to the problem has been to clean the insulators periodically with water jets. However, often this approach is not adequate due to the very large number of insulators and the limited availability of cleaning teams. Additionally, as environmental action is not constant, it sometimes happens that insulators are cleaned unnecessarily.

An alternative solution is to have a system that forecasts the imminent flashover occurrence using the predicted value of the surface resistance in sufficient time for the insulators to be cleaned before failure. To demonstrate that this method is effective and to determine the relationship between surface resistance and environmental conditions, equipment to measure environmental variables every hour was installed in an electrical substation. The variables recorded are wind velocity and direction, barometric pressure, precipitation, atmospheric and dew temperature, and time. From these data, the absolute and relative humidity are calculated.

Also, equipment to measure and record surface resistance (in kilohms) on the polluted insulator was installed, such that a low resistance measurement indicates a much greater probability of flashover occurrence. A technical discussion of the problem can be found in [Gon93, Ham64, FRM95]. The measurement data are stored in files (one column for each variable), and chronologically ordered; these data are the basis for the knowledge discovery algorithm computation.

Before attempting to accomplish the surface resistance forecast three fundamental questions have to be answered:

a) The environmental variables that are being registered are adequate, so that a true relationship can be established between them and the surface resistance ?

b) Which variables are the most influential over the surface resistance behavior ?

c) What behavior pattern of the environmental variables causes that the surface resistance be low ?

Given the characteristics of the flashover problem, the three previous questions can be solved using artificial intelligence techniques, specifically the induction of decision trees by data mining, as it will be shown in the following section.

DESCRIPTION OF THE PROPOSED TOOLS

In this section the four artificial intelligence tools developed at the Informatic Systems Unit of the Electrical Research Institute (IIE) are described, the data mining algorithm for knowledge extraction is based on a combination of the ID3 algorithm and a case-based nearest neighbor approach.

Decision Tree Generator Tool (DTGT)

This tool is used in the learning stage or extraction of knowledge and builds the decision tree from historic data. The data set is organized in a table with rows and columns, where the rows are the examples, and each example is composed of a fixed number of attributes (the number of columns), one of them is the class to which the example belongs. The tool is based on the ID3 algorithm presented in [Win94] and it performs, on the data set, the steps of Table 1.

Table 1 ID3 Algorithm

A. Find the attribute that better divides the data set into homogeneous subsets: for each attribute, calculate the disorder or entropy according to the following formula:

$$E = \sum_r [Nr/Nt] \left[\sum_c \{-(Nrc/Nr) \log_2 (Nrc/Nr)\} \right]$$

where
Nr = number of examples in branch r
Nt = total number of examples in all branches
Nrc = total of examples in branch r of class c

B. The attribute which has the smallest value of E is taken as the root node of the tree (attribute-node) and there will be one branch for each value that the attribute has[1].

C. For each value of the attribute-node, select all the examples (rows) with the same attribute value. For each subset do the following:

C1. If all examples belong to the same class, the branch is labeled with the class.

C2. If the subset is empty, the branch is labeled as "unknown class".

C3. If the examples in the subset belong to different classes, go to step A, with this subset as the new data set.

D. If there are branches without labels, go to step A, otherwise finish.

This algorithm is called "ID3 complete with unknown classes" or simply ID3-UNK, and step C2 has been modified in the way shows by Table 2.

This modified algorithm is a combination of the ID3 algorithm and of the case-based nearest neighbor approach, and is called "ID3 complete to the nearest neighbor" or simply ID3-NEA.

The two algorithms were coded in C language. The tools take as input data a text file that contains: a) a dictionary with the classes, attributes and the attribute values, and b) a data set (examples) organized as a table.

In order to get faster response in the classification process and to use less amount of main memory, a routine reads this file and transforms all the values that are 'character strings' into numbers. For example, if the first class of the dictionary is "low", the routine assigns the number zero "0", "1" to the next class, and so on. With this transformation the process time is reduced since the

searches are accomplished with numbers instead of character strings. Also, the space in main memory is reduced since an integer number uses less space than a character string. Furthermore, the memory arrays (matrices and vectors) necessary to store the information are dynamically created by the tool.

The current version, that runs on a personal computer, has the following maximum parameter values:

Maximum number of characters per attribute value: 20
Maximum number of class names: 127
Maximum number of attribute names: 127
Maximum number of values per attribute: 127
Maximum number of nodes: 9,000

At the end of the process the tool generates a text file with the classification tree (decision tree) in a special format with links that are interpreted by the three tools that are described in the following sections.

---

[1] The idea of a *complete* tree is taken from [Guz95].

Table 2 Modified ID3 Algorithm

C2.   If the subset is empty, find the most similar example (smaller distance) to the current branch; if the distance is acceptable (according to certain threshold previously defined), label the branch with the class of the most similar example, otherwise, label the branch as "unknown class".

Discovered Knowledge Visualization Tool (DKVT)

The DKVT tool serves to visualize the discovered knowledge in the examples as production rules: it takes as input data the file with the decision tree and transforms it into production rules which are stored in another file.

Expert System Consulting Tool (ESCT)

This tool takes as input data the file with the tree generated by DTGT, then, it starts a session where the user interacts with the expert system until a conclusion is reached. In this fashion the decision tree is used as the knowledge base of the expert system.

Discovered Knowledge Efficiency Measurement Tool (DKEMT)

This tool takes a set of test cases in 'batch' without intervention of the user: takes as input the file with the tree generated by DTGT, a file with test cases, and shows the results on the screen.

The test cases file does not include the dictionary section, only the table with the test cases. The tool will calculate the classification accuracy, indicating the proportion of cases that were correctly classified. This tool has an additional feature: if the table does not include the class column, then the tool will classify (or assign a class to) each one of the test cases.

EXPERIMENTAL RESULTS AND COMPARISONS

Results Using the Proposed Tools

To perform the experiments, data from an N-120P high-voltage insulator were registered during 21 days (504 examples of meteorological and surface resistance values). The attributes used by DTGT are: hour of the day, wind direction, wind velocity, temperature, precipitation, dew temperature, barometric pressure, relative humidity and absolute humidity. The class attribute is the surface resistance.

The values of each column in the data set are substituted by class values taken from a set of 10 values that represent class intervals; for example, for the hour of the day column the first class interval goes from 0:00 to 2:20 hours, the second class interval goes from 2:21 hours to 4:40 hours, etc. Three classes were assigned to the surface resistance: "low" (for values between 1,013 and 3,489 kilohms), "medium" (for the interval 3,490/7,992) and "high" (for the interval 7,993/11,482).

Environmental Variables and Surface Resistance Relationship

Using the ID3-NEA algorithm and the data set, DTGT obtained a decision tree in 3.68 seconds using a PC with a 80486 processor at 33 MHz. Next, the classification tree was loaded to the DKEMT tool, then, with a new data set containing 9 days (216 test cases) of randomly selected records, DKEMT obtained a classification accuracy of 82.87% (179 cases correctly classified and 37 wrongly classified). Figure 1 shows the comparison of the real surface resistance and the one obtained by the decision tree (ID3-NEA).

This result establishes that the recorded environmental variables determine the surface resistance behavior in more than 80% of the time.

The Most Influential Variables

To determine the most influential meteorological variable in the behavior of the surface resistance, the DKVT tool was used to generate the production rules from the extracted knowledge. Analyzing these rules, the number of times that a variable appears in the whole set of rules can be determined and is shown in Table 3.

From this table, it is found that the relative humidity variable appears in all the rules (except in default rule), so that, it is clear that this variable is important in the surface resistance behavior. Additionally, the precipitation and temperature variables hardly appear in the rules (in less than 6% of the rules), that is, they do not affect to a large extent the surface resistance.

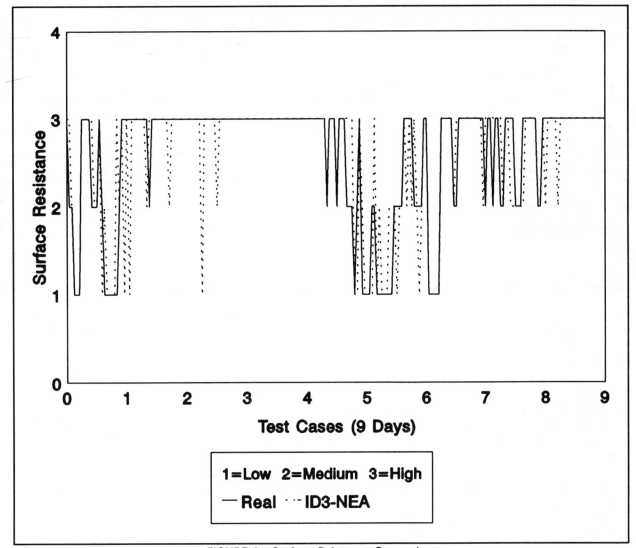

FIGURE 1   Surface Reistance Comparison

Low Surface Resistance Characterization

From the rules that have as a conclusion surface resistance "low", the circumstances that induce toward the flashover can be observed. For example, the following rule,

IF	Relative Humidity = 60.5/65.9
and	Dew Temperature = -6.0/-3.9
and	Hour of the Day = 0:00/2:20
THEN	Surface Resistance = 1013/3489

indicates that the surface resistance is "low" when the relative humidity is high, the dew temperature is low and

the hour of the day is between 0:00 and 2:20 hours.

Finally, using the ESCT, the domain expert can outline hypothetical cases and observe how the surface resistance behaves.

Comparison against C4.5, FOIL, CN2 and OC1

With the same data sets described above, the tools C4.5 [Qui89], FOIL [QC93], CN2 [CN88] and OC1[MKS94] were used; the results obtained are summarized in Table 4.

From this table it can be observed that:

Table 3 Number of Times that a Variable Appears in the Rule Set

VARIABLE	FRECUENCY	PERCENTAGE
Relative Humidity	495	99.79
Hour of the Day	391	78.83
Wind Direction	282	56.85
Dew Temperature	174	35.08
Barometric Pressure	160	32.25
Wind Velocity	88	17.74
Absolute Humidity	63	12.70
Temperature	25	5.04
Precipitation	10	2.01

● The classification accuracy to classify the same examples used in the learning stage is better for ID3-UNK and ID3-NEA, since these algorithms generate a complete decision tree.

● The accuracy to classify known test cases is similar for ID3-NEA, C4.5 and CN2 (82%), though ID3-NEA requires less time to generate the decision tree, and it can be a better option for real time environments, like our problem.

● C4.5 generates less number of rules, and apparently it is easier to understand the knowledge extracted with this tool. However, for the particular case of the high-voltage insulators problem, it is not explicitly clear what are the conditions when the surface resistance takes on the "low" values since C4.5 defines the class "low" as the default class.

● Although ID3-NEA obtains the biggest number of rules (496), this is an advantage since it is possible to analyze how frequent a variable appears in the rules (as shown above) and on the other hand, the number of rules can be reduced to only 224 if the class "high" is considered as the default class or can be reduced still to 89 rules using simple similarity properties, but this was not done to preserve the detail of the extracted knowledge.

CONCLUSIONS

Data mining tools are valuable algorithms to extract the scattered knowledge hidden in large volumes of information and help to detect patterns and behavior trends.

For the particular case of the high-voltage insulator flashover problem we could successfully determine:

● The relationship between the environmental variables and the surface resistance.

● The variables that have more impact on the phenomenon, and the variables that hardly affect the surface resistance behavior.

● The circumstances that cause the surface resistance to be low.

In the near future we will work in the simplification and reduction of the number of rules generated for the DKVT tool, to facilitate the interpretation of the discovered knowledge.

ACKNOWLEDGEMENTS

Authors wish to thank Marco A. Ponce and José L. Fierro for their support during the project and for his valuable suggestions.

REFERENCES

[CN88] P. Clark, T. Niblett. *Manual for CN2 release 1*, The Turing Institute, 1988.

Table 4 Machine Learning Tools Comparison

TOOL	ID3-NEA	ID3-UNK	C4.5	FOIL	CN2	OC1
TEST-Acc	82.9 %	72.2 %	82.9 %	62.9 %	82.4 %	80.1 %
TRAIN-Acc	99.6 %	99.6 %	85.3 %	96.8 %	92.0 %	81.5 %
# Rules	496	253	9	61	52	NA
Default	Unknown	Unknown	Low	NA	High	NA
Time(secs)	3.68	3.13	143	10.7	74	643

where:

TEST-Acc =   Accuracy to classify known cases.
TRAIN-Acc =   Accuracy to classify the same examples used in learning stage.
Rules =   Number of generated rules.
Default =   Default class (if no rule applies).
Time (secs) = Seconds required to generate the results.
NA =   Characteristic not available by the tool.

[FRM95]   J. Fierro, I. Ramírez, G. Montoya. Monitoreo en línea de la corriente de fuga de cadenas de aisladores para 400 kV en zonas de contaminación. In *Memorias de la Reunión de Verano del Capítulo de Potencia* (IEEE Sección México), Acapulco, México, 82-88, July 1995.

[Gon93]   J. González. La Resistencia Superficial como Criterio de Diagnóstico del Aislamiento Externo expuesto a Contaminación. In *Memorias de la Reunión de Verano del Capítulo de Potencia* (IEEE Sección México), Acapulco, México, 9-13, July 1993.

[Guz95]   A. Guzmán. Sustitución de Sistemas Expertos por Arboles k-d. In *Congreso de Reconocimiento de Patrones, CIMAF-95*, La Habana, Cuba, January 1995.

[Ham64]   B. Hampton. Flashover Mechanism of Polluted Insulation. In *Procceedings IEE*, Vol 111, No. 5, 986-992, May 1964.

[Her95]   V. Hernández. Sistema de Descubrimiento de Conocimiento en Bases de Datos. Tesis de Maestría, ITESM-Morelos, 1995.

[Lee86]   W. Leech. A Rule Based Process Control Method with Feedback. In *Proceedings of the International Conference and Exhibit, Instrument Society of America*, 1986.

[MKS94]   S. Murthy, S. Kasif, S. Salzberg. A System for Induction of Oblique Decision Trees. *Journal of Artificial Intelligence Research 2*, 1-33, 1994.

[MSR+91] M. Mejía, E. Sánchez, G. Rodríguez, H. Cuevas. EXEB: A Varnishes Expert System. In *Operational Expert Systems Applications in Mexico*, F. Cantú eds., Pergamon Press, 40-47, 1991.

[QC93]   J. Quinlan, R. Cameron. FOIL: a Midterm Report. In *Proceedings European Conference on Machine Learning*, Springer Verlag, 3-20, 1993.

[Qui79]   J. Quinlan. Discovering Rules by Induction from Large Collections of Examples. *Expert Systems in the Micro-Electronic Age*, D. Michie eds., Edinburgo, Escocia, Edinburgh University Press, 1979.

[Qui86]   J. Quinlan. Induction of Decision Trees. *Machine Learning*, Vol 1, Num 1, 81-106, 1986.

[Qui89]   J. Quinlan. *C4.5: Programs for Machine Learning*, Morgan Kaufmann Publishers, 1989.

[Sha⁺91]   G. Piatetsky-Shapiro, et al. Knowledge Discovery in Databases: An Overview. In *Knowledge Discovery in Databases*, G. Piatetsky-Shapiro eds., Cambridge, MA, AAAI/MIT, 1-27, 1991.

[SS96]   R. Sasisekharan, V. Seshadri. Data Mining and forecasting in Large-Scale Telecommunications Networks, *IEEE Expert*, Vol. 11, No. 1, 37-43, February 1996.

[Win94]   P. Winston. *Inteligencia Artificial*, Addison-Wesley Iberoamericana, 455-475, 1994.

A FRAMEWORK FOR INTEGRATING DSS AND ES
WITH
MACHINE LEARNING

Saleh M. Abu-Soud
Department of Computer Science
Princess Sumaya University College for Technology
Royal Scientific Society
Amman - Jordan 11941
{abu-soud@mars.rss.gov.jo}

ABSTRACT

Knowledge Acquisition is a major problem in developing ES, it is costly and time consuming. An alternate to the acquisition of knowledge through the dialog with an expert is to convert a history of previous cases into a set of IF-THEN rules. DSS are considered as rich sources of such cases and can be integrated with ES to build an integrated decision expert system with automatic knowledge acquisition capability. In this paper, three contributions have been presented: (1) A framework for integrating DSS and ES has been proposed, (2) In this framework, a new inductive learning algorithm (called ERIC) for extracting rules from DSS has been presented Experiments showed that the results of this algorithm are better than the results of some well-known algorithms in the field of inductive learning, and (3) A DSS and an ES have been built and integrated with each other using the proposed framework.

Keywords: *Artificial Intelligence, Machine Learning, Decision Support Systems, Expert Systems, Inductive Learning, Knowledge Acquisition.*

1. INTRODUCTION

Expert Systems (ES) are considered as one of the most significant branches of AI, since they have been shown to be successful in a wide range of commercial products as well as an interesting research tools. An ES is a special kind of a system that employs human knowledge captured in a computer to solve problems that ordinarily require human expertise in a specific domain.

[RK91,Tur88,LS93,Pat90] Decision Support Systems (DSS) on the other hand, are interactive systems built to assist managers in their decision making process when they deal with semi and unstructured tasks.[Gri80, DS83,SW87, Thi88]

Most ES and DSS are usually built separately from each other and unrelated. However, integrating both systems may enhance the quality of both systems or one of them depending on the aim and the way in which they are integrated. Each system can be fitted into the other to enhance one or more of the capabilities of the other system. For example, as it is known, DSS can answer *what if?* questions, but merging an ES into the components of the DSS will make the explanatory facilities of the DSS more powerful, i.e. it will be able to answer the *why?* and *how?* type of questions. While the integration of DSS into the components of ES, will give the ES an opportunity to use the huge DSS's database.

On the other hand, the output of one system can be the input of the other. For example, ES produces decisions that can be used as input to a DSS to help it in identifying the problem and to produce suitable solutions for this problem. On the contrary, the output of a DSS can be fed into an ES to produce a good decision out of the several suggestions resulted from the DSS.

The idea of integrating DSS and ES was originally appeared in 1976 by [Min76]. Since that time, some other attempts and suggestions have been made to integrate the two systems.[Bon[+]81,Bon[+]84,Gou[+]84, TW86]

The quality of knowledge plays a great role in the success or the failure of ES. So, high attention must be given to the process of acquiring knowledge.

Unfortunately, knowledge acquisition is considered as the bottleneck and even the most exhaustive phase in building ES in both time and cost.[MD95]

This problem can be eliminated by the fact that DSS and ES can be connected in a way in which the results of the quantitative analysis that are produced by the DSS are compiled, revised and then converted to a form (rules for example) that is acceptable by the ES as an input. In this case, the DSS will act as an expert, that is, the source of knowledge. With this process, the task of acquiring knowledge will be automated, which means that there will be no need to the whole traditional knowledge acquisition process. This will yield to a cheaper and faster implementation of ES.

A DSS-ES interconnection framework can be built with machine learning techniques, more specifically with learning by induction. Inductive learning algorithms are usually built to extract IF-THEN rules from a history of previous cases (called training examples in machine learning literature). DSS can be considered as a rich source of such examples, this is because the output of DSS are in the form of iterative/interactive reports in which a plenty of decision cases can be found, which in turn, can be transformed to production rules that can be fed into an ES.

There exist several approaches for extracting rules from training examples. The most common used approach is to form a decision tree from training examples. This approach is efficient, especially for large sets of examples. ID3 [Qui83] is considered as the mainstay of 'decision-tree' based algorithms so far. It is simple and very efficient especially when the set of training examples becomes large. Many derivatives of ID3 were proposed, for example, ID4 [SF86], ID5 [Utg88], GID3 [Ira+93], and C4.5 [Qui93]. The algorithms based on this approach do not always produce

the most general rules, because the resulting rules usually contain unnecessary conditions that make the rules more specific, in addition to that, the resulting decision trees are always too large.

To overcome the problems associated with dec sion-tree algorithms, many non decision-tree algorithms are proposed by other researchers. Some known algorithms from this category may include CN2 [CN89], BCT [Cha89], RULES [PA95], and AQ family such as AQ11 [ML78] and AQ15 [MMH+86].

In this paper, a new non decision-tree inductive learning algorithm (called **ERIC**) is presented. This algorithm generates rules that are more general and simpler than rules obtained from some well-known algorithms in the field of inductive learning, such as ID3, AQ family, CN2, and C4.5. Then a framework for integrating DSS with ES is constructed using this algorithm. The proposed algorithm converts the output of the DSS into a set of production rules that constitutes the knowledge base of the ES. At the end, a real DSS for Students Advisory System called DSSAS, in addition to an ES for the same purpose called ESSAS, have been built and then integrated in one system called *Integrated Decision-support Expert Advisory System* **(IDEAS)**, using the proposed framework.

2. THE BASIC STRUCTURE OF ES
2.1 COMPONENTS OF ES

An Expert System usually consists of the following basic components: 1) *The Knowledge Base*, which contains *facts* and *rules* that are necessary for understanding, formulating, and solving the problem. This component can be built as a separated part in the ES. If this is the case, then it can be replaced with other knowledge bases. This situation gives us the ability for using the ES for different purposes. 2) *The Inference Engine*: It is considered as the *brain* of the ES. It provides a methodology for reasoning about information in the knowledge base, and for formulating conclusions. 3) *Knowledge Acquisition System*, which is defined as the process of accumulating, transfer, and transformation of expertise from some knowledge source, such as books, experience, data base, reports, or any other sources, to a computer program for constructing or expanding the knowledge base[Tur88], 4) *User Interface:* ES must have a user friendly interface to ease the communication between the ES and the user. Figure 1 shows the basic components of an ES.

2.2 THE BOTTLENECK IN DEVELOPING ES

Knowledge Acquisition process is considered as one of the major bottlenecks in developing ES. This process

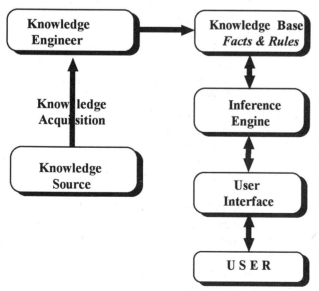

FIGURE 1. The Basic Structure of an ES.

is time consuming, slow, and costly. An alternate to acquiring knowledge through an interaction with an expert is to convert an existing database or a history of cases into a set of IF-THEN rules[Qui83].

Unfortunately, until now no automatic systems are available to automate this process. All what we have are some simple systems aim at helping in extracting rules or decision trees from a set of training examples.

Some of these systems are: RULE-MASTER from Radian Corporation, EXPERT-EASE from Export Software International, TIMM from General Research Corporation, and XiRULE from Attar Software Corporation. Most of these systems are independent products that accept examples and produce rules or decision trees that can be used as knowledge for ES.

3. THE STRUCTURE OF DSS

A DSS should have at least three major components, which are: 1) *A Data Base* : this components constitutes a rich source of data for the DSS. Most successful DSS have found it necessary to create a DSS data base that is logically separated from other data bases. This data base must be characterized with its flexible nature in the extraction process and in the quick response to unanticipated user requests. 2) *A Decision Model Base*: A promising characteristic of DSS is their ability to form decision models available to decision makers. 3) *A Software System*: This component consists of three subsystems : a) Dialog Subsystem, which is the interface between the user and the DSS. It collects the user's inputs to the DSS and produces the output of the DSS in the form interactive/iterative reports. b) Decision Management: This part is responsible for generating new and updating the existing decision models. It is also responsible for generating the interactive reports which constitute the output of the DSS, and c) Data Base: The users of DSS need to interact repeatedly and creatively with a relatively small sets of data. So, data extraction that allows additions and deletions to a DSS data base from large transactional data base is the critical area. Figure 2 depicts the basic components of a traditional DSS.

4. THE *'EXAMPLES-TO-RULES'* INDUCTIVE CLASSI FIER (ERIC)

ERIC generates rules from a set of training examples. Examples must be organized in a two dimensional array. A row in this array corresponds to an example and each column contains an attribute. As an initial requirement, all examples are not marked. Figure 3 shows the flowchart of ERIC.

4.1 EXPLANATION OF ERIC

As an illustration of the operation of ERIC, consider the training set for Human Identification Problem [Qui83] which is shown in Table 1. This set consists of three attributes: Height with values (short, tall), Hair with values (blond, red, dark), Eyes with values(blue, brown), and a class attribute Class with values (+,-).

Applying the steps of the algorithm, the class value '+' is considered first, and then all combinations, with one attribute each, are formed. These combinations are: {Height}, {Hair}, and{Eyes}. For this class value, the probabilities of the values of all combinations that belong to the class '+' but not to other classes are computed. It is noted that the Hair value of 'Red' is the only value that satisfies this condition, with probability of 0.333. So the following rule is formed: *1- IF Hair is red THEN Class is '+'.* and the examples that contain the value 'Red' are marked as classified. Since there are still two unmarked examples in the class '+', and no single values exist that satisfy the previous condition, we will increase k by one and form combinations of attributes with two values. These combinations are: {Height, Hair}, {Height, Eyes}, and {Hair, Eyes}. Now, we will search for the values under each of these combinations that appear under the class '+' but not '-'. If we take a look at the training examples, it is easy to note that the combination of values {blond,blue} which appear under the combination of attributes {Hair,Eyes} is the only combination that satisfies the condition, with probability 1.0. So the following rule will be added: *2- IF Hair is blond AND Eyes is blue THEN Class is '+'*, and the related examples are marked as classified.

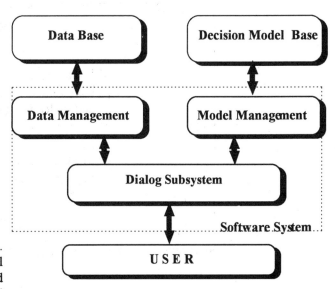

FIGURE 2. The Basic Components of a DSS.

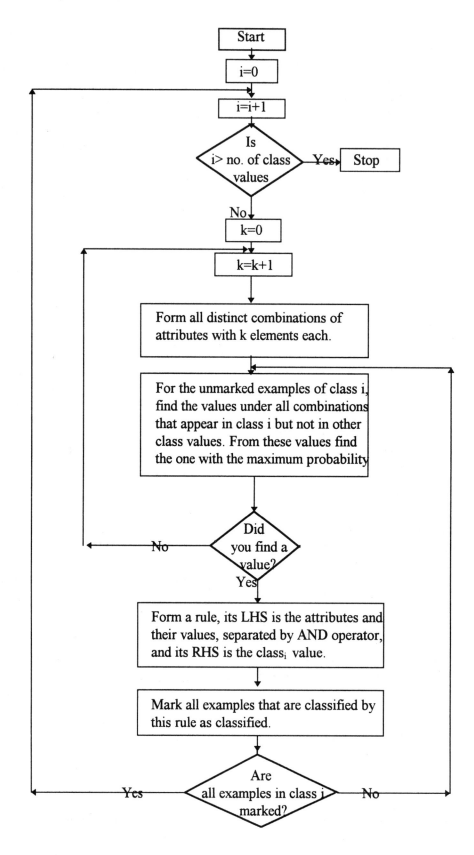

FIGURE 3. The Flowchart of ERIC.

Since all examples under the class '+' are marked as classified, we will move to the second class '-', and repeat the same steps as done for the examples in class '+'. We will find that the value 'dark' of the attribute Hair is repeated three times with probability 0.6, and the value 'brown' of Eyes is also repeated 3 times with the same probability. Since both values have the same probabilities we can take any one of them. Let us consider the value 'brown', then the following rule is extracted: *3- IF Hair is dark THEN Class is '-'*, and the related examples that are classified by this rule are marked as classified. Two examples are still not classified, namely, the fourth and fifth examples. If we repeat the previous process on these examples, it will be found that the value 'dark' of Hair is repeated twice with probability 1.0 and occurred in class '-' but not '+'. So the following rule is added: *4- IF Hair is dark THEN Class is '-'*, and these two examples are marked as classified. Now, all examples are marked as classified. This will cause the algorithm to halt with the extracted four rules as its output.

4.3 EXPERIMENTS WITH ERIC

Two training sets are used for testing ERIC: 1) Balance training set with 625 examples and 5 attributes, and 2) Tic-Tac-Toe training set with 958 examples and 10 attributes. These training sets are obtained from the University of California Irvine Repository of Machine Learning Databases and Domain Theories.

For comparison purposes, the results of applying ERIC on these training sets are compared with applying two well-known inductive algorithms, namely ID3 and AQ, on the same sets. Figure 4 summarizes the results of the three algo rithms using the above training sets.

It is noted from Figure 4 that the results obtained from ERIC are much better than those obtained from ID3 and AQ. Figure 4-a shows that ERIC produced less number of rules than other algorithms, also it produced

Table 1. The Training set for Human Identification Problem

Height	Hair	Eyes	Class
short	blond	blue	+
tall	blond	brown	-
tall	red	blue	+
short	dark	blue	-
tall	dark	blue	-
tall	blond	blue	+
tall	dark	brown	-
short	blond	brown	-

rules with less number of conditions as depicted in Figure 4-b. Rules with less number of conditions are better because they are more general and can classify more unseen examples (examples that are not listed in the table of training examples). For example, one can easily note from the figure that for Balance training set, ERIC extracted 303 rules with 3.41 average conditions in the L.H.S. of the rules, while ID3 extracted 401 rules with average conditions of 3.85 and AQ produced 312 rules with 3.53 average condition.

5. THE DSS-ES INTEGRATION FRAMEWORK

The integration framework, as depicted in Figure 4, consists of three components: A DSS, an ES, and an integration interface between the two systems. These two systems can work independently from each other. The user of DSS needs not to run the ES unless he likes to do so, and vice versa concerning the ES. In addition, the user of one system can run the other to enhance or to refine the obtained output. The integration interface consists of three stages. In the first stage, when the output of the DSS is displayed to the user on the computer screen, a copy of this output is logged into a file. This process is repeated whenever there is a user working with the DSS. The logged output is not limited to the final decision or advise, but contains also all intermediate results that the user accepts during the interaction with the DSS.

The second stage is to convert the DSS output into training examples that are going to be appended to the training file that contains the examples previously converted by this system, if any. So, the training file contains an accumulation of examples, that is it growths as someone runs the DSS. This is good because as the number of examples gets large as the extracted rules become more general.

The format of training file is a two-dimensional table, rows correspond to examples, and columns represent attributes, the last column corresponds to the class attribute or decision.

So the process of converting the DSS output into training examples is naive in the sense that it is a way of filling slots with data, in other words; to fill the attributes with their values.

It is known that ES are domain specific systems, this means that it is difficult to build very general ES that can deal with a wide range of problems. If it happens that DSS are more general than ES, the general domain of a DSS can be divided into a number of specific domains, each domain has different attributes and different class, that can be stored in a separate file. This is the reason that the training sets, as shown in Figure 4, are divided

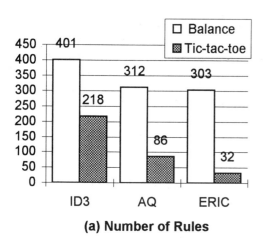

(a) Number of Rules

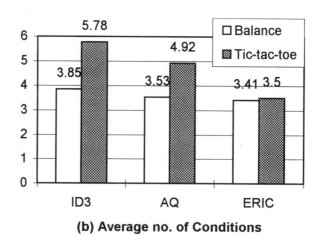

(b) Average no. of Conditions

FIGURE 4. Summary of Results of The Three Algorithms.

into n sets, and according to the required ES d-main, the related set is considered to be the input of the third stage, ERIC.

The classification algorithm ERIC is the last stage of the interface. It accepts one of the 'training examples' sets, according to desire of the ES user, as its input and produces the related rules in their general form, as discussed in the previous section. These rules are placed in the knowledge base component of the ES as its rule base for the considered domain, the fact base of the ES is the set of the attributes with their properties.

ERIC may be executed whenever the user feels that it is necessary to update the ES knowledge base. Actually, this is the powerful aspect of the framework, that is, the ES knowledge can be updated easily and frequently without the need for Knowledge Engineer or the existence of an expert. In other words, there is no need for the traditional Knowledge Acquisition process at all. This will influence the over all building process of an ES, in a way that it will be done with less time and cost.

6. THE INTEGRATED DECISION-SUPPORT EXPERT ADVISORY SYSTEM (IDEAS)

6.1 THE STUDENTS ADVISORY DSS (DSSAS)

DSSAS has been built for the aim of providing the students' advisors in the university with information needed in their decision making process that is related to

the academic situations of the students. DSSAS has been implemented on a PC with Pentium processor using dbase iv. It consists of more than 12,000 lines of code.

DSSAS is composed of two main parts: the undergraduates' part and the graduates' part. Each part has its own functions, but some functions are common for both parts. The main functions of both parts are:
1) *The students system*, this system is responsible for providing the decision maker with general information and suggestions which are related to students, courses, and rules and regulations in the university, 2) *The data/model bases management system*, which is responsible for performing the emergence updating the data in the DSS data base and model base (the traditional updating processes is the responsibility of management information systems and traditional DBMS in the university), and 3) *Planning and Analysis system*, this is the major part of DSSAS, it is responsible for performing tasks which demand some kind of analysis, such as analyzing the student's academic situation.

6.2 THE STUDENTS ADVISORY ES (ESSAS)

ESSAS is a rule-based expert system. It has been built to be used as a student advisor. This system can be used by the students themselves to get information required for their registration of courses or information regarding their academic situations in the university.

ESSAS has been implemented with Texas Instrument's Personal Consultant Plus Shell.

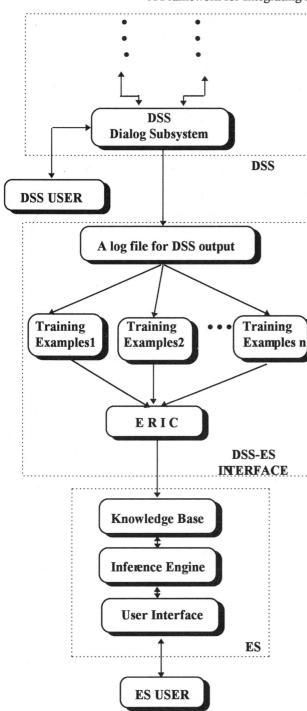

FIGURE 4. The DSS-ES Integration Frame - work.

6.3 THE DECISION-SUPPORT EXPERT AD-VISORY SYSTEM *(IDEAS)*

The main source of ESSAS's knowledge is mainly the university regulations. These regulations are well structured and can be easily converted to rules. ESSAS was started initially with 126 rules. But, it is noted that there exists a wide array of situations that are not covered completely or expressed explicitly in these regulations, these situations depend mainly on judgment and intuition of the advisor him self and on the way in which he understands the student's situation.

DSSAS, on the other hand, works as an assistant for the student's advisors. It provides them with necessary information needed in their decision making process regarding students. DSSAS has the ability of analyzing many situations that are not mentioned explicitly in the regulations. So, it produces a huge number of cases that can be converted to rules for future references for similar situations.

IDEAS is an integrated system that combines ESSAS with DSSAS in one system. IDEAS can be used as a separate ES or a stand alone DSS. In this system, DSSAS acts (besides its original function) as a wealthy source of new cases and situations that are revised and converted to rules using the inductive algorithm ERIC. This process increases the power of ESSAS, in a way that it makes its knowledge always updated and capable of dealing with the new student cases and situations that may ap pear.

6.4 THE TRAINING EXAMPLES IN IDEAS

A training example in IDEAS contains four attributes plus one class attribute. These attributes may be summarized as follows :

1. DURST for duration of studying years with values 1 to 7.

2. UOG for undergraduate or undergraduate with values U and G.

3. NOCREDITS for number of credits the student has completed with values from 0 to 132.

4. CGPA for cumulative GPA with values between 0 and 4.

5. SITUATION: This is the class attribute. It shows the student situation with values : D, P, S, R, H, and G for dismissed, in probation, successful, repeat, honor, and graduated respectively.

6.5 A CASE EXAMPLE

As an illustrative example of the process of trans-forming the output of the DSS into training examples, let us consider the following DSS output case :

'This is the first semester for this undergraduate student. He finished the five 3-credit courses he registered in this semester successfully with GPA of 3.2. He can register for the courses of the second semester.'.

This output case is one of hundreds of output situations of the DSS. It can be easily transformed into a training example just by filling the values of the six attributes. The training example that corresponds to this case may be as follows :

DURST	UOG	NOCREDITS	CGPA	SITUATION
1	U	15	3.2	S

This process can be repeated with all DSS output cases to produce the corresponding training examples which are going to be stored into a new file or appended to the file that contains previous examples with eliminating all duplicates.

7. CONCLUSIONS

A framework for integrating DSS and ES has been developed. In this framework, an inductive learning algorithm (called ERIC) has been used to transform the output of DSS into production rules that constitute the knowledge base of an ES. It has been shown that ERIC produces rules that are less and more general than those produced by some well-known inductive algorithms.

This approach helps in decreasing the time and the cost of building ES considerably. This is because the acquisition process of the ES knowledge (which is considered as the major bottleneck in developing ES) can be done easily and quickly using the proposed framework.

Using this framework, the knowledge base of the ES can be updated repeatedly. This process allows the ES to be able to cope with any changes that may be necessary to be applied to its rule base. In addition, this approach gives the ES an opportunity to deal with the growth of the new situations and cases that may occur in the domain of the ES under consideration.

REFERENCES

[Bon⁺84] Bonczek R. et al., *Developments in Decision Support Systems*, Advances in Computers, vol. 23, M. Yoritz, ed. New York: Academic Press, 1984.

[Bon⁺81] Bonczek R. et al., *Foundations of Decision Support Systems*, New York, Academic Press, 1981.

[Cha89] Chan, P.K. *Inductive Learning with BCT*, Proc. Sixth International Workshop on Machine Learning. Cornell University, Ithaca, New York, 104-108, 1989.

[PN89] Clark, P. & Niblett, T., *The CN2 Induction Algorithm*. Machine Learning, 3, 1989.

[DS83] Davis D. and Steen J., *Computing the Advantages of Detailed Data Systems* (for contract administration), Personnel Journal 62, 1983.

[Gou⁺84] Goul M. et al., *Designing the Expert Component of a Decision Support System*, Paper delivered at the ORSA/TIMS National Meeting, San Francisco, 1984.

[Gri80] Grindlay A., *Decision Support Systems*, Business Quarterly, 45, 1980.

[Ira93] Irani K. et al. *Applying Machine Learning to Semiconductor Manufacturing*, IEEE EXPERT, February, 1993.

[LS93] Luger G. and Stubblefield W., *Artificial Intelligence: Structures for Complex Problem Solving*, Second Edition, Addison Wesley, 1993.

[MD95] Metica V. and Dolnicar D., *Evaluation of Automatic Rule Induction Systems*, Expert Systems with Applications, vol. 8, no. 1, pp 77-87, 1995.

[ML78] Michalski, R.S., & Larson, J.B., *Selection of most representative training examples and incremental generation of VL1 hypothesis: The underlying methodology and the descriptions of programs ESEL and AQ11*. (Report No. 867). Urbana, Illinois: Department of Computer Science, University of Illinois, 1978.

[MMH⁺86] Michalski, R.S., Mozetic, I., Hong, J., & Lavrac, N. *The Multipurpose Incremental Learning System AQ15 and Its Testing Application to Three Medical Domains*, Proc. of the Fifth National Conference on Artificial Intelligence, Philadelphia, PA: Morgan Kaufmann, 1041-1045, 1986.

[Min76] Mintzberg H., *Planning on the Left Side and Managing on the Right*, Harvard Business Review, 1976.

[MKS94] Murthy, S.K., Kasif, S., & Salzberg, S., *A System for Induction of Oblique Decision Trees*, Journal of Artificial Intelligence Research, 2, 1-32, 1994.

[Pat90] Patterson D., *Introduction to Artificial Intelligence & Expert Systems*, Prentice Hall, 1990.

[PA95] Pham, D.T. & Aksoy, M.S., *RULES: A Simple Rule Extraction System*, Expert Systems with Applications, 8(1), 59-65, 1995.

[Qui83] Quinlan J., *Learning Efficient Classification Procedures and their Application to Chess End Games*, In R.S. Michalski, J.G. Carbonell & T.M Mitchell, Machine Learning, an Artificial Intelligence Approach, Palo Alto, CA: Tioga, 1983.

[Qui93] Quinlan, J., *C4.5: Programs for Machine Learning*. Philadelphia, Morgan Kaufmann, 1993.

[RK91] Rich E. and Knight K., *Artificial Intelligence*, Second Edition, McGraw-Hill, 1991.

[SF86] Schlimmer, J.C. & Fisher, D., *A Case Study of Incremental Concept Induction*. Proc. of the Fifth National Conference on Artificial

Intelligence, Philadelphia, PA: Morgan Kaufmann, 496-501, 1986.

[SW87] Spargue R. and Watson H., *Decision Support Systems: Putting Theory into Practice*, Prentice Hall, 1987.

[Thi88] Thierauf R., *User-Oriented Decision Support Systems: Accent on Problem Finding*, Prentice Hall, 1988.

[Tur88] Turban E., *Decision Support and Expert Systems*, Macmillan Publishing Company, New York, 1988.

[TW86] Turban E. and Watkins P., *Integrating Expert Systems and Decision Support Systems*, MIS Quarterly, 1986.

[Utg88] Utgoff, P.E. *ID5: An Incremental ID3*, Proc. of the Fifth National Conference on Machine Learning, Ann Arbor, MI, University of Michigan, 107-120, 1988.

PARTITIONING THE REPRESENTATION SPACE:
A METHOD FOR SIMPLIFYING THE DISCOVERY PROCESS

Ibrahim F. Imam

Machine Learning and Inference Laboratory
George Mason University
Fairfax, VA 22030
email: iimam@aic.gmu.edu

ABSTRACT

Most discovery systems face difficulties when discovering knowledge from large databases. Adapting the discovery systems to cope with large databases is the traditional solution to overcome these limitations. Another interesting solution is to partitioning the representation space, and applying the discovery system in parallel on data from each subspace, then combining the discovered knowledge if necessary. This paper introduces a new methodology for partitioning the representation space. The method selects an irrelevant attribute, using a utility function, to partition the representation space. Since irrelevant attributes are not needed to describe the concepts discovered from the data, the knowledge discovered from all subspaces should be identical. In such cases, discovery can be done only in one subspace. If the representation space is partitioned by a relevant attribute, the knowledge discovered from all subspaces can be combined simply using information about that attribute. The method is analyzed using two learning systems, AQ15c for learning decision rules from examples and C4.5 for learning decision trees from examples.

1. INTRODUCTION

Machine learning systems are very effective tools for discovering hidden, previously unknown knowledge from data. Some symbolic learning systems perform better when discovering knowledge from small databases than from very large databases. Reasons of these limitations are due to memory limitation, complexity of the search algorithm, loss of control over the degree of generalization, inefficient methodologies of handling different attributes' domains (e.g., continuous or structured), learning from sample (window) of the training data, etc. Most attempts to overcome these limitation were focused on improving the learning system to handle very large databases. Also, approaches for discovery from portions of the data have difficulties in combining knowledge learned from all portions [CS93].

This paper investigates the claim that *"selecting an irrelevant attribute for partitioning the representation space results in producing knowledge from each subspace that is identical to the knowledge obtained from the whole space"*. Because of the fuzziness of the relevancy of an attribute to the decision classes, combining knowledge learned from different subspaces is very complex [Ima94]. Such complexity of combining knowledge proportionally depends on the degree of relevancy of the attribute(s) used in partitioning representation space. The <u>representation space</u> of a given database is the cross product of all attribute domains. By <u>partitioning the representation space</u> it is meant that generating a set of sub-spaces with smaller dimensionalty such that the union of all these sub-spaces gives the original representation space, and there is no intersection between any two subspaces. The proposed methodology differs from any classification system in that it selects the most *irrelevant* attribute to partition the representation space instead of using the most *relevant* attribute, and the representation space is partitioned once (either using one attribute or a combination of two or more attributes). This research has high potential which could lead to great simplification of the process of discovering knowledge from databases.

There are two main issues concerned with this approach. The first issue is the relationship between the generalization of any set of examples with respect to the whole representation space and with respect to a subspace. This issue is of great interest to be analyzed, however, having different but correct generalization of the examples should be acceptable. In other words, it is necessary to obtain a consistent and complete concept description. The second issue is concerned with the density of the available examples in the original representation space. The performance of most discovery systems may change after partitioning the representation space if the ratio between all possible examples (i.e., the number of all possible combinations of values of the attribute domains) and the number of available examples is very high. In such cases, the discovery systems are expected to differently generalize the learned concept before and after partitioning the representation space.

The paper proposes a solution to this problem by quantizing all attribute domains with continuous or large number of values. Quantizing any attribute domain shrinks the representation space which reduces the number of possible examples. The number of available examples may be slightly reduced due to possible redundancy or ambiguity. However, the ratio between the two numbers will be highly reduced.

Determining the best quantization for continuous domains is another difficult problem. Usually continuous domains are quantized either by an expert, by a threshold or by fitting a statistical distribution. This process is supposed to be repeated many times to find the optimal quantization. An optimal quantization of a continuous domain minimizes the number of ambiguous and redundant examples (i.e., examples belonging to more than one decision class; and duplication of examples in the same decision class). In this paper, attribute domains with a large number of values are quantized according to a uniform distribution of the number of examples per an interval.

The paper includes an analysis of different patterns and their relationship before and after partitioning the representation space. Also, it presents an experiment on artificial data in which the relevant and irrelevant attributes are known. The experiment uses two systems AQ15c [WKB95] for learning deicsion rules from examples and C4.5 [Qui93] for learning decision trees from examples. Also, some experiments on real world databases are presented to show the relationship between the density of the data and the proposed approach.

2. RELATED WORK

The proposed approach is concerned with symbolic approaches of learning. Most symbolic learning systems (e.g., decision tree systems) use search techniques either to select the most relevant attribute for classification or to determine the set of attributes to be used in generalizing the training examples. An attribute is relevant if it is needed for describing the learned concept. Different learning approaches could have different ranking of relevant attributes for the same data.

The proposed methodology for partitioning the representation space is related to the basic idea of learning decision trees. The decision tree learning approach is based on repeatedly selecting the attribute which best classifies the given decision classes. Such an attribute is selected according to a given criterion. This attribute is used to divide the representation space into subsets, each corresponding to one of the attribute values. Then, the same process is repeated for each subset of the data unless all examples in one subset belong to one decision class [Qui79, BFO*84]. In the proposed approach, the best attribute for partitioning the representation space is usually the worst attribute for classifying the decision classes.

2.1. Criteria for Evaluating Attributes from Data

The attribute selection criteria can be divided into three categories. These categories are logical-based, information-based, and statistical-based. The logical-based criteria for selecting attributes use logical relationships between the attributes and the decision classes to determine the best attribute to be a node in the decision tree, such as the MAL criterion, Minimizing Added Leaves, [Mic78]. The information-based criteria are based on the information theory. Basically, these criteria measure the information conveyed by dividing the training examples into subsets. Examples of such criteria include the Information Measure IM, the entropy reduction measure, and the gain criteria [Qui83], the gini index of diversity (BFO*84), and Gain-ratio measure [Qui86]. The statistical-based criteria measure the association between the decision classes and other attributes using statistical distributions. Examples of statistical criteria include the Chi-square method and the G statistic [SR81; Har84; Min89].

2.2. Parallel Discovery

Parallel discovery is one of the main approaches for discovering knowledge from very large databases. Parallel discovery can be divided into two categories. In the first category, the discovery system is paralleled and applied to the whole data [ZMM*89]. One of the main problems with this category is utilizing the multiple input/output channels. The second category is to apply the discovery system on subsets of the data [CS93]. In the work by Chan and Stolfo, the data was randomly divided into subsets. Then, the discovery system was applied on each subset independently. Also, the process of learning was separated from the process of combining the discovered knowledge. They used a meta-learning methodology to combine the knowledge discovered from all subsets. The meta-learning used was a binary tree of arbiters. The proposed approach does not require combining the discovered knowledge when partitioning the representation space with an irrelevant attribute.

3. PARTITIONING THE REPRESENTATION SPACE

3.1. The Proposed Approach

Relevancy in most discovery systems is concerned with searching for relationships among attributes. Discovery, in such cases, is the determination of non trivial, previously unknown, and frequently occurring relationships [FI92, Sch93, Zia91]. The proposed approach benefits from irrelevant attributes to partition the data into smaller subsets. Partitioning the representation space using irrelevant attributes generates a set of subspaces where any pattern that can be discovered from the whole space can also be discovered from any generated subspaces. However, since some real world databases do not contain irrelevant attributes, partitioning the data with the least relevant attribute is desirable if the learning system can provide consistent and complete concept description. Failure to provide a consistent and complete concept description results in over-generalizing the data differently before and after any possible partitioning of the representation space. The following is a description of the proposed method:

Input: A database and preferred decision attribute.

Output: The most irrelevant attribute, and knowledge obtained from subspaces partitioned by this attribute.

Step 1: A large sample of the data is selected to determine the most irrelevant attribute.

Step 2: Attributes with continuous domains or with many values (e.g., > 20) are quantized according to a uniform distribution of the number of examples per each interval.

Step 3: Ambiguous and redundant examples are removed from the data.

Step 4: Evaluate the relevancy of all attributes to the decision classes.

Step 5: Select the most *irrelevant* attribute, say x. For each value, say v_i, (or quantized interval) of x, combine all records from the original database that contain that value. This set of records represents examples in a subspace corresponding to the value v_i.

Step 6: Apply the discovery system on each subset of records in Step 5.

Step 7: Compare the knowledge obtained from each subset.

When partitioning the representation space using relevant or less relevant attributes, decision trees learned from all subspaces can be combined together by generating a root node with the number of branches equal to the number of subspaces, and connecting each branch with one of those trees. The root of the decision tree should be assigned to the attribute used in partitioning the representation space. Each branch should be assigned to a value of that attribute corresponding to the subspace.

With rule learning systems, add to all rules learned from each subspace an extra condition representing the attribute used in partitioning the representation space and the value corresponding to that subspace. Unlike the decision tree learning approach, adding an irrelevant attribute (adding a condition containing a disjunction of all of its values) to all learned rules does not affect the representation for the semantics of the learned concept. Also, it does not change the predictive accuracy of the learned concept.

3.2. Analyzing Patterns Before and After Partitioning the Representation Space

This section introduces a set of artificial patterns that are generated before and after partitioning the representation space. There are two categories of patterns: those that can be described without the partitioned attribute and others that can not be described without the partitioned attribute. Experiments show different patterns from each category. Also, this section illustrates how partitioning the representation space may affect the generalization process.

First, let us define the density of a pattern. Figure 1 shows a visualization diagram of two different representation spaces. Each cell of the diagram represents a possible combination of the cross product of the attributes' domains. Any example can be represented as one cell in this diagram. Consider the rule [x2=1] represented by the shaded area in Figure 1 (a and b). In both figures, the rule [x2=1] is covering 12 positive examples.

Definition 1: The *density* of a database is the ratio between the number of distinct records in the database to the number of all possible records of the attribute domains.

To explain the definition, assume there are a database with only 12 examples, Figure 1. The number of *possible examples* in Figure 1-a is 36, and in Figure 1-b is 72. The density of this database in the representation space in Figure 1-a is 1/3, and 1/6 for Figure 1-b. Consider a high density database that consists of four attributes, x1 (4 values), x2 (2 values), x3 (3 values) and x4 (3 values). Assume that x1 is selected to partition the representation space. The following examples show only these patterns before and after partitioning the representation space.

Patterns described without attribute x1: Figure 2 shows a visualization of the pattern [x2 = 1 v 2] & [x3 = 1] & [x4 = 1] before and after partitioning the representation space using x1. The same pattern can be obtained from each subspace.

Patterns described using attribute x1: Figure 3 shows the two patterns [x1 = 1 v 3] & [x3 = 1] & [x4 = 1], and [x1 = 1 v 2] & [x3 = 1] & [x4 = 3]. Those two patterns are mapped into the following three patterns: [x3 = 1] & [x4 = 1 v 3] which exists in the first subspace, [x3 = 1] & [x4 = 3] which can be obtained from the second subspace, and [x3 = 1] & [x4 = 1] which appears in the third subspace.

Illustrating the importance of having high density data by an example: In the case of partitioning the representation space with relevant attribute, learning concept description from subspaces may result in generating different but consistent description. Consider the five positive examples given in Figure 4. In the original space, these examples may be generalized to the following patterns:

[x1 = 1 v 3] & [x2 = 1] & [x3 = 1] & [x4 = 1 v 3]
[x1 = 3 v 4] & [x2 = 1] & [x3 = 3] & [x4 = 1]

The following patterns obtained when partitioning by x1:

[x2 = 1] & [x3 = 1] & [x4 = 1 v 3]   when (x1=1)

[x2 = 1] & [x3 = 1 v 3] & [x4 = 1]   when (x1=3)

[x2 = 1] & [x3 = 3] & [x4 = 1]   when (x1=4)

It is clear that the cell (x1=3, x2=1, x3=1, x4=3) is covered by one of the patterns obtained from the original data, while it is not covered by any pattern of the third subspace. Also, the second pattern can be generalized to the pattern [x2 = 1] & [x4 = 1] which covers the cell (x1=3, x2=1, x3=2, x4=1) which was not covered by any pattern in the original space.

4. ANALYZING OF THE APPROACH

This subsection introduces an experiment using the learning systems AQ15c [WKB95] and C4.5 [Qui93] to discover knowledge from the MONK-1 problem [TMC91]. The MONK-1 problem has six attributes, three of which are highly relevant to the concept and the other three are irrelevant. A sample size of 60% of the data (which is considered high density for the MONK-1 problem) is randomly selected to determine the relationship between the knowledge learned before and after partitioning the representation space.

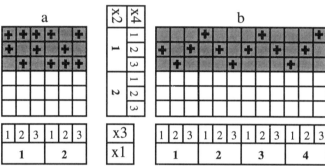

Figure 1: An example to illustrate the density of a rule.

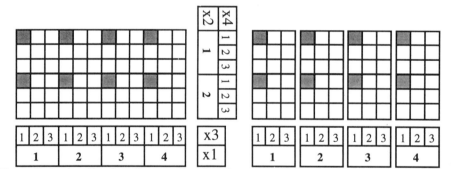

Figure 2: A visualization diagram showing the pattern [x2 = 1 v 2] &[x3 = 1]&[x4 = 1].

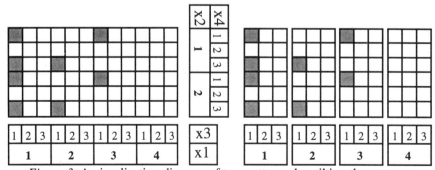

Figure 3: A visualization diagram of two patterns describing the patterns [x1 = 1 v 3]&[x3 = 1]&[x4 = 1], and [x1 = 1 v 2] & [x3 = 1] & [x4 = 3].

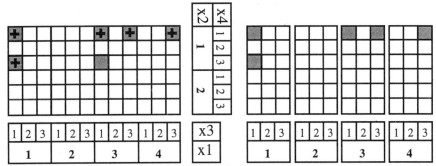

Figure 4: Different generalizations before and after partitioning the representation space.

The MONK-1 problem describes robot-like figures. It uses six attributes to classify each robot to either friendly (positive) or not friendly (negative). The concept can be described using attributes x1, x2 and x5 only, see Figure 5. This concept can be summarized as follows: it is positive if [x5=1] or [x1=x2], and negative otherwise.

AQ15c learned the correct concept (Figure 5) when applied to the training data. Table 1 presents the subspaces generated when each attribute was used for partitioning the representation space, and the decision rules learned by AQ15c from each subspace. In the case of partitioning with x1, x2 or x5 (relevant attributes), when adding one condition corresponds to the subspace, to each rule, a concept description similar to the one in Figure 5 should be obtained.

Positive rules	Negative rules
1 [x5 = 1]	1 [x1 = 1][x2 = 2..3][x5 = 2..4]
2 [x1 =3][x2 = 3]	2 [x1 = 2][x2 = 1..3][x5 = 2..4]
3 [x1 = 2][x2 = 2]	3 [x1 = 3][x2 = 1..2][x5 = 2..4]
4 [x1 = 1][x2 = 1]	

Figure 5: Decision rules describing the target concept of the MONK-1 problem.

When C4.5 was used to learn decision trees from the same data it produced the decision trees in Figure 6. For all experiments with C4.5, the window size of the training examples was set to 100%. This was necessary to increase the density of the training examples. The results obtained by C4.5 (see Table 2) support those obtained by AQ15c.

5. EXPERIMENTING ON DATABASES WITH DIFFERENT DENSITIES

This section presents analysis of the proposed approach on different databases with different density. These experiments evaluate the average increase in the error rate when learning from one subspace partitioned by an attribute from the database. These results are illustrated in set of tables 3, 4, 5, and 6. Attributes are ordered in each table from the most relevant to the least relevant.

The MONK-3 and the Congressional Voting Records of 84, the density was relatively high. The other two databases are concerned with the classification of the Iris Plants and diagnosing the liver Disorders. Both databases have low density. Only the C4.5 learning program is used to determine the increase of the average error rate of learning concepts from the whole space and from a subspace.

Table 1: **Rules learned by AQ15c for the MONK-1 problem from different subspaces.**

Attribute	Subspace	Rules	
x1	1	Pos: [x2=1] v [x5=1]	Neg: [x2=2..3] & [x5=2..4]
	2	Pos: [x2=2] v [x5=1]	Neg: [x2=1 v 3] & [x5=2..4]
	3	Pos: [x2=3] v [x5=1]	Neg: [x2=1..2] & [x5=2..4]
x2	1	Pos: [x1=1] v [x5=1]	Neg: [x1=2..3] & [x5=2..4]
	2	Pos: [x1=2] v [x5=1]	Neg: [x1=1 v 3] & [x5=2..4]
	3	Pos: [x1=3] v [x5=1]	Neg: [x1=1..2] & [x5=2..4]
x3	1, 2	Same as in Figure 5	
x4	1, 2, 3	Same as in Figure 5	
x5	1	Positive for all examples	
	2	Pos: [x1=1][x2=1] v [x1=2][x2=2] v [x1=3][x2=3]	
		Neg: [x1=1] & [x2=2..3] v [x1=2] & [x2=1 v 3] v [x1=3] & [x2=1..2]	
	3, 4	Same as subspace 2 of x5	
x6	1, 2	Same as in Figure 5	

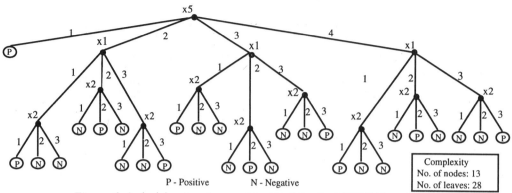

Figure 6: A decision tree Learned by C4.5 for the MONK-1 problem.

Table 2: **The results of using C4.5 for learning decision trees from different subspaces.**

Attribute	Subspace	Rules
	1	Tree similar to the one in Figure 7-b, but with x2 instead of x1
x1	2	Tree similar to the one in Figure 7-c, but with x2 instead of x1
	3	Tree similar to the one in Figure 7-d, but with x2 instead of x1
	1	Tree similar to the one in Figure 7-b
x2	2	Tree similar to the one in Figure 7-c
	3	Tree similar to the one in Figure 7-d
x3	1, 2	Tree similar to the one in Figure 6
x4	1, 2, 3	Tree similar to the one in Figure 6
x5	1	One leaf tree with a Positive decision
	2, 3, 4	Tree similar to the one in Figure 7-a
x6	1, 2	Tree similar to the one in Figure 6

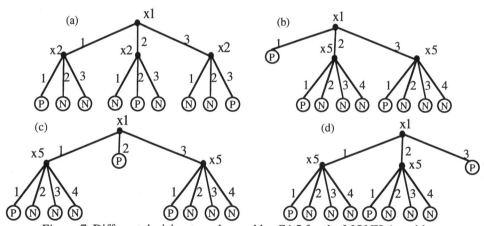

Figure 7: Different decision trees learned by C4.5 for the MONK-1 problem.

The MONK-3 Problem: Table 3 shows the attributes of the MONK-3 problem ordered from the most relevant first to the irrelevant last according to the Gain criteria. The first attribute x5 is the most relevant. In this experiment, the average error of concepts learned from a subspace generated by an attribute is proportionally depends on its relevancy.

The Congressional Voting 84 Problem: The U.S. Congress Voting Records-1984 contains two decision classes "Democratic" and "Republican". Each voting record of a Democrat or a Republican is described in terms of 16 binary attributes. Even though the density of the training data was low, the results in Table 4 support the claim that learning concepts from a subspace generated by an irrelevant attribute (all attributes in the third row) is equivalent to learning from the whole data.

Classification of Iris Plants: The Iris plants database contains three decision classes describing different kinds of irises, and four attributes describing the length and width of the sepal and petal. All attributes have continuous domains. In this experiment, all attribute domains were quantized into three or four intervals with equal number of examples in each interval. Table 5 shows the four attributes ordered according to their relevancy, and the average increase in the error rate of concepts obtained from a subspace generated by each attribute. It is very clear that the more relevant the attribute the higher the increase in the error rate.

Diagnosing Liver Disorders: The liver database diagnose whether or not there is any disorder in the liver. The decision is made based on five blood tests (with continuous domain) and the number of drinks per day. The first five attributes were quantized according to a uniform distribution of the number of examples per an intervals. The data contains 345 available examples, only 272 examples were used for learning (less than 3% of the cross product of the attribute domains, 9375). The concept obtained is described with all the attributes. In other words, there was no total irrelevant attributes. Only $x6$ produced unusual results which could be due to the number of its values. Also, $x4$ produced better results from its second subspace.

6. CONCLUSION

This paper introduced a methodology for simplifying the discovery process in large databases. The paper is only concerned with analyzing the proposed methodology on well known databases. The method divides the representation space into subspaces, discovers knowledge from only one subspace. A representation space of a dataset is the cross product of all attribute domains. The method uses utility functions (e.g., the gain criterion), for evaluating the relevancy of attributes.

Table 3: **Results of the MONK-3 problem.**

Attributes	Change in the error rate for subspaces				Average Errors
	Sp. 1	Sp. 2	Sp. 3	Sp. 4	
x5	19.4%	19.4%	25.0%	52.8%	29.2%
x2	22.2%	22.2%	44.4%	-	29.6%
x1	0%	0%	0%	-	0%
x4	5.6%	2.8%	2.8%	-	3.7%
x3	0%	0%	-	-	0%
x6	0%	0%	-	-	0%

Table 4: **Results of the Congressional Voting 84 problem.**

Attributes	Change in the error rate		Average Error
	Space 0	Space 1	
x4	34.5%	55.9%	45.2%
x5	34.5%	0%	17.3%
x12, x3, x8, x9, x14, x13, x15, x6, x7, x11, x1, x16, x10, x2	0%	0%	0%

Table 5: **Results from the Iris Plant dataset.**

Attributes	Change in the error rate for subspaces				Average Errors
	Space 1	Space 2	Space 3	Space 4	
x4	63.6%	42.4%	8.1%	45.4%	39.87%
x3	63.6%	36.4%	5.6%	45.4%	37.75%
x1	33.3%	46.4%	5.7%	42.4%	31.95%
x2	12.1%	6.1%	27.3%	--	15.17%

Table 7: **Results of the liver disorders problem.**

Attributes	Change in the error rate for subspaces					Average Errors
	Sp. 1	Sp. 2	Sp. 3	Sp. 4	Sp. 5	
x5	9.6%	20.5%	8.2%	9.6%	9.6%	11.5%
x3	9.6%	13.7%	8.2%	15%	9.6%	11.2%
x2	9.6%	12.3%	6.8%	13.7%	4.1%	9.3%
x6	8.2%	17.8%	20.5%	--	--	15.5%
x4	13.7%	-1.4%	10.9%	6.8%	9.6%	7.9%
x1	9.6%	4.1%	16.4%	0%	12.3%	8.4%

Having high density data is a major requirement of the proposed methodology. The density of a database is measured by the ratio of the number of available (present) examples to the maximum number of examples (that can be generated in its representation space). Having higher density data increases the probability that the same patterns discovered from the original data can be discovered after partitioning the representation space using an irrelevant attribute. One approach to increasing the density of the data is by quantizing the attribute domains. Attribute' domains are quantized according to a uniform distribution of the number of examples per interval. Another way to overcome the density problem, specially with rule learning systems, is to learn characteristic rules. The methodology requires also the learned concept to be consistent and complete. Failure to provide consistent and complete concept description results in learning different knowledge before and after partitioning the representation space regardless of the partitioning method.

This paper introduced a set of explanatory examples showing high density patterns before and after partitioning the representation space. An example using the MONK-1 problem was introduced for learning decision rules and decision trees from examples. Also, other experiments show that having high density of the given database is very important for the success of the approach.

REFERENCES

[BFO*84] Breiman, L., Friedman, J.H., Olshen, R.A., and Stone, C.J., "Classification and Regression Trees", Belmont, California: Wadsworth Int. Group, 1984.

[CS93] Chan, P., and Stolfo, S., "Toward Parallel and Distributed Learning by Meta-Learning", *Proceeding of the AAAI-93 Workshop on Knowledge Discovery in Databases*, Washington D.C., July 11-12, 1993.

[FI92] Fayyad, U.M., and Irani, K.B., "On the Handling of Continous-Valued Attributes in Decision Tree Generation", *Journal of Machine Learning*, Vol. 8, No. 1, pp. 87-102, 1992.

[Har84] Hart, A., "Experience in the use of an inductive system in knowledge engineering", *Research and Developments in Expert Systems*, M. Bramer (Ed.), Cambridge, Cambridge University Press, 1984.

[Ima94] Imam, I.F., "An Experimental Study of Discovery in Large Temporal Databases", *Proceedings of the Seventh International Conference on Industrial and Engineering Applications of Artificial Intelligence and Expert Systems (IEA/AIE-94)*, Anger, F., Rodriguez, R., Ali, M. (Eds.), pp. 171-180, Austin, Texas, June, 1994.

[Mic78] Michalski, R.S., "Designing Extended Entry Decision Tables and Optimal Decision Trees Using Decision Diagrams", *Technical Report No.898*, Urbana: University of Illinois, March, 1978.

[Min89] Mingers, J., "An Empirical Comparison of selection Measures for Decision-Tree Induction", *Machine Learning*, Vol. 3, No. 3, pp. 319-342, Kluwer Academic Publishers, 1989.

[Qui83] Quinlan, J.R., "Learning efficient classification procedures and their application to chess end games" in R.S. Michalski, J.G. Carbonell, and T.M. Mitchell, (Eds.), *Machine Learning: An Artificial Intelligence Approach*. Los Altos: Morgan Kaufmann, 1983.

[Qui86] Quinlan, J.R., "Induction of Decision Trees", *Machine Learning* Vol. 1, No. 1, pp. 81-106, Kluwer Academic Publishers, 1986.

[Qui93] Quinlan, J.R., "C4.5: Programs for Machine Learning", Morgan Kaufmann, Los Altos, California, 1993.

[Sch93] Schlimmer, J., "Using Learned Dependencies to Automatically Construct Sufficient and Sensible Editing Views", *Proceeding of the AAAI-93 Workshop on Knowledge Discovery in Databases*, Washington D.C., July 11-12, 1993.

[SR81] Sokal, R., and Rohlf, F., "Biometry", Freeman Pub., San Francisco, 1981.

[TMC91] Thrun, S.B., Mitchell, T., and Cheng, J., (Eds.) "The MONK's Problems: A Performance Comparison of Different Learning Algorithms", *Technical Report*, Carnegie Mellon University, October, 1991.

[WKB*95] Wnek, J., Kaufman, K., Bloedorn, E., and Michalski, R.S., "Selective Induction Learning System AQ15c: The Method and User's Guide", Reports of the Machine Learning and Inference, No. 95-4, George Mason University, 1995.

[Zia91] Ziarko, W., "The Discovery, Analysis, and Representation of Data Dependencies in Databases", *Knowledge Discovery In Databases*, Shapiro, G., Frawley, W., (Eds.), AAAI Press, 1991.

[ZMM*89] Zhang, X., Mckenna, M., Mesirov, J., and Waltz, D., "An Efficient Implementation of the Backpropagation Algorithm on the Connection Machine CM-2", Technical Report RL89-1, Thinking Machine Crop., 1989.

COMPARATIVE EVALUATION OF THE DISCOVERED KNOWLEDGE

Mondher Maddouri and Ali Jaoua
Department of Computer Science, Faculty of Sciences of Tunis,
Campus Universitaire, Le Belvédère, 1060, Tunis, Tunisia

ABSTRACT

Rule induction is the branch of machine learning that remedies to the mutually exclusivity of the rules generated by decision tree methods. The known rule induction methods (PVM, CHARADE) can be applied only to little data sets since they use costly procedures [WK91, DK91]. In this paper, we present a faster rule induction method, the Incremental Production Rule based method (IPR method). The empirical evaluation in many real world applications proves that IPR is more precise than the rule induction method PVM and the decision tree methods ID3 and SIPINA. Other empirical results focus on the usefulness of the rule conflict resolution algorithm used in the IPR system.

Key words : machine learning, rule conflict, precision, disease diagnosis, classification.

1. INTRODUCTION

Inducing IF-THEN rules from data is an interesting aim of machine learning. In many real world applications, the rules discovered are used as a classifier to assign an appropriate class label to a new coming [DK91]. Medical diagnosis, pattern recognition and other applications can be considered as classification tasks, where the input data is shaped by a relational data base containing the classified patterns. The rule induction methods (a branch of machine learning) try to discover a consistent set of rules from this data base. The decision tree methods try to induce a decision tree from the data (a powerless knowledge representation alternative). Whereas decision tree methods are generally faster, precision capabilities measured in real world applications remain a sharp criterion to judge the consistency and the reliability of each method [HCK95, WK91].

The Incremental Production Rule based method (IPR method) is a rule induction method. It was used for meta-knowledge discovery to organise the rule base access. It allows the incremental maintenance of the rule base, each time the data base is updated [MJ96]. It gives tools for detecting the poor predictive attributes as well as generating multi-decisional rules with a minimal overhead [MJ95]. Recently, the IPR method includes a punishment/award algorithm to resolve the rule conflict problem. In this paper, we focus on the real world applications of the IPR method in order to evaluate its performance and to compare it to existing machine learning methods. To do this, we consider two rule induction methods (the IPR method and the PVM method) and two decision tree methods (the method ID3 and the method SIPINA). These methods were carried on four applications : the classification of the IRIS flowers, the classification of brandy solutions, the diagnosis of the Wisconsin breast cancer disease and the diagnosis of the heart diseases. We use different randomly resampling estimators to evaluate the precision of these methods. A detailed complexity analysis is, also, given to estimate the time requirements of the methods.

In section 2, we present the mathematical concepts of the IPR method and we introduce an incremental heuristic solution for covering a binary relation by a set of its optimal rectangles. In section 3, we present a data representation method that transforms the input data base into an imaginary binary relation. We explain the knowledge acquisition process and we present the punishment/award algorithm. In section 4, we compare the rule induction methods. We estimate their precision in IRIS flower classification and we give a comparative complexity analysis. In section 5, we compare the IPR method with the decision tree methods when they were applied to the four applications. In section 6, we give other empirical results to compare four data representation methods and to measure the usefulness of the punishment/award algorithm.

2. BASIC CONCEPTS

In this section, we present the basic concepts of the IPR method and we give the main definitions.

2.1. Mathematical concepts

Let O be a set of objects and P be a set of properties.

Definition 2.1: A *binary relation* R between O and P is a subset of the Cartesian product OxP. An element of R is denoted by (x, y). For any binary relation R, we associate the following definitions :

* The *domain* of R is defined by :

Dom(R) = {e | ∃ e' : (e, e') ⊆ R},

* The *codomain* of R is defined by :

Cod(R) = {e' | ∃ e :(e, e') ⊆ R},

* The *cardinality* of R is defined by :

Card(R)= number of pairs in R♦

Definition 2.2: Let R be a binary relation defined between O and P. A *rectangle* of R is a Cartesian product of two sets (A, B) such as A ⊆ O, B ⊆ P and AxB ⊆ R, where A is the domain and B is the codomain ♦

Definition 2.3: A rectangle (A, B) of R is *maximal* if

AxB ⊆ A'xB' ⊆ R ⇒ A=A' and B=B'. ♦

Definition 2.4: A rectangle containing (a, b) of a relation R is *optimal* if it gives the greater gain value among all the maximal rectangles containing (a, b). The gain of a rectangle RE = (A, B) is measured by :

g(RE)= card(A)*card(B)-[card(A)+card(B)] ♦

Definition 2.5: The union of all the maximal rectangles containing a couple (a, b) of R is called the *pseudo-rectangle* of (a, b) in R and denoted ∅$_R$(a, b)♦

Definition 2.6: A *coverage* of a relation R is a set of rectangles CV ={RE1, ..., REn} of R where any element (a, b) of R is contained in at least one rectangle of CV♦

Definition 2.7: A coverage CV defined by {RE1, ..., REn} of a relation R is *minimal* if it is made up by a minimal number of optimal rectangles ♦

Example: In figure 1, we present a binary relation R and we give two maximal rectangles. RE1 is the optimal rectangle containing the couple (o1, p1) with a gain value of '1' and RE2 is the optimal rectangle containing the couple (o2, p2) with a gain value of '0'. RE1 and RE2 constitute a minimal coverage of the relation R. Each couple from R belongs at least to one rectangle of the coverage. If we consider the rectangle R3={o1, o2, o3,

o4, o5}x{p3}, the coverage CV' ={RE1, RE2, R3} is not minimal since the rectangle R3 is redundant.

2.2. Heuristic Concepts

The problem of covering a binary relation by a minimal set of optimal rectangles is known as the problem of covering a binary matrix by a minimal number of complete sub matrix [C69]. A complete sub matrix is a matrix where all its entries are equal to '1'. This was found as an NP-hard problem [KJ97]. In the IPR system, we use a polynomial heuristic solution that was successfully applied in many fields such as signature files and documentary data bases [KJ97].

Let R be the binary relation to cover. The heuristic solution consists in dividing R into p packages: P1, ..., Pp. Each package represents one or more couples. The idea is to construct step by step the minimal coverage of R. In the first step, we cover the relation R₁=P₁ by CV₁. In the kth step, let R$_{k-1}$=P₁ ∪ ... ∪ P$_{k-1}$ and let CV$_{k-1}$ be its minimal coverage, the task is to construct the minimal coverage CV$_k$ of R$_k$=R$_{k-1}$ ∪ P$_k$ using only the initial coverage CV$_{k-1}$ and the package P$_k$. In the pth step, we obtain a set of rectangles covering the relation R. This incremental aspect was found very economical [MJ96].

```
Begin Algorithm :
1. Let R be partitioned to p packages P₁, ..., Pₚ.
2. FOR k=1 to p DO
2.1. Sort the couples of Pₖ by the value of the gain
     function in decreasing order
2.2. While  (Pₖ≠∅) Do
       * Select a couple (a, b) in Pₖ by the sorted order
       * Search PR : the pseudo-rectangle containing
         (a, b) within Rₖ=CVₖ₋₁∪Pₖ
       * Search RE: the optimal rectangle containing
         (a, b) within PR
       * CVₖ = (CV ₖ₋₁- {r∈CV ₖ₋₁/ r⊆RE } ) ∪{RE}
         (Delete all the redundant rectangles from CVₖ)
       * Pₖ= Pₖ - { (X, Y )∈Pₖ / (X, Y)∈RE}
       End While
End FOR
End
```

FIGURE 1: Example of minimal coverage of a binary relation R.

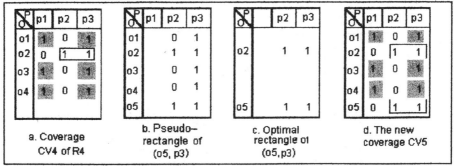

Figure 2: Incrementation steps for generating the minimal coverage of R.

Example: Let R be the relation of figure 1. R is partitioned into five packages : P1={o1}x{p1, p3}, P2={o2}x{p2, p3}, P3={o3}x{p1, p3}, P4={o4}x{p1, p3} and P5={o5}x{p2, p3}. Figure 2 presents the incrementation steps when adding P5. Initially, R is an empty relation and in each step we add a package. In this step R4 encloses the four rows P1, ..., P4. The initial minimal coverage CV4 encloses the rectangles RE1={o1, o3, o4}x{p1, p3} and R2={o2}x{p2, p3} (figure 2.a). The package P5 encloses only two couples. The pseudo rectangle (figure 2.b) containing the couple (o5, p3) represents the union of two maximal rectangles {o1, o2, o3, o4, o5}x{p3} and {o2, o5}x{p2, p3}. So, the optimal rectangle containing (o5, p3) is the rectangle RE2={o2, o5}x{p2, p3}. The rectangle R2 is redundant since it is included in the rectangle RE2. Thus, the final coverage of R contains only the rectangles RE1 and RE2.

3. THE IPR METHOD

In this section, we present the different procedures of the IPR method : the data representation, the rule extraction and the rule conflict resolution tools [MJ95].

3.1. Data representation

In our context, the information from which we will extract rules is modelled by a relational data base. Each column of the data base represents an attribute. The attributes can be qualitative or quantitative. For qualitative attributes, the tests (in the condition part of the rules) will be built using the attribute label, a qualitative value and the operator "equal to : =". For the attribute « eyes colour » that can be blue, black, ..., a valid test can be « eyes colour=blue ». For the quantitative attributes, we should find the necessary thresholds that divide the attribute into few intervals called modalities. The tests will be built using the attribute label, the thresholds and the operators "less than : <" or "greater than : >". For the attribute « age », a valid test can be « age less than 18 ». The thresholds and the optimal number of modalities constitute the major difficulties of the existing approaches [C90].

The method « modal », relevant to the IPR system, resolves this problem [MJ96]. The idea is that, for every quantitative attribute, we build the histogram representing the distribution of its values. Then, we divide it into elementary modalities (intervals). An elementary modality is limited by two attribute values related to two (not null) disjoint local minima of the appearance number. Each occurrence in the histogram can be coloured by the natural class of the object associated to it. Thus, to each elementary modality, we can assign a class of objects as the most frequent class that appears within the modality. Finally, if two disjoint

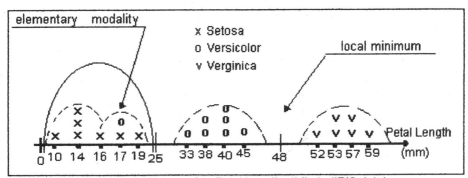

Figure 3: Histogram for Petal Length attribute (IRIS data)

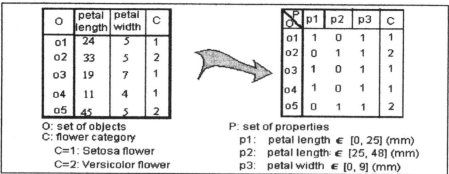

Figure 4: Transformation of continuous attributes related to 5 flowers (IRIS Data base).

modalities have the same class, we join them in one modality. Here, we note that the number of modalities and the choice of thresholds can be modified after each addition of new objects to the data base. For example, we consider the attribute "Petal Length : PL" that characterises the flowers of IRIS [WK91]. These flowers constitute three classes : "Setosa", "Versicolor" and "Verginica". To divide this quantitative attribute into modalities, we build the histogram of figure 3. In this example, there are eighteen flowers for which the petal length varies between 10 mm and 59 mm. In the histogram, we can find seven (not null) local minima (for the number of appearance) related to the values : 10 mm, 16 mm, 19 mm, 33 mm, 45 mm, 52 mm and 59 mm. So, we can determine four elementary modalities : [10, 16], [16, 19], [19, 45] and [45, 59]. If we consider the class of each object, we assign the class Setosa to the modalities [10, 16] and [16, 19], the class Versicolor to the modality [19, 45] and the class Verginica to the modality [45, 59]. The modalities [10, 16] and [16, 19] will be joined in one modality [10, 19] since they have the same class. After adjusting the thresholds, we obtain these three modalities : [10, 25], [25, 48] and [48, 59].

The data sets are represented by an n-ary relation joining a set of objects "O" with a set of attributes "A" (relational data base). This n-ary relation can be transformed into a binary relation that joins the set of objects "O" with a set of properties "P". Each property illustrates a possible test that may appear in the condition part of a rule. In fact, each column in the data base will be replaced by its different modalities if the attribute is quantitative. Whereas, if the attribute is qualitative, it will be replaced by its different values. Then, the value "1" (joining an object o and a property p) in the binary relation expresses that the object o has the property p and the value "0" expresses the opposite.

In figure 4, we illustrate this transformation. The n-ary relation is made up by five objects (flowers) and two attributes : "petal length : PL" and "petal width : PW". Once the thresholds are calculated, we generate a binary

relation. The attribute "petal length" is replaced by its three modalities of figure 3. Here, the third modality [48, 59] does not appear because we cannot find a value of petal length greater than 48 mm. For the attribute "petal width", only one modality appears [0, 9], since we cannot find a value of petal width greater than 9 mm.

3.2. Rule extraction

With the modal representation of pattern data base, an optimal rectangle RE =(A, B) is a Cartesian product of the subset A of objects (domain of RE) by the subset B of properties (codomain of RE). The Cartesian product reflects the fact that all the objects of the domain have simultaneously, all the properties of the codomain of RE [MJ95]. Then, we cannot find another property p (not belonging to B : $p \notin B$) that is checked by all the objects of A. Similarly, we cannot find an other object o (not belonging to A : $o \notin A$) that has all the properties of B. This is due to the maximality of the optimal rectangle. We have a big similarity between the objects of A based on the properties of B. Each optimal rectangle RE of R is given a class label. It is the label of the class to which belongs the majority of the objects constituting its domain.

From the definition of optimal rectangle and its semantic characteristics, we assume that the properties of the codomain of an optimal rectangle RE constitute the conditions of the rule associating the rectangles category (the most frequent class) to the objects of the rectangle domain. So, each optimal rectangle RE of the binary relation R is an equivalent representation of a production rule. The codomain of RE constitutes the action part of the rule and the domain cardinality of RE can be used to ponder the rule (rule conflict problem).

Figure 4 shows the petal length and petal width of five IRIS flowers. For all the flowers, we know the associated class. We want to extract the rules that designate the classification of these flowers. The method consists of generating the binary relation and the minimal coverage of the set of data (figure 3). Once the two optimal

Table 1 : Example of production rules extracted from the relation R of figure 4.

The optimal rectangles	The associated rules	
RE1={o1, o3, o4}x{p1, p3}/C=1	*IF* Petal Length $\in$ [0, 25] *AND* Petal Width $\in$ [0, 9]	*THEN* 'Setosa' flower.
RE2 ={o2, o5}x{p2, p3}/C=2	*IF* Petal Width $\in$ [0, 9] *AND* Petal Length $\in$ [25, 48]	*THEN* 'Versicolor' flower.

rectangles were obtained (RE1, RE2), we deduce the rules. Table 1 shows the rule associated to each rectangle. The codomain of RE1 encloses the properties "p1 : petal length $\in$ [0, 25]" and "p3 : petal width $\in$ [0, 9]". So, the condition part of the rule joins these two properties by the logical word "AND" (disjunction form). The class of the three objects o1, o3 and o4 enclosed in the domain of RE1 is "C=1 : Setosa". So, the action part of the rule expresses the assignment of the class label "Setosa".

In the whole paper, the words "optimal rectangle" and "production rule" mean the same thing. The problem of extracting an optimal number of consistent rules from a data base is considered as a problem of research for a minimal coverage of the binary relation related to the data base. The binary relation is deduced from the data base as presented in section 3.1. The set of rules is deduced from the minimal coverage as presented in table 1. The consistency of the rule base derives from the minimality of the coverage. The consistency of each rule derives from the optimality of the associated rectangle.

3.3. Rule conflict resolution

In the literature, we consider the rules that are not mutually exclusive as a more powerful representation than mutually exclusive rules. However, in this representation, more than one rule can fire simultaneously and the knowledge base will have conflict problems when these rules yield different classes. The task of conflict resolution, usually, involves weights or confidence factors that will be assigned to each rule [G94]. In our case, we present a punishment/award algorithm that calculates the rule weights using the resolved cases (classified objects).

The principal idea is simple. We apply the knowledge base to the resolved cases. For each case, we look for the applicable rules. We decrease the weights of these rules with the same quantum. Then, we apply the rule having the great weight. If this rule yields the true class, it will be awarded (its weight increases). Otherwise, it will be punished (its weight decreases). This iteration is repeated unless we achieve a stable state. Let O be the set of resolved objects and r_i be the ith rule. The algorithm states

```
Begin Algorithm:
1. Initialise the rule weights
2. While (the state is not stable) DO
    2.1. FOR each resolved object oi from O DO
        * select the applicable rules for oi
```

```
        * FOR each applicable rule ri DO
            weight(ri)= weight(ri) - quantum * weight(ri)
            End FOR
        * choose the best applicable rule rc
        * IF rc yields the true class of oi
            THEN   weight(rc)= weight(rc) + award
            ELSE weight(rc)= weight(rc) - punishment
            End IF
        End FOR
    End While
End.
```

4. COMPARISON OF RULE INDUCTION METHODS

In this section, we present the method PVM (Predictive Value Maximisation method [WK91]) and we give the complexity and precision comparisons with IPR.

4.1. The PVM method

The goal of the PVM method is to find the single best rule of length less than or equal to a fixed value. For each variable, a set of interesting threshold and constant values are first determined. These cut-off points are local maxima of the predictive values for the single variable being considered. Logical expressions with variables are generated (in disjunctive normal form) and instantiated with the interesting constants and thresholds. These are tested for discriminatory performance, and a relatively small table of the most promising expressions is kept. Then, combinations of the stored expressions are used to generate longer expressions, and the most promising of these are in turn stored in the table and are used to generate even longer expressions. The method is an approximation to the exhaustive generation of all possible rules of a fixed length or less [WK91].

4.2. Complexity analysis

Let c be the number of attributes, d be the number of objects of the data base and n the number of couples in the binary relation representing the data base (n $\approx$ c * d). The total complexity of the PVM method is about :
$$O\left[\sum_{k=1}^{c} (n * d)^{2^k}\right]$$ [MJ96]. Whereas, the cost of the IPR

method is pessimistically approximated by $O[n^2*(c+d)^2]$ [MJ95]. Here, we accentuate the fact that the PVM method is exponential in complexity. This is due to the part of the algorithm that looks for the combinations of simple conditions to build complex ones. In the IPR system, we use a heuristic that enables us to identify a complex condition rule without testing a combinatory number of possibilities. The complexity analysis proves that the IPR is faster than the PVM. PVM cannot be applied to large data bases [WK91].

4.3. Precision estimators

The precision of the system is measured by the true error rate. It is the error rate tested on the true distributionof cases in the population - which can be approximated by a very large number of new cases gathered independently from the cases used to design the system [WK91]. In our work, we estimate the true error rate by the apparent error rate and the leaving-one-out error rate.

The apparent error rate of a system is its error rate when it is tested on the same sample cases used to build the knowledge base. It is known as the resubstitution or reclassification error rate.Since we are trying to extrapolate performance from a finite sample of cases, the apparent error rate is the obvious starting point in estimating the performance of a system on new cases. It is a biased estimator.

Leaving-one-out is the simplest estimator based on randomly resampling. Suppose we have a data base of n resolved objects and we want to estimate the true error rate of a learning system. We extract our rules from (n-1) objects and we leave only one object by which we test the system. This iteration is repeated n times. In each time, we choose a new object for testing. The error estimated is the average of the n iterations. Leaving-one-out is a very costly method, but it is a non biased estimator.

4.4. IRIS flower classification :

The precision measures of the methods are done on the IRIS data base. It is a one hundred fifty patterns data base with three flower categories : 'Setosa', 'Verginica' and 'Versicolor'. Each class is represented by fifty patterns. Each pattern is described by four continuous attributes : petal length (PL), petal width (PW), sepal length (SL) and sepal width (SW). The problem is to extract from the data base, the rules with which we can assign a certain flower class. These rules will be used, later, to predict the class of a new coming flower using only the measures of its petal length, petal width, sepal length and sepal width.

The PVM method generates only two rules with a total of three variables. The IPR method generates more rules pondered by the punishment/award algorithm. The IPR method gives well interesting results (Figure 5). The apparent error rate indicates that it performs well in the learning task. The Leaving-one-out error rate proves that the IPR system will be more precise on the new cases than the PVM system. The results of the PVM method are taken from [WK91]. The values on the other data sets

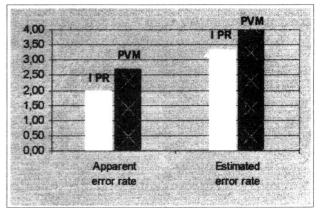

FIGURE 5 : Comparative performance on IRIS data base

are not given. That is why we compare IPR and PVM only on the IRIS data.

5. COMPARISON WITH DECISION TREE METHODS :

Here, we compare the IPR method with the decision tree methods ID3 and SIPINA.The results of IPR are relevant to the data representation method 'modal' and the ponderation via the punishment/award algorithm initialisad by zero values. More interesting results can be atchieved (separately on each data base) using other data representation methods, other rectangular decomposition heuristics or other rule ponderation procedures. As a fact, the results given here are differents from those of section6.

5.1. The methods :

ID3 is a well known machine learning method, where the knowledge is represented by a non binary decision tree. It derives from the works of Quinlan in the later seventeen's [DK91]. It has many successfully industrial applications. Let O be the set of objects belonging to various classes. The algorithm chooses the more predictive attribute [HN96] and partitions the objects of O into a set of disjoint subsets C_1, C_2, ..., C_n. Where C_i contains the objects of O having the same i^{th} value of the chosen attribute. Then each subset C_i will be partitioned using the same strategy, unless the objects that it contains belong to the same class. In the resultant decision tree, each node represents a tested attribute and each leaf is assigned a class. For a new coming object, we cross the decision tree. In each node, we look for the associated attribute value of the object and we branch to a second node. We repeat this testing task since the root node, to a terminal node. When achieving the terminal node, we assign its associated class to the new object. ID3 is also an incremental method.

The method SIPINA (Système Interactif pour les Processus d'Interrogation Non Arborescente) derives from the work of A. Zighed in the later eighteen's [ZAD92]. It was applied in many classification applications. The main idea is to induce a decision tree where we can join similar nodes to improve a certainty factor. In other words, let S_i be a partition of the data set O. Using S_i, the algorithm looks for a new partition S_{i+1} of O that is more certain than S_i. The certainty of a partition is measured by an entropy function. SIPINA gives many choices of the known entropy measures. To generate the partition S_{i+1} from S_i, we try the possibilities of dividing a subset of objects in S_i or joining two subsets in S_i. Each possibility gives a new partition S_i^k and the chosen partition that makes S_{i+1} is taken as the partition S_i^k having the best certainty value.

5.2. Complexity analysis

Let c, d and n be the parameters defined in section 4.2. The complexity of the method ID3 is found to be about $O(c^2*d^2)$. The complexity of the method SIPINA is also proven to be about $O(c^2*d^2)$. The IPR method is not faster than these two decision tree methods since its complexity is about $O(c^2*d^2*(c+d)^2)$. In general, the rule induction methods are more complex than the decision tree ones.

5.3. Precision estimators

When the number of objects increases, the leaving-one-out estimator is no more appropriate since it is a very costly method. The true error rate is estimated by the apparent error rate and the 10-fold cross-validation error rate. The idea of the 10-fold cross-validation estimator is to partition the whole data base into 10 subsets. We take 9 partitions as the learning set and we keep the tenth set as a test set. The iteration is repeated 10 times. The errors obtained are averaged [HN96].

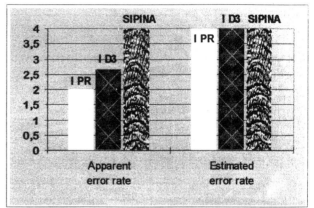

FIGURE 7 : Comparative performance on IRIS data

5.4. Diagnosis of the W. B. Cancer Disease

This data base is relevant to the works on medical diagnosis of the "Wisconsin breast cancer". The problem is to extract a set of rules and to use these rules to decide whether a patient has a dangerous cancer or a benign cancer. The samples consist of 683 objects (patients), nine attributes and two classes. For 444 of the patients, it was a dangerous cancer. For the others, it was a benign cancer.

Figure 6 summarises the results. The apparent error rate proves that the IPR method fits well the learning examples. Only 1,96 % of the patients has been miss-classified within the 683 examples of the data base. ID3 and SIPINA yield a same error rate (3 %). When diagnosing new patients, SIPINA can miss-classify seven patients per cent. The method IPR miss-classify only 4,73% of the patients. It is more reliable than SIPINA and ID3.

5.5. IRIS Flower Classification

The data set related to this application is described in section 4.4. In general cases, when a learning system gives best results on learning samples, it gives bad results

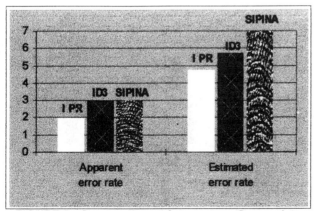

FIGURE 6 : Comparative performance on Cancer data

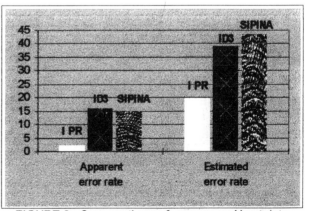

FIGURE 8 : Comparative performance on Heart data

on testing samples because it over-specialises the learning data. The IPR method gives best results on both learning and testing samples (Figure 7). Whereas SIPINA makes 4 % of mistakes on the learning set as well as the testing set, IPR gives only 2 % of mistakes on learning data and 4 % on testing data. Here the difference between IPR and the other methods is not enormous since the data base is not very large (150 examples).

5.6. Diagnosis of the Heart disease

This data base is relevant to the works of R. Detrano in the V. A. Medical center (Long Beach and Cleveland Clinic Foundation - USA). We try to extract a set of rules that designates the possible diagnosis of a patient having a heart disease. The samples consist of 297 patients. For each patient, we have 14 clinic tests. An additional attribute states the disease of each patient.

Figure 8 shows that the IPR method fits well the input data and generalises well the discovered knowledge since it gives better results on both learning and testing samples. We have not over-specialisation problems. In the reclassification task, the IPR method makes only 2,35 % as an error rate, whereas the ID3 method gives 16% and SIPINA gives 15 %. In the prediction task, IPR makes only 19,86 % of mistakes. ID3 makes 39,05 % and SIPINA makes 43,34 % of mistakes. 39,05 % and 43,34 % are enormous rates and prove that ID3 and SIPINA cannot be reliable tools to diagnose such dangerous diseases.

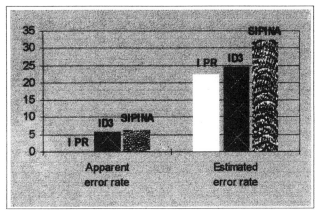

FIGURE 9 : Comparative performance on Brandy data

5.7. Brandy Classification

This data set is obtained from the laboratory of DGCCRF in Strasbourg. It deals with the classification of the alcoholic drinks (brandy). There are three classes of brandy : « Kirsch », « Mirabelle » and « Poire ». This drink is made up by nine chemical solutions with different dosages. The data base contains 158 examples of brandy and nine attributes. A tenth attribute shows the brandy class. Given the different dosages of a brandy solution, the learning system should give the appropriate brandy class.

While the data base is not very large, the precision of the IPR method is extremely better than those of the ID3 and SIPINA methods. Figure 9 shows that the IPR

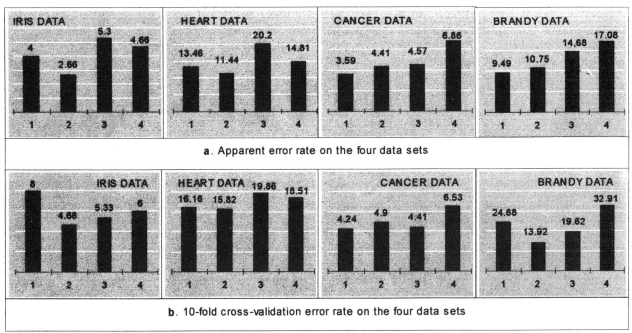

a. Apparent error rate on the four data sets

b. 10-fold cross-validation error rate on the four data sets

FIGURE 10 : evaluation of the data representation methods : (1) modal, (2) fusinter, (3) MDLPC and (4) chi-merge.

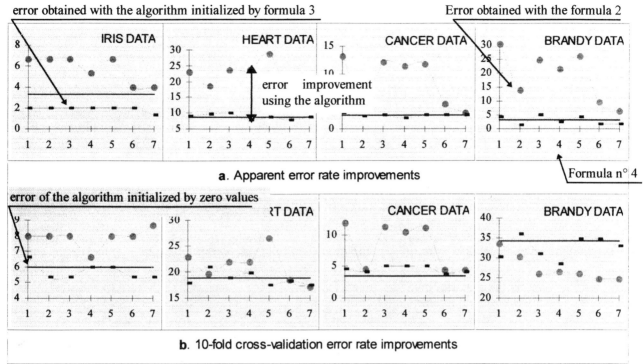

FIGURE 11 : Improvement of the IPR error rate using the punishment/award algorithm (black points)

method performs well on this data set. It does no mistakes and it reclassifies correctly all the learning samples. The ID3 method does better than SIPINA on these samples. Also, on the new cases, ID3 does better than SIPINA. But, the error rates of the two methods (ID3 : 24,68 %, SIPINA : 32,27%) are greater than IPR error rate (22,29).

6. OTHER EMPIRICAL RESULTS

Other than the comparison of the presented methods, we give an empirical evaluation of some tools in the IPR system. We evaluate the data representation method : modal and the punishment/award algorithm.

6.1. Data representation methods

In section 3.1, we presented the method « modal » that identifies the interesting cut-off points of a quantitative attribute. These points are used with logical operators to build the tests of a production rule. Many other methods exist and were used with well known learning systems. Here, we consider four methods : « modal », « fusinter », « MDLPC » and « Chi-merge ». « fusinter » is the method proposed with the SIPINA system. We study the effect of data representation methods on the performance of the IPR system. In each time, we use one data representation method with the IPR system and we measure its precision. We measure the apparent error rate (figure 10.a) and the 10-fold cross-validation error rate (figure 10.b), on the four presented data bases. Figure 10.a shows that modal and fusinter are the most promising methods since they give the best results. Modal gives the best precision values on cancer and brandy data. It gives the second best values on iris and heart data. Whereas, fusinter gives the best precision values on iris and heart data and gives the second best values on the other data sets. Figure 10.b proves that fusinter is more reliable since it gives best values on three data sets : iris, heart and brandy. Modal gives the best value on cancer data and the second best value on the heart data. It is more interesting than MDLPC and Chi-merge.

6.2. Rule conflict resolution

Rule induction methods are considered as promising methods since they generate non mutually exclusive rules. Nevertheless, this powerful knowledge representation leads to the problem of conflict knowledge. Assigning weights to the rules is the widely used way for conflict resolution. Here, we propose seven formulae to calculate the rule weights. The numbers 1, 2, ..., 7 in figure 11 denote the seven formulae. As an example, we state that the formula 1 takes the number of objects enclosed in the associated rectangle as the rule weight. The other formulae are more complex to describe here. So, we give only the precision obtained with them (the grey points of figure 11). The punishment/award algorithm use the formulae resultant weights as initial weights. It tries to improve the precision of the IPR system by adjusting these weights. Figure 11 shows the effect of this on the error rate of the IPR system, measured respectively on the iris data base, the heart data base, the cancer data base and the brandy data base. The black bold points denote the values related to the punishment/award algorithm when initialised by the formulae weights. The thick horizontal line shows the precision value relevant to the application of the algorithm when initialised by zero values.

In general cases, the punishment/award algorithm improves well the precision of the IPR system. It improves the apparent error rate more than the 10-fold cross-validation error rate since it tries to fit the learning data. This is clear in the different figures. On the brandy data, the algorithm over-specialises the learning samples. It improves the apparent error rate and damages the testing error rate. To remedy this problem, we should iterate the algorithm for the necessary time. Finally, note that the precision values obtained by the punishment/award algorithm are not very different. The algorithm tries to achieve the same precision level independently of the formulae used to initialise the weights. The precision values relevant to the algorithm when initialised by zero values are generally reliable. They are not the best values, but they are always near to them. This accentuates the reliability of the punishment/award algorithm and its stability.

7. CONCLUSION

The Incremental Production Rule based method is a rule induction method that relies on the concept of rectangular relation. In this work, we introduced a punishment/award algorithm for rule conflict resolution and we focused on the empirical evaluation of the precision capabilities of the IPR method. The comparison with the PVM method proves that IPR is a faster and a precise rule induction method. The comparison with widely used decision tree methods like ID3 and SIPINA, proves that non mutually exclusive form is not only a powerful aesthetic knowledge representation, but also, it leads to a well precise classifier. The other empirical results focused on the usefulness of fusinter, the data representation approach of SIPINA, and emphasised the contribution of the punishment/award algorithm.

So far, we do not know a rule induction method faster and more precise than decision tree methods. In the IPR method, since it is the heuristic algorithm of rectangular decomposition that induces the great complexity of the method, many studies are interested in searching a more suitable solution. We hope to achieve the complexity order of the decision tree methods and to keep an interesting precision level.

ACKNOWLEDGEMENT

This work is financed by the Tunisian state secretary of scientific research as a national research project (PNM'93). We thank particularly the students S. Tlili, L. Khediri, M. Saadana and J. Boughizan for their useful contribution to this work.

REFERENCES

[C69] M. Chein, "Algorithme de recherche de sous matrices premières d'une matrice", Bulletin de Math (R.S. Roumanie), p.21-25, 1969.

[C90] P. Cazes, "Codage d'une variable continue en vue de l'analyse des correspondances", Rev. Statistiques Appliquées, p.35-51, 1990.

[DK91] E. Diday, Y. Kodratoff, « Induction symbolique et numérique à partir de données », Cépadues-édition, France, mai 1991.

[G94] D. E. Goldberg, « Algorithmes génétiques : exploration, optimisation et apprentissage automatique », Edition of Addision-Wesley, Franch, June 1994.

[HN96] T. Ho, T. Nguyen, "Evaluation of attribute selection measures in decision tree induction", In proc. of the ninth international conference IEA/AIE'96, Fukuoka (Japon), June 4-7, 1996.

[HCK95] HSU, Chun-Nan and Knoblock, «Estimating the robustness of discovered knowledge», In Proc. of Inter. Conf. (KDD-95), Montreal, 1995.

[KJ97] R. Khcherif, A. Jaoua, "Incremental rectangular decomposition of documentary data base", will appear in the journal 'Information Science', 1997.

[MJ95] M. Maddouri & A. Jaoua, "Incremental learning : proposition and evaluation of methods", in proc. of the Joint International Conference on Information Science, JICIS'95, Durham (USA), September 28, 1995.

[MJ96] M. Maddouri & A. Jaoua, "Incremental rule production : towards a uniform approach for knowledge organisation", In proc. of the International Conference IEA/AIE'96, Fukuoka (Japon), June 4-7, 1996.

[WK91] S. M. Weiss & C. A. Kulikowski, "Computer Systems that learn", Morgan Kaufmann Publishers, USA, 1991.

[ZAD92] A. Zighed, J. P. Auray, G. Duru, "SIPINA : méthode et logiciel", Edition Alexandre Lacassagne, Lyon (France), 1992.

IDENTIFICATION OF A LOSS OF COOLANT ACCIDENT ON A SUBMARINE NUCLEAR STEAM RAISING PLANT USING AN INTELLIGENT KNOWLEDGE BASED SYSTEM

Richard J Carrick

Department of Nuclear Science & Technology, Royal Naval College
Greenwich, London, England, SE10 9NN.

ABSTRACT

Submarine Nuclear Propulsion Plant Supervisors operate under different constraints to the civilian organisations. Not only must they ensure the safe intelligent operation of the equipment and systems, but they must also provide continuous propulsion availability and electrical generation, without which the submarine can not manoeuvre and fight effectively. This paper presents an investigation into the use of a Knowledge Based System in identifying, diagnosing and correcting a particularly disabling fault - A Loss of Coolant Accident - on a Nuclear Steam Raising Plant.

KEY WORDS

Expert System, Knowledge Base, IKBS, Loss Of Coolant Accident, LOCA, Pressurised Water Reactor, PWR, Nuclear Steam Raising Plant, NSRP.

NUCLEAR STEAM RAISING PLANT

The safe, reliable and continuously available operation of the Nuclear Steam Raising Plant (NSRP) is the fundamental aim of all nuclear plant managers, supervisors, operators and maintainers. Loss of performance can have severe penalties on operational time and in extreme cases place the vessel in potential danger. Reduction in plant availability on a submarine can have amplified problems when the operational environment is considered. The sea is unforgiving and during hostile activities the loss of propulsion that may result from a Loss Of Coolant Accident (LOCA) could prove catastrophic.

The LOCA potentially poses one of the most severe challenges to the operator and to a Pressurised Water Reactor (PWR). On current classes of submarine the responsibility for immediate protection of the reactor following a LOCA lies solely in the hands of the supervisor of the propulsion watch, consisting of a Charge Engineer and various Engineer Technicians and Mechanics. The human is the most important factor in the safe operation of the NSRP, ensuring the integrity of the reactor is not compromised and being obliged to take any necessary action required to prevent predetermined protection limits, and consequently core thermal constraints, from being violated. In addition, the overall submarine safety must be considered and the supervisor must endeavour to provide the Command with propulsion availability at all times. In doing this he must have due regard for Reactor Safety and Nuclear Safety. This is the state achieved when the probability (and with Nuclear Safety the consequence in addition) of a Nuclear Reactor Accident has been reduced to an acceptably low level and the potential for radiation exposure, whether routine or accidental, has been reduced to a level as low as reasonably practicable, by sound organisation and good discipline in **operation**, design, construction, maintenance, repair and training. Nothing is achieved in endangering the submarine to prevent possible, but not certain, core damage The need for accurate diagnosis and responsibility in decision making now becomes more apparent.

NEED FOR OPERATOR AID

To provide a reasonable chance of success in dealing with a Fast Primary Coolant Leak (the most severe classification of leak) the supervisor must ensure operators respond rapidly to leak symptoms, including diagnosis, responding according to the relevant operating procedures. This requires rapid recognition of the fault condition and swift response to rectify the condition, including the correct operation of all required emergency actions. The less severe classifications of Slow Primary Coolant Leak and Leak Search allow considerably longer diagnostic times. The operators of NSRP have a considerable number of individual parameters concerning the primary, secondary and auxiliary supporting equipment and plant to use as a basis for decision making. Experience of reactor plants such as Chernobyl and Three Mile Island have highlighted the weaknesses of the human element in any control system. The frequency of human errors on the NSRP remains relatively low [AEA92], however, the potential consequence arising from human error may be high and hence the stimulus, together with surveillance IKBS work [CBMC95], for the study into the use of Intelligent Knowledge Based Systems as an operator/supervisor aid.

Intelligent Knowledge Based Systems (IKBS) were considered as potentially useful tools to provide a source of advice to supervisors which are under great stress at the time critical and rapid decisions are required. The provision of a suitable "expert base" was available within the Department of Nuclear Science & Technology in the form of propulsion plant operators and relevant academic experts in a range of fields within the nuclear industry.

DEVELOPMENT OF THE LOCA APPLICATION

Knowledge Elicitation. The correct courses of action in the case of many plant faults are laid down in Emergency Operating Procedures (EOP's). These provide the basis for the correct course of action and hence the guidance the IKBS should be able to provide the supervisor. EOP contain only the immediate actions to place the plant in a safe situation, any remedial action to correct the situation is not included and this was obtained from the sources detailed in a previous paragraph.

System development. The construction of the overall LOCA application followed the procedure detailed below:

- Identification of the type of problem and recognition of any constraints on the project including time, computing power, expertise and costs. In this case the IKBS was PC Based. The budget was minimal but the use of a wide range of expertise was freely available within DNST.

- Highlight the key concepts of the problem. This was to identify the existence of a primary coolant leak, classify the size and then identify the leak site.

- An understanding of the nature of the type of search required within the IKBS. In this case the decisions needed to be highly reliable and rapidly achieved. Any doubt in the decision should produce at least some advice to the supervisor.

- Production of an executable program. This is the building of clear, complete and unambiguous rules, then creating a logical rule structure.

- A testing and validation stage to ensure the correct response is reached and that information provided is reliable and, importantly, helpful.

The above process is iterative with redesign and refinements taking place as the application evolves.

THE IKBS

XI-Plus™ was the PC based IKBS tool used. It utilises "know-how" programming to form an advisory tool. It consists of three main parts:

- A Knowledge Base containing the rules or "know-how", collected from the expert sources.

- An Inference Engine or rule interpreter capable of processing the knowledge base. This is the link between the users and expert, understanding the methodology used by the Knowledge Engineer in constructing the knowledge base.

- A database which holds the current status of consultation, providing a means of checking the logic behind a derived solution.

The IKBS has an in-built rule checking facility and a window facility that gradually encourages the user to use more sophisticated tools. In this instance the shell facility was used, together with the built-in toolkit and menu functions. This inevitably led to some limitations in the construction method but was time efficient. After the collection of knowledge from the relevant experts the rule construction for development of the knowledge base could commence. Three main methods were employed:

- Use of the "is" clause, where if an identifier is the required value then the rule is the consequential action. The text has no intrinsic meaning to the IKBS and was chosen to have meaning to humans. The command language includes words such as "is", "if" and "then" to link the statements together and where necessary user defined clauses may be stipulated.

- The use of "Assertions" that are meaningless patterns to the package but are helpful to the user.

Only true or false statements are found in assertions.

- The use of "Demons" were used to remain
  dormant until a condition was satisfied and then
  fire immediately to generate the required response;
  that is, the action that an experienced operator
  would take.

Rule Sequencing and Search Patterns. On
information being provided by the user the expert system
will backward chain to find the answer to the query, the
package having found a rule statement to match the
required goal as a consequence. It then examines the
conditional clause of the rule, which could in turn be set
by the consequences of another rule. Alternatively the
system could forward chain where the inferences
propagate forward to a conclusion that may not be known
in advance. The IKBS package used can be forced to
carry out particular sequences or left to establish its own.
In practice a mixture is used and with experience the
Knowledge Engineer will achieve the most responsive
application by careful selection of sequencing. This
selection also extends to the search pattern, which may be
depth or breadth first. The depth method tends to reach
the required rule quicker but only if guided in some way.
Breadth-First searching is similar to the admissible
properties of algorithms, and will take the shortest path
available if one exists, and may eliminate redundant
decision paths at an early stage [Jac90]. The LOCA
application used a combination of search patterns.

RULE GENERATION

From the enormous amount of information available
to the Manoeuvring Room Watchkeepers it was necessary
to extract all the relevant parameters for identification and
classification of a primary coolant leak. The bases for
this information were existing operating rules,
supplemented by operator experience. The LOCA
problem was considered in two parts; determine if a leak
exists and then establish the size and location of leak site.

Five fundamental parameters were used in the initial
task of establishing the existence of a leak. The various
combinations were then used to classify the leak as
"probable", "possible" or "unlikely". The parameters
were:

- Core Average Temperature - Tav.

- Reactor Power.

- Pressuriser Level.

- Loop Pressure.

- Steam Generator (SG) Activity.

All of these parameters are monitored on current
classes of plant. Normal operation may, with the
exception of SG Activity, see a change in any of the
above parameters and must not be construed as a primary
coolant leak. However combinations of the above
parameters can lead to the increased likelihood of a leak
and the application reflects this by guiding the operator in
his next course of action. Decision Trees were used to
identify the possibilities and link the conclusions to the
next course of action. In this application the Decision
Trees were formulated from the Operating Rules to cover
every considered eventuality and to identify the required
response from the application. 18 separate Decision
Trees were formulated, containing over 3000 separate
outcomes. The separate use of the trees eased
construction, matching the Decision Trees to discrete
parts of the identification and searches, and separate
knowledge bases constructed based on the information
provided.

When a probable conclusion is reached the next
requirement is to determine the rate of leak, the correct
emergency actions can then be suggested to the operator.
Different emergency actions are required depending on
the leak rate:

- Fast Leak Isolation.

- Slow Leak Isolation.

- Leak Search.

The first knowledge base encountered contained rules
to determine if a leak existed, the next classifies the leak
rate using the following parameters:

- Pressuriser Level.

- Make-up Pumps state.

- Pressuriser Level fall rate.

- Discharge state.

Sufficient time must be available to complete the slow
leak drill when classified. If insufficient time exists then
the fast leak drill must be chosen, this is the reason for the
Pressuriser level parameter. The state of the Make-Up

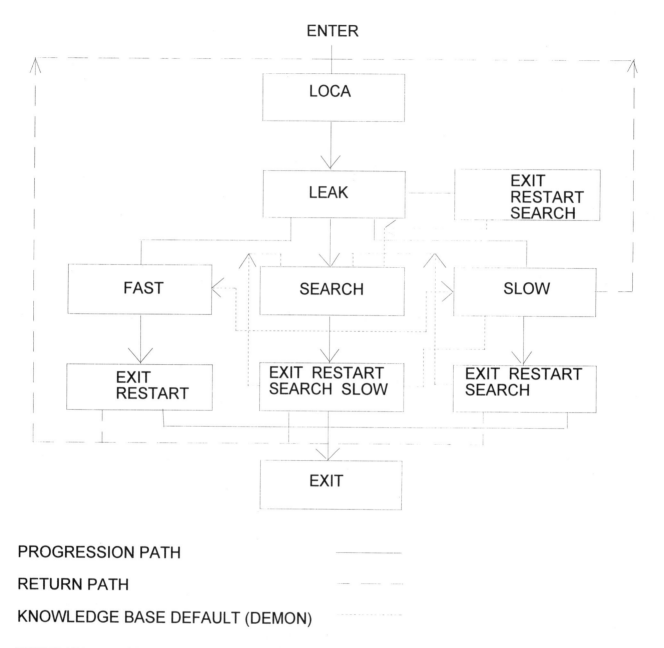

PROGRESSION PATH

RETURN PATH

KNOWLEDGE BASE DEFAULT (DEMON)

FIGURE 1 Diagram of Knowledge Base Interrelationship.

Pumps and Discharge System would effect the rate of fall and must therefore be considered when establishing the leak rate. If the level of the Pressuriser falls below a level when it is considered that insufficient time exists to complete Slow Leak Isolation then the Fast Drill must be commenced immediately. The use of a "Demon" for this rule ensures that the application can advise the operator of the correct action should the level fall below a predetermined threshold.

After establishing the leak rate the application then link to the next required Knowledge Base. From Figure 1 the next level can be seen, and the application would have picked either "Fast", "Search" or "Slow" depending on the rate. The action required of the operator differs in each case, balancing plant availability with the isolation of the leak. Clearly the action required to find and isolate a very small leak will differ from isolation of a leak rate that will quickly result in a lack of cooling capability.

When the correct course of action has been taken the application leads the operator through a series of questions, but in a working solution these would be automatically provided as plant configuration changes.

After the relevant path has finished, resulting in the leak being identified and isolation action taken, the application will allow the operator to reset the system or to continue the search pattern. The latter is in the event that he symptoms of a leak still exist and continued action is required. It is worth noting that the leak search actions are considerably more detailed than the fast and slow leak emergency actions.

APPLICATION TESTING

Ensuring that the application performs as intended is crucial if the expert system is to prove a useful diagnostic aid. Testing was carried out in a number of discrete steps.

- On-going testing. The construction of a series of knowledge bases allowed their individual checking as the application developed. A query was built into the knowledge base and executed in isolation to ensure the correct conclusion was reached.

- Checking facilities. XI-Plus™ contains a number of built-in check facilities detailed below:

    -- Names and values - lists the identifiers with the relations and values, including any assertions used.

    -- Unused consequences - Inferences cannot be made due to the knowledge base being set of rules are not being used.

    -- Missing values - conditions that cannot be satisfied. Linking between the Inference Engine and the knowledge bases cannot not be completed due to incomplete information.

    -- Circular reasoning - reports any identifiers or assertions that could result in looping in the chaining method employed. This can be avoided by good construction.

    -- Conflicting values - possible conflicts between value identifiers and assertions - confusion in the inference cycle can result.

    -- Trace. This is a built in command that proved valuable in resolving failures in the chaining process, or how the application arrived at the wrong conclusion. The "thought" process is shown in full and allows the constructor to step through and check that rules are written correctly.

    -- The final stage of the testing was to systematically input the information contained within the Decision Trees, ensure the application links correctly and the conclusion reached is in accordance with the experts views.

REFINING

Careful design and forcing of the application to use a particular inferencing technique can increase the speed of the system in arriving at a conclusion. In the LOCA application the Inference Engine was forced to predominately forward chain by the use of demons.

However, in certain sections of the application it was advantageous to limit the inferencing to backward chaining, in doing so breaking the rules down to solve them individually. This prevents the application from interrogating a complete set of rules when the correct conclusion has been found during the first set.

The sequence order is also controlled. Normally the system will work through the rules in the order they appear in the knowledge base. Unfortunately some rules have a dependence on a set of sub-rules that must have provided a conclusion in order to proceed. The construction of the application took this into account to limit the number of times a knowledge base needed to be interrogated.

The multiple firing of rules and demons allows each rule set containing like statements to be accessed concurrently. This can reduce enquiry time by preventing the application from attempting to satisfy the rule conditions with incomplete information. In essence, instead of the system attempting to use one piece of information at a time and then trying to reach a conclusion; the system accesses each rule set with enough information to satisfy all rules in each set the first time.

The original application consultations involve the use of the query tool. A particular identifier, relation and value is set into the knowledge base and then fired to check a conclusion can be reached. The system automatically asks the questions after initiation and then prompts the user for inputs. This method is unfriendly to the user and it was at this point that "forms", an example is shown at Figure 2, were designed to limit the inputs and ask for relevant information at the correct time, limiting the value ranges (to sensible figures). In total this application used eighty-four forms containing over two-thousand rule statements to arrive at a conclusion.

The response from lines of enquiry were made on "reports" which were linked to a number of "forms". The contents reflect the conclusion an engineer in charge of the situation would arrive at if provided with the same information as the application. Careful linking of the Form and Report libraries allows a response to be given and the next line of enquiry pursued.

APPLICATION ASSESSMENT

The LOCA application was based on a single operator's knowledge, and as such may contain incomplete information and errors in judgement. Validation of the this system utilised two independent operators and an Artificial Intelligence Engineer. They were used to comment on the accuracy and completeness of suggested courses of action, the ease of use and the usefulness of information provided. No restrictions were placed on the assessor's inputs and any values that led to the application being unable to find a response were solved on refining the knowledge bases. It was found that to a large extent that the suggested course of action was in agreement with the assessors own thoughts. It is considered that substantially more assessors, or practitioners, from varying disciplines, would be used tin developing the full expert system. The system was considered easy to use with little training required to grasp the fundamentals of use. The full assessment of the usefulness of such an application would require a real time system in a "live" Manoeuvring Room situation.

Correct diagnosis of the Loss of Coolant Accident requires a considerable number of plant parameters to be monitored in order that accurate advice may be provided. Human operators may be selective when faced with vast quantities of information, helped by their experience, but the knowledge based application must be exhaustive to arrive at a credible solution. The Knowledge Engineer must take extreme care to include complete information for the application, without which the application will be unable to arrive at a suitable diagnosis. This situation may prove to be highly confusing for the operator and could possibly be worse than having no advice system at all. Worse still could be an application that arrives at a conflicting decision with the operator's conclusion, leaving the operator with the choice of believing the application or the conclusion reached via his own thought train.

"INPUT PLANT PARAMETERS"

Tav	[ ]	Celsius
Power	[ ]	Percent
Pressuriser Level Fall	[ ]	Visible ?
Primary Coolant Discharge	[ ]	Yes/No
Loop Pressure	[ ]	Bar
Fall in Loop Pressure	[ ]	Bar/Sec

"Press Return when finished"

FIGURE 2 Example of "Form" questions.

Knowledge Base Systems such as this one would prove extremely useful in providing the corrective action to remedy a leak situation. The initiation of a course of corrective action would be the operators independent decision. These could also act as the impetus to start the diagnostic process in the case of the more controlled Slow and Leak Search Drills.

An extremely useful function of the system developed and the IKBS software package is the ability to display how the conclusion is reached. Indeed the user can be lead through on a step by step basis in slow time, analysing the questions and responses to reach a conclusion. Although time would not allow this feature to be used in a "live" situation, the benefit during training is evident.

CONCLUSIONS

A workable Knowledge Based application has been created, capable of offering identification, diagnostic and corrective advice to operators. The current application is dependent on the user inputting information manually, and it is considered that the application can only be of real use with real-time inputs due to the speed of response required by the operator. The completeness of information for the LOCA has proved difficult due to the large number of possible outcomes and numerous parameter inputs.

The investigation into the usefulness of expert systems in the NSRP environment has thus far not considered multiple failures, where plant parameters may lead to conflicting conclusions. Rule statements would need to be written extremely carefully to avoid this scenario, and when all possible credible failures have been considered. Currently operators can use experience and their heuristic skills to deal with undocumented failures. In the case of the IKBS this is not possible and the knowledge must already have been imparted in order to generate a solution. Doubt exist as to the suitability of such systems in multi-scenario situations, indeed this was the stimulus for further investigation of neural network application of an Automatic Manoeuvring Room Advisory System (AMRAS).

The application has potential in the training field due to the inherent capability of the application to allow investigation of how the relevant decisions were reached, and hence giving an insight into the experts knowledge.

Current philosophy within the UK means that an operator must have absolute control of the reactor plant, and the responsibility should not be delegated to an operator aid [Be89, Wa95]. This limits the use of expert systems to an advisory role and precludes the use of such devices in automated control of emergency actions. Additionally submarine safety may carry equal importance to reactor safety in some situations, and may require the operator to consider alternative remedies to the normal emergency actions in order to maintain propulsion or electrical generation capability.

The creation of a custom built environment may help to alleviate problems occurring as the application becomes more involved, as it would when more failure scenarios are included. However, it is considered that the use of knowledge based systems would become unmanageably large, and contain insoluble problems, due to incomplete information, when all fault conditions are included.

The application built and tested in this investigation has adequately demonstrated the usefulness and limitations of IKBS. Similar studies have reached parallel conclusions [BMD95, HL89, RO95], but have been on plants of an entirely different nature with fundamentally different operating constraints.

Any views expressed are those of the author and do not necessarily represent those of DNST or HM Government.

REFERENCES

[AEA92] AEA Technology, Safety and Reliability Directorate, Human Reliability Document, Dec 1992.

[Be89]   Bernard J, Expert System Applications within the Nuclear Industry, Library of Congress, 1989.

[BMD95]   Berry H L, Menell S J, Davies J, An Introduction to Artificial Intelligence, Nuclear Energy, Vol 34, Feb 1995.

[CBMC95]   Cadas C N, Bowskill J, Mayfield T, Clarke J C, An IKBS approach to surveillance for Naval Nuclear Submarine Propulsion, Journal of the British Nuclear Energy Society, Feb 1995.

[HL89]   Hassberger J A, Lee J C - A Simulation-Based Expert System for Nuclear Power Plant Diagnostics, Nuclear Science and Engineering, 1989.

[Jac90]   Jackson P, Introduction to Expert Systems, second edition, Addison-Wesley Publishing Company, 1990.

[Ro95]   Rowe A, Applications of Artificial Intelligence in the Nuclear Industry, Journal of the British Nuclear Energy Society, Feb 1995.

[Wa95]   Wainwright, N, A Regulators Viewpoint on the use of AI in the Nuclear Industry, Nuclear Energy Vol. 34, 1995.

UTILIZATION OF AN EXPERT SYSTEM FOR THE ANALYSIS OF SEMANTIC CHARACTERISTICS FOR IMPROVED CONFLATION IN GEOGRAPHICAL SYSTEMS

Harold Foley III, Fred Petry, Maria Cobb, Kevin Shaw
Naval Research Laboratory
Stennis Space Center, MS, USA 39529
{foley,petry,cobb,shaw@nrlssc.navy.mil}

Keywords: geographical information systems, conflation, feature matching, deconfliction

Abstract

One of the major contributions of geographical information systems (GIS) is their ability to efficiently manage geographical information. Geographical information can be represented and stored in a variety of ways, each with its advantages. Thus, a challenging problem facing GIS is the ability to effectively utilize the various types of geographic data. The process by which these different information sources are merged in order to yield a more comprehensive dataset is referred to as *conflation*. In this paper, we will fully describe the issues and problems related to conflation, as well as discuss how an expert system is utilized to assist in the process.

1 Introduction

The need to handle imprecise and uncertain information concerning spatial data has been widely recognized in recent years [Goo90, Bur96] particularly in the field of geographical information systems (GIS). GIS is a rather general term for a number of approaches to the management of cartographic information. Most definitions of a GIS [Mag91] describe it as an organized collection of software systems and geographic data able to represent, store and provide access for all forms of geographically referenced information. At the heart of a GIS is a spatial database. The spatial information describes the location and shape of geographic features in terms of points, lines and areas. This is essentially a map represented in the database by means of cartographic features and techniques. The loca-

tion data has corresponding descriptive information about the geographic features.

There has been a strong demand to provide approaches that deal with inaccuracy and uncertainty in GIS work. The issue of spatial database accuracy has been viewed as critical to the successful implementation and long-term viability of GIS technology [Ver89]. The value of GIS as a decision-making tool is dependent on the ability of decision-makers to evaluate the reliability of the information on which their decisions are based. Users of GIS technology must therefore be able to assess the nature and degree of error in spatial databases, track this error through GIS operations and estimate accuracy for both tabular and graphic output products. There are a variety of aspects of potential errors in GIS encompassed by the general term "accuracy." However here we are mostly interested in those aspects that lend themselves to modeling by fuzzy set techniques.

Many operations are applied to spatial data under the assumption that features, attributes and their relationships have been specified a priori in a precise and exact manner. However, this assumption is generally not justifiable, since inexactness is almost invariably present in spatial data. Models of uncertainty have been proposed for GIS information that incorporate ideas from natural language processing, non-monotonic logic and fuzzy set, evidential and probability theory. For example in [Sto87] there are reviews of four models of uncertainty based on probability theory, Shafer's theory of evidence, fuzzy set theory and non-monotonic logic. Each model is shown as appropriate for a different type of inexactness in spatial data. Inexactness is classified as arising primarily from three sources. "Randomness" may occur when an observation can assume a range of values. "Vagueness" may result from imprecision in taxonomic definitions. "Incompleteness of evidence" may occur when sampling has been applied, there are missing values, or surrogate variables have been employed.

Robinson [Rob85, Rob88, Rob90] has done extensive research on fuzzy data models for geographic information. Robinson has considered several models as appropriate for this situation, including two early fuzzy database approaches using simple membership values in relations [Gia79, Bal83] and a similarity-based approach [Buc82].

Active recent research has extended these approaches to current GIS software [Wan90], object-oriented approaches [Geo92] and other GIS features [Boo85]. Another current development involves the use of fuzzy logic in topological queries for the relationships between objects (represented by minimal bounding rectangles) found in typical geographic data [Daw95].

As GIS become increasingly more popular, efficient methods for performing **conflation** will have to be developed. Conflation can be defined as the process of merging multiple geographic data sets for the purpose of developing a more accurate geographic data set(map). The first successful attempt at automating the conflation process occurred in the mid-1980's. The United States Geological Survey, United States Bureau of Census (USGS-USBC) collaborated in an effort to consolidate their respective digitized map files of various US regions [Saa88]. This effort serves as the foundation for much of the conflation research occurring today. Conflation, also referred to as map compilation, entails two major processes: (1) **feature matching** and (2) **deconfliction.**

Features in GIS are representations of real world entities; and the characteristics describing these features are known as attributes. Thus, feature matching can be defined as the method for matching corresponding features in separate maps, where each map contains different representations of the same real world objects. In the past, this task was done by a human who had to painstakingly match individual objects by overlaying separate maps. This process is subject to considerable use of human judgment relative to the similarity of geometric features and linguistic terms in the feature matching process. In this paper, we will focus on an approach that utilizes the inherent imprecision of feature matching for conflation.

During the process of feature matching, it is almost inevitable that conflicts in the data will be found. There are a number of factors which may

contribute to these conflicts, such as imprecision in data, differing data representations and formats, missing information, etc. The overall conflation process cannot be successful unless there exists a method by which such conflicts can be satisfactorily resolved. The process of resolving these types of conflicts due to data integration is referred to as *deconfliction.*

In this paper, we fully describe the issues related to conflation and discuss the ways in which this process can be done. In particular, we present a method that combines algorithmic and knowledge processing techniques for performing conflation. Finally, we provide a brief discussion of a knowledge acquisition tool currently under development which will operate on the World-Wide-Web.

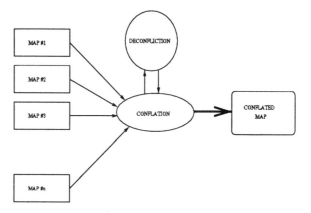

Figure 1: Conflation Process

2   Background

In this phase of the research, we have restricted our analysis to the four types of Vector Product Format(VPF) products [Def93] developed by the Defense Mapping Agency (DMA). The four VPF datasets are World Vector Shoreline (WVS), DNC (Digital Nautical Chart), Vector Smart Map(VMap), and Urban Vector Smart Map(UVMap)[Def93]. Each one of the data formats serves a different purpose and contains different types of information. Thus, it would be beneficial to have a system that integrates multiple datasets for the simple reason that it is better to analyze data in conjunction with each other rather than in a standalone fashion.

The technique for solving this problem combines both a computational and rule– based module. The computational module employs a statistical and probabilistic technique for managing the many uncertainty aspects of the problem. The expert system, on the other hand, manages the reasoning processes associated with the conflation process.

Based upon the premise that all aspects of the geographical data are to be used for efficiently performing conflation, we offer a framework for managing and interpreting the various semantics associated with both spatial and non-spatial data. Spatial characteristics are classified as either geometric or topologic. The geometric characteristics may include length, width, height, etc. associated with a particular feature. The topological structure we use in this research is the winged-edge topology [Def93]. The winged-edge topology manages the orientation of line segments in linear-based features. Non-spatial attributes, on the other hand, include the descriptive properties associated with a particular feature, such as building, river, railroad, number of tracks, type of road(e.g. highway, city street, etc.) All of these characteristics contribute to the overall behavior of the conflation process.

As with any expert system, there must exist a method for eliciting expert knowledge. We propose the use of a fuzzy knowledge acquisition tool for acquiring such expert knowledge [Gai93].

In general, we consider that the solution to the conflation problem requires the development of a complete environment as opposed to the development of individual systems. The environment in current development is known as ICE (Intelligent Conflation Environment). ICE represents an environment for managing all aspects of the conflation process, including rubber-sheeting, feature matching, and deconfliction. Figure 2 provides a conceptual design of ICE.

ICE : Intelligent Conflation Environment

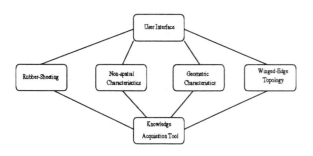

Figure 2: ICE: Intelligent Conflation Environment

3 Conflation Process

The first stage of conflation employs a rubber-sheeting technique, which utilizes a computational geometry method called *triangulation*. Rubber-sheeting entails conceptually aligning the maps over one another and performing feature matching based primarily on spatial proximity and orientation of the features. Features from the first map are selected and candidate features from the second map are chosen to be evaluated as possible matches.

Often in digitized maps, there exist multiple characteristics describing the various objects. The characteristics can be classified as either spatial or non-spatial. Considering, say, only non-spatial information for feature matching would represent an incomplete analysis of the problem. In contrast, the consideration of all characteristics would indicate a more comprehensive approach to solving the problem. Thus, we feel a good solution requires the utilization and exploitation of all aspects of the data for performing conflation, as demonstrated in Figure 2.

A major problem is the consideration of how various aspects of the geographical data are to be both analyzed and combined in order to make sound judgments during the feature matching phase of conflation. What criteria are most important? What if the feature labels are equivalent, yet they are not positionally similar? These questions, as well as a host of others, have to be considered when performing conflation. In this paper, we focus on these aspects of utilizing the various types of characteristics for the matching of geographical features.

4 System Architecture

In order to exploit the semantic nature of the data, we have developed an expert system that can perform heuristic knowledge processing. The expert system was developed in NEXPERT by Neuron Data, Inc. NEXPERT is a expert system shell that supports forward, backward, and mixed chaining. It allows for the creation of expert system rules and classes & objects. Rules are of the standard IF-THEN-ELSE production rule format. NEXPERT's data organization scheme adheres to the standard object oriented paradigm. Since NEXPERT code is precompiled to C, it integrates easily with the other system modules. C, in turn, is used to facilitate an interface between the Display module, which was developed in *Smalltalk*, and other internal modules. Figure 3 below provides a high-level design of the overall system.

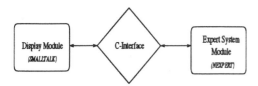

Figure 3: System Architecture

4.1   Data Organization

Since the expert system supports the object oriented design, it's data organization scheme remains consistent with that of the other modules. In addition, knowledge processing can be invoked at any time during the execution through the C calls, thus allowing for dynamic object instantiations. For example, two objects, **Railroad-1 & Railroad-2**, are considered to be two candidate matching features. Suppose both are instances of the superclass **Railroad**, which contains the following attributes.

ctl	Cumulative Track Length
rra	Railroad Power Source
rrc	Railroad Categories
ltn	Number of Tracks

As mentioned above, objects can be created dynamically by simply invoking the NEXPERT module. Thus, the two instantiations, **Railroad-1** and **Railroad-2**, are created during execution. Remaining consistent with the object oriented paradigm, all attributes of the superclass are inherited by the objects. The attribute values, in turn, are retrieved from the **ODBMS** via the *Smalltalk* component.

Once all preprocessing of the information is finished, knowledge processing can begin. The rules are written so as to allow *inexact* matching and reasoning to occur. For all practical purposes, requiring corresponding feature attributes to completely match would be too restrictive a constraint.

4.2   Knowledge-Base Structure

The knowledge-base(KB) represents an encoding of the reasoning process of an expert(e.g. cartographer or geographer) familiar with the domain, namely VPF products and various data merging techniques. In constructing the KB, we partitioned the rule set into distinct rule subsets where each

subset analyzes a specific domain. Such a decomposition has proven beneficial in modularity, understandability, and analysis.

Continuing with the railroad example, a sample rule subset would consist of rules relevant to analyzing the matching of railways. In Figure 4, a snapshot of the NEXPERT development shell demonstrates a rule subset that analyzes railroad features.

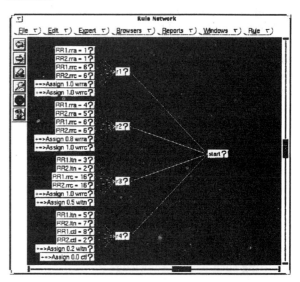

Figure 4: Rule Structure

4.3   Imprecise Feature Matching

Since human reasoning techniques are quite vague and the representation of geographical information is typically inconsistent with respect to both the content and structure, it is usually true that facts and rules are neither totally certain nor consistent [Neg85]. Thus, we have developed a technique for effectively capturing and structuring this inexact and uncertain information. Since it is rare to have a case where corresponding features match identically, it would be impractical to have a conflation system that simply selects matching features on an identical basis. Furthermore, it would be just as infeasible to match features with very little relevancy. Thus, a system that has the ability to consider various degrees and types of evidence is more desirable and practical.

Consider two objects, **Railroad-1** and **Railroad-2**. Suppose the attribute values for *ltn* do not match. Yet, the features may, in fact, correspond. That is, they are the correctly matched features. We have developed a technique for reasoning with the types of cases where there exist inexact matches. In addition, the technique is also able to accommodate the cases where the attribute sets themselves are not equivalent.

For matching, we consider each feature object as a set of attribute-value pairs:

$$((a_{11}, v_{11}), (a_{12}, v_{12}), ..., (a_{1n}, v_{1n}))$$
$$((a_{21}, v_{21}), (a_{22}, v_{22}), ..., (a_{2m}, v_{2m}))$$

The key is the degree of matching of the individual values for attributes. We call this matching "similarity" as this term has been used commonly in GIS. This is not the same as the formal similarity relationship in fuzzy set theory. There are two types of attribute domains for which we must specify the degree of matching, numeric and linguistic. In general, we have formulated membership matching functions for numeric domains and tables for the matching degree for linguistic domain elements. A major difficulty is acquiring these for each domain in the VPF description as there are several hundred such domains.

The similarity table contains a value in the range [0,1] for each attribute domain value, whereby each similarity value represents a degree of matching between attribute values. The domain values, in many cases, are numeric encodings of some linguistic characteristic. Thus, the similarity values in the table can also represent a degree of similarity between linguistic terms.

To illustrate this principle, consider the attribute *RRA*(Railroad Power Source). Also, *RRA* is restricted to the following values.

0	Unknown
1	Electrified Track
3	Overhead Electrified
4	Non-electrified
999	Other

The following table contains the similarity values for the attribute *RRA*.

RRA	0	1	3	4	999
0	.2				
1	.2	1			
3	.2	.6	1		
4	.2	.1	.1	1	
999	.2	.2	.2	.2	.2

Such tables are expected to be generated with the use of the knowledge acquisition tool.

The matching cannot simply be based on the tables alone as the attributes for each feature may not be independent, but may have semantic interrelationships. We have represented these interrelationships as rules in our expert system.

In Figure 4, a sampling of the rules used to analyze the matching of railroad features is offered. Translating the production rule for rule3(r3), we get the following.

IF ((RR1.ltn = 3 *and* RR2.ltn = 2) **and**
(RR1.rrc = 16 *and* RR2.rrc = 16))
THEN $w_{rra} \leftarrow 1.0$
$w_{ltn} \leftarrow 0.5$

The return values from the expert system are weights associated with the attributes. These weights can either strengthen or weaken the matching criteria. Considering the code segment from above, we can see this demonstrated. The weight, w_{ltn}, for *ltn* is weakened since the attribute values are not equal. Thus, *ltn* has less significance to the overall matching process than *rra*, whose respective weight is significantly larger. In short, this phase of the matching routine attempts to capture the inherent semantic relationships that exist in the non-spatial attributes.

Given the values returned from the expert system and those retrieved from the similarity table, an overall matching score can be computed. The matching score provides some indication as to what is the "best" matching feature for a given feature. The overall matching score for two features is given as:

$$MS_{i,j} = (\textstyle\sum_{k=1}^{N} [simA_k(F_i, F_j) \times ESW_{A_k}])/N$$

where

$A_k = k$th attribute in both F_i and F_j, where $0 \leq k \leq N$.

N = number of attributes that describe **both** F_i and F_j.

ESW_{A_k} = weight associated with A_k computed by expert system.

5   Shape Analysis of Linear Features

In the previous sections, a method for exploiting the nonspatial characteristics were explored. In this section, a technique for utilizing the geometric components for improved matching of linear features is proposed. The approach represents a means for quantitatively measuring the concept of shape similarity. Combining this with the feature matching technique discussed above, the use of shape similarity measurements greatly enhance the accuracy of the matching process since both spatial and non-spatial characteristics are considered. The shape similarity technique that we propose is algorithmically constructed.

In order to accomplish this, the authors apply a set of transformations. These include *translation, rotation,* and *scaling.* The objective of utilizing these transformations is to bring linear features into a *standard position* [Saa88]. Once in the standard position, shape similarity analysis between linear features can be performed.

5.1   Transformation

As stated earlier, three transformations are required in order for features to be *standardized* [Saa88]. The first operation performed is translation, where the features' starting nodes are translated to the origin. The next operation required is rotation, where the features' end nodes are rotated to the x-axis. The last transformation is scaling, where the starting and end nodes in the linear features are scaled to [0,0] to [1,0] on the x-axis. It is important to note that because of the linear transformations, the algorithm is able to accomodate linear features of different sizes since only shape is considered. Once these operations are completed (ie. features are standardized), the shaping analysis algorithm can proceed. The following figure depicts the standardization of two linear features by applying all of the necessary transformations.

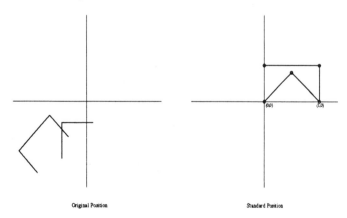

Figure 5: Standardization of Linear Features

5.2   Node Merging

The next phase in the shape similarity process is *node merging.* Node merging requires that all nodes (i.e., points) for one linear feature are mapped onto another linear feature, and vice versa. This mapping occurs with respect to each node's distance "in" the linear feature. Consider the two features, F_1 and F_2, comprised of the nodes $(n_{11}, n_{12}, ..., n_{1j})$ and $(n_{21}, n_{22}, ..., n_{2k})$, respectively. The associated ratios of the nodes are given as $(r_{11}, r_{12}, ..., r_{1j})$ and $(r_{21}, r_{22}, ..., r_{2k})$ where $r_{11} = r_{21} = 0.0$ and $r_{1j} = r_{2k} = 1.0$. Thus, two new features, $F_1\prime$ and $F_2\prime$, are created such that

$$F_1\prime = \aleph(F_1) \oplus \aleph(F_2)$$
$$F_2\prime = \aleph(F_1) \oplus \aleph(F_2)$$

where

$\aleph(F_i)$ is the number of nodes contained in F_i.

$\oplus$ - operation for merging nodes based upon their ratios.

The following figure demonstrates how the nodes are merged. The darkened nodes represent nodes that have been mapped onto a given feature, and the opened nodes represent existing nodes. The dotted segments connecting corresponding nodes, both mapped and existing, represent the distances to be measured.

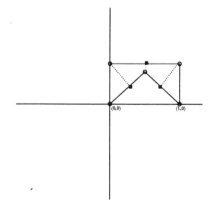

Figure 6: Merging of nodes in the linear features

5.3   Similarity Measure

Once the linear features have been standardized and nodes have been merged, the last phase of the shape similarity analysis can begin. The step simply measures the distances between the corresponding nodes and normalizes the result over the sum of the distance of both linear features. This process is similar to measuring the distance between two curves. Thus, our measurement represents more of an approximate, yet accurate, measure. With this, we have the following formula for quantitatively measuring the similarity between linear features.

$$\text{SHsim} = \left(\sum_{i=1}^{j+k-2} Dist(n_{1i}, n_{2i}) \, (D_1 + D_2) \right)$$

where
$j + k - 2 =$ the maximum number of distances to be computed since ratios of $n_{11} = n_{21}$ & $n_{1j} = n_{2k}$. D_i represents the distance of the linear feature F_i.

Thus, the linear features that have closer similarity with respect to shape represent the best candidates for feature matching.

This measure of shape similarity is based on Frechet's [Saa88] distance for measuring distance between polygonal arcs(i.e. linear features). The measurement is a natural distance measure, $L_2 - Distance$ which deals with all potential parameterizations. Suppose we have two functions:

$$P : [0,1] \rightarrow R^2$$

$$Q : [0,1] \rightarrow R^2$$

Also, $(p_0, p_1, ..., p_n)$ and $(q_0, q_1, ..., q_m)$ are the ordered points sets for P and Q, respectively. Similar to node merging, the set of ratios is computed. This is to be used for explicitly computing the distance between P and Q at all of the r_i's. Hence, the following is used for computing the $L_2 - Distance$.

$$\begin{aligned} L_2 &= \left(\int_0^1 \| P(r) - Q(r) \|^2 dr \right)^{\frac{1}{2}} \\ &= \left(\sum_{i=1}^{n+m} \int_{r_{i-1}}^{r_i} \| P(r) - Q(r) \|^2 dr \right)^{\frac{1}{2}} \end{aligned}$$

If we have (x_{P_i}, y_{P_i}) for the coordinates of $P(r_i)$, and, similarly (x_{Q_i}, y_{Q_i}) for the coordinates of $Q(r_i)$, and if we let $\Delta_{x,i} = x_{P_i} - x_{Q_i}$, and if we let $\Delta_{y,i} = y_{P_i} - y_{Q_i}$, which are the measured separations of the linear features at the bending points of one or the other polygonal line, then evaluating the integrals for the $L_2 - Distance$ gives:

$$\{ \tfrac{1}{3} \sum_{i=1}^{n+m} ((r_i - r_{i-1})(\Delta_{x,i-1}^2 + \Delta_{x,i-1}\Delta_{x,i} + \Delta_{x,i}^2 + \Delta_{y,i-1}^2 + \Delta_{y,i-1}\Delta_{y,i} + \Delta_{y,i}^2)) \}^{\frac{1}{2}}.$$

This represents the closed form for the evaluation of the integral.

6 Knowledge Acquisition

The real challenge in developing an expert system is actually acquiring expert knowledge [Gon93]. As knowledge engineers, we have the responsibility of eliciting such knowledge from an expert, in this case a cartographer, geographer, etc. In addition, the acquired knowledge must be structured so as to allow for easy encoding. This phase of the development process is commonly referred to as "knowledge acquisition bottleneck."

Typically, knowledge acquisition has been performed by either interview or observing an expert. Such a means of acquiring knowledge has proven problematic and error-prone. Both processes are rigorous and tedious requiring numerous sessions. In addition, it demands that the knowledge engineer have an in-depth understanding of the problem domain. Lastly, efficient knowledge acquisition requires that both the expert(s) and knowledge engineer(s) be in the same location.

For these reasons, we have decided to automate the knowledge acquisition process by constructing a knowledge acquistion tool(KAT) that has the ability to operate over the internet. Thus this greatly reduces the need for both the expert and knowledge engineer to be in the same location. It will be developed in *Java*, and will utilize *CGI* scripts. As a result, the KAT will be platform independent, and modifications to the KAT, which are quite common, can be reflected immediately.

Another motivation for developing a KAT is that it will greatly enhance construction of the knowledge base. The KAT will have the added ability to translate the knowledge into expert system rules automatically, thus this greatly reducing the time and effort required in writing expert system rules.

The foundation of the KAT is Kelly's [?] repertory grids. Repertory grids have been adapted for use in many systems [Boo85, For92b]. Such systems query the expert about the domain. This, in turn, assists the user in the construction of a comprehensive KB. The repertory grid is a matrix that correlates attributes with the constructs that represent the expert's ability to evaluate the degree of presence or absence of these elements [For92a]. Typically, the rating assigned by an expert to an element for a given construct is either 0 or 1. For our implementation, we use a multivalued rating for an element that represents a degree of membership to a fuzzy set [Boo85]. Specifically, we measure the degree of similarity among attribute value pairs. As a result, our repertory grids resemble those described in Section 4.3.

7 Conclusion

Our prototype of this system has shown that this approach is indeed feasible. We are continuing development of the full system(ICE), in particular now focusing on geometric characteristics. Additionally, as we realized in our prototype for non-spatial data matching, there is a tremendous problem of knowledge acquisition. Thus, we are currently developing a knowledge acquisition interface for the non-spatial, geometric, and topological characteristics of maps which will be able to operate over the internet.

8 Acknowledgments

We wish to thank the Office of Naval Research, and specifically the Command and Control program under the direction of Mr. Paul Quinn, for funding this work.

References

[Bal83]   J. Baldwin, Knowledge Engineering Using a Fuzzy Relational Inference Language, *Proc IFAC Symp. on Fuzzy Inf. Knowledge Rep. and Decision Analysis*, 15-21, 1983.

[Boo85]   Boose, J.H. A Knowledge Acquisition Program for Expert Systems Based on Personal Construct Theory. *Int'l Journal of Man-Machine Studies*. vol. 23, no. 5, pp 495-525.

[Buc82]   B. Buckles and F. Petry, A Fuzzy Model for Relational Databases,*Int. Jour. Fuzzy Sets and Systems*, 7, 213-226, 1982.

[Bur96]   P. Burroughs and A. Frank, eds, Geograhic Objects with Indeterminate Boundaties, *GISDATA Series, Vol. 2*, Taylor and Francis , London UK, 1996.

[Cob96]   M. Cobb and F. Petry, Integration of a Fuzzy Query Framework with Existing Spatial Querying Languages, *Proc. of 1996 IEEE Int. Conf. onFuzzy Systems*, pp93-99, 1996.

[Cob95]   Chung, M., Cobb, M., Arctur, D., and Shaw, K. (1995). An Object-Oriented Approach for Handling Topology in VPF Products. *Proc. GIS/LIS 95, Nashville, TN, 163-174.*

[Def93]   Defense Mapping Agency (DMA). Military Standard: Vector Product Formats, (1993) Draft Document No. MIL-STD-2407, (1994) Draft Military for Digital Nautical Chart, MIL-D-89023, (1994)Draft Military for Urban Vector Smart Map, MIL-U-89035, (1994) Draft Military for Vector Smart Map, MIL-U-89039,(1995) Draft Military for World Vector Shoreline, MIL-W-89012A. Defense Mapping Agency, Fairfax, VA.

[Daw95]   T. Dawson and C. Jones, Representing and Analyzing Fuzzy Natural Features in GIS, 405-412, *Ninth Annual Symp. on Geographical Information Systems*, 1995.

[For92a]   Ford, K.M. (1992) An Approach to the Automated Acquisition of Expert System Rules." *Proc. of the 1st Florida Artificial Intelligence Research Symposium*. St. Petersburg, FL: FLAIRS,pp. 86-90.

[For92b]   Ford, K.M., H.Stahl, J.R. Adams, et.al. ICONKAT: An Integrated Constructivist Knowledge Acquisition Tool. *Knowledge Acquisition*, vol. 3. pp. 215-236.

[Gia79]   C. Giardina, Fuzzy Databases and Fuzzy Relational Associative Processors, *Tech. Rep, Stevens Inst. of Technology*, Hoboken NJ, 1979.

[Gai93]   Gaines, B. and M. Shaw. (1993). Eliciting Knowledge and Transferring It Effectively to a Knowledge-Based System. *IEEE Trans. on Know. and Data Eng.*. vol 5, no 1.

[Geo92]   R. George, B Buckles, F Petry , and A. Yazici Uncertainty Modeling in Object-Oriented Geographical Information Systems,1992 *Proceed.Conf Database & Expert System App.(DEXA)*, 77-86, 1992.

[Goo90] M. Goodchild and S. Gopal, eds. *The Accuracy of Spatial Databases* , Taylor and Francis, London, UK. 1990.

[Gon93] Gonzalez, A. and Dankel. (1993). *The Engineering of Knowledge-Based Systems: Theory and Practice.* Prentice Hall, London.

[Kel55] Kelly, G.A. (1955) *The Psychology of Personal Constructs, Vol. 1–A Theory of Personality.* New York, NY: W.W. Norton.

[Mag91] D. MaGuire, An Overview and Definition of GIS, *Geographical Information Systems: Principles and Applications, VOL 1 - Principles,* (eds. D.MaGuire, M. Goodchild, and D. Rhind), 9-20, Longman, Essex GB, 1991.

[Neg85] Negoita, Constantin. (1985). *Expert Systems and Fuzzy Systems.* The Benjamin/Cummings Publishing Co.

[Rob85] V. Robinson and A. Frank, About Different Kinds of Uncertainty in Geographic Information Systems, *Proc. AUTOCARTO 7 Conference,* 1985.

[Rob88] V. Robinson, Implications of Fuzzy Set Theory for Geographic Databases, *Computers, Env, Urban Systems,* 12, 89-98, 1988.

[Rob90] V. Robinson, Interactive Machine Acquisition of a Fuzzy Spatial Relation, *Computers and Geosciences,* 6, 857-872, 1990.

[Saa88] Saalfeld, A. (1988). Conflation: Automated Map Compilation. *Int'l Journal of Geographical Information Systems 2(3).*

[Sto87] D. Stoms, Reasoning with Uncertainty in Intelligent Geographic Information Systems, *Proc. GIS 87 - 2nd Annual Int. Conf on Geographic Information Systems,* 693-699, American Soc. for Photogrammetry and Remote Sensing, Falls Church VA, 1987.

[Ver89] H. Veregin, A Taxonomy of Error in Spatial Databases, *Technical Report 89-12, National Center for Geographic Information and Analysis,* Santa Barbara, CA, 1989.

[Wan90] F. Wang, G. Hall and S. Subaryono, Fuzzy Information Representation and Processing In Conventional GIS Software: Database Design and Application, *Int. Jour. Geographical Information Systems,* 4, 261-283, 1990.

COOPERATIVE SOLVING OF URBAN-TRAFFIC PROBLEMS[1]

L.A. García, F. Toledo
Department d'Informàtica, Universitat Jaume I
12071 Castelló, Spain
Email: garcial@inf.uji.es toledo@inf.uji.es

ABSTRACT

In this paper we present a multiagent environment suitable to be applied on a system for controlling urban-traffic. This work is an evolution of the expert system prototype developed inside the ESPRIT II EQUATOR project[EQ94], which incorporates reasoning frameworks of the application domain theories by means of a qualitative/temporal model of the system behaviour. The expert system is composed by six functional agents which respectively perform pre-processing of data, detection of current problems, diagnosis of problems, short term prediction, generation of solutions and refinement of solutions. For each one of the system objects (streets, intersections, etc.), the temporal knowledge base (TKB) contains the past, present and future events that are important in their own evolution. This paper describes the structure of this TKB and proposes a new and efficient, but application oriented, heuristic algorithm for doing the reasoning maintenance process.

1. URBAN TRAFFIC-CONTROL SYSTEMS

Urban Traffic-Control Systems (UTCs), like many other control systems [Bell91], are generally composed of two layers: a first one "algorithmic" and a second one in which the decisions are taken by an human expert. The predictive aspect, that is, to apply decisions of control earlier than traffic jams occurs, is one of the more important control aspects taken by the human expert. However, the complexity of this problem has makes that this two layers UTC system be insufficient for an effective control. This has led to several authors [Amb91], [For87], [Bell91], to introduce knowledge-based systems inside the UTC system architectures. This knowledge should incorporate the reasoning structures according to the application domain theories: knowledge about the inner structure, behaviour and/or composition of the study objects (that is, deep knowledge). Therefore, the knowledge is structured in two levels: knowledge for explaining the system behaviour, and knowledge about how to find the solution to a given problem (control knowledge).

In our approach the first level has been built by means of a qualitative model that extends the knowledge representations of traditional qualitative parameters [DeK84] for supporting bidimensional qualitative parameters [Mor93]. The main idea is to model the spatio-temporal street evolution by defining a discrete quantic space in which the density take values, and setting-up that the spatio-temporal limit that divide these two regions with constant density is a straight line [Mor93]. In figure 1 it can be seen an example of this representation model. The vertices of each qualitative region are called *events* and can represent phenomena as it can be the appearing or disappearing of a queue or the colour change of a traffic-light. The system behaviour is modelled by a set of objects (streets, intersections, traffic-lights, inputs, outputs, etc.) that have associated parameters whose evolution is restricted by means of constraints among them.

These qualitative data are built from scratch taken the data collected by the traffic sensors, which at moment are traffic loop sensors. This kind of sensors gives the measured density, speed and flow over a delimited section of a street at certain time periods. These data, unfortunately, are not sufficient to fully determine the state of a street. To deal with this uncertainly is needed a model of the system to be controlled.

---

[1] This work is partly supported by the FUNDACIÓ CAIXA-CASTELLÓ under a PLAN DE PROMOCIÓN DE LA INVESTIGACIÓN grant number P14A94-21

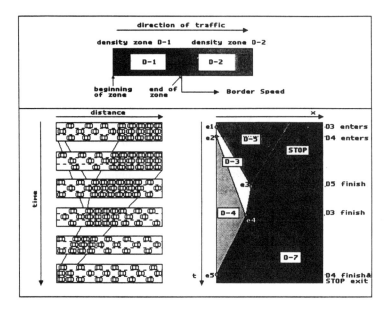

FIGURE 1. Temporal Qualitative Representation of traffic evolution

The project ESPRITT II EQUATOR had been the frame in which these characteristics have been evaluated. Since the end of this project we are working in the developing of parallel techniques for this kind of systems. The first step we are solving consists on the reimplementation of the EQUATOR demonstrator by taking into account, as philosophy of design, the blackboard architecture [Nii86], but by relaxing some of their main characteristics. The choice of this kind of system architecture was done due to the kind of problem we are dealing to --that is, a continuously interpretation problem (as the DVMT [Les89] system for the continuous identity and position of vehicles under a controlled scene, or the REAKT [Bar94] system for real-time control of industrial applications). Some of these systems have been evolved incorporating additional characteristics, for example, uncertainly, imprecise reasoning, parallel techniques, scheduling heuristics, etc. This growing evolution of the systems developed shown the suitability of the blackboard paradigm for their reimplementation or for its facility to experiment with different scheduling heuristics. The weakest point of it is its "lack of efficiency", but this last question is not generally accepted [Cor91][Vel92].

This paper is organised as follows. Section two describes each one of the agents --the application agents and the control agent--, its functionalities and relationships. The structure of the Temporal Knowledge Base, which is built as a classical blackboard is also described in this section. In section three it is described the temporal behaviour of the system. It proposes a new and theoretically efficient, but application oriented,

heuristic algorithms for doing the reasoning maintenance process. In section four the conclusions and future work to be done are exposed.

2. FUNCTIONAL ARCHITECTURE.

The architecture is composed of six application agents plus one control agent (for organisation of the overall system). All the data needed for representing the system under control are on the Temporal Knowledge Base (TKB). This TKB is organised by dividing it into levels and panels (following a classical blackboard organisation).

The application agents do different, but related, works:

Pre-processing agent: This agent translates the input data from sensors, collected each five minutes, into the temporal/qualitative representation described before. This module is complex mainly due to the data lack for fully determining the traffic state. This is because there are not information from real sensors in every temporal instant of the period and because the information given by the sensors are not enough to reconstruct the current state. These problems are solved using qualitative simulation from the previous state, instead of using direct analysis of sensor data. The qualitative simulation from the previous state will give a whole image of the state of the system assuming that the previous state is correct, that the prediction process is correct, and that no unpredictable events have happened during the prediction. Sensor data are used to verify these facts.

Detection problem agent: It examines the world instant state for picking the existing traffic problems. This work is done by applying the knowledge of what means a problem at each situation. This knowledge is rule codified and it is applied for problem prediction at different levels of spatial and temporal granularity [Mar93] -- a punctual problem it can not be considered as it and, in the other side, a problem belonging to an overall zone it can be detected as it, it must not be considered as the sum of the problems over the spatial components over the zone.

Diagnosis agent: It analyses if the existence of a problem it is due to bad traffic adjustment or if it is originated by an incident (bad parking, an accident, .etc.). This distinction is important because there are different actions to take on (at present an algorithmic system it reacts absurdly for a queue due by an accident by assigning more green in that direction). This work is done by reading qualitative state data from the objects that are involved into the zone where the supposed problem is detected and from the objects related to them These data are the input data for a data minimisation process that finish when it can be determined if the actual situation represents a problem or it represents an incident.

Prediction agent: It predicts the possible future problems in function of the traffic behaviour (represented into the model of the system). This functionality is very important for an UTC system because it should anticipate to the problems (particularly, trying to avoid that a primary congestion-increasing of a queue by an excess of the demand versus the capacity of a street can turn into a secondary one blockage due to a growing queue of an intersection located afterward, with the consequent extension of the blockage along the network).

Solution generator agent: It generates changes into the traffic lights as response to the present problems detected. This agent needs data from the zone where the problem is detected and from its neighbour zones, in order to find a (sub) optimal (heuristically based) solution for the problem detection. The main objective of this agent is not turn on these problems over other zones.

Solution Refinement agent: It acts over the previous generated solution for getting better if that is possible (e.g. if a time before, as a response to a given problem, the green time of an itinerary has been reduced, then it can be profited for incrementing the green time over certain traverses, not as a response of a given problem, but as an improvement of the obtained solution). Figure 2 shows this functional architecture.

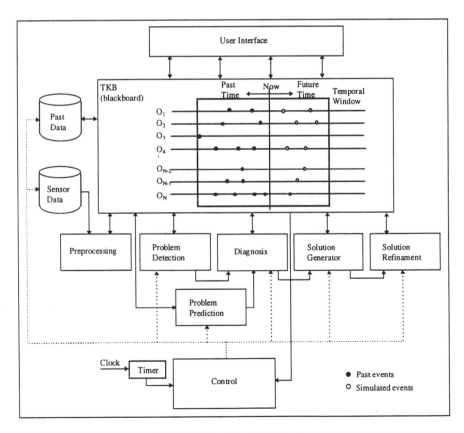

FIGURE 2. Functional Architecture

All of these agents are communicated among them by using a temporal knowledge base (TKB) that works like a blackboard. For each one of the system objects (streets, intersections, etc.) the TKB contains the past, present and future events that are important in their own evolution.

This TKB is hierarchically organised into levels and panels. There are two panels in the TKB, one for representing the world instant status, and the other one for representing the control problem. This second panel permits to experiment with different control heuristics for scheduling. These control heuristics can be initially stated or dynamically changing. The panel is divided into three levels. The first level stores the atomic constraints for the system objects. The second level uses the information stored into the first level and its own stored information -about different control heuristics, for example, to prefer for execution the demands that are closest in time. The last level works with the information from the lowest levels (first and second levels) and its own spatial preferences to help the scheduler to choose the next agent to execute.

The other panel, used for representing the evolution of the system under control, is divided into four levels. The lowest level corresponds to the scratch data (qualitative data provided for the pre-processing agent that can be real data --coming directly from sensors-- or simulated data). The next level is used for representing the states of street sections (using data from this same level and from the lower level). Data for representing streets, intersections, inputs or outputs are stored into the next level and it is constructed by using data from this same level and from the lowest levels. The top level is used for representing the state of sets of streets and street-sections, that is, zones over the city. Figure 4 shows the structure of the panel for representing the evolution of the system.

The application agents work over a temporal window open on the TKB. Each of the agents take the input data from the adequate level (or levels) of the simulation panel to reason about them. The pre-processing agent uses real data and past results to provide the present traffic status. The present detection problem and diagnosis agents work over the past and present parts. The problem prediction agent works over the full window. The solution generator and refinement agents work over the present and future parts.

The six agents use the TKB as communication channel, but the direct message interchange among them is allowed too using high level information (present or future problems, traffic plans generated, etc.). The detection problem and the prediction problem agents send a message to the diagnosis agent when a problem it is detected (based of both real sensor data or simulated data). When the diagnosis agent receives a message, then it looks for the information that needs to work on this problem. If this agent finds that there is a real problem, then it sends a message to the solution generation agent. The solution generation agent, as response to the reception of a message determines where to look for the information needed to solve it. When it finds a solution then it is communicated to the refinement solution agent. All the work that must be realised by the application agents is regulated by the control agent, that includes the agenda where the demands for execution are stored.

3. TEMPORAL BEHAVIOUR OF THE SYSTEM

As it has been previously described, for each one of the actual system objects, it is defined an object in which, at every moment, there are stored its past states, the actual state and the probably future states (obtained by simulation).

At a given moment there is a value for each one of these objects, that it can be a simulated or a real value (that is, directly collected by the sensors of the system under control, or simulated with values of other sensors). The real value is determined by the qualitative reasoning process previously described.

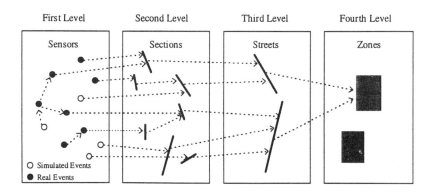

FIGURE 4 TKB Simulation Panel Structure

The real data of the system are collected at certain periods of time, at present each five minutes. During this period there are no more available real data. That is the reason because it is needed to apply a reasoning process based on both, the past behaviour of certain objects of the system and with the present traffic-light regulations (that is, the red and green period times of each one the traffic-lights of the system) to rebuild the present time state.

When real data arrive then it is starting a process for ensuring the consistence among the simulated values of the objects at this moment and the data received from the sensors. If these values are not greatly differ, then it is assumed that the previously simulated process (the qualitative evolution of this object) was correct. Therefore, the system begins a new simulation process taken these real data (that is, data collected directly by the sensors) as starting data. If the values received by sensors are greatly differ with the simulation values, then this disagreement can be due for one of two reasons: 1) anyone of the situations of the objects had changed, for example, it could be the appearing of a vehicles queue where it was not there before (for whatever reason: the time of the day, sports event, accident, etc.), in that case there is to do a reasoning maintenance process for this last time simulated period, or 2) if the most of the objects related among them (for example, the associated with a determined section of a street) have values that it can be considered consistent with the simulated data, except for the value of a concrete object (for example, data from a sensor of the same section of the street) then, it could be possible that this sensor was malfunctioning. The action to do in this situation is to begin a new simulation period with the real data as starting data. There is no reasoning maintenance process in this case, instead of it this sensor it marked as probably incorrect. If the problem persists, then it can be deduced that this sensor is really incorrect. In another case, it must be due to a punctual error (and, therefore this abnormal situation is forgotten).

This behaviour allows the creation of a set of sensors to revisited because repeatedly their real data were not corresponding with a "normal" behaviour of the system. This information can be used by the personal maintenance team of the traffic system for repairing them. The next figure (figure 4) shows several examples of this kind of behaviour.

In case (a) of figure 4, there are a little disagreement among the simulated values and the real sensor ones, but this situation is easy to explain with the behaviour of the other sensors of the street (for example, it could have been that the vehicles of this street have taken a greater speed that the speed expected by the control system). In this case it is started a new simulated period with the real sensor data as the basis for it. It is not executed any adaptation process.

Case (b) presents a great disagreement among the simulated values and the values collected by the sensors. As response to this situation a reasoning maintenance process is followed over the set of hypothesis taken by the system during the simulating period. The main purpose of this process is to locate the points in which these erroneous hypothesis were taken and to continue with the correct path of the behaviour tree that it can be directed to the final real situation collected by the sensors.

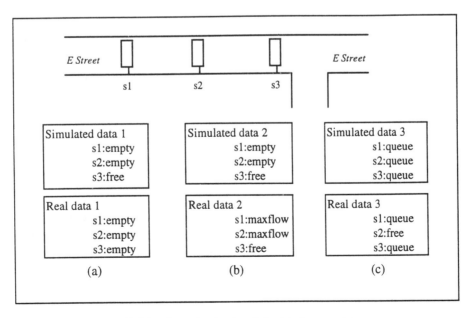

FIGURE 4. Example of real and simulated sensor data.

Last case (c) represents that an abnormal situation has happened. Sensor 2 determines that the street is into a *free* state, but sensors 1 and 3 determine that this street is *collapsed*. As between the sensors 1 and 3 there is not any intersection or any output, then this disagreement can be due to one of two reasons: 1) an incident (bad parking, traffic-cut, etc.) or 2) this sensor is out of order. If this abnormal situation is presented several times into the future then sensor 2 is assumed to be out of order. This sensor is deleted of the active sensors list ant it is inscribed into the list of faulty sensors.

When there are to revisited the set of hypothesis taken during a period of simulation time, then the behaviour tree is traverse into an inverse direction, that is , from the leaf that represents the real state of the data sensors to the root (that fall in with the initial state of this finished period). By this way, the reasoning maintenance process in this system is needed only for this last period simulation time. That is a very important characteristic because in this way this process has a moderated computational cost (polynomial in time and with respect to the number of objects whose values have to be rebuilt) and it maintains a great compactly of the stored system knowledge.

This inverse traverse of the path on the tree must be done as fast as possible. It is assumed that the transition over the traffic states of the qualitative data is -continuous, that is, from a given state only two possible transitions are possible: 1) to a state representing the immediate upper degree of traffic density, or 2) to a state representing the immediately lower degree of traffic density. The discontinuities between qualitative traffic states are not allowed (by example, it is not allowed that the next traffic state to a qualitative state *free* could be the qualitative state *congestive*). With this assumption, the set of all the possible paths on the tree can be represented as a regular language. The *lineal left regular grammar* that generates this language is the one defined by the following production set:

Start Symbol: S; P = {(S, Ee), (S, Ff), (S, Mm), (S, Cc), (S, Qq), (E, Ee), (E, Ff), (E, λ), (F, Ff), (F, Mm), (F, Ee), (F, λ), (M, Mm), (M, Cc), (M, Ff), (M, λ),(C, Cc), (C, Qq), (C, Mm), (C, λ), (Q, Qq), (Q, Cc), (Q, λ)}

Each of the terminal symbols represents one of the several qualitative states that a street can follow at every moment: *empty, free, maximum flow, congestive* and *queue*. By this way there are defined the valid transitions between the different states. The process for determining which are the correct choices of hypothesis in the tree, can be guided by this *LL* regular grammar. The system has a word that represents the choices of the qualitative states that it were taken during the last finished time period. The last symbol of this word is the one given by the sensors of the system each five minutes. The problem now is to build a new word belonging to the language generated by this left linear grammar into the minimum number of steps. This new word should be as close as possible to the one obtaining by the simulation process.

To illustrate this method the following example is proposed. Lets suppose that the word obtained by the simulation process was *eefffmfff* (note that the time period finished with a qualitative state *free*) and that the qualitative state given to this object by the sensors was *congestive*, then there is a strong disagreement so the reasoning maintenance process is started. The first step of the algorithm to rebuilt the path over the tree is to apply the production rule that represent this qualitative state, i.e. *(S, Cc)*. As it is not possible to transition from a *congestive* state to a *free* state then it must be selected as many intermediate productions as needed. In the example, these intermediate rules are *(C, Mm), (M, Mm)*. The system, therefore, obtain the (one of the possible) correct word that represent the evolution of the system during this last period, *eefffmmmc*.

$$S \Rightarrow Cc \Rightarrow Mmc \Rightarrow Mmmc \Rightarrow Mmmmc \Rightarrow$$
$$Ffmmmc \Rightarrow Ffmmmc \Rightarrow Ffffmmmc \Rightarrow$$
$$Eefffmmmc \Rightarrow Eeefffmmmc \Rightarrow eefffmmmc$$

Then, the system assumes that the set of hypothesis taken for determining the state of the world over this time period was correct until the sixth state. So, we must revise the set of hypothesis that were related with the seventh, eighth and ninth symbols of the word.

4. CONCLUSIONS AND FUTURE WORK

In this work we are mainly dealing with the knowledge structuring problems and with the reasoning maintenance problems into a Knowledge Based System. The techniques and characteristics described are currently been applied to an Experimental Traffic Control System for its evaluation and for measuring its performances respect to all of the components.

The main advantages of this approach are (1) the clearness of the options taken for representing the knowledge of the system, (2) the integration of the events, which incorporate cognitive meaning, into the basic elements of knowledge stored into the system, (3) the good performances that it can be obtained, that is because the simplicity of the reason maintenance process implemented, (4) the suitability of the proposed architecture for experimenting with different control heuristics, and (5) the mixed approach for agent communication, agent to agent and TKB based (i.e.

shared memory), which permits to use the, apparent, easiness for system definition of the classical blackboard architectures without loosing efficiency (the agents are informed via messages sending by other application agents as well as by the scheduler).

We are currently working in to implement this overall architecture into a parallel computer for evaluating it as support for knowledge based application systems [Gar96]. The system is being written using the parallel version of ECL^{i}PS^e [Mei95] in two parallel computers, a Sun SuperSparc with two processors, and a SILICON GRAPHICS Power Challenge with 16 processors.

5. REFERENCES.

[Amb91]   Ambrosino G., Bielli M., Boero, Mastretta M., *A blackboard model for traffic control operations*, in Advance Telematics in Road Transport, Proc. Of the Drive Conference, ELSEVIER pub. Pp 596-614, Brussels, 1991.

[Bar94]   F. Barber, V. Botti, E. Onaindía, A. Crespo, *Temporal Reasoning in REAKT (An environment for Real-Time Knowledge-Based Systems)*, AI Communications, Vol7, Num. 3, Sep, pp 175-202.

[Bell91]   Bell, M.C., Scenama G., and Ibbetson L.J., *CLAIRE: An Expert System for Congestion Management*, in Advance Telematics in Road Transport, Proc. Of the Drive Conference, ELSEVIER pub. Pp 596-614, Brussels, 1991.

[Cor91]   D. D. Corkill, Blackboard Technology Group, Inc. *Blackboard Systems* in: AI Expert 6(9):40-47, September 1991.

[DeK84]   De Kleer J. And Brown J.S., *A Qualitative Physics Based Confluences*, Artificial Intelligence, 1984, 24.

[EQ94]   EQUATOR Final Report, ESPRIT II project EEC 1994.

[Gar96]   L.A.García, F. Toledo, S. Moreno and E. Bonet, *An Expert System with deep knowledge for Urban Traffic Control based on a parallel architecture.* Proc. of EXPERSYS-96, Expert Systems Applications & Artificial Intelligence, J. Zarka, E. Mercier-Laurent, D.L. Crabtree and M. Narasipuram editors, I.I.T.T. International, París 1996.

[Les89]   Lesser V.R. and Corkill C., *The Distributed Vehicle Monitoring Testbed: A Tool for Investigating Distributed Problem Solving Networks*, in: R.S. Engelmore and A.J. Morgan, eds., Blackboard Systems, chapter 6, pp. 135-157, Addison-Wesley, Reading, MA, 1988.

[For87]   Forasté B. And Scemana G. *Expert System Contribution for control system: the traffic case*, Preprints IMCAS International Symposium on AI, Expert Systems and Languages in Modelling and Simulation, Barcelona, June 1987.

[Mar93]   Martín G., Toledo F., Barber F., Forradellas R. and Ibáñez F., *Qualitative Modelling of the Urban Traffic Behaviour Using Constraint Logic Programming*, Proc. 13th International Workshop on Expert Systems & their Applications, Avignon 1993, pp 391-409, De. EC2 Paris 1993.

[Mei95]   M. Meier, *ECL^{i}PS^e 3.5 ECRC Common Logic Programming System. User Manual*, December 1995. European Computer Industry Research Centre, Arabellastr. 17, 81925 Munich, Germany, December 1995.

[Mor93]   Moreno S., Toledo F. And Martín G., *An application of (lambda, T) formalism for handling systems with continuous change: The many tanks problem case-study*, in Qualitative Reasoning and Decision Technologies, CIMNE de., pp 419-428, 1993.

[Nii86]   H. P. Nii, *Blackboard Systems: the blackboard model of problem solving and the evolution of blackboard architectures*, AI Magazine, 7:38-53, 1986.

[Sha89]   Shanahan M., *Representing Continuous Change in the Event Calculus*, TR. Dep. Of Computing, Imperial College, 1989.

[Vel92]   H. Velthuijsen, *The Nature and Applicability of the Blackboard Architecture.* PTT Research, 1992.

INTEGRATING NEURAL NETWORKS AND EXPERT SYSTEMS FOR FAULT DIAGNOSIS OF THE MGC-20 CYCLOTRON

Mostafa M. Syiam, Osman M. Badr, M. N. H. Comsan
and Mamdouh M. A. Dewidar.
Department of Computer Science & Eng., Faculty of Elect. Eng.,Menoufia Univ., Menouf,
Menoufia,Egypt. Email: syiam@frcu.eun.eg

ABSTRACT

A hybrid expert system has been designed and implemented for fault diagnosis of the ATOMKI MGC-20 cyclotron. Two artificial intelligence methodologies, multilayer feedforward neural network and the rule based expert system, are integrated to build the proposed hybrid expert system. The developed hybrid expert system consists of two levels. The first level is two feedforward neural networks and the second one is a rule based expert system. The two Neural networks are used for isolating the faulty parts of the cyclotron. The inputs of the networks are the indicators conditions of the cyclotron control panel, symptoms, where the outputs correspond to the status of the five main parts of the cyclotron. A rule based expert system is used for troubleshooting the faults inside the faulty part. It uses inputs and outputs of the neural networks and also use questions and answers from the user to define precisely the faults in the faulty part. The Performance evaluation of the developed hybrid expert system indicated that it has a high level of diagnostic performance compared with the diagnosis of a human professional expert.

Keywords: Hybrid Expert systems, Neural Networks, Faults diagnosis, Cyclotron, Integrated Systems.

INTRODUCTION

Diagnosis is a very complex and an important task for determining the condition of faults or finding the root cause of faults. Accurate diagnosis can lead to a high system reliability and save maintenance costs, since the fault can be diagnosed before a sever damage occurs. Artificial intelligence strategies have been successfully applied to develop expert systems for fault diagnosis and similar problems. The two AI methodologies which have consistently proven themselves to be useful in the development of diagnostic systems are expert systems and neural networks.

Each of these techniques has its own advantages and disadvantages when applied to a diagnostic task. Expert systems with both shallow and deep knowledge are able to diagnose novel faults. However, many of them are running in consulting mode, and need large amount of computation. Thus, they are not suitable for real time diagnosis. On the other hand, neural networks can classify condition patterns of a system as normal or faulty, for fault detection and diagnosis at real time speed. Neural networks are highly efficient in detecting patterns and regularities in the input data. The application of neural networks for fault isolation problems requires less restrictive assumptions as to the structure of the input. Also the inherent parallelism of these networks allows very rabid parallel search and best match computations. Integrating these two methodologies into a single hybrid expert system resulted in greatest potential benefits in many applications. An integrated expert system with neural network is used to improve the fault diagnosis in particle physics detector [Ant92]. A hybrid expert system with embedded neural network within the expert system, as a separate module from the knowledge base and the inference engine, is built for making diagnosis for chemical process plant [Bec92]. Many neural network architectures are used for orbit control in accelerators and storage rings [BF94]. An integrated pattern classifier, using three different techniques, neural networks, nearest neighbor classifier, and expert systems, was used for condition monitoring and diagnosis of power circuit breakers [CV92]. An integrated expert system with shallow and deep knowledge was built for condition monitoring and diagnosis of high voltage circuit breakers [VC92]. A connectionist expert system, using neural network as the expert system knowledge base, was built for the transient identification in nuclear power plants [CC93]. Many neural network architectures are used for fault diagnosis in a realistic heat exchanger - continuous stirred tank reactor system, the best architecture is the multilayer perceptron network.

Neural nettworks are used for cells classification for cancer diagnosis as well cell or not well [Moa91]. A combination of a rule based approach with the neural network was used for classification and probability estimation [DHM92]. A combined neural network with knowledge base system are used for large scale systems dynamic control [HS91]. Neural networks are used for sensor signal analysis for surveillance in nuclear reactors [KR92]. Core monitoring models have been developed with the use of neural networks for prediction of the core parameters for pressurized water reactors [KKC93].

In this paper, a hybrid expert system is developed to diagnose the faults of the ATOMKI MGC-20 cyclotron. The architecture of the developed hybrid expert system consists of two levels, the first level contains two neural networks which are used for fault isolation and defining the faulty main parts of the cyclotron. The inputs of the neural networks are the 27 indicators conditions of the control panel. The outputs of the neural networks are five outputs express the five main parts condition (faulty or normal) of the cyclotron. The second level is a rule based expert system, which uses the five outputs of the two neural networks, the two neural network inputs, and other inputs from the user for troubleshooting the faults in the isolated faulty parts.

THE MGC-20 CYCLOTRON

The MGC-20 cyclotron operating in institute of nuclear research, Debrecen, Hungary, is an azimuthaly varying field cyclotron with spiral sectors. It accelerate protons, deuterons, alpha particles, and helium-3 to different energies. The applications are isotope production, nuclear reaction, and nuclear spectroscopy studies. The MGC-20 cyclotron has two hollow metal accelerating electrodes, called two dees, between them an oscillating electric field generated by a radio frequency generator [LB62]. The charged particles are produced by the ion source, which are located centrally between the two dees. The magnetic field of the cyclotron main magnet causes the particles to move in an approximately circular orbit. The radius of the orbit is a function of the particle velocity υ, therefore the radius increases with time, so the particles follow a spiral path from the ion source to the edge of the magnet, where they pulled out from the cyclotron by an electrostatic deflector. The extracted beam is guided by the beam transportation system to the users. The five main parts of the MGC-20 cyclotron are, The ion source, the deflector, the beam transport system, the concentric and harmonic coils, the radio frequency

system. In addition, the cyclotron contains a control and operation panel which contains a very large number of indicators to diagnose the faults of the cyclotron.

CONVENTIONAL METHOD FOR FAULT DIAGNOSIS OF THE CYCLOTRON:

Conventional method for fault diagnosis of the MGC-20 cyclotron depends on the experience of the human experts, who are the group responsible for the cyclotron operation and maintenance. Due to the cyclotron control panel which contains a large number of indicators, and the effects of the faulty part on the operation of other parts of the cyclotron. This makes the diagnosis process for faults of a certain part very complex and tedious for the normal users and operators. Thus the existence of the human expert is necessary all the time of operation, but this situation can not be verified really. To overcome these situations, a hybrid expert system is proposed in this paper for cyclotron fault diagnosis, to help the users and operators to diagnose easily the faults during the operation.

The symptoms which are used as inputs for the developed system to diagnose any fault in the cyclotron parts are the conditions of the control panel indicators. Each indicator has two conditions. For red indicators (ON=0, OFF=1), but for green indicators (ON=1, OFF=0). The symptoms are listed in table 1.

NEURAL NETWORKS AND RULE - BASED EXPERT SYSTEM - A REVIEW AND A COMPARISON

Artificial Neural Networks (ANN) are inspired by understanding of biological neural networks [ML94]. ANN is composed of basic units called artificial neurons, which is the processing elements in the network [Zur92]. Each neuron receives input data, process it and delivers a single output. The input can be raw of data or output of other neurons. The output can be the final product or it can be an input to another neuron. The learning algorithm used for training the multilayer feedforward neural network is the Back Propagation training algorithm (BP) [ML94, Zur92,]. The training procedure involves the presentation of a set of pairs of input and output patterns, the BP algorithm first uses the input pattern to produce its own output pattern and then compares this with the desired output, if the error between the desired and the actual outputs is sufficiently small, then no training take place, otherwise training continue. Numerous expert systems have been developed for diagnostic applications in a wide variety of domains.

Table 1. the symptoms used for cyclotron fault isolation.

No	symptom name	No	symptom name
1	-The gas valve red indicator	17	-The magnetic correctors AC supply red indicator
2	-The ion source cooling red indicator	18	-The switching & bending magnets (2,3,4,5,6) red indicator
3	-The ion source p.s. cooling red indicator		
4	-The ion source AC supply red indicator	19	-The beam transportation system green indicator condition
5	-The vacuum of the cyclotron vacuum chamber red indicator	20	-The concentric & harmonic coils AC supply red indicator
6	-The cyclotron main magnet red indicator	21	-The concentric & harmonic coils p.s. cooling red indicator
7	-The ion source green indicator condition		
8	-Deflector P.S. overload red indicator	22	-The concentric & harmonic coils green indicator condition
9	-Deflector P.S. cooling red indicator		
10	-Deflector cooling red indicator	23	-The cathode heating of the RF system green indicator.
11	-Deflector AC supply red indicator		
12	-The beam transportation system red indicator	24	-The RF system high voltage green indicator
13	-The deflector green indicator condition	25	-The final stage anode voltage green indicator
14	-The magnetic lenses p.s. cooling red indicator	26	-The pre-final stage anode voltage green indicator
15	-The magnetic lenses AC supply red indicator		
16	-The magnetic correctors power supply cooling red indicator	27	-The Radio Frequency system green indicator condition

As catalogued by Warren [Bec92]., the expert system approach for diagnosis is intuitively attractive, as symptoms can be linked to causes explicitly. In a rule - based knowledge representation scheme, a distinct symptom-cause linkage is expressed without any deep knowledge of the system's structure, function, or principle of operation. The use of this compiled shallow knowledge of a diagnostic expert in the form of heuristic rules has been shown to provide good results in domain areas where the underlying knowledge of the symptom-cause linkage is not clearly defined, or where an adequate model of the system to be diagnosed is not readily available. The limitations of rule based expert systems are revealed when they are confronted with novel fault situations, for which no specific rules exist. If the knowledge base does not contain the necessary information about a particular fault situation, in this case the expert system will be unable to diagnose the fault, however. In the domain of physical system diagnosis, this diagnostic brittleness can be overcome through the inclusion of deep knowledge of the process in the knowledge base, usually in the form of models of the process structure and function. The expert system can reason from the first principles of the system to diagnose the novel faults.

Neural networks and expert systems involves fundamentally different approaches to the diagnostic task, which can be expressed as being numeric and symbolic. Each approach contains its own strengths and weaknesses. Neural networks are fast, handle noisy data well, learn from experience, and generalize in novel situations if sufficiently structured and trained. However, they are black box, operators unable to explain its own reasoning methodology, incorrectly generalize novel faults if improperly trained, and forget past training if retrained on new data.

Expert systems have explicit representations of knowledge which eases the modification and validation of the systems, able to generate explanations for its reasoning methodology, and can use deep knowledge to reason about novel events. However they are unable to learn from experience, hard to maintain if the knowledge base becomes extremely large, and require extensive computational time if a deep model of the process must be consulted. To take the advantages of each technique, as well as avoiding the weakness of either individual technique. Thus the integration of neural network and expert system could be an excellent approach for an efficient diagnosis [Bec92,ML94].

DESIGN OF THE DEVELOPED HYBRID EXPERT SYSTEM

The developed hybrid expert system uses two diagnostic tools, a neural network for making fault isolation, and a rule based expert system to perform a precise fault diagnosis for the isolated part. We used a multilayer feedforward neural network as the model of neural network, with three layers, input layer, hidden layer, and output layer.

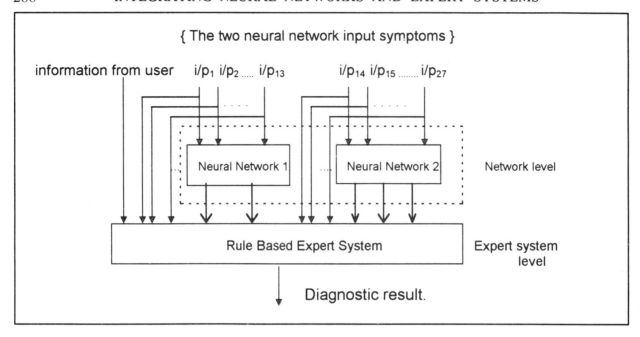

FIGURE 1 Schematic diagram for the developed hybrid expert system

In orde ro decrease the complexity of the neural network, increasing the classification accuracy, and to reduce the diagnosis time, two neural networks are used in the developed hybrid expert system. The first neural network has 13 input neurons, 6 hidden neurons and 2 output neurons corresponding to two cyclotron main parts, the ion source, and the deflector. The second neural network has 14 input neurons, 5 hidden neurons and 3 output neurons corresponding to three cyclotron main parts, the first part is the beam transportation system, the 2nd part is the concentric & harmonic coils, and the 3rd part is the radio frequency system.

The development of a hybrid expert system involving both neural networks and expert systems integrates the high level reasoning capabilities of the expert systems with the low level processing capabilities of the neural networks. The complete hybrid expert system as shown in Figure 1. The first level consists of two multilayer feedforward neural networks which have 27 inputs and 5 outputs. The 2nd level is the rule based expert system, its inputs consists of the two neural networks 5 outputs, 27 inputs, and other informations from the user, for troubleshooting the fault in the faulty part.The level of neural networks consists of two multilayer feedforward backpropagation neural networks. The typical structure for the two neural networks as shown in Figure 2.

Neural network 1:

We used 3-layer network for this network, the input layer has 13 neurons. The output layer has 2 neurons,

every output neuron corresponds to a cyclotron neurons, every output neuron corresponds to a cyclotron part. The first output neuronrepresents the status of the ion source, the second one represents the status of the deflector. The output neuron which has output value $\cong 1$ means that the corresponding part is faulty, and which has output value $\cong 0$ (zero) means that no fault is in the corresponding part.The learning is performed using different values of the learning factor η and the momentum factor α, also for different numbers of hidden layer neurons. The holdout method was used for estimating the error rates for each architecture. The data set used to train and test the network architectures consisted of 100 patterns.

The data set was divided into two equal groups. The first group of data was first used to train the network architectures, while the second group of data was used to test it. Eight network architectures were trained and tested. Secondly, the training and testing sets were swapped and training and testing process were repeated. The estimated error rate for each architecture is the average of the error rates for the testing results of the two learning phases of such architecture. The efficiency of classification is **96%**. The training results are shown in Table 2-a. and indicate that the optimal architecture for the fitrst neural network is a three layer network with a hidden layer of **8** neurons, learning factor η=**0.85**, momentum factor α=**0.7**. The number of iterations for learning completed is **6199** iteration for final error < 0.01.

FIGURE 2. The typical two neural networks used in the hybrid expert system.

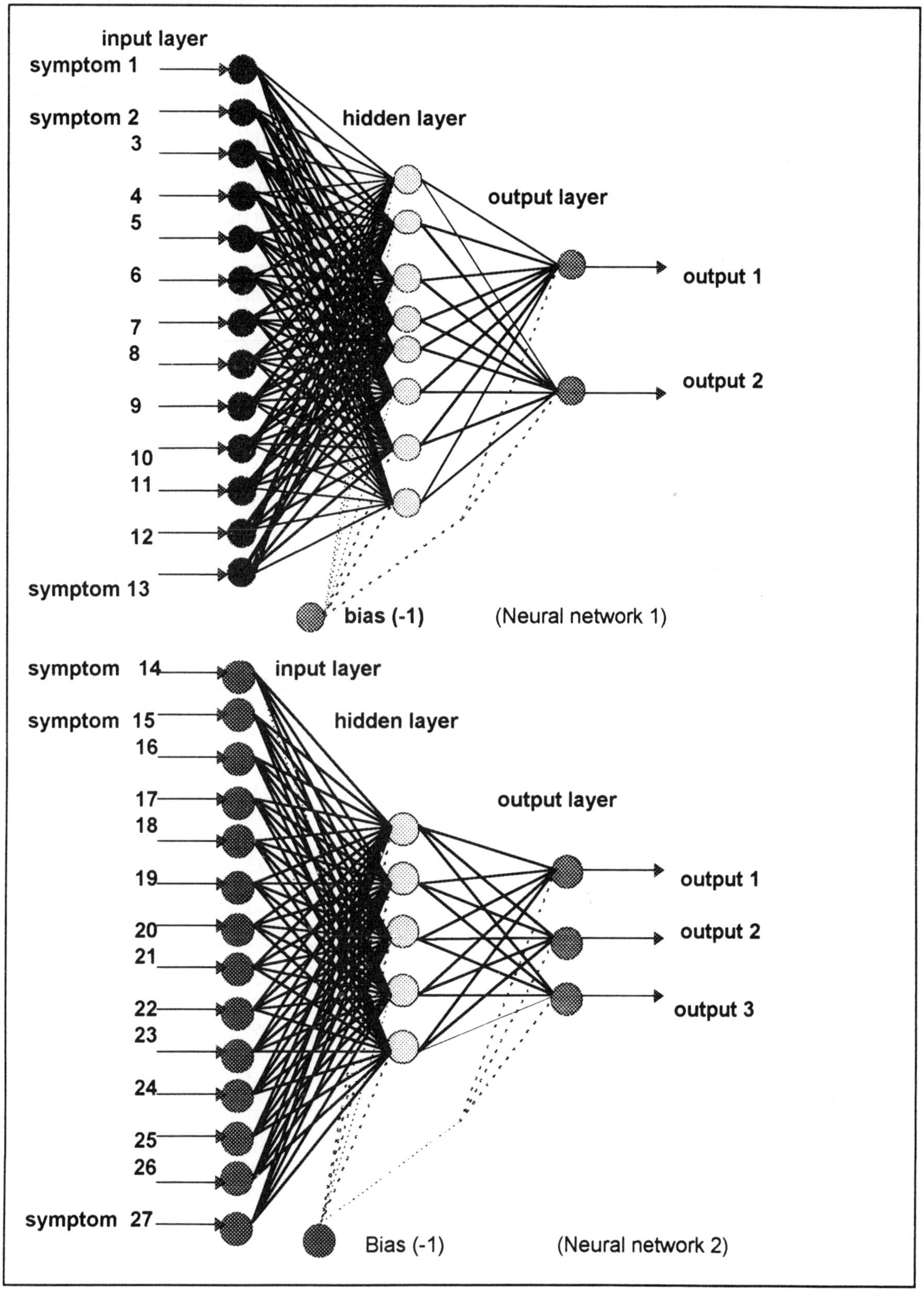

Neural network 2:

We used 3-layer network for this network, the input layer has 14 neurons. The inputs are the symptoms 14 to 27 shown in Table 1. The output layer has 3 neurons. Every output neuron corresponds to a cyclotron part. The first output neuron corresponds to the status of the beam transportation system. The second output neuron corresponds to the status of the concentric & harmonic coils. The third output neuron corresponds to the status of the radio frequency system. The output neuron which has output value $\cong$ 1 means that a fault in the corresponding part related to this neuron, and which has output value $\cong$ 0 means that no fault in the corresponding part,

We made learning by different values of the learning factor η and the momentum factor α, also for different numbers of hidden layer neurons. The holdout method was used for estimating the error rates for each architecture. The efficiency of classification is **100%**. The training results are shown in Table 2-b. and indicate that the optimal architecture for the second neural network is a three layer network with a hidden layer of **5** neurons, learning factor $\eta = 0.6$, momentum factor $\alpha = 0.7$. The number of iterations for learning completed is 7071 iteration for final error less than 0.01. Table 3. shows 20 patterns for testing the developed neural networks.

Neural Network Testing

Each testing pattern has a different set of 27 symptoms. The output which is nearly 1 shows that the fault exist in the corresponding cyclotron part, but the output which is nearly 0 shows that the fault free in the corresponding cyclotron part. The first pattern recall shows that the problem in the ion source and in the beam transport system, the second pattern recall shows that the problem in the ion source only, the third pattern recall shows that the problem in the deflector and also the fourth pattern, the 5th and 6th patterns recall shows that the problem in the beam transport system, the 7th and the 8th patterns recall shows that the problem in the concentric and harmonic coils, the 9th pattern recall shows that the problem in the RF system, the 10th pattern recall shows that the problems in the ion source and the RF system, and so on for the other ten patterns.

The Rule based Expert System:

The expert system is built as a borland C ++ program, which uses the rule base reasoning and object - attribute representation. The rule that is applied to an attribute states that the object either has or has not the attribute. The expert system accepts the two neural networks outputs directly into the general knowledge base.

Table 2a The first neural network training results.

Learning Factor η	Momentum Factor α	Hidden Nodes Number	Iteration Number of Training		Classification Rate
0.5	0.7	2	19849	11149	96 %
0.5	0.7	3	14749	10149	96 %
0.5	0.7	4	12449	9049	96 %
0.5	0.7	5	13799	8949	96 %
0.5	0.7	6	10549	8899	96 %
0.5	0.7	7	9749	8649	96 %
0.5	**0.7**	**8**	**9399**	8199	**96 %**
0.5	0.7	9	10849	9899	96 %
0.5	0.45	8	32049	21749	96%
0.5	0.5	8	27799	17299	96%
0.5	0.6	8	17549	14099	96%
0.5	0.65	8	11999	10149	96%
0.5	**0.7**	**8**	**9399**	8899	**96 %**
0.5	0.8	8	10399	6699	96%
0.4	0.7	8	13149	11049	96%
0.5	0.7	8	9399	8899	96%
0.6	0.7	8	8599	6699	96%
0.7	0.7	8	6599	5899	96%
0.8	0.7	8	6899	5799	96%
0.85	**0.7**	**8**	**6199**	5299	**96%**
0.9	0.7	8	7549	4149	96%

Table 2b. The second neural network training results

Learning Factor η	Momentum Factor α	Hidden Nodes Number	Iteration No. of Training		Classification Rate
0.6	0.7	2	10997	9931	100 %
0.6	0.7	3	10451	9399	100 %
0.6	0.7	4	9229	9151	100 %
0.6	**0.7**	**5**	7253	**7071**	**100 %**
0.6	0.7	6	8137	8683	100 %
0.6	0.7	7	8527	8475	100%
0.6	0.7	8	8189	7201	100%
0.6	0.55	5	21787	23763	100 %
0.6	0.6	5	13909	13597	100 %
0.6	0.65	5	8735	7331	100 %
0.6	**0.7**	**5**	7253	**7071**	**100%**
0.6	0.8	5	14715	15417	100 %
0.6	0.85	5	15807	20487	98.5 %
0.4	0.7	5	16171	16145	100 %
0.45	0.7	5	15079	15053	100 %
0.5	0.7	5	9281	9229	100 %
0.55	0.7	5	8969	9463	100 %
0.6	**0.7**	**5**	7253	**7071**	**100%**
0.63	0.7	5	7565	7929	100 %

Table 3. The 20 patterns recalls for testing the trained networks

No	Neural network 1 input symptoms (1 2 3 4 5 6 7 8 9 10 11 12 13)	Neural network 2 input symptoms (14 15 16 17 18 19 20 21 22 23 24 25 26 27)	net1 out 1	net1 out 2	net2 out 3	net2 out 4	net2 out 5
1	1 1 1 1 0 1 0 1 1 1 1 0 0	1 1 1 1 1 0 1 1 1 1 1 1 1 1	**1.00**	.000	**889**	.000	.001
2	0 0 1 0 0 1 1 0 1 1 1 1 1	1 1 1 1 1 1 1 1 1 1 1 1 1 1	**1.00**	.000	.055	.045	.041
3	1 1 1 1 1 1 1 1 0 1 1 1 0	1 1 1 1 1 1 1 1 1 1 1 1 1 1	.002	**.860**	.055	.045	.041
4	1 1 1 1 1 1 1 1 1 1 0 1 0	1 1 1 1 1 1 1 1 1 1 1 1 1 1	.002	**.954**	.055	.045	.041
5	1 1 1 1 1 1 1 1 1 1 1 0 0	0 1 0 0 1 0 1 1 1 1 1 1 1 1	.052	.059	**.989**	.000	.000
6	1 1 1 1 1 1 1 1 1 1 1 0 0	0 0 1 0 0 0 1 1 1 1 1 1 1 1	.052	.059	**.990**	.000	.000
7	1 1 1 1 1 1 1 1 1 1 1 1 1	1 1 1 1 1 1 0 0 0 1 1 1 1 1	.072	.069	.001	**.981**	.023
8	1 1 1 1 1 1 1 1 1 1 1 1 1	1 1 1 1 1 1 0 1 0 1 1 1 1 1	.072	.069	.001	**.977**	.019
9	1 1 1 1 1 1 1 1 1 1 1 1 1	1 1 1 1 1 1 1 1 1 1 1 0 0 0	.072	.069	.004	.024	**.978**
10	0 1 0 1 0 1 0 1 1 1 1 1 1	1 1 1 1 1 1 1 1 1 1 0 0 0 0	**1.00**	.000	.004	.024	**.978**
11	1 1 0 0 1 1 0 1 1 1 1 1 1	1 1 1 1 1 1 1 1 1 1 1 1 1 1	**1.00**	000	062	.055	.054
12	1 0 1 1 1 1 0 1 1 1 1 1 1	1 1 1 1 1 1 1 1 1 1 1 1 1 1	**1.00**	.000	062	.055	.054
13	1 1 1 1 1 1 1 0 1 1 1 1 0	1 1 1 1 1 1 1 1 1 1 1 1 1 1	.007	**.955**	062	.055	.054
14	1 1 1 1 1 1 1 1 0 0 1 1 0	1 1 1 1 1 1 1 1 1 1 1 1 1 1	.008	**.957**	062	.055	.054
15	1 1 1 1 1 1 1 1 1 1 1 1 1	0 0 1 1 1 0 1 1 1 1 1 1 1 1	.531	.005	**.990**	.001	.000
16	1 1 1 1 1 1 1 1 1 1 1 1 1	1 1 0 0 1 0 1 1 1 1 1 1 1 1	.531	.005	**.988**	.001	.000
17	1 1 1 1 1 1 1 1 1 1 1 1 1	1 1 1 1 1 1 0 1 0 1 1 1 1 1	.531	.005	.002	**.977**	.007
18	1 1 1 1 1 1 1 1 1 1 1 1 1	1 1 1 1 1 1 1 0 0 1 1 1 1 1	.31	.005	.002	**.976**	.007
19	1 1 1 1 1 1 1 1 1 1 1 1 1	1 1 1 1 1 1 1 1 1 1 1 1 0 0	.531	005	.002	.001	**.976**
20	1 1 1 1 1 1 1 1 1 1 1 1 1	1 1 1 1 1 1 1 1 1 1 1 0 0 0	.531	.005	.003	.012	**.959**

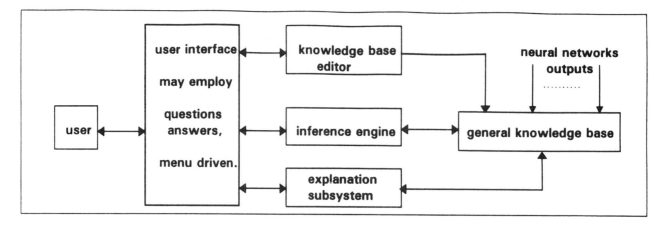

FIGURE 3. The architecture of the developed rule based expert system.

In addition it acquires the required neural networks inputs and other informations from the user via questions and answers as inputs for doing a complete diagnosis. If the expert system reaches a conclusion, then a message will appear for the user. The developed rule based expert system as shown in Figure 3, it has user interface module, inference engine module, knowledge base module, knowledge base editor, and explanation subsystem. The inference engine is the core of the expertsystem. It generates diagnosis and faults reasons. The goal driven backward chaining method is used for constructing the inference engine. At any time the expert system begins with a problem and tries to verify it by verifying all the condition parts of its rule.

The knowledge base is built up by capturing domain expert expertise "the shallow knowledge", by cyclotron design and manufacturer manuals and service guides "the deep knowledge". The expert system knowledge base consists of five independent knowledge bases. The first one stores the ion source fault diagnosis cases, it has 31 rules. The second knowledge base stores the deflector fault diagnosis cases, it has 16 rules. The third knowledge base stores the beam transport system fault diagnosis cases, it has 17 rules. The fourth knowledge base stores the concentric & harmonic coils fault diagnosis cases, it has 8 rules. The 5th. knowledge base stores the radio frequency system fault diagnosis cases, it has 75 rules. Each rule made up of a conditional part, which is preceded by the word "IF" and a conclusion part, which is preceded by the word "THEN". Each clause within these parts is made up of an attribute which is a keyword or phrase. The knowledge base is represented by the rules method. It can also expanded by acquiring new rules from experts.

The expert system can explain its answers, by using the explanation module. The knowledge base editor used to enter a new knowledge base or append a new rules.

The user interface applies menu driven and question - answer methods for dealing with the program.
Example for the rule based expert system diagnosis for the pattern number 1 in table 3., which shows a problem in the ion source and a problem in the beam transport system. For the ion source problem, the expert used for diagnosing the faults of pattern number 1. For example in rule (11), the first condition part of the rules which is "the problem in the ion source " is defined automatically from the neural networks outputs, the second condition part is" the vacuum of the cyclotron vacuum chamber lamp ON " is a symptom from the neural networks input symptoms, its condition is defined by the user during the user answers for the symptoms conditions. The third condition part defined when the expert system asks the user questions and the user answer it. Depending on the user answer of the third condition part,the expert system complete the diagnosis. and the user can overcome this faults.

For the beam transport system fault, the expert system uses rule number 63 for diagnosing the fault. The fault from rule 63 produced due to the fault happen in the cyclotron vacuum chamber vacuum, and the automatic valve closed automatically to try keeping the vacuum.

Performance Evaluation:

To evaluate the performance of the developed hybrid expert system, the 10 patterns recalls that are used for testing the developed hybrid expert system, as shown in table 3, are given to a professional expert to diagnose the faults of each pattern. Comparing the results of the developed hybrid expert system with the diagnosis of the professional expert indicates that the developed system has a high level of diagnostic performance. Table 4 shows the diagnosis of the human expert .

Table 4 The human expert diagnosis for the same previous 20 patterns

No	Input patterns (symptoms)					Expert decision				
	1 2 3 4 5 6 7	8 9 10 11 12 13	14 15 16 17 18 19	20 21 22	23 24 25 26 27	1	2	3	4	5
1	1 1 1 1 0 1 0	1 1 1 1 0 0	1 1 1 1 1 0	1 1 1	1 1 1 1 1	✓	x	✓	x	x
2	0 0 1 0 1 1 0	1 1 1 1 1 1	1 1 1 1 1 1	1 1 1	1 1 1 1 1	✓	x	x	x	x
3	1 1 1 1 1 1 1	1 0 1 1 1 0	1 1 1 1 1 1	1 1 1	1 1 1 1 1	x	✓	x	x	x
4	1 1 1 1 1 1 1	1 1 1 0 1 0	1 1 1 1 1 1	1 1 1	1 1 1 1 1	x	✓	x	x	x
5	1 1 1 1 1 1 1	1 1 1 1 0 0	0 1 0 0 1 0	1 1 1	1 1 1 1 1	x	x	✓	x	x
6	1 1 1 1 1 1 1	1 1 1 1 0 0	0 0 1 0 0 0	1 1 1	1 1 1 1 1	x	x	✓	x	x
7	1 1 1 1 1 1 1	1 1 1 1 1 1	1 1 1 1 1 1	0 0 0	1 1 1 1 1	x	x	x	✓	x
8	1 1 1 1 1 1 1	1 1 1 1 1 1	1 1 1 1 1 1	0 1 0	1 1 1 1 1	x	x	x	✓	x
9	1 1 1 1 1 1 1	1 1 1 1 1 1	1 1 1 1 1 1	1 1 1	1 1 0 0 0	x	x	x	x	✓
10	0 1 0 1 0 1 0	1 1 1 1 1 1	1 1 1 1 1 1	1 1 1	1 0 0 0 0	✓	x	x	x	✓
11	1 1 0 0 1 1 0	1 1 1 1 1 1	1 1 1 1 1 1	1 1 1	1 1 1 1 1	✓	x	x	x	x
12	1 0 1 1 1 1 0	1 1 1 1 1 1	1 1 1 1 1 1	1 1 1	1 1 1 1 1	✓	x	x	x	x
13	1 1 1 1 1 1 1	0 1 1 1 1 0	1 1 1 1 1 1	1 1 1	1 1 1 1 1	x	✓	x	x	x
14	1 1 1 1 1 1 1	1 0 1 1 1 0	1 1 1 1 1 1	1 1 1	1 1 1 1 1	x	✓	x	x	x
15	1 1 1 1 1 1 1	1 1 1 1 1 1	0 0 1 1 1 0	1 1 1	1 1 1 1 1	x	x	✓	x	x
16	1 1 1 1 1 1 1	1 1 1 1 1 1	1 1 0 0 1 0	1 1 1	1 1 1 1 1	x	x	✓	x	x
17	1 1 1 1 1 1 1	1 1 1 1 1 1	1 1 1 1 1 1	0 1 0	1 1 1 1 1	x	x	x	✓	x
18	1 1 1 1 1 1 1	1 1 1 1 1 1	1 1 1 1 1 1	1 0 0	1 1 1 1 1	x	x	x	✓	x
19	1 1 1 1 1 1 1	1 1 1 1 1 1	1 1 1 1 1 1	1 1 1	1 1 1 0 0	x	x	x	x	✓
20	1 1 1 1 1 1 1	1 1 1 1 1 1	1 1 1 1 1 1	1 1 1	1 1 0 0 0	x	x	x	x	✓

Table 5 A part of the set of rules used to diagnose the faults in pattern number 1.

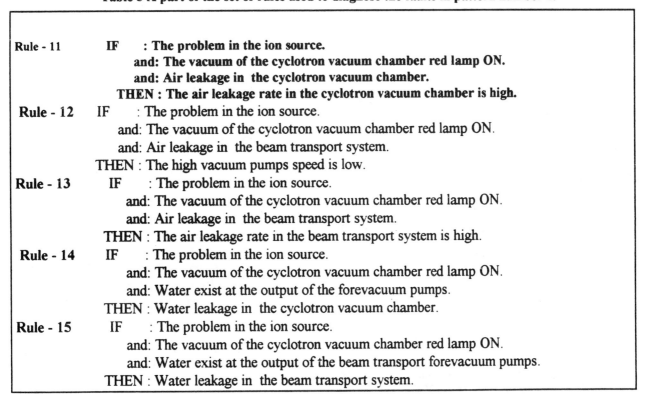

Rule - 11   IF   : The problem in the ion source.
    and: The vacuum of the cyclotron vacuum chamber red lamp ON.
    and: Air leakage in the cyclotron vacuum chamber.
    THEN : The air leakage rate in the cyclotron vacuum chamber is high.

Rule - 12   IF   : The problem in the ion source.
    and: The vacuum of the cyclotron vacuum chamber red lamp ON.
    and: Air leakage in the beam transport system.
    THEN : The high vacuum pumps speed is low.

Rule - 13   IF   : The problem in the ion source.
    and: The vacuum of the cyclotron vacuum chamber red lamp ON.
    and: Air leakage in the beam transport system.
    THEN : The air leakage rate in the beam transport system is high.

Rule - 14   IF   : The problem in the ion source.
    and: The vacuum of the cyclotron vacuum chamber red lamp ON.
    and: Water exist at the output of the forevacuum pumps.
    THEN : Water leakage in the cyclotron vacuum chamber.

Rule - 15   IF   : The problem in the ion source.
    and: The vacuum of the cyclotron vacuum chamber red lamp ON.
    and: Water exist at the output of the beam transport forevacuum pumps.
    THEN : Water leakage in the beam transport system.

CONCLUSION

The integration of two different AI methodologies, feed forward back propagation neural networks and the rule based expert systems, into a single hybrid expert system, shows an extreme promise in producing useful tools for diagnostic applications. The diagnostic process of linking symptoms to causes is speeded and enhanced using a hybrid expert systems. The advantages of each technique can be combined while avoiding the disadvantages of using either neural networks or expert systems singulary. A hybrid expert system is developed to diagnose the faults of the ATOMKI MGC-20 cyclotron. The developed hybrid expert system contains two levels, the first level contains two feedforward back propagation neural networks which are used for isolating and defining the main part which is faulty. The second level is the rule based expert system, which used the five outputs of the two neural networks, the two neural networks inputs, and other inputs from the user for troubleshooting the faults in the isolated faulty part. The optimal architecture for the two neural networks are three layer network, the two neural networks used the backpropagation algorithm for training. 20 patterns are used for testing the developed neural networks. Each testing pattern has a different set of 27 symptoms. Testing results indicated that the neural networks can identify the faulty part with a very high correct rate. Performance evaluation for the developed hybrid expert system indicated that it has a high level of diagnostic performance compared with the diagnosis of a professional human expert.

ACKNOWLEDGMENT

The authors would like to acknowledge Dr. A. Valek the head of the cyclotron department, Institute of Nuclear Research, Debrecen, Hungary, for his support and cooperation.

REFERENCES

[Ant92]   I. D' Antone, " Neural Network in An Expert Diagnostic System", IEEE Transaction Nuclear Science, pp. 58-62, 1992.

[Bec92]   Warren R.Becraft,"Neural Network/Expert System Integration for Diagnostic Systems ", Proceedings of The 2nd Pacific Rim International Conference on Artificial Intelligence, Vol. 1, pp. 831- 837, 1992.

[BF94]   Eva Bozoki, Aharon Friedman, " Neural Networks and Orbit Control in Accelerators ", The Fourth European Conference For Particle Accelerators, pp 1589-159, 1994,.

[CC93]   Se Woo Cheon, Soon Heung Chang, " Application of Neural Networks to A Connectionist Expert System for Transient Identification in Nuclear Power Plants ", Nuclear Technology, Vol. 102, pp. 177-191, 1993.

[CV92]   Lianhui Chen, Paul Voumard, " An Integrated Pattern Classifier for The Condition Monitoring & Diagnosis of Power Circuit Breakers", pp. 1221-1227, 1992.

[DHM92]   Rodney M. Doodman, Charles M. Higgins, Ihon W. Miller, " Rule - Based Neural Network for Classification and Probability Estimation ", Neural Computation 4, pp. 781-804, 1992.

[HS91]   Alistair D.C. Holden, Steven C. Suddarth, " Combined Neural Net / Knowledge Based Adaptive Systems for Large Scale Dynamic Control ", International Journal of Pattern Recognition and Artificial Intelligence, Vol. 5, Number 4, pp. 1-20, October 1991.

[KKC93]   Bon Hyun Koo, Hyong Chol Kim, And Soon Heung Chang, " Development of Real Time Core Monitoring Systems Models With Accuracy Enhancement Neural Networks ",IEEE Transactions on Nuclear Science, Vol. 40, No. 5, April 1993.

[KR92]   Shahla Keyvan And Luis C. Rabelo, " Sensor Signal Analysis by Neural Networks for Surveillance in Nuclear Reactors ", IEEE Transactions on Nuclear Science, Vol. 39, No. 2, April 1992.

[LB62]   M. Stanley Livingston, Jhon P. Blewett, " Particle Accelerators ",Mc Graw-Hill Book Company, 1962.

[ML94]   Larry Medsker, Jay Liebowitz, " Design and Development of Expert Systems and Neural Networks ", Macmillan Publishing Company, New York, 1994.

[Moa91]   Ciamac Moallemi, " Classifying Cells for Cancer Diagnosis Using Neural Networks", IEEE Expert, pp. 8-12, December 1991.

[Sch87]   Herbert Schildt, " Artificial Intelligence Using C ",Mc Graw - Hill, Inc., 1987, USA.

[VC92]   Paul Voumard, Lianhui Chen, " Expert System in The Condition Monitoring & Diagnosis of High Voltage Circuit Breakers ", Proceedings of The 2nd Pacific Rim International Conference on Artificial Intelligence, Vol. 1, pp 631- 637, 1992.

[Zur92]   Jacek M. Zurada, " Introduction To Artificial Neural Systems ",West Publishing Company, 1992.

INTELLIGENT USER-INTERFACES FOR EXPERT SYSTEMS: REQUIREMENTS & IMPLICATIONS

Abbes Berrais

Abha College of Technology, POB 238, Abha, Saudi Arabia

ABSTRACT

Intelligent user-interface is a necessity for the success of expert systems. Issues such as *user requirements*, *user model*, and *system model* have been given insufficient attention in the development of majority of many such systems - particularly those for engineering applications. This paper discusses the requirements, the benefits, and the problems of building intelligent user-interfaces for expert systems. The notions of a *user model* and *system model* are discussed and their impact on the success of intelligent user-interfaces considered. The paper concludes with a description of the user-interface developed for an expert system prototype for preliminary seismic design of reinforced concrete buildings.

Key words: Knowledge-based expert system, user-interface, user model, system model, user requirements.

INTRODUCTION

The user-interface stands between the user and the application program, and facilitates the communication and interaction between the two. It can be considered as the channel in which the user can interact with the internal processes of the application program (and vice versa). The user-interface may be considered at three levels:

Semantic: internal granulity and inter-relationships of the system's functions

Conceptual: structure of individual interactions with the system

Syntactic: the operation of input and output commands

As conventional computer-aided design (CAD) systems have become larger and more complex, the need for an appropriate and effective intelligent user-interface has become more critical. At the same time the expectation of users has been raised by widespread exposure to windows based applications software. The requirements for an intuitive and easily mastered user-interface is equally applicable to expert systems (ESs) and must be taken into account early in the development process. The design of the user-interface is influenced by many factors, amongst these: the cognitive model of the user's thought processes, aspects of usability, the type and capabilities of the programming tools, the hardware environment, and the intended function of the ES.

The majority of prototype ESs for civil engineering applications reviewed in recent papers[1,2,3,4,5]- particularly those relating to design applications - do not include reference to a *user model* or tackle the problem of *user requirements* in detail. Indeed, in many cases the intended users of the system are often not adequately defined and considered. Development of the user-interface should take into account the global needs of users and the tasks the system is intended to support. Thus, an understanding and analysis of *user requirements* will facilitate design of the user-interface. In this context the term "user-interface" embraces far more than simply the visual appearance of the screen. The success of the user-interface depends largely on the completeness of the *user model*.

This paper addresses some of the issues of establishing *user requirements* and the problems faced when implementing these requirements within an ES. It does not go deeply into the theory of cognitive psychology. The concepts, the requirements and the theoretical definitions of *user model* and *system model* are presented, and an attempt is made to implement some user requirements with reference to an expert system prototype by the author for preliminary seismic design of reinforced concrete buildings.

USER REQUIREMENTS

Before attempting to establish the *user model*, the anticipated users of the system must first be defined. To what extent is the user himself knowledgeable about the task domain? When the user is defined, the requirements of user and system can more easily be established. User requirements should take into account the following [7]: type of users, the tasks performed, and the social and dynamic environment in which they interact with the system. This will lead us to identify what will be the system requirements such as: what are the areas of the task in which the user is supported (diagnosis, planning, design, etc.), and what are the limitations of the system.

User requirements can be extracted from an analysis of the tasks the system is intended to support, user characteristics and characteristics of the information and its use. In order to have compatibility between the user's understanding of the system and his skill of knowledge, the design of the user-interface should take into account the followings [8]: (i) classify the user in terms of his level of skill or knowledge, (ii) analyze the functionality of user tasks, their important procedural characteristics and, (iii) the type of information use.

It has been suggested that it is a good idea to separate between the semantics of the system application and the user-interface [9]. This separation will lead to the following benefits:

- Portability, to allow the same application to be used on different systems

- Reusability, increases the likelihood that component can be reused in order to cut development costs
- Customization, the user interface can be customized by both the designer and the user to increase its effectiveness without having to alter the underlying applications.

A proposed user-interface model is illustrated in Fig. 1. As shown in this Figure, both the user requirements and system requirements are taken into account which in turn affect the *user model* and *system model* during the process of developing the user interface.

USER MODEL

The development of a *user model* is a difficult task which requires information about the specific users and the responses to their reaction with the system. User modelling can help matching the facilities that a system provides to the needs of the user; improve user learning; guide design decisions and make design choices and assumptions explicit. A *user model* is required by the ES to help to: identify what needs to be explained, determine the depth and complexity of the explanation, and establish the knowledge necessary to assist the user in achieving his goals and understanding the solution.

Sparck Jones [10] sees three advantages in the inclusion of user models in ESs:

1. *Acceptability*. The form in which information is elicited and explanations given need to be tailored to the intended user, be they novice or expert.
2. *Efficiency*. The most efficient mode of system-user interaction will usually vary according to the user's level of skill.
3. *Effectiveness*. More effective task performance through more accurate interpretation of user behaviour and by making the system's requirements more comprehensible to a particular user.

Philip Slatter [11] argues that a good match between the computer system and the user is vital for several reasons:

- Without this cognitive compatibility, the system's behaviour can appear surprising and unnatural to the user.

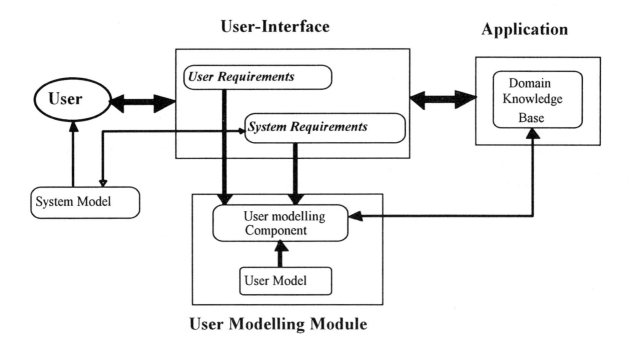

Fig. 1 User-Interface Model

- To counter the potential dehumanising influence of expert system technology (through ensuring the knowledge and reasoning of the system are understandable to the user, user-controlled dialogues, etc.)
- To ensure the system will be acceptable within its intended social and organizational context of use.

Different artificial intelligence researchers use different definition of a *user model* [8,12,13,14,15,16,17]. Several *user models* have been proposed in the field of human-computer interaction [12]. However, these *user models* deal more generally with user's interactions with interactive computer systems and are more concerned with the cognitive modeling of the user rather than the interaction of the user with the application domain embedded within the computer. Examples of *user models* that have been proposed include: *conceptual* and *quantitative* models[13]. *Conceptual* models are mainly concerned with representing human cognitive processes and the strategies involved in human computer interactions. *Quantitative* models are concerned with the interaction of the user with the computer and how to improve this interaction (independent of application domain).

In the *user model*, some aspects of the user's understanding, knowledge, and processing are modeled. Majchrzak *et al.* [18] divided cognitive models into three categories. The first described the user's task and goal structures. The second category is concerned with linguistic and grammatical models which emphasized the user's understanding of the user-system interaction. The third category was based on the more solid understanding of the human motor system.

The *user model* can represent certain aspect of the user which is implicit within the system and used to adapt the system to suit the user. It seems difficult to have a unified *user model* which embodies all the user requirements (because not all the users even in the same field - think or process information in the same mode or have the same level of skill).

In this paper we define the *user model* as being the model of a typical user which resides in the KBES. This *user model* may encompass the following characteristics: user understanding and expertise of the application domain, user acceptability, and user confidence. The building of a specific *user model* depends on the system designer's understanding of these factors. This *user model* can be considered as a *static* model. Static model may represent a specific user and can then be updated to the changes occurred with the user.

SYSTEM MODEL

The *system model*, in contrast, is the model of the computer system as seen by the user. The *system model* structures the interaction of the user with the system. The user may (or may not) have a model, which may be incomplete, of the system (structure, functionality, and interaction of the system). This model may differ from a novice user to an expert user. This difference poses different problems for the designer of the user-interface.

The *system model* may be considered as a conceptual model which represent the system, and in which the system developer might wish to represent to the user (*system model* is different from the mental model as seen by the user). In other words *system model* is external and explicit, whereas mental model cannot directly expressed. Browne *et al.* [19] described a conceptual model as a model invented by a system developer, while a mental model is a naturally evolving internal representation of a system.

While interacting with the computer, a user progressively builds an internal model of the system. This model is a representation of how the system works. It has been suggested that users construct such a cognitive model through a process of anchoring and adjustment [20]. Generally, users create an initial *system model* based on their understanding of the system, and they adjust this model on the basis of interactive experiences that are not consistent with their current model. More detailed information about adjustment of the cognitive behavior of users during the interaction with computer systems is given by Lehner and Kralj [20].

The *system model* is likely to be characterized by ease of use, capabilities, friendliness, complexity, and speed. The success of the user-interface to explain internal system actions and reasoning processes to the user will be reflected in the *system model*. The *system model* may also reflect the task domain of the ES which may affect the way how the user interacts with the system. Figure 2 shows the relationships between the user, computer system, *user model* and *system model*, via the user-interface, and how the design of a good user-interface should take into account both these two models. It is crucial that the *user model* and the *system model* match, because this will lead to better user-interface and will increase user acceptability and adaptability. If, however, the *user model* and the *system model* do not align then the user's performance can rapidly deteriorate.

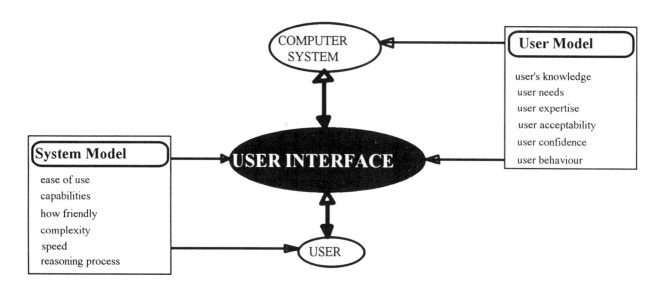

**Figure 2 Relationship between the user and the computer system
via the user interface**

KNOWLEDGE ACQUISITION FOR USER MODEL

Much research has been done on the process of emulating human expert's cognition but little on the emulation of users cognitive task [11]. Since the user requirements are to be considered as a factor in developing KBESs, it is advisable that some acquisition about the user's knowledge be performed in a similar way to the acquisition of domain knowledge from a human expert. Kidd [21] argues that the user is an active agent in the problem-solving process and he recommends that knowledge acquisition about the user's knowledge should include the followings:

● Identifying the different classes of users likely to use the system and their needs.

● Analyzing user requirements: what are the common classes of problems and question, what advice does the user require and in what form (e.g. Does he usually have his own idea of a solution and only require a critique from the ES, or does he need to have a set of alternative solutions ?) and what type of justification does he require.

● Analyzing what type of knowledge the user brings to bear on the problem-solving process. Recent research indicates that these include the user's goals within the domain, his constraints on acceptable solutions (e.g. time, availability, cost , etc.), his own model of the type of problems that the system can solve for him.

The acquisition of user requirements is a difficult process and many methods have been developed with varying approaches [7]. The most widely used method is the *Contextual Inquiry technique* [22]. In this technique, the user is observed interacting with

the system, and asked some specific questions, then the analysis is carried out.

Stelzner and Williams [23] suggested that the roles of the user and the system must complement and supplement each other. When subjective decision are to be taken it is important that the user remains in control and the system act as an experiment consultant. For objective decisions, the system should provide more control of the process, freeing the user from unnecessary errors. They also identified five major requirements for the user-interfaces to ESs:

1. The user-interface should represent the domain in the user's natural idiom (the way of expressing things in a person's way of understanding).

2. The user-interface should provide immediate feedback to the user on the effect of changes to system state by explicitly maintaining and displaying complex constraints and interrelationships.

3. The user must be able to recover easily from trying different alternatives.

4. The user-interface must support the user at different granularities, or level of abstractions.

5. User-interface must be implemented in such a way that it is possible to have multiple interfaces to the same knowledge.

Figure 3 shows a simplified model of how user requirements can be applied to the ES development cycle. The design cycle is based on the prototyping approach which can allow the evaluation of the system in an iterative manner. User requirements can be obtained by structured interviewing and observations of the user during his interaction with the system. These user requirements can help tailoring of the initial user-interface specifications to the cognitive and behavior characteristics of the user, and the range of the decisions he has to take.

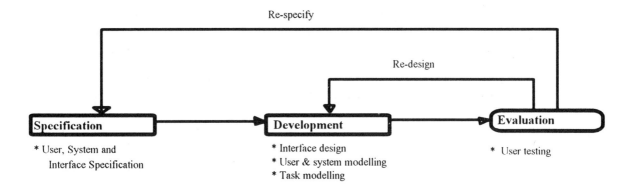

Fig. 3 User requirements specifications for KBES design cycle

Based on the type of design problem performed, the user-interface can be hierarchically organized as a set of design modules (agents) where each module accomplishes a portion of the design at a certain level of abstraction. The upper levels of the hierarchy are modules in the general aspects of the design problem, while the lower levels deal with more specific decisions. In the next section, the implementation and development of the user-interface for the SDA system is described and discussed in detail.

ES IN CIVIL ENGINEERING

Many ESs have been developed to assist civil engineers in various tasks [1,2,3,4,24]. None of the ES prototypes, published in the literature, has tackled or discussed the problem of the user-interface from the *user model* point of view. And they were only restricted on the external appearance and form of the user-interface at the surface level. This lack of research and investigation of the user-interface based on *user model* and *system model* is due mainly to the shortage of established practical information, from artificial intelligence field, on *user models* and *system models*. The *user model* could be taken into account by: collecting information about the user of the system, this information could be presented as the user's expertise and knowledge about the task domain and computer manipulation. In addition different domain tasks may require different requirements for man-machine interaction. Construct user group taxonomies, for which for each user group different explanations can be defined, both for the domain and the problem solving knowledge.

SDA SYSTEM

The SDA [6] system is an expert system to assist structural engineers in the analysis and design of reinforced concrete (RC) buildings subjected to earthquake forces. Figure 4 shows the architecture of SDA. This architecture has six main components:

- *Knowledge base*: comprises of seven main modules as shown in Figure 4, each module being responsible for a specific task.
- *Context*: contains the collection of facts which represent the current state of the problem at hand.
- *Inference Mechanism*: controls the system by modifying and updating the context.
- *Explanation facility*: provides the user with the necessary explanation about the task being performed.
- *External analysis programs*: contain the finite element analysis programs.
- *User-interface*: facilitates communication between the user and the modules of the system. The user-interface is described in more detail in the next section.

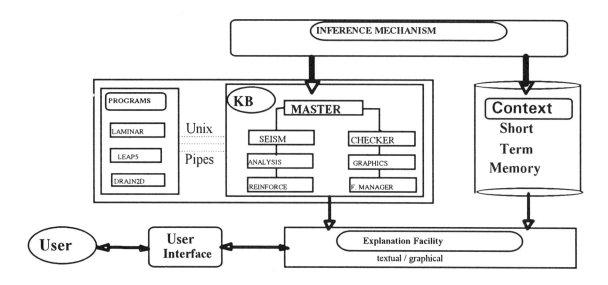

Figure 4   Macro level architecture of the prototype

SDA USER-INTERFACE

The aim of this section is not to develop a model for the users cognitive tasks, but to implement and validate some of the user requirements discussed previously. The development, architecture and characteristics of the user-interface are also described. In the SDA system the user-interface is separated from the application domain. This distinction is very important in many aspects of development, allowing the system developer to present different, independent interfaces to the same application. The user-interface can be customized to increase its effectiveness without having to alter the underlying application.

The system in question is a design tool to assist in preliminary seismic design of RC buildings and the type of users who can use SDA are classified in two types: a structural engineer, who is knowledgeable about the wider perspective of structural engineering, but without depth knowledge of seismic design aspects; and a structural engineer with expertise of seismic design aspects, but little knowledge of computer applications. The first type of user needs some assistance and advice on how to design reinforced concrete buildings to resist seismic forces. With this type of user, the system will initiate the dialogue. The second type of user is helped to carry out the seismic design in an efficient way with a critique, and justification from the SDA system. In this case, the user takes the initiative and controls his interaction in a flexible manner to suit his needs. In both cases, the user has his own model of the type of problem that the system can solve for him.

The knowledge about the two types of users was coded using Quintec-Flex production rules. This included building a help module, which determines when the user is making errors and to offer advice on how to perform commands. The help module contains information which are conveyed to the user, about how to use and communicate with the SDA system. In addition, the help module contains meta-level reasoning in the context of explanation (e.g. why certain inferences had been drawn in preferences to others). Production rules have also been used to model the interaction between the user and the system, and to set the menu display according to the user wishes. The user errors were identified and classified into two categories: computing errors, and task errors. Computing errors are concerned with system commands and functions, whereas task errors are concerned with seismic design methods. The user errors were specified in terms of Quintec-Flex production rules.

Since the types of the users are defined and the task concisely described, the *user model* can be considered as a static model[28] in which the system interaction is based around that model. In this model the user and the system are required to work as a partnership, the interaction is initiated by both parties. This is *called Mixed Initiative Mode* and this can be fulfilled by the combination of wide range of system options coupled with natural language. The *user model* is conveyed to the user through the SDA display representation, which direct the user to interact with the system in a certain desired way. The user must form a mental model of the task which is as close to the *user model* as possible.

The user-interface within SDA incorporated the following types of knowledge: (i) knowledge about seismic design methods & terminology; (ii) knowledge about the system operations being used; (iii) knowledge about the different tasks the user might want to carry out (static and dynamic analysis & design) ; and (iv) knowledge about the user such as: his level of understanding and expertise, and his preferred method of interaction.

The following requirements were investigated and implemented within the SDA system:

General requirements

- the system is easy to use, takes the initiative and questions the user
- informs the user about the limitations of the system
- allows the user to invoke the system at any level of abstract
- gives brief help for each task required by the user
- warns the user about any unpredictable actions

Specific Task Domain Requirements

Other more specific task requirements arise from a consideration of the engineering aspects of the system domain:

- Provide the user with better understanding of the actual seismic behavior of RC buildings.

- Provide the user with clear distinction between elastic and inelastic analysis of a building.
- The use of passive and active graphics to integrate the seismic design process, from conceptual design to the final structural design.

The general screen display (Master Menu) of the SDA user-interface is shown in Fig. 5. The Master Menu is considered to be as a dynamic screen menu, where the memorization of commands is overcome, and the interface is based on a hierarchical structure of modes, each corresponding to a functional subgroup.

USER-INTERFACE MENUS

The Master Menu is divided into five windows as shown in Fig. 5. The top and left hand windows are used for the high-level options which contain all the major functions (of SDA). In addition to the five main windows, the user-interface uses other windows which are opened and closed dynamically while performing a task. The large window in the middle is the main input/output window where most of the dialogue between the user and SDA is carried out while performing a task. The right hand side window is a graphical window which deals with the graphical explanation of the input as well as the results. The bottom window is a message/warning window which displays warnings about unusual actions by the user. For each option the user is guided through a pull-down menu with multiple sub-options, each sub-option corresponds to a specific task. The high-level options are divided into two types based on their

function (as shown in Fig. 5): *engineering options* and *system options*. On the left side are the *engineering options*, while across the top are the *system options*. These options are described briefly below:

Engineering options:

Engineering options are concerned with the engineering capabilities of the SDA and are divided into five options:

HELP : provides help on the SDA menus, seismic design terminology and background information.

GEOMETRY CHECK : concerned with eccentricity requirements of a typical building.

CODE LOADS : concerned with evaluating the base static shear force and the lateral static forces.

ANALYSIS : concerned with the modeling and elastic & inelastic analysis of buildings .

DISPLAY : concerned with the graphical and textual display of building and earthquake information.

System options:

System options are concerned with tasks related to the management of the system (see Fig. 5):

RESTART : clears the windows and return the user to the first display in a defined sequence.

FILES : concerned with the management of data and system files.

UNIX : gives temporary access to UNIX operating system.

EXIT : to exit SDA completely.

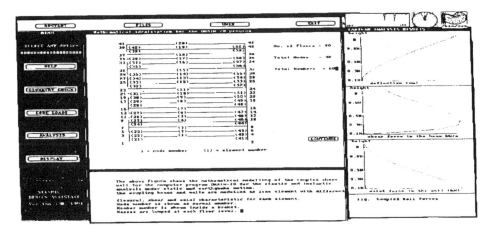

Figure 5   Screen display of the SDA user interface

The use of menus can display different commands organized in tree-like structures. The user simply chooses the command from a menu of possible commands language. In other words, a menu might not only contain commands, but also routes to other menus. The dialogue design within the menus has provided the following advantages: syntax and commands are kept simple; number of commands are limited in format; information passed by a command is limited; sequence of dialogues are organized into logical groups; simple error handling; user should not be expected to remember too much.

The philosophy behind using these different menu options is that the user can have the opportunity to initiate SDA at any level of abstraction he wishes with some degree of control on SDA behavior. Menu displays are beneficial because they always display the possible range of commands, research has shown [18] that humans are much better at recognizing a correct choice rather than recalling that choice from memory. The user is able to volunteer information concerning his problem easily and rapidly to the system.

EXPLANATION FACILITY

The ability of the ES to provide an explanation to its reasoning is an important aspect of its intelligibility to the user. The explanation offered by the user-interface are not limited to the conventional 'Why' or 'How' explanation, deep explanation about a specific task can be provided with how it derived the ordering of its hypotheses. This can be activated by an *Explain* or *More Details* button associated with the textual window of the required task. In addition, some of the textual explanation are coupled with passive and active graphical explanations, the utility of this is that the user will have more understanding of the meaning of different variables used by SDA. As an example, the mathematical model for the finite element method can be displayed in a graphical form showing all the different parameters with their explanation (see Fig. 5). Complex interrelationships can be explicitly conveyed using graphics by presenting them as spatial relationships. This exploit human highly developed spatial awareness and ability to extract meaning from spatial patterns. Information provided by the system to the user is minimum and in usable form that is essential to making decision or performing an action. In addition, SDA maintains a

record of the decision it makes. It uses this record to explain and justify its decisions and conclusions on request. This ability of SDA to provide an explanation of its reasoning is obviously an important aspect of its acceptability to the user.

SUMMARY AND CONCLUSION

The paper has discussed the benefits of including *user models* and *system models* when building user-interfaces for ES. *User models* and *system models* have been largely ignored by developers of ES for engineering design applications. In this paper, an attempt has been made to investigate *user requirements*, *user model*, and *system model* for user-interfaces. Although many of the requirements of *user models* have been investigated, comprehensive realization remains beyond the reach of the current research. Problems include the limitations of the programming tools used, the complexity of building user-interface knowledge which take more than one specific type of user into account, and the difficulties associated with the simulation of user knowledge (because this knowledge is dynamic thorough time and must be continuously followed and updated). User-interfaces should be built and tested during development of ESs, and not "tacked on" at the end. A *user model* is important for the design of an appropriate user-interface, and there are no standard user models. A *user model* has to be established or derived on the basis of the user type, the task to be computerized, and the particular human-machine interaction under consideration. The *user model* should be taken into account early on in the development of any ES. A poor match between the *user model* of the system and *system model* of the user will lead to systems which are difficult to use and user-error prone.

The user-interface developed in this research was aimed at a specific type of user, with a particular body of knowledge and particular goals. The type of *user model* adopted in this research is a static model, and human-machine interaction is determined by that model. More initiative is taken by the system, thus minimizing its passive role. Even though the static model has its drawbacks, but the investigation has paved the way to understand and research more advanced user models, such as dynamic models, which could reflect user behaviors more accurately, and improve the interaction between the user and the system. The user-interface has demonstrated the advantages of using both display menus and graphical windows to enhance the seismic design process and user-interaction with the system. The research has demonstrated that menus have great advantages over keyboard commands in that memorization of complex sequences of commands is eliminated; interactions are simplified; and response times of both the user and the system are minimized. The user-interface could be enhanced by developing a *multiple user model* which would provide better interaction than the existing static model. The development of a *multiple user model* would require a multidimensional map with many aspects, reflecting the different user types. These user types would range from the novice user to the experienced user. The *multiple user model* would allow for switching between different user models. The problem of building a *multiple user model* stems from the difficulty of defining user types. It is unclear how to elicit this information and more difficult to build the information into a computer system. Building such a model would require the system designer to have a detailed knowledge of human behavior, of learning patterns, and of task requirements. This complex task requires sophisticated artificial intelligence techniques to work well. Additionally, there is a need for new effective cognitive based techniques to allow the user to develop an accurate mental model of the computer system. Moreover, knowledge about users, such as intelligence, educational background and job skills, can be inferred by studying how users use computers commands and deriving user characteristics through the use of stereotypes.

REFERENCES

1. A. Berrais and A. S. Watson, "Expert Systems for Seismic Engineering: the State of the Art," *Engineering Structures*, 15(3), 146-154 (1993).

2. IABSE. Proceeding of the IABSE Colloquium *Expert Systems in Civil Engineering*, International Association for Bridges and Structural Engineering, Bermago, Italy, 1989.

3. CIVIL-COMP93, The Third International Conference on the Application of Artificial Intelligence to Civil and Structural Engineering, Cambridge, England, August 1993.

4. B. H. V. Topping, *Developments in Artificial Intelligence for Civil and Structural Engineering*, Civil-Comp Press, Edinburgh, UK, 1995.

5. C. J. Anumba, "Considerations in User-Interface Design for Knowledge-Based Systems", Artificial Intelligence and Object-Oriented Approaches for Structural Engineering, Topping, B. H. V. and Papadrakakis, M. (eds.), Civil-Comp Press, Edinburgh, 1994, pp. 47-51.

6. A. Berrais and A. S. Watson, "A Knowledge-Based Design Tool to Assist in Preliminary Seismic Design," *Microcomputers in Civil Engineering.*, 9, 199-209 (1994).

7. R. C. Jones and E. Edmonds, "Knowledge-based requirements", Human Aspects in Computing: Design and Use of Interactive Systems and Information Management, H. J. Bullinger (ed), Elsevier Science Publishers, 1991, pp. 796-800.

8. I. Cole, M. Lansdale, and B. Christie, "Dialogue Design Guidelines", in *Human Factors of Information Technology in the Office*, B. Christie (ed.), John Wiley & Sons Ltd., 1985, pp. 212-241.

9. A. Dix, J. Finlay, A. Gregory, and R. Beale, *Human-Computer Interaction*, Prentice Hall, UK, 1993.

10. J. K. Sparks, "Issues In User Modelling for Expert Systems", *Proceedings of AISB85*, University of Warwick, April, 1985.

11. P. E. Slatter, *Building Expert Systems : Cognitive Emulation.*, Chichester : Ellis Horwood, 1987.

12. R. M. Young, A. Howes, and J. Whittington, "A Knowlegde Analysis of Interactivity, in *Human-Computer Interaction"* - *INTERACT 90*, Proceedings of the IFIP TC 13 Third International Conference on Human-Computer Interaction, Cambridge, (D. Diaper, D. Gilmore, G. Cockton, and B. Shackel, eds), U.K, 1990, pp. 115-120.

13. R. C. Williges, "The Use of Models in Human Computer Interfaces Design", Ergonomics Society Lecture, Swansea, wales, 6-10 April 1987.

14. Brown G. M., Cognitive Models in Human-Computer Interaction, in An Introduction to Human-Computer Interaction ed. by Paul Booth, Lawrence Erlbaum Associates, UK, 1989, pp. 65-101.

15. K. Gregory, "Methodology for Designing a Normalized User Interface", in Advances in Human Factors/Ergonomics - Cognitive Engineering in the Design of Human-Computer Interaction and Expert Systems, ed. by Salvendy G., Elsevier Science Publisher, Netherland, 1987, pp. 139-146.

16. P. Johnson and S. Cook, *People and Computers: Designing the Interface*, Proceedings of the Conference of the British Computer Society Human Computer Interaction Specialist Group, British Information Society Ltd., Cambridge, 1985.

17. K. R. McKeown, "User Modeling and User Interfaces", AAAI-90, Proceedings of the Eighth National Conference on Artificial Intelligence, American Association for Artificial Intelligence, July 29-August 3, 1990, Vol. 2, pp. 1138-1139.

18. A. Majchrzak and Tien-Chien Chang et al., *Human Aspects of Conputer-Aided Design*, Taylor & Francis Ltd, 1987.

19. D. Browne, M. Norman, and E. Adhami E., "Methods for Building Adaptive Systems", in Adaptive User Interfaces, edited by Browne D., Totterdell P., and Norman M., Academic Press, 1990, pp. 85-130.

20. P. E. Lehner and M. M. Kralj, "Cognitive Impacts of the User Interface", in Expert Systems: The User Interface, (J. A. Hendler, ed), Norwood, NJ, 1988, pp. 307-318.

21. A. L. Kidd, *Knowledge Acquisition For Expert Systems : A Practical Handbook.*, Plaenum Press, New York, 1987.

22. D. Wixon, K. Holtblatt., and S. Knox, "Contextual design: an emergent view of the system design", Proceedings, CHI'90, ACM , 1990, pp. 329-336.

23. M. Stelzner and M. D. Williams, "The Evolution of Interface Requirements for Expert Systems", in Expert Systems: The User Interface, James A. Hendler (ed.), 1988.

24. M. Maher, *Expert Systems for Civil Engineers: Technology and Application*, American Society of Civil Engineers, 1987.

25. Quintec System Ltd., QUINTEC-PROLOG, Programmers Reference, Unix version, 1989.

26. Quintec System Ltd., QUINTEC-PROLOG, System Predicates, Unix version, 1989.

27. Quintec System Ltd., QUINTEC-FLEX, User Manual, Unix version, 1989.

28. D. M. Cleal and N. O. Heaton, *Knowledge-Based Systems: Implications for Human-Computer Interfaces*, Ellis Horwood Ltd., England, 1988.

29. K. R. Howey, M. R. Wilson, and S. Hannigan, "Developing a User Requirements Specification for IKBS Design", Proceedings of the Fifth Conference of the British Computer Society Human-Computer Interaction Specialist Group, University of Nottingham, Sept. 1989, pp. 277-289.

AN ARCHITECTURE FOR GENERATING INTELLIGENT INTERFACES

Bhardwaj VVSR and Hrushikesha Mohanty
Artificial Intelligence Laboratory, University of Hyderabad,
Hyderabad - 500 046, India.
Email : hmcs@uohyd.ernet.in

ABSTRACT

Though numerous User Interfaces of various kinds are being produced every year, there has been a consistent demand for more active and versatile interfaces. Separating the interface from the application has proved to be useful in a large number of cases. An intelligent user interface will be capable of independently handling all the I/O routines and controlling itself - a feature that reduces the burden on the main application. In this paper, we present such a self-controlled and intelligent interface and describe an architecture that generates it. This interface can react dynamically to the changing environment and to handle the I/O process according to specifications and will be generated by the UI generator when the generator receives a request from an application for an interface using its Interface Repository, which contains information about the interfaces already generated and Knowledge base, which contains the necessary rules and constraints for interface generation.

Keywords : Intelligent Interface, Interface definition model (IDM), Interface Specification Language (ISL), self-controlled User Interface, User-Interface Management System.

1.0 INTRODUCTION

User interfaces have come a long way since the character-based I/O of 80's till the GUI of 90's. The common primary objective of any interface obviously happens to be User-friendliness - the more the interaction, the more it is user-friendly. The research works in [Cou91, ES92, Lee90, RW92] establish this fact. Yet, the demand for more user-friendly interfaces is on the rise. Though today's interfaces are maintainable and reusable, most of them lack the flexibility because of their dependence on the code in which they are embedded. Undoubtedly, an interface separated from the application distributes the job among the application developer and the interface developer, thus making it more convenient to develop a Software system. Besides, this kind of separation makes the interface self-controlled i.e., all the operations on the interface will be performed by itself, but not by the application for which this interface is used. Most of these operations are carried out as the result of the requests received by the interface from the application. The efficiency of this kind of interface will be enhanced if it is also provided with the ability to react dynamically according to the situations, for e.g., deciding "What kind of User screen is required from the current operation?". This needs the User Interface be equipped with a Knowledge base, i.e., the interface being "Intelligent". Needless to say, the main advantage of using such an intelligent interface is that only relevant and required data items are communicated between the application and the interface. The ability to decide about the I/O formats depending on the operation makes the interface more versatile. The works [Med95, Pfa83, PW92, Rob95, RW92, Tay96, Tay95, Whi95] do talk about some of the above properties, but the emphasis in those works was on reusability and not intelligence.

In this paper, we propose a model called Interface Definition Model (IDM) for Intelligent User Interfaces. We also present an architecture for the development of Intelligent User Interfaces. These Object Oriented User interfaces are just not only self-controlled but also Knowledge-based. One of the salient features of this sort of UI happens to be its decision making capability, i.e., reacting dynamically according to the situations. Also, this interface happens to be application independent and can be composed with relevant programs that require identical interface behavior. Attachment of the interface to the application is done by the Interface Generator, which will be described later in the paper, upon receiving a request from the application. We call the language used to communicate the requests for interface generation as Interface Specification

Language (ISL).[†] The user Interface Management System (UIMS) handles the tasks of managing and controlling the UI. Precisely, we propose the architectural design for the generation of an intelligent GUI and propose a model for it. The motivation for this work comes mainly from the works [ES92, Lee90, Pfa93, RW92, Tay96] and the terms Interface, User Interface, UI and GUI have been used as synonyms throughout the paper.

The organization of the paper is as follows. Section 2 describes the IDM in detail, Section 3 gives an account of the structural and behavioral aspects of the interface model. The architectural aspects of the Interface generation are presented in Section 4. Section 5 explains the behavior of the interface as well as the process of interface generation. The paper ends with concluding remarks in section 6.

2.0 THE MODEL

In this section we define an abstract model of the proposed user interface. This model, which we call as Interface Definition Model (IDM) is a 5-tuple (I, S, K, P, T) where I represents the interface scheme to be used (or precisely, the structure of the user interface), S represents the situation i.e., the conditions under which this interface is used and K represents an index for the requesting application program. The set of parent models, if any, from which the current model inherits is represented by P. If the current model happens to be a base-class, then the value of P is assigned to NULL or ϕ. Obviously, each $p \in P$, which represents a single parent class, is also a 5-tuple with the same parameters mentioned above, perhaps with different values. If there exists only one element in P, then it is singular inheritance and it is a case of multiple inheritance if there are two or more elements in P. The term T represents the set of allowed operations on the interface. No operation other than these is allowed to be performed on the interface. In other words, this model is encapsulated. Typically, I consists of information regarding user screens, control routines and attributes like Interface id, type, size, color, positioning, no. of windows, etc., S contains the constraints like Task id, type of the task, whether an interface is readily available etc. With the values of K and P being obvious, T may contain operations like identify caller id, recognize problem type etc. Fig. 1 gives an illustration for a typical problem of reading a matrix of size 3x3.

[†] We do not present all the aspects of ISL in this paper, but describe only those statements of ISL that are used in the paper. However, the reader is free to contact the authors for specific details regarding ISL.

Formally, I can be defined as a set { n, A_i, R_i }, where n represents the number of user screens for the current application, A_i is the set of attributes and R_i is the set of control routines (i.e., the routines that control the user screens), for the ith screen. Thus A_i = (a_1, a_2, ..., a_n) where each a_j, $1 \leq j \leq n$, represents an attribute or subattributes of the ith screen and R_i = { r | r is a control routine for the ith screen }. Here the number of tuples in A_i varies from one application to another.

For a simple task like reading an integer value 4 from the keyboard, n=1 and R_i = { Open, Close, Read }. The attribute set A_i may be given as (Display = Standard, No. of windows =1, No. of Data items read =1, Type = Integer, Value = 4) Here, Display represents commonly used screen attributes, which in this case, are Standard attributes.

3.0 STRUCTURAL & BEHAVIORAL ASPECTS

This section presents the intelligent behavior and self-controlling nature of the User interface pertaining to IDM. The intelligence of the interface is established by its ability to respond adaptively to a dynamically changing environment, i.e., controlling I/O processes even in those situations which are not explicitly specified during the design & development stage. For example, if the end user enters an integer where a floating point type should be read, the interface automatically converts the entered type to floating point and passes it on to the program. In other words, some sort of "Fault Tolerance" is achieved and the User Interface makes sure that only valid data is sent to the application.[†] This means that the UI interprets what the user thinks- a sign of intelligent behavior. As the interface is object oriented, all the operations like creation, destruction and manipulation of user screens are performed by the UI itself. Moreover, all the controls like interface retrieval from repositories, interface repository making, inheriting interfaces etc., rest with the UI, but not with the application program. Thus we termed this interface as "self-controlled UI". Rest of the section explains these concepts in detail.

The structure of the IDM interface is depicted in Fig. 2. It is an interface that is separated from the application program. As indicated, the main components of the system are UI controller, User screen, Selector and

---

[†] Obviously, this capability is limited. In case the UI encounters such a situation where it can not do anything, the user will be asked to correct the mistake. The interesting point to be noted in this context is that the UI remembers this situation (Just as a Case-based reasoner does in some intelligent systems) and acts accordingly next time, when it faces a similar situation.

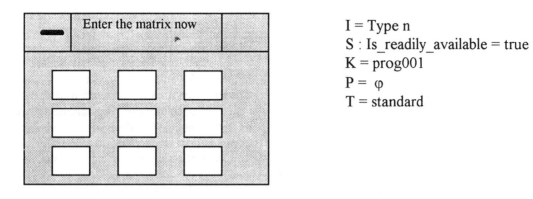

I = Type n
S : Is_readily_available = true
K = prog001
P = φ
T = standard

Fig 1 : Reading a matrix

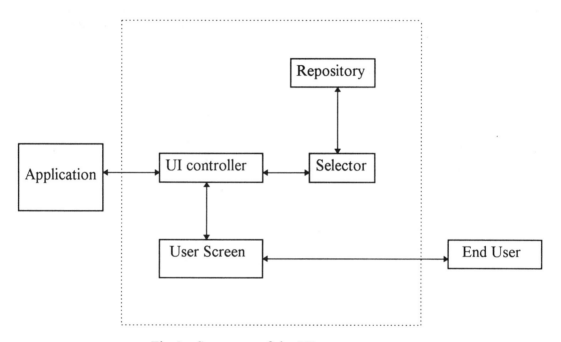

Fig 2 : Structure of the UI

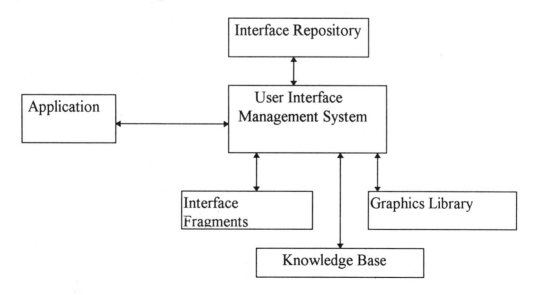

Fig 3: Architecture for User Interface Generation

Repository. The user communicates to the interface through the user screen and the application program, through the User Interface controller.

3.1 UI Controller

This controls the entire User Interface and communicates with the application program. All the routines to manipulate the user screens are stored in this component. The basic operations that will be undertaken by this component are Opening and Closing the user screens and the other operations include stacking up the screens, manipulating dialog boxes, buffering the Data that are read either from the application (Output) or from the user (input) etc. On encountering an unexpected situation (i.e., a situation which is not explicitly defined before) during processing, it seeks the assistance of the selector component and takes appropriate action. The communication between the UI controller and the Selector is through keywords. The controller generates the appropriate Keywords when an unexpected event takes place (for e.g., UI is requested by the application for a problem which was encountered never before). The selector processes the keywords and returns to the controller, the decisions taken previously in similar situations (Two situations are said to be similar if more than 75 percent of their parameters match). From this, the controller decides the future course of action and this event will be stored in the repository. Currently, we are exploring the possibility of using ISL or a similar language for the communication between various components of the User Interface.

3.2 User Screen :

This is the actual display meant for the end user. It is totally under the control of the UI controller. Its typical attributes include Length, Breadth, Color, Background etc. This component physically takes input and gives output to the user.

3.3 Selector :

This component is activated by the UI controller when an unfamiliar situation arises during the processing. It consists of an embedded rulebase to generate the situation parameter.. As mentioned, the UI passes on a set of key words to this component. After receiving them, the selector generates the situation using its rule base and compares it with the situation structure in the repository. The rules of this rulebase guide the selector in formulating the situation that suits keywords passed on by the UI controller. If more than 75 per cent parameters of situation generated by selector match a situation in the repository, then that old result will be retrieved and sent to the UI controller. If less than 75 per cent parameters match, then the Human Intervention is sought and this case will be stored in the repository. We decided to fix the ratio of above comparison at 75 instead of a common-sense driven value of 50 percent just to make sure that the best situation that matches the

current is matched. Even otherwise, the designer is free to fix this value according to his/her convenience.

3.4 Repository :

This consists of two parameters viz., Situation Parameter Set and Decision Set. The situation parameter set consists of situations described by their attributes, for e.g., status of the printer, Amount of Discrepancy in the entered input etc. The decision set contains the decisions taken in a particular situation. Each of these decisions implicates the selection of a user screen that is to be generated. All the situations are mapped to appropriate decisions.. The selector passes on situation to repository and repository returns to selector, all those which match the current situation. When a new result comes, it will be communicated by controller to repository through the selector and repository keeps the result for future reference.

Thus, in a typical process, the Application requests the interface for some I/O operations, the controller takes charge, manipulates user screens and performs required operation. In case of a strange situation, it consults the selector (Which, in turn consults the repository) and gets a previous decision, if possible, and follows it up with necessary action.

4.0 THE DESIGN

In this section we present the architecture for the IDM. While fig.3 depicts the architecture graphically, the rest of the section is dedicated to presentation of the details of various components involved in UI design.

The main components of the model apart from the calling program are the UI management system (UIMS), Knowledge base, Interface fragments, Repository and a Graphic Library.

The UIMS plays a managerial role in maintaining and providing the interfaces. Up on receiving a request from a program for an appropriate interface, it interacts with other components and generates appropriate interface and serves it to the requesting program. If an interface is readily available for the current task, then it will be recovered, else a new interface would be generated. The interaction between the program and the interface occurs through a new interface specification language, named ISL developed by us exclusively for this purpose.

The knowledge base contains the necessary rules, constraints etc. that help the UIMS in deciding the type of interface that should be supplied to the program. The moment a request comes from the UIMS along with the details of the current task, it specifies the type of interface that is required for the current task. The knowledge structure, (apart from the rules) in this knowledge base can be abstracted as a pair (T, I) where t represents the task type and I represents the interface type. In simple terms, the UIMS sends the task specifications to the Knowledge base and gets back the interface specifications. Once again the interaction occurs by means of ISL

Interface fragments are nuts & bolts of the interface, in the sense that they are the basic independent components of an interface which will be picked up by the UIMS as per the task specifications.

Interface Repository is a set of readily developed interfaces, generated either at the design time or in a previous case. If the request from the UIMS matches any of the existing interfaces, then the corresponding interface is returned to UIMS. If no interface is readily available then a null value is returned to UIMS and UIMS then proceeds with the construction of new interface.

UIMS also makes use of the graphics library that contains the graphical objects (e.g.. an icon) that are to be displayed on the screen, at the time of interface generation.

5.0 THE DEVELOPMENT

Up on receiving a request from the program, the UIMS identifies the type of request and consults the knowledge base for the type of interface that it should have. After getting the interface specifications, it checks with the repository, whether an interface is readily available for the current specifications. If available, then it is supplied to the program, otherwise it either generates a new interface from Fragments and Graphics library or adds on extra features to an existing interface in the repository to derive a "Child Interface" and this interface is supplied to the program.

6.0 CONCLUSIONS

In this paper, we propose the architecture for the generation of an intelligent interface depending on the task specified in a program. The advantage of having such a self-controlled intelligent UI is that in case of un expected situations, the interface can behave intelligently to avoid most of the situations wherein the system would have crashed. Moreover, due to fast changing standards, if a need arises to change the UI then it will not be a difficult job because of the separation of interface from the application. Also, because of the ability to store the attributes of the interface, the interface becomes highly reusable.

7.0 ACKNOWLEDGMENTS

This work is supported in part by the Council for Scientific and Industrial Research, New Delhi, India.

REFERENCES

[Cou91] Coutaz J, "Architectural Design For User Interfaces", Proc. Third European Software Engg Conf., ESEC'91, Milan, Italy, Oct 1991.

[ES92] Ege, R & Stary, C., "Designing Maintainable, Reusable Interfaces", IEEE SW, Nov 1992

[Lee90] Lee Ed, "User Interface Development tools", IEEE Software, May 1990.

[Med95] Medvidovic, N. , " Formal Definition of the Chiron - 2 Architectural Style ", UCI - ICS Technical Report UCI-ICS-95-24, Dept of Information and Computer Science, Univ. of Calif., Irvin, July 1995.

[Pfa83] Pfaff G E, ed. "User Interface Management Systems", Seeheim, FRG: Eurographics, Springer-Verlag, Nov. 1983.

[PW92] Perry & Wolf DL, "Foundations for the Study of Software Architecture", ACM SIGSOFT Software Engg Notes, 17(4), Oct. 1992.

[Rob95] Robbins J E et. al, " A software architecture Design Environment for Chiron - 2 style architectures", Arcadia Technical report UCI-95-01, University of Calif., Irvin, Jan 1995.

[RW92] Rudolf J & Waite C, "Completing the job of Interface design", IEEE SW, Nov 1992.

[Tay96] Taylor, R. et. al, " A Component and message based Arch. style for GUI Software", IEEE SE, June 1996.

[Tay95] Taylor R N et. al " Chiron1 : A software Architecture for User Interface Development, Maintenence and Run-Time support", ACM Trans. Computer- Human Interaction, 2(2), Jun 1995.

[Whi95] Whitehead, EJ et.al, "Softw re Architectures: Foundation of a Softw re Component Marketplace". D Garlan, Ed., Proc. 1st Intl workshop on Architectures for Software Systems, Apr. 1995.

AN INTELLIGENT TUTORING SYSTEM AND ITS PEDAGOGICAL STRATEGIES

MSc. Lidia L. Elias Hardy and Jose M. Yáñez Prieto
Department of Engineering, Institute of Nuclear Sciences and Technology,
Plaza, Havana, Cuba, AP 6163.
E-mail: lauren@isctn.edu.cu

ABSTRACT

This paper describes some characteristics of an Intelligent Tutoring System oriented to train on the functioning and operation of industrial equipment and written in C++ v. 3.1. Data structures able to organize the expert's knowledge and others which allow to make a distribution of the tutoring knowledge was designed. Based on these structures, the algorithm able to solve the problem of the analysis of the learner's answers, allowing the detection and correction of the errors, and the algorithms, which the system can present the appropriate task for the learner in accordance to his knowledge in each moment, were elaborated. Control strategies applied on the Learning Management Model were defined to allow the introduction of the strategies on the Learning Process. The paper reports the used pedagogical strategies in the Intelligent Tutoring on the functioning and operation of the pressurizer for Nuclear Power Plant with PW reactor type WWER 440.

Keywords: Intelligent Tutoring Systems, pedagogical strategies, progression graph, knowledge base, operation of the pressurizer.

INTRODUCTION

The construction of the new industry and its subsequent operation with quality and efficiency, performing the established norms, requires an appropriate qualification of the personnel who will work in it.

Nowadays different computing systems for the training and certification of the operation personnel are in use. These systems are very expensive and complex and have one main disadvantage: they are unique.

During the last few years simulators and Computer Based Systems without adaptability for the user have been developed [DEF $^+$ 92]. Furthermore, the Artificial Intelligence techniques have been introduced to make these systems more powerful [DEF $^+$ 92, MDL $^+$ 92].

We would like to discuss the issues that have been raised in our attempt to build our ITS. The main goal is to discuss the pedagogical strategies (progression and error strategies) used in the system. The paper reports how we use the pedagogical strategies in an example of the Intelligent Tutoring on the functioning and operation of the pressurizer for Nuclear Power Plant.

GENERAL OVERVIEW

Starting from the idea that each ITS can be divided in two parts: a general and an specific part, we built our model. The general part of the ITS model consists of a set of basic components.

The construction of the ITS model is based on inheritance. The general part consists of a base classes set which respond to the basic components. These classes include every necessary method and structure required for the confection of application. Thus, creating the derivative classes at these base classes, we can elaborate the specific part of this ITS.

The advantages of this model is that, if the derivative classes inherit from the basic components, then the structure and interrelated connections between different components are inherited.

The fundamental characteristics of our model are:
- It solves the tasks with expertise.
- It selects the more adequate pedagogical strategies considering the learner's knowledge.
- It has a graphical environments, which allows the communication of learner with tutoring system by

means of several learner's actions on equipment graphic.
- It contains the student modeling.

LEARNING MANAGEMENT MODEL

Tasks as Basis Units for the Learning Process

The Learning Process is based on propositing the learner exercises for its solution. These exercises we known as TASK. During the design of the specific part, TASK must be defined. For solving the TASK, the learner should use his own knowledge which we want to verify through TASK.

The ITS model allows theoretical exercises as well as practical exercises. The former could be implemented as questions (true/false question, matching question, multiple-choice question and marking question [AS85], while the last one is related with the equipment operation. These practical exercises generate the possible states of the equipment while the learner, considering the initial state and observing the behavior of the parameter values (pressure, temperature, flow, etc.), needs to change the states of the auxiliary devices (valves, heaters, pumps and others) to drive the equipment to stationary state. The exercises have different level of complexity depending on the combination of parameters and equipment states.

The TASKS can be oriented to goal or to states. In the first case, they allow to know if the learner masters the execution of a determined objective whereas in the second case, they allow to know if the learner masters the operation form of the equipment.

In order to offer as many tasks as the learner needs to obtain an appropriate level and for selecting such according the learner's requirements, it is necessary to use a data structure, which allows to make a distribution of tutoring knowledge.

Based on the Levels Graph structure [PL89], a new structure was designed, which is called PROGRESSION GRAPH. This is a specific graph showing several (N,E) pairs, where N is a set of graph nodes, and E is a set of graph links.

The nodes represent the TASKS and we denominate them as TASK-NODES. During the Learning Process, the quantity of TASKS can change allowing more adaptability to each learner. This Progression Graph property is denominated Dynamicity.

The links indicate possible transitions between different tasks, revealing a minimum level of knowledge that a learner has to master in the task-source to receive instruction on task-destination. Graphically this can be represented by an arrow which begins in task-source and finishes in task-destination (see figure 1).That is to say, to receive instruction in T_d, the learner has to solve correctly n_1, n_2, ..., n_k task level T_o. The quantity of links in Progression Graph can also change during the Learning Process.

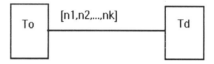

Figure 1. Links between two tasks

Control Strategies

Control strategies are applied in the Learning Management Model to facilitate the guidance of the learner through Progression Graph. This strategies are classified into:
    PROGRESSION STRATEGIES,
    ERROR STRATEGIES
With their combinations during the Learning Process the Pedagogical Strategies defined for each application will be implemented.

Progression Strategies

The progression strategies are established for each node-task, except for terminals nodes and allow to decide the ways to be followed by the learner to advance from Progression Graph's node-base to node-end.

The strategies can be classified into:
    strategy "SOME"
    strategy "EVERY"
    strategy "NEITHER"
The strategy "SOME" requires to overcome the levels of one of possible transitions from current task to anyone preceding task. Firstly the system will find a way not overcome by the learner, and then it will proposes him to solve the corresponding levels. If levels are overcame then he will solve the new task.

The strategy "EVERY" requires to solve corresponding levels for every transition, from current task to every preceding one and will allow the execution of all of them.

The strategy "NEITHER" stops the Learning Process in the current task and it can be used when we want to interrupt or finish the Learning Process in no terminal task.

For example, if T_{i1} is the task which we want to execute (figure 2) and the strategy "SOME" was defined, then it is necessary to overcome levels 1, 2 only. Otherwise if the strategy is "EVERY", then the learner should overcome every T_{ik} levels (at 1 to 6) to continue to T_{j1}, T_{j2} and T_{j3} tasks. If the strategy is "NEITHER" the Learning Process is stopped.

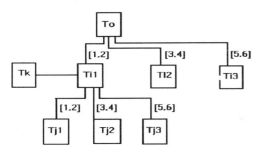

Figure 2. Example of a Progression Graph

Error Strategies

These strategies are established for each node-task and indicate what action the system has to perform when the learner has solved the task incorrectly.

The error strategies can be classified into:

    strategy "LEVEL"
    strategy "ABANDON"
    strategy "DEVIATE"
    strategy "REPEAT"

"LEVEL" allows to find another level not solved of the same task.

"ABANDON" will leave temporally the current task, which was incorrectly answered and will find another Progression Graph's way, which allows to continue to terminal node.

"DEVIATE" allows the deviation of instruction. For this, the system searches for a level not solved of the same task. If this level is not found, then the system searches another task. If this search fails, then the process is stopped.

"REPEAT" performs the current task repetition at the same level, but using a new set of data.

Besides, the model allows the definition of new strategies, joining the existing strategies or creating new ones.

CPTUTOR, AN INTELLIGENT TUTOR FOR TRAINING ON OPERATION OF THE NPP's PRESSURIZER.

The knowledge base is performed on physics and thermodynamics. This theoretical knowledge is represented by backward chaining rules. Those rules describe different kinds of knowledge:

1. Knowledge of the physical relations between the different parameters, which modelize the thermodynamics of the plant;
2. Knowledge of the different automatism that can be triggered, as well as the knowledge that can trigger them;
3. Knowledge of the influences of the different actions and automatism on the various parameters of the plant.

<u>Parameter-Parameter</u> <u>Rule</u> (first type of knowledge)
IF   the reactor coolant temperature increases
THEN the reactor coolant specific volume increases
<u>Automatism</u> <u>Rule</u> (second type of knowledge)
IF   the primary pressure goes to 12,8 MPa
THEN valve 1 should be triggered
<u>Action-Parameter</u> <u>Rule</u> (third type of knowledge)
IF   the spray is started
THEN the primary pressure should decrease

<u>Pedagogical strategies</u>

The goal of CPTUTOR is the training of the learner on the operation of the pressurizer. The learner masters the pressurizer operation if he solves the tasks of functioning the level control system and the different pressurizer systems:

- five groups of sprinkler system's valves
- two groups of blow-off valves

- two groups of safety valves
- five groups of heaters systems.

The tasks and its control strategies for each system were defined.

CPTUTOR trains the learner on the operation form of the equipment considering the parameter values and the states of the auxiliary devices. To reach this goal the learner must master the operation of all auxiliary devices. This is the reason why we have selected "EVERY" as the progression strategy for all tasks, while the error strategy is different for each task. Figure 3 shows the structure of the Progression Graph for CPTUTOR.

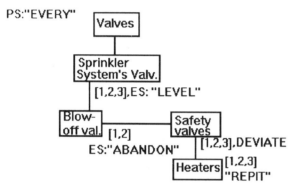

Figure 3. CPTUTOR's Progression Graph

For the sprinkler system's valves, the error strategy "LEVEL" was selected, considering that we have five group of valves and each is triggered at different pressure values. Furthermore, we defined three levels of complexity, where the first level shows the incorrect state only for one group of valves depending of the pressure values; the second level shows the incorrect state of two groups; and the third level shows for three groups of valves.

The rest of the tasks have been designed with the same point of view.

RESULTS

The CPTUTOR was used to evaluate the performance of 10 students of Nuclear Engineering in the Institute of Nuclear Sciences and Technology, in Havana.

CPTUTOR has as a goal to train the learner on operation of the pressurizer systems (pressure and level control systems) when the nuclear power plant works in normal functioning.

The TASKS defined were:

- Depending of the main parameter values and states of the auxiliary devices, keep the pressure on 12,5 MPa paying attention to:
  - Sprinkler system's valves
  - Blow-off valves
  - Safety valves
  - Heaters
- Depending of the main parameter values and states of the auxiliary devices keep the level in the pressurizer.

The results appear in Table 1, where the column Sessions shows the number of working sessions (2 hours duration) where the students had worked to master each task:

Table 1 Results of the students' evaluation using CPTUTOR

Sessions	Number of students mastering the level control system	Number of students mastering the pressure control system			
		Sprinkler system's valves	Blow-off valves	Safety valves	Heaters
1	2	5	5	1	2
2	4	3	2	3	2
3	3	2	3	2	2
4	1	0	0	1	3
5	0	0	0	0	1
6	0	0	0	2	0
7	0	0	0	1	0

Table 1 shows that the students had more difficulties to solve the tasks on operation of the heaters and safety valves. To master the last one, 30% of the students needed more tan five working sessions with CPTUTOR, while the majority of the students worked during three sessions to solve the tasks for each control system.

CONCLUSION

The Progression Graph structure facilitates the distribution of tasks. The algorithm of task's generation and execution, allows to raise the task, that goes with the learner's knowledge state.

To allow the introduction of pedagogical strategies, a group of Control Strategies was defined .

The Control Strategies selected for CPTUTOR have resulted corrects for the defined goal. The students' work shows that the number of sessions necessary to master the pressurizer's systems, should be 3 as minimum.

The ITS Model which has been developed, will allow to increase the quality of operation personnel training.

REFERENCES

[AS85]   Alessi M., Stanley R. Computer-Based Instruction: Methods and Development. Prentice Hall, USA, 1985.

[DEF⁺92] Díaz A., Elorriaga J., Fernández I., Gutiérrez J., Vadillo J. Diagnóstico y estrategias de recuperación en sistemas ICAI aplicados a entornos industriales. *In proceedings of III Congreso Iberoamericano de Inteligencia Artificial*. La Habana, Cuba, Febrero 1992.

[MDL⁺92] Mercier V., Delmas D., Lonca P., Moreau J. SEPIA: An Intelligent Training System for French Nuclear Power Plant operators. *In Proceedings of the NATO Advanced Research Workshop "The use of computer models for explication, analysis and experiental learning"*. Bonas, France, October 1992.

[OT89]   O'Shea T., Self J. Enseñanza y aprendizaje con ordenadores: Inteligencia Artificial en educación. Editorial Revolucionaria, La Habana, 1989.

[PL89]   Prieto, M. Manuel, Loret de Mola, Gustavo. Consideraciones sobre la construcción de Entrenadores basados en Sistemas Expertos Universidad de la Habana, 1989.

THE TRAJECTORY DETERMINATION AND THE KINEMATIC PROBLEM SOLVING FOR A HUMAN ARM MOVEMENT: A NEURAL NETWORK APPROACH

*M. Costa, *P. Crispino, **A. Hanomolo, *E. Pasero
*Politecnico di Torino, Dipartimento di Elettronica
24, Corso Duca degli Abruzzi, Torino, Italy
**Université Libre de Bruxelles, Faculté des Sciences Appliquées
Service d'Automatique, Av. F.D. Roosevelt 50,
1050- Bruxelles, Belgique
e-mail: ahanomol@labauto.ulb.ac.be

ABSTRACT

The article is the result of a preliminary study concerning the possibility of simulating human arm movements. The research work will be continued on a larger scale in the frame of an european project -ANNIE: Artificial Neural Networks for Integrated Ergonomics. The objective of the work was to implement a neural network based methodology for simulating human movements in order to check and ensure ergonomical properties in different environments like work cells, car and spacecraft interiors etc. If most of the studies are concerned with robot like movements, only a few have dealt with human movements. The neural networks potentiality is proved in this case, too: feed-forward nets are employed and the resilient propagation is used as training algorithm.

INTRODUCTION

CAD tools for 3-D modelling can be used to simulate working environments and study ergonomic parameters for a correct positioning of the objects. The best way to verify the man-object interaction is to have a human-like mannequin moving in the environment. Robotic mannequins are available to simulate human actions. The robotic kinematics is well defined by several equations system, but the human movements are not easy to model. The simulation of the mannequin-objects interaction is therefore affected by the lack of precision of this approach. Some works suggest to use Artificial Neural Networks (ANN) to simulate the human movements.

ANN demonstrated their efficiency when used to emulate specific human jobs, such as handwritten recognition, voice recognition etc. The main advantage of these systems is that they don't need a model of the job they have to emulate. An example based training is sufficient to have a network able to replicate the human behavior. The main drawback is that the generalisation of these systems is related to the number of examples you use to train the network. The more examples used the better performance.

The paper is part of a complex work consisting in the simulation of a human arm movement. The objective is to build up a methodology for simulating human arm movements in order to check the ergonomic features of various environments. The work consists in the accomplishment of a procedure which, using experimental data, provides a neural interface able to control a model built in a CAD. In our case the CAD is ROBCAD implemented by Tecnomatix which allows the simulation of 3 dimensional robots.

STATE OF THE ART IN NEURAL NETWORKS APPROACH FOR THE ARM TRAJECTORY FORMATION

Experimental observations of human unconstrained point-to-point reaching movements have indicated that these movements are characterized by straight hand paths [Cru90],[CB87] and symmetric bell-shaped velocity profiles that tend to remain invariant, despite variations in movement direction, speed and initial position.

In order to control voluntary movements one must solve on one hand a kinematic problem and on the other hand, a

dynamic one. The central nervous system must solve the following computational problems:
1. the determination of a desired trajectory in the visual coordinates;
2. the transformation of its coordinates to the body coordinates: THE INVERSE KINEMATICS PROBLEM;
3. the command generation (what muscle forces should be generated): THE INVERSE DYNAMICS PROBLEM [WK93].

When classical methods are used, several differential equations have to be build in order to better simulate the arm movement. Besides the difficulties arising from the finding of the correct equations parameters, one has to deal also with solving equations system which involves huge computations, work with huge matrices and their inverses. Here lies the advantage of neural networks approach which avoids these incoveniences.

Most of the proposed methods are concerned with the resolution of certain models that try to simulate in a better way the human arm behavior. As a matter of fact the work is carried out for the research of the right physical measurements and models which can justify the bell-shaped human arm velocity.

A minimum torque-change model was proposed in [UKS89] which was then improved in [Dor94] by a minimum muscle-tension change model. In [WK93] a model is presented which uses both a forward dynamic model (FDM) as well as an inverse one (IDM), plus a minimum torque-change trajectory generation mechanism. It proved to be very efficient for via points movement. In [FJ95] a cascade of neural networks was proposed in which each network takes care of one aspect of the movement at a time.

In all the studies little is said about the types of the neural networks involved and their characteristics. The best performance index seems to be the one proposed in [WK93]. Only a few have used real data (monkey's arm movement [Dor94]); the methods are rather model driven instead of data driven. Nothing or little [MB89] is said and done about the generalization capabilities of the neural networks and all movements studied are planar.

With all these results in view our first concern was to avoid completely the physical laws and consequently the burden of difficult computations, and to take advantage of the capabilities of the neural networks. First a movement without via points was simulated by means of feed-forward networks, and then a procedure able to use the same neural structures for via point movement was proposed.

MOVEMENT WITHOUT VIA POINTS

Following the computational problems mentioned in the introduction to the paper the first concern was to build the trajectory. The movement was temporally discretized in ten equal intervals. In each interval, the coordinates of the end effector are acquired so that at the end we have the coordinates of 11 points (including the starting and the ending point) - the discrete representation of the desired trajectory.

For what concerns the kinematics of the movement, the angular values of the joints are collected at each time interval, by means of ROBCAD.

These data are obtained from thousands of movements and used as training examples for the neural interface. The interface is made of two neural networks [Cri96]:

- the first one computes the 9 intermediate points having as inputs the starting point, the ending point and the time;
- the second one computes the poses corresponding to the position of the hand joints in the 10 subsequent points of the movement having as inputs the outputs of the first network and the pose of the initial position.

As real data is not yet available, the data used for the training phase of the networks was artificially generated so as to fulfill the two main features of the human arm movement as shown in the introduction. As velocity profile a cosinus was chosen with the particularity that it keeps the peak value for a certain time (T_3).

The maximum arm velocity was chosen $v_{lim} = 3m/s$, the maximum time $T_{max}=5s$ and the maximum admissible trajectory length 2m (the arm length was taken 1m). The expression of the velocity is the following:

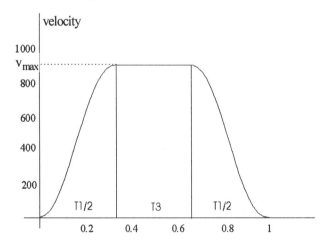

FIGURE 1 Example of a velocity profile

$$v(t) = \begin{cases} \dfrac{v_{max}}{2}\left[1+\cos\left(\dfrac{2\pi}{T_1}\left(t-\dfrac{T_1}{2}\right)\right)\right] & 0\le t\le T_1/2 \\ v_{max} & \\ \dfrac{v_{max}}{2}\left[1+\cos\left(\dfrac{2\pi}{T_1}\left(t-\dfrac{T_1}{2}\right)-T_3\right)\right] & T_1/2+T_3\le t<T \end{cases} \quad (1)$$

The velocity profile for a certain movement will be determined by imposing:

1) the total distance -S- equal to the integral of the velocity:

$$S = \int_0^T \frac{v_{max}}{2}\left[1+\cos\left(\frac{2\pi}{T_1}\left(t-T_1/2\right)\right)\right]dt + T_3 v_{max} \quad (2)$$

Thus, we obtain:

$$v_{max} = \frac{S}{T_1/2+T_3} \quad (3)$$

2) the following relation between time space and velocity:

$$T_3 = \frac{(S/T)^2}{5(S/T)^2+v_{lim}^2} \quad (4)$$

Both for the trajectory building and the kinematics problem feed-forward networks with one hidden layer were employed with sigmoidal activation functions and with the resilient propagation as training algorithm known for its fast convergence. The algorithm consists in defining an update value $\Delta_{k,i}$ (5).

$$\Delta_{k,i}(v) = \begin{cases} \eta^+\cdot\Delta_{k,i}(v-1) & \text{if } \dfrac{\partial F^{(p)}(v-1)}{\partial w_{k,i}}\cdot\dfrac{\partial F^{(p)}(v)}{\partial w_{k,i}}\rangle 0 \\ \eta^-\cdot\Delta_{k,i}(v-1) & \text{if } \dfrac{\partial F^{(p)}(v-1)}{\partial w_{k,i}}\cdot\dfrac{\partial F^{(p)}(v)}{\partial w_{k,i}}\langle 0 \\ \Delta_{k,i}(v-1) & \text{otherwise} \end{cases} \quad (5)$$

where:

$$0\langle\eta^-\langle 1\langle\eta^+$$

and

$$F = \frac{1}{2\cdot P}\cdot\sum_{p=1}^{P}\sum_{i=1}^{m}(d_i^{(p)}-y_i^{(p)})^2 \quad (6)$$

is the cost function, with d -the desired outputs, y - the network outputs, m - the number of outputs and P - the number of training samples.

The weight update rule becomes:

$$w_{k,i}(v+1) = w_{k,i}(v)+\Delta w_{k,i}(v)$$

with:

$$\Delta w_{k,i} = \begin{cases} -\Delta_{k,i} & \text{if } \dfrac{\partial F^{(p)}(v)}{\partial w_{k,i}}>0 \\ \Delta_{k,i} & \text{if } \dfrac{\partial F^{(p)}(v)}{\partial w_{k,i}}<0 \\ 0 & \text{otherwise} \end{cases}$$

The results were tested on an industrial model from the ROBCAD library: a robot with 6 joints The coordinates and the total time were represented as values between 0 and 1. One thousand sets of values were used for the training phase and other several thousands for the validation phase.

For the first network, days were needed for the training phase while for the second one, good results were obtained in weeks. For the first one, the influence of initial conditions, number of hidden layers and of hidden neurons was tested. A network with two hidden layers with 27 and 20 neurons provided a cost function (6) of 152 after 1000 iterations. The validation error was 156. For a network with one hidden layer with 27 neurons 3 training runs were carried on, in which the initial weights were different. In all experiments the validation error was very close to the training error and the best result was obtained after 10000 iterations with a cost function of 22 (figure 2).

For the second neural network, because of the time constraints only one structure was tested: with one hidden layer and 60 neurons. After 3000 iterations the cost function was 64 as was the value of the validation error (figure 3).

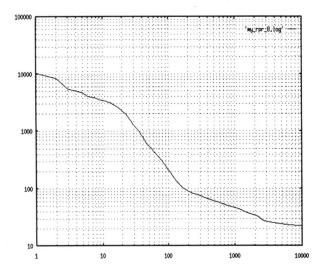

FIGURE 2 The cost function for the trajectory generation

Simulations were carried out on ROBCAD by means of the ROSE environment and the generated movements (with the neural networks) were very close to the real ones.

MOVEMENT WITH VIA POINTS

When obstacles are involved the movement becomes more complicated. Certain space coordinates have to be avoided. In ROBCAD, such a movement " brakes " at the via points (the velocity becomes zero) which is not at all the case in the real life.

As we had promising results from movement without via points it was interesting to find out a way to use the same neural networks to generate more complicated movements. We suppose only one via point whose coordinates are known and through which the hand has to pass. In reality there is a neighborhood of this point that can be passed through so as to avoid the obstacle.

As the neural networks built were provided only for complete trajectories and not for " pieces of them " the idea was to divide the movement in two unconstrained movements. The only known parametres are: the starting, the via and the ending points and the total time. Thus we know the two distances: s_1 - between the initial and the via point - and s_2 - between the via and the final point. We also know that the velocity at the via point (v_v) is the same for both trajectories.

As the available neural network can provide intermediate points only for entire movements and not only for parts of them, we had to build from the known parameters two complete trajectories characterized by the total length: $S_i = s_i + d_i$ and the total time T_i. These values allow the computation of the final coordinates of the first subtrajectory and of the initial coordinates for the second subtrajectory.

As the determination of the movement parameters leads to the solving of an undetermined system of equations different conditions had to be imposed. To relations (1) to (4) we have added the space constraint: both movements have to end, respectively, begin inside the hemisphere in which the movement is allowed (the radius = the arm length). More precisely, it was imposed that the trajectories end or begin on the border of the definition area. The total times T_1 and T_2 are determined by means of the Newton-Raphson method.

As data was created artificially, no real data being available at that moment, we had to check the constraints connected to velocity: v_{lim} and time: T_{max}. Obviously, some of the movements proved to be inconsistent with all these and we had to eliminate them.

Once we had the total description of both trajectories - initial and final coordinates and total time- we could use two neural networks trained for simple movements, for the trajectory determination. The intermediate points generated by both networks represented the trajectory followed by the end effector.

From the outputs of the network which has as inputs the initial, final coordinates and total time for the trajectory until the via point, all the intermediate points beginning with the first and ending with the one closest to the via point were kept. From the outputs of the network which has as inputs the initial, final coordinates and total time for the trajectory from the via point, all the intermediate points beginning with the one closest to the via point and ending with the last were kept. For the total movement all the intermediate points from both networks are used. The error was very small. Better results can be obtained if the network is trained so as to generate more intermediate points. In this case there will be more chances of falling into a point closer to the via point.

CONCLUSIONS

In the context of human arm movement simulation by means of neural networks a strategy was proposed that uses less computation effort and less knowledge about the movement itself than previous studies did. Also, the research was no longer reduced to planar movements, but spatial movements were considered. Two networks were trained for simple unconstrained movements, one for the trajectory building, the other for the solving of the kinematics problem. A resilient propagation algorithm was used as training algorithm and the generalisation power was tested with success on several thousands of examples. More neural structures should be tried, especially for the kinematics problem.

The same networks were used for a movement with one via point. In order to build a whole trajectory from partial

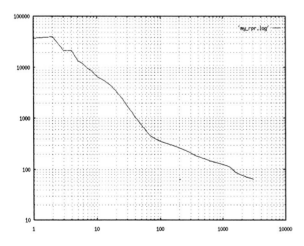

FIGURE 3 The cost function for the kinematics problem

trajectories, the movement was divided in two unconstrained movements and a geometrical approach was employed. As little is known about the control algorithms used by the human brain in performing the obstacle avoidance certain constraints were imposed in order to obtain from partial movements entire movements (characterised by the bell-shaped velocity profile (fig.1)). But when these constraints were added to those connected to velocity, time and space some of the generated movements proved to be impossible to accomplish.

As stated in the beginning, this work is to be continued in the frame of an European project where real data will be available. It is expected that a better strategy will be implemented for via point movements. The idea of using the same neural structures as for simple movements should remain but a better method should be sought for the determining whole trajectories in order to avoid unrealistic situations. Up to now, the results are encouraging; once trained, the networks provide the correct trajectories and if more intermediate points are generated performance will surely improve.

REFERENCES

[CU93] A. Cichocki, R. Unbehauer - Neural networks for optimisation and signal processing, J. Wiley & sons Publ. 1993

[Cri96] P. Crispino - Progetto e realizzazione di un'interfaccia neurale per la simulazione del movimento con tecniche CAD, Tesi di laurea, 1996

[Cru90] H. Cruse, et al - On the cost functions for the control of the human arm movement, *Biol. Cybernetics*, 62, pp. 519-528, 1990

[CB87] H. Cruse, M. Bruwer - The human arm as a redundant manipulator: the control of path and joint angles, *Biol. Cybernetics*, pp.137-144, 1987

[Dor94] M. Dornay et al - Simulation of optimal movements using the minimum-muscle-tension-change model, *NIPS*, Morgan Kaufmann Publishers, pp.627-635, 1994

[FJ95] Tamar Flash, M.I. Jordan - Computational schemes and neural network models of human arm trajectory control

[MB89] L. Massone, E. Bizzi - A neural network model for limb trajectory formation, *Biol. Cybernetics* 61, pp. 417-425, 1989

[UKS89] Y. Uno, M. Kawato, R. Suzuki - Formation and control of optimal trajectory in human arm movement, *Biol. Cybernetics,* 61, pp. 89-101, 1989

[WK93] Y. Wada, M. Kawato - A neural network model for arm trajectory formation using forward and inverse dynamics models, *Neural Networks*, vol. 6, pp. 919-932, 1993

A Neural Network Velocity Estimator for Discrete Position Data

S. P. Chan

School of Electrical and Electronic Engineering
Nanyang Technological University
Block S1, Nanyang Avenue
Singapore 639798
Republic of Singapore
e-mail: espchan@ntu.edu.sg

ABSTRACT

A method of using neural network to estimate the velocity signal of robotic joint from discrete position versus time data is proposed and evaluated. The architecture of the neural net and the training methodology are presented and discussed. Based on computer simulations, comparison of the accuracy of the neural network estimator with two other well established velocity estimation algorithms are made. The neural net approach is able to maintain good performance even in the presence of measurement errors.

INTRODUCTION

In the industry, many motion control systems are controlled digitally using DSP, microprocessor or computer. With these systems, optical incremental encoders are widely used to monitor the output position for feedback control purpose. Some of the most important areas of incremental motion control systems are in computer peripherals, such as printers, disk drives, and numerical control equipment such as NC machines and robots.

One technique of obtaining position and velocity information requires the use of two separate sensors: an encoder and a tachometer. An obvious disadvantage of this approach is the extra hardware, the tachometer. It will increase the total system cost. In robotic, two extremely important considerations often dictate against using a separate tachometer. They are the weight and space requirements which affect the size and cost of the manipulator mechanical structure. In certain instances, it is not possible to mount additional sensors on the joint motor shaft.

Mathematically, velocity is the time derivative of position. Instead of using direct measurement, velocity can be estimated from discrete position versus time information provided by an incremental encoder. The problem of generating derivatives of measured data when appropriate transducers and/or observers are not available has been reviewed by Harrison [HM93]. Six methods or schemes by using the central finite-difference methods are described. Each scheme is subject to two principal sources of error: noise or quantized data and the presence of high-order derivatives of motion.

In essence, the velocity is estimated by performing an approximate derivative operation on the discrete data. Many designs of discrete-time derivative filters exist today. Unfortunately, most of these are unsatisfactory for control applications as delay inherent to these derivative filters adversely affects stability. Furthermore, it is well known that derivative operations tend to magnify errors. However, several methods have been developed to estimate velocity from discrete data while reducing estimate errors and minimizing the delay. Some existing velocity estimator algorithms are reviewed in [BSM92]. With the aid of computer simulations, the accuracy of these algorithms are evaluated, including the dynamic range and transient response. The major conclusion to these analyses is the choice of "best estimator" to use is, at best, application dependent.

Research in artificial neural networks has recently been active as a new means of information processing. Artificial neural networks try to mimic the biological brain neural networks into mathematical model. In general, they consist of a collection of simple nonlinear computing elements whose inputs and outputs are tied together to form a network.

The main advantage of neural networks for robotics is their ability to adaptively learn nonlinear functions whose analytic forms are difficult to derive and whose solutions are hard to compute. The most dominant forms of neural networks used in robotics are the multi-layer perception and the Hopfield network. The neural network has been used for the calculation of the kinematics and the inverse kinematics statically and nonlinearly [GA88]. The neural network is also used for the control dynamically as a controller [MKS+88, KUI+88], a compensator for uncertainty and nonlinearity [Cha95], and a tool for identification. In robotic sensing, the use of neural networks in vision has been quite extensive. However, not much have been done on the use of neural networks in other areas of sensing.

In this work, it is proposed that a neural network can be used to estimate velocity information from the joint position data as measured by an optical incremental encoder. That is to say, the neural network is trained to perform the equivalent mathematical operation of calculating the time derivative of measured data. The estimated velocity signal can then be incorporated into a suitable control law which enables the robot to track the desired motion trajectory.

ROBOTIC ASSEMBLY

Precision position control is a prerequisite for the application of robot to assembly operations in industry. The most common assembly task involves the insertion of one part (the peg) into another part (the hole). Typical example is the insertion of electronic components onto a circuit board. The robot is taught the desired points and the sequence of motions. The reproducibility of the task depends upon the repeatability of the robot motion and the equipment setup used. To facilitate automatic assembly, the parts should be designed for automation such as holes with chamfers and mating parts with proper tolerance. Interactions often take place between the parts being assembled. Factors such as changes in the workspace, and variations in the pose and dimensions of the workpieces introduce possible positioning errors of peg and hole alignment.

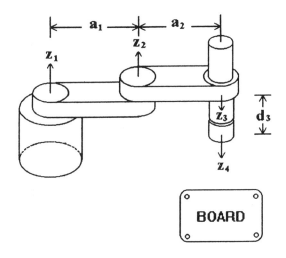

FIGURE 1 Schematic of a SCARA robot

Some applications also require a downward force to be exerted so as to ensure proper locking of the parts. These potential interactions make accurate position control in assembly more difficult compared to the simple pick-and-place operations.

A reliable robotic assembly system has to be able to accommodate uncertainties in the environment. SCARA (selective compliance assembly robot arm) configuration is a special type of robot developed primary for assembly tasks [MF82]. A SCARA robot has four degrees of freedom (Figure 1). Horizontal rotations of the first two joints determine planar position of the gripper. The third joint controls the vertical (z-axis) motion of the end-effector. The last degree of freedom decides the orientation of the gripper about the z-axis. Since this configuration provides compliance in the horizontal plane and substantial rigidity in the vertical direction, this robot is ideal for many light-duty assembly tasks. The dynamic equation of motion in the joint space can be written in a component form [CL94]

$$(1) \quad \begin{bmatrix} P_1 + P_2 c_2 & P_3 + 0.5 P_2 c_2 & 0 & -P_5 \\ P_3 + 0.5 P_2 c_2 & P_3 & 0 & -P_5 \\ 0 & 0 & P_4 & 0 \\ -P_5 & -P_5 & 0 & P_5 \end{bmatrix} \begin{bmatrix} \ddot{q}_1 \\ \ddot{q}_2 \\ \ddot{q}_3 \\ \ddot{q}_4 \end{bmatrix} + \begin{bmatrix} -0.5 P_2 s_2 \dot{q}_2 & -0.5 P_2 s_2 (\dot{q}_1 + \dot{q}_2) & 0 & 0 \\ 0.5 P_2 s_2 \dot{q}_1 & 0 & 0 & 0 \\ 0 & 0 & 0 & 0 \\ 0 & 0 & 0 & 0 \end{bmatrix} \begin{bmatrix} \dot{q}_1 \\ \dot{q}_2 \\ \dot{q}_3 \\ \dot{q}_4 \end{bmatrix}$$

$$+ \begin{bmatrix} P_6 \dot{q}_1 \\ P_7 \dot{q}_2 \\ P_8 \dot{q}_3 \\ P_9 \dot{q}_4 \end{bmatrix} + \begin{bmatrix} P_{10} \text{sgn}(\dot{q}_1) \\ P_{11} \text{sgn}(\dot{q}_2) \\ P_{12} \text{sgn}(\dot{q}_3) \\ P_{13} \text{sgn}(\dot{q}_4) \end{bmatrix} + \begin{bmatrix} 0 \\ 0 \\ -P_4 g \\ 0 \end{bmatrix} = \begin{bmatrix} \tau_1 \\ \tau_2 \\ \tau_3 \\ \tau_4 \end{bmatrix}$$

where q_i and τ_i are the i-th components of the joint displacements and joint torques/forces respectively. The parameters $P_1, ..., P_{13}$ are physically related to the mass, moment of inertia, mass centre, length, coefficient of viscous friction, and Coulomb friction associated with each link. The term g is the acceleration constant due to gravity and sgn is the sign function. The terms c_2 and s_2 are the functions $\cos(q_2)$ and $\sin(q_2)$ respectively.

An assembly operation is performed by first moving the gripper to a position directly above the nominal insertion point with proper orientation. The insertion stroke requires the gripper to be lower from an initial point along the z-axis according to the planned motion trajectory until it reaches the goal point where the component is properly seated. During this step, a trajectory consisting of three segments, namely, the approach phase, the insert phase, and the set-down phase can be adopted. From the initial point, the gripper rapidly approaches an assembly location which is just above the insertion point. During the insert phase, the gripper travels with a reduced constant velocity so that the components can be inserted successfully. Finally it comes to a complete stop at the terminal position in the set-down phase.

The present generation of robotic manipulators use hard control algorithms at the direct digital control level. As robot consists of a finite number of mechanical chains, it is a multi-variable nonlinear coupled system. The solution of this problem is difficult because even the simplest movement requires sophisticated and computational intensive mathematics [LWP80]. The well known computed torque control [Mar73] and other model based control techniques [AAG+89] require joint velocity information. Unfortunately, most industrial robots are not equipped with sensors to measure the actual joint velocities.

NEURAL NETWORK VELOCITY ESTIMATOR

There has been considerable research interest in incorporating "intelligent" into robot control systems. Intelligent can be introduced directly into the controller of a robot, or may be distributed in components such as sensors, end-effectors, and programmable fixtures. Compare to other velocity estimation techniques, a neural network has several desirable characteristics. It can learn from training data and can map nonlinearly. Hence it is highly plausible that a neural network can be trained to learn the mathematical function of time derivative operation. Furthermore it is well known that the accuracy obtained from performing derivative operation on measured data is sensitive to measurement noise. Nevertheless, the structure of a neural network is

robust for noise and it has generalization capability. Therefore its performance will not be adversely affected by measurement noise and it can even interpolate data which it may not have learnt before. The approach proposed here is to teach a neural network until it learns the relation between successive position data which are separated by equal sampling period, T and the corresponding time derivative of the position data. Once trained, the neural network can then process very efficiently the computation of time derivative on measured position data, i.e. the velocity.

The structure of the velocity estimator consists of a multilayer network which is composed of input, hidden, and output layers as shown in Figure 2. A preconditioning module together with normalizing and de-normalizing modules serve as input and output buffers. The network has four input cells and one output cell. The number of nodes in the hidden layer is chosen to be five. Units with nonlinear sigmoid transfer characteristic are used for both the output and hidden layers. Learning proceeds with the error propagation algorithm of the generalized delta rule according to Rumelhart et al [RHW86].

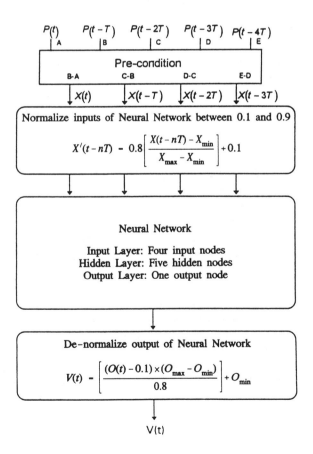

FIGURE 2 Structure of the neural network estimator

The most straightforward approach for training the neural network is to use directly the few latest absolute positions as it inputs and the corresponding expected value of velocity as the output. Unfortunately, this approach requires the network to learn not only the time derivative operation but also the dependency of the data on the absolute position of the joint in the entire workspace. To obtain precision with such a network, a tremendous amount of training data would be required. To avoid this difficulty and reduce the training time, only the variations of the position data are fed into the network inputs. The preconditioning module is inserted between the position encoder output and the network inputs. It computes the position variations between the latest sampled data. Hence the network is totally independent of the absolute position of the joint and memorizes only the fundamental relationship containing time derivative.

Furthermore, the data processed by the neural net should be normalized such that the maximum normalized value is lower or equal to the activation function maxima. For this study, a unipolar sigmoidal activation function covering the range from 0 to 1 is used. Each displacement data computed by the preconditioning module is normalized to stay within the range of 0.1 and 0.9. The normalized values are fed directly to the inputs of the network. The normalizing process prevents using the extreme values of the sigmoid function. This helps to reduce the time required during training to converge to a required error level. A similar de-normalizing process is used to convert the result of the network output back to real world amplitude.

NEURAL NETWORK TRAINING SCHEME

The neural network is trained to estimate velocity from discrete position data. To train the network, a set of input patterns containing discrete position data and output pattern of the corresponding velocity must be available. The training data are specially designed to train the neural network in a generalize form. If the network is not trained to generalize, it may not predict the correct output if the input conditions are different. Thus, the neural network needs to be re-trained again with the modified condition.

Training of the network is conducted off-line using data obtained from a specially designed motion trajectory. It is desirable that the training motion is smooth, i.e., it is continuous with a continuous first derivative (velocity), and a continuous second derivative (acceleration). Furthermore, the training data must cover the ranges of position, velocity and acceleration which are within the normal operating condition of the robot.

Taking these factors into consideration, a position trajectory which consists of six segments of quintic polynomials [Cra89] is adopted. For each path segment, the position trajectory is of the form

$$P(t) = Pos(t)$$
$$= a_0 + a_1 t + a_2 t^2 + a_3 t^3 + a_4 t^4 + a_5 t^5 \qquad (2)$$

The corresponding velocity and acceleration along this path segment are

$$\dot{P}(t) = Vel(t)$$
$$= a_1 + 2a_2 t + 3a_3 t^2 + 4a_4 t^3 + 5a_5 t^4 \qquad (3)$$

and

$$\ddot{P}(t) = Accel(t)$$
$$= 2a_2 + 6a_3 t + 12a_4 t^2 + 20a_5 t^3 \qquad (4)$$

The path constraints are

Initial position: $Pos(0) = P_0 = a_0$

Final position: $Pos(t_f) = P_f = a_0 + a_1 t_f + a_2 t_f^2$
$$+ a_3 t_f^3 + a_4 t_f^4 + a_5 t_f^5$$

Initial velocity: $Vel(0) = \dot{P}_0 = a_1$

Final velocity: $Vel(t_f) = \dot{P}_f = a_1 + 2a_2 t_f + 3a_3 t_f^2$
$$+ 4a_4 t_f^3 + 5a_5 t_f^4$$

Initial acceleration: $Accel(0) = \ddot{P}_0 = 2a_2$

Final acceleration: $Accel(t_f) = \ddot{P}_f = 2a_2 + 6a_3 t_f$
$$+ 12a_4 t_f^2 + 20a_5 t_f^3$$

$$(5)$$

Each path segment will be evaluated over a time interval from 0 to t_f. Using a quintic polynomial, it is possible to specify the position, velocity, and acceleration at the beginning and at the end of each path segment so as to achieve desired continuous position, velocity, and acceleration trajectories. The entire training motion is of 2 s duration. The specifications at the beginning and at the end of each path segment are given in Table 1.

The path equations are evaluated at discrete sampling periods to generate consecutive discrete position data points and velocity data points. A sampling period of 2 ms is adopted and hence there are 1000 sets of training data. Each set of training data consists of an ordered arrangement with five position data points and one velocity data point. At any sampling time, the current position together with the four previous position points which are each separated by one successive sampling period serve as the inputs to the estimator. The current velocity is taken as the expected estimator output. Figure 3 shows the learning scheme for training the neural network.

Table 1 Specifications of path segments

Time (s)	0	0.25	0.75	1.0	1.25	1.75	2.0
Accel (mm/s²)	0	500	-500	0	-500	500	0
Vel (mm/s)	0	62.5	62.5	0	-62.5	-62.5	0
Pos (mm)	0	4.6875	60.9375	65.625	60.9375	4.6875	0

The neural network is trained using the backpropagation algorithm. The sets of training data are presented to the network sequentially. Training of the proposed neural net with the entire 1000 data sets require a large number of epochs to obtain the acceptable error level. Since network learning is conducted off-line, there is no practical constraint on the training time. In this study, training terminates when the mean square error (MSE) is less than 0.000009.

APPLICATION TO ROBOTIC ASSEMBLY

The application under study requires the robot to insert a PCB into an edge connector. The edge connector used in the study is a 31-way connector. It consists of two rows of spring contacts which are designed to hold an inserted board firmly for good conductivity. The connector height $h = 15.5$ mm, width $w = 9.14$ mm, overall length $l_o = 85.3$ mm, board aperture $l_a = 81.3$ mm, board insertion depth $h_d = 7.6$ mm, and inner width of the socket $w_i = 2$ mm. The gap w_g between the two rows of spring contacts is approximately 0.6 mm and the thickness of the PCB to be inserted is 1.6 mm.

The PCB which is attached to the gripper is positioned vertically above the nominal position of the edge connector. The insertion task requires the gripper to be lowered from an initial point along the z-axis according to a planned motion trajectory until the PCB reaches the goal point at the bottom of the socket in the connector. The desired assembly motion trajectory consists of three segments of quintic polynomial

$$z_d(t) = \begin{cases} 974.33t^3 - 1794.91t^4 + 884.05t^5 & :0.0 \leq t < 0.81, \\ 53.40 + 5.0(t - 0.81) & :0.81 \leq t < 1.86, \\ 58.65 + 5.0(t - 1.86) - 272.21(t - 1.86)^3 \\ \quad - 1241.36(t - 1.86)^4 + 1438.82(t - 1.86)^5 & :1.86 \leq t \leq 2.22, \end{cases}$$

(6)

where $z_d(t)$ is in unit of mm. From the initial position (0 mm), the end-effector moves vertically downward to the goal position (61 mm) in 2.22 seconds. The first segment from 0 mm to 53.40 mm is the approach phase having the fastest motion with a maximum velocity of 121.44 mm/s and a maximum acceleration of 458.27 mm/s^2. This is followed by the insert phase from 53.40 mm to 58.65 mm during which the PCB pushes open the spring contacts. A slow constant velocity profile is selected to avoid producing a large impact force which might lead to oscillation and result in loss of control. In the set-down phase from 58.65 mm to 61 mm, the board comes to a stop at the bottom of the socket.

It is interesting to note that the neural network has never been trained with the desired trajectory (6). However, this trajectory consists of velocity and acceleration which are within the ranges of values in the special training data. If the network learnt the mapping properly, its generalizing ability should enable it to give the correct output. Hence it is reasonable to believe the network is capable of providing the appropriate velocity estimation during the actual insertion process.

To investigate the performance of the neural network velocity estimator, the assembly motion trajectory is evaluated at sampling intervals of 2 ms to generate discrete position data points. At a particular sampling time, the current position point together with the four position points obtained at the pervious four successive sampling are fed to the neural network estimator as

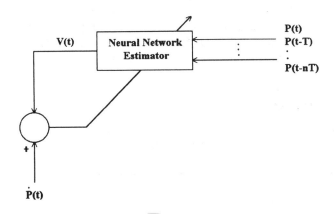

FIGURE 3 Training of the neural network estimator

inputs. Output of the estimator which is the estimated velocity is compared with the value of velocity calculated at the same sampling time using the desired velocity trajectory equation. The different between the calculated velocity and the estimated velocity is taken as the estimation error at that sampling time. The root mean square (RMS) value of the estimation errors for the 2.22 s trajectory is then computed which is equal to 4.79444×10^{-4} m/s.

PERFORMANCE COMPARISON

Performance of the neural network velocity estimator is compared with two other well established velocity estimation algorithms, namely the finite difference method [BSM92] and the differential filter technique [Kho88]. The finite difference method estimates the velocity using

$$Vel(t) - \frac{P(t) - P(t-T)}{T} \qquad (7)$$

where T is the sampling time, $P(t)$ and $P(t-T)$ are two position points separated by one sampling period. The differential filter is described by the difference equation

$$Vel(t) - \frac{-2P(t-4T) - P(t-3T) + P(t-T) + 2P(t)}{10T} \qquad (8)$$

where four position points, $P(t)$, $P(t-T)$, $P(t-3T)$, and $P(t-4T)$ are used.

The position data points obtained from the assembly path trajectory (6) serve as the test data. Estimates of velocity are then obtained from the application of (7) and (8). The actual velocity is calculated using the velocity trajectory equation which is obtained by differentiating (6). The RMS errors between the calculated velocity and the estimated velocity is then computed. The results are summaries in column two of Table 2 for the case without noise. The error comparison tends to suggest that finite difference method

provides the best velocity estimation under ideal condition. However, the actual position data obtained from an optical incremental encoder will be affected by measurement errors due to noise and quantization. The most significant error source is position measurement truncation. Position measurement truncation errors occur because position can only be measured as an integer number of lines, thus causing the truncation of the position measurement. Additional position measurement errors can occur due to imperfections in the encoder, i.e., the encoder lines are not equally spaced in position. Typically, deviation of the measured position data from the ideal case will not exceed one position count.

For the z-axis of the SCARA robot under consideration, the spacing between two adjacent encoder lines corresponds to a displacement of 0.0025 mm. Since the maximum error which can occur in the position counts is considered to be ±1 count, the encoder error is translated to a position error of ±0.0025 mm. In this study, a random number generator is utilized to generate random numbers which are between -0.0025 mm and 0.0025 mm. At each sampling period, the random number generated will be added to the position data calculated from the motion trajectory equation to simulated the effect of measurement noise. The noisy position data points are then fed to the three velocity estimators for comparison. The RMS velocity error for each estimator is computed and the results are summaries in column three of Table 2.

The results obtained are quite interesting. The neural network estimator produces the best estimation result with the smallest RMS error. Whereas differential filter technique is not sensitive to measurement noise and maintains more or less the same error performance. This feature is probably due to its build-in filter characteristic which can minimize the effect of noise. On the other hand, performance of the finite difference method is adversely affected. Its RMS error becomes the largest, as the measurement errors directly affect the velocity estimation. For the neural network estimator, there is a small degradation in the performance. It is important to realise that the neural network estimator has never been trained with the noisy position data. However its

Table 2 Velocity estimation errors

Velocity Estimator	RMS Error (m/s) Without Noise	RMS Error (m/s) With Noise
Neural Network	4.7944×10^{-4}	6.1959×10^{-4}
Finite Difference	1.9864×10^{-4}	10.0×10^{-4}
Differential Filter	7.9716×10^{-4}	8.1778×10^{-4}

structure is quite robust to noise and hence it is able to maintain reasonable performance.

CONCLUSION

In this paper, neural network is applied to estimate the velocity of robotic joint using discrete position versus time measurements. An interesting approach which consists of making the neural network process only relative variations instead of absolute position is introduced. It has proved to be effective in reducing the training effort. The computer simulation study for a typical assembly motion trajectory shows the neural net can make velocity estimation with good accuracy. It is capable to generalize even for un-train inputs. Furthermore, performance of the proposed neural net estimator is found to be robust to measurement errors.

REFERENCES

[AAG+89]   C. H. An, C. G. Atkeson, J. D. Griffiths, and J. M. Hollerbach. Experimental evaluation of feedforward and computed torque control. *IEEE Trans. Robot. Automation.*, vol. 5, pp. 368-373, June 1989.

[BSM92]   R. H. Brown, S. C. Schneider, and M. G. Mulligan. Analysis of algorithms for velocity estimation from discrete position versus time data. *IEEE Trans. Industrial Electronics*, vol. 39, no. 1, pp. 11-19, 1992.

[Cha95]   S. P. Chan. A neural network compensator for uncertainties in robotic assembly. *J. of Intelligent and Robotic Systems*, vol. 13, pp. 127-141, 1995.

[CL94]   S. P. Chan and H. C. Liaw. A parameter estimation technique for SCARA robots. In *Proc. 3rd Int. Conf. Automation, Robotics and Computer Vision (ICARCV'94)*, Singapore, Nov. 1994, pp. 1487-1491.

[Cra89]   J. J. Craig. *Introduction to Robotics*, 2nd ed. Addison-Wesley, Reading, MA, 1989.

[GA88]   A. Guez and Z. Ahmad. Solution to the inverse kinematics problem in robotics by neural networks. In *Proc IEEE Conf. on Neural Networks*, San Diego, July 1988, vol. II, pp. 617-624.

[HM93]   A. J. Harrison and C. A. McMahan. Estimation of acceleration from data with quantization errors using central finite-difference methods. *Proc. Instn. Mech. Engrs.*, vol. 207, part 1, pp. 77-86, 1993.

[Kho88]   P. K. Khosla. Estimation of robot dynamic parameters: Theory and application. *Int. J. Robot. Automat.*, vol. 3, no. 1, pp. 35-41, 1988.

[KUI+88]   M. Kawato, Y. Uno, M. Isobe, and R. Suzuki. Hierarchical neural network model for voluntary movement with application to robotics. *IEEE Control Systems Magazine*, vol. 8, pp. 8-17, 1988.

[LWP80]   J. Y. S. Luh, M. W. Walker, and R. P. C. Paul. On-line computational scheme for mechanical manipulators. *J. Dynamic Systems, Measurement, and Control*, vol. 102, pp. 69-76, June 1980.

[MAR73]   B. Markiewicz. Analysis of the computed torque drive method and comparison with conventional position servo for a computer-controlled manipulator. *Technical Memo 33-601*, Jet Propulsion Laboratory, Pasadena, CA, 1973.

[MF82]   H. Makino and N. Furuya. SCARA robot and its family. In *Proc. 3rd Int. Conf. Assembly Automation*, W. Germany, May 1982, pp. 433-444.

[MKS+88]   H. Miyamoto, M. Kawato, T. Setoyama, and R. Suzuki. Feedback-error-learning neural network for trajectory control of a robotic manipulator. *IEEE Trans. Neural Networks*, vol. 1, pp. 251-265, 1988.

[RHW86]   D. Rumelhart, G. E. Hinton, and R. L. Williams. Learning internal representations by error propagation. In D.E. Rumelhart and J.L. McClelland (eds), *Parallel Distributed Processing: Explorations in the Microstructure of Cognition. Vol. 1: Foundations*, MIT Press, Mass. 1986.

USING AN ON-LINE ARTIFICIAL NEURAL NETWORK TO DIAGNOSE REAL TIME TRANSIENT CONDITIONS IN A CRITICAL, COMPLEX SYSTEM

Peter Weller[1,2], Paul Hutchison[3], Alex Thompson[2], Rick Carrick[2]
1. Department of System Science, City University,
Northampton Square, London, EC1V 0HB, UK.
email. p.r.weller@city.ac.uk

2. Department of Nuclear Science & Technology,
Royal Naval College, Greenwich, London, UK.

3. Directorate of Nuclear Propulsion,
MOD (Procurement Excutive),
Abbeywood, Bristol, UK.

ABSTRACT

This paper reports on an investigation [Hut96] into using an on-line artificial neural network (ANN) to diagnose real time transients as part of an advisor to operators of a complex system, a nuclear reactor. An ANN is first developed to successfully diagnose six plant conditions using a selected input set of twelve reactor variables for three time steps. Once developed this network is embedded in a reactor simulator program to produce on-line diagnosis. The results show that not only can the system successfully diagnose transients quicker than a human operator but it can produce some unexpectedly accurate diagnosis of multiple faults.

KEYWORDS

Artificial Neural Networks, Diagnosis, Complex Systems, Nuclear Reactors

INTRODUCTION

In many complex, critical systems, typified by a nuclear reactor, a potentially catastrophic situation can develop very quickly. It is important that such systems are constantly monitored and fault situations identified promptly so action can be taken before serious damage occurs. The operators of such systems are themselves under constant pressure and factors such as stress and reaction time can be seriously impaired by extraneous conditions [Hut96]. A tool that can support the operators in their role must be of benefit and worthy of investigation. Artificial Intelligence techniques, and Artificial Neural Networks (ANNs) in particular, would seem ideal for this application.

Research in ANN applications for the nuclear industry has progressed steadily over a number of years. Several of the applications reported concern the diagnosis of nuclear plant condition. Plant variables are used as inputs to an ANN which has been trained on a series of transient conditions. The output of the network can provide a guide to the present state of the reactor [BB94, BU92]. Due to the nature of nuclear reactors, computer simulations of the plant are used to gather data and test results. The output from these simulators is free from interference such as sensor accuracy. It is therefore normal to introduce a random noise into the data to simulate a more realistic environment.

This paper reports on a continuation of this idea with the development of an operators advisor [Hut96]. An ANN is first developed to diagnose a small number of reactor transients. The program code of the resulting network is then embedded into a computer simulation of a nuclear reactor to provide an on-line, real time diagnostic tool. This system is rigorously tested for a range of conditions, not included in the original ANN training data, to explore the versatility and response times of the system and the applicability of ANNs to on-line transient identification.

The remainder of this paper is as follows. A brief introduction to pressurised water reactors (PWRs) is followed by a description of the transients and reactor variables selected for use in the ANN. The methods used to develop the ANN are described and initial results are given. The process for embedding the final ANN into the PWR simulator is explained. Results are reported for the testing of this system. Finally further developments of the work are discussed.

THE PRESSURISED WATER REACTOR (PWR)

The Pressurised Water Reactor (PWR) is one of the world's most popular design for nuclear reactors. A simplified diagram, Fig 1, showing the main components is given below.

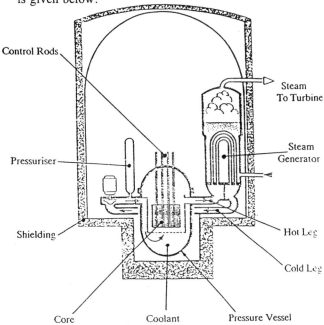

Fig 1: Diagram of PWR

The whole assembly is enclosed in heavy shielding. The main structure is a large pressure vessel made of welded steel. It has a lid which is secured to the vessel by a ring of heavy bolts. The pressure vessel contains the reactor core and the control rods. The core is composed of a lattice of fuel elements containing enriched Uranium 235. Control rods are mounted in the top of the vessel. In an emergency these can be driven down into the core and control the fission process. The remaining volume in the pressure vessel is filled with water under high pressure. The water is used as coolant and a moderator. Heat is generated by the fission of the Uranium and is absorbed by the circulating water.

The water leaves the pressure vessel through pairs of heavy pipes welded to the vessel. One pipe carries heated water away from the core and is referred to as the 'hot leg'. The other pipe carries water back to the core and is called the 'cold leg'. Each reactor has two or more loops of these pairs of pipes. The hot leg takes the heated pressurised water to a steam generator. The generator is a heat exchanger composed of many small tubes surrounded by water. These tubes are filled with heated water from the core. The pressurised primary circuit water inside the tubes cannot boil, but the water in the shell surrounding the generator tubes does. The steam produced is then processed and used in a turbine to produce electricity or propulsion power. The cooler pressurised water is then pumped back to the pressure vessel through the cold leg.

One of the pressurised water loops contains a pressuriser which evaporates a quantity of coolant to maintain the water pressure in the primary vessel. This component is also used to compensate for the unexpected pressure changes resulting from a transient or fault condition. Other very important components of a PWR are the safety systems. The pressurised nature of the system means that if a leak occurs in one of the primary pipes, a large amount of the cooling water may be lost very quickly. Continued cooling must be maintained to prevent potential core damage.

DEVELOPMENT OF THE DIAGNOSTIC ANN

The diagnostic ANN was developed in the following manner. A range of PWR transients was first chosen for the ANN to diagnose. Secondly, a set of reactor variables were selected that could be used to determine change of conditions in the PWR. The ANN training, test and validation sets were then produced and these were used to develop a range of back-propagation ANNs. The best of these networks, in terms of lowest RMS error, was used for the remainder of this investigation.

A bespoke computer simulator of a nuclear reactor was used to produce a selection of six transient conditions on which to develop the diagnostic ANN. These scenarios were chosen to provide a range of both similar and contrasting results on which to test an ANN. The steady state condition was included as one of the transient conditions in order to report a no fault output. The final group of transients chosen were:
     a) Steady State Condition
     b) Fast Primary Coolant Leak
     c) Downstream Steam Leak
     d) Throttle Opening Transient
     e) Single Rod Drop
     f) Group Rod Drop

These conditions were each represented as a binary value for the output of the ANN. A '1' signified the presence of a transient while a '0' indicated absence. Each transient was initiated from identical steady state conditions.

The PWR variables used for the ANN input set for each of the transients was selected by a domain expert as the most relevant for diagnosing plant condition. From a maximum of 108 measurable plant variables, 12 were identified as the most important. The final set of selected variables included temperatures, pressures and other information on the pressuriser, steam circuit and reactor core.

The following graphs, Figures 2 to 7, show the relationships and rate of change of the selected variables for each transient considered. The numbers on the right hand side of each graph identify the PWR variable.

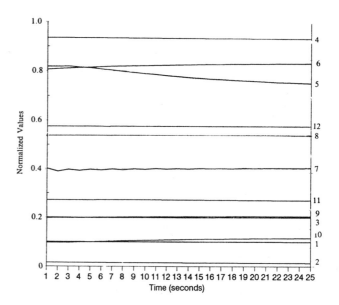

Fig 3: Fast Primary Coolant Leak

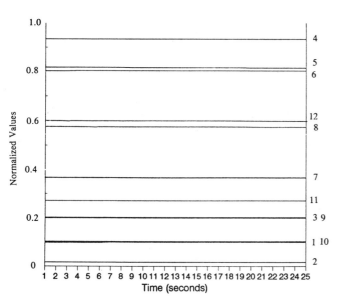

Fig 2: Steady State Conditions

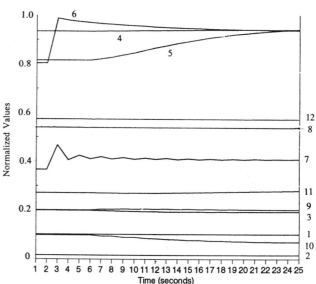

Fig 4: Downstream Leak

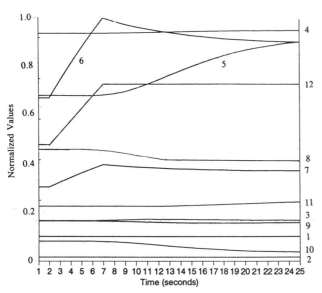

Fig 5: Throttle Opening Transient

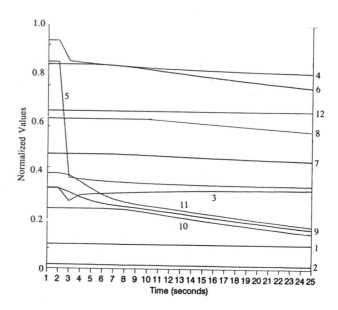

Fig 7: Group Drop

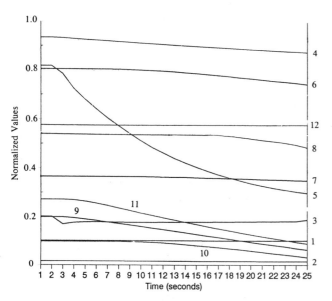

Fig 6: Rod Drop

The recent history of the reactor was felt to be important for the diagnosis of condition. Data from the last three time periods was used in the input set. Three time steps were felt to offer sufficient plant information to enable accurate results yet to allow rapid diagnosis of potentially serious conditions requiring prompt action. The final ANN structure was therefore thirty-six inputs and six outputs.

Data from the first twenty-five time periods from the simulation of each condition were used for producing the training, test and validation sets. For each transient, sets of three successive outputs from the simulator were copied and concatenated with the corresponding output to produce a valid ANN input entry. Two of these entries were incomplete, due to the three time step input, and were discarded. The combined large set was then randomly divided into independent training, test and validation sets in the approximate ratio of 60:30:10. The final training set consisted of 84 entries, the test set of 42 cases and the validation set of 12 examples.

These data sets were then used to train a range of ANNs. A maximum of two hidden layers was considered with a varying number of nodes in each layer. The back-propagation algorithm was used for training and the RMS error of the test set was the measure of an ANN's performance. Each network was trained for 20,000 cycles, the test set was presented every 1000 cycles with the best network, based on RMS error, being saved. The validation file was presented to the best ANN saved during the training run to determine the accuracy of the diagnosis. The best ANN developed was composed of two hidden layers of 18 and 15 nodes. The presentation of the test set to this network produced an RMS error of 0.014. All the transients were correctly identified within the previously defined limit of > 0.9 for transient existence.

INTERFACING WITH THE PWR SIMULATOR

A facility in the ANN development software package was used to produce a 'C' module of the best network. This code was then embedded in the simulator program to give a real time response to the output of the reactor simulator. The output from the ANN was translated into a graphical form so that a constant display of the ANN result could be compared to that of the simulator. The input to the ANN was composed of results from calculations in the program and operator initiated entries on the computer keyboard to represent transient selection and occurrence.

RESULTS

The above system was used to explore the accuracy and speed of diagnosing reactor transients. Each transient was individually initiated on the simulator and the response time for an accurate (>0.9) diagnosis explored. The results for each condition are given below in Figures 8 and 9.

The convention used for all the results is as follows:
1 - Throttle Opening Output Node
2 - Steady State Output Node
3 - Primary Leak Output Node
4 - Downstream Leak Output Node
5 - Group Drop Output Node
6 - Rod Drop Output Node

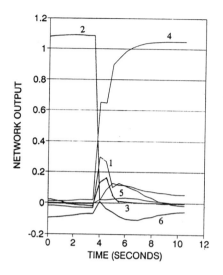

Fig 8: Downstream Steam Leak

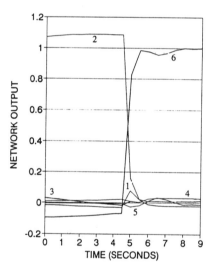

Fig 9: Rod Drop

The response times are very good and quicker than a human operator. In each case the simulator program was allowed to reach steady state conditions before the transient was initiated. This is shown in the figures by the value of the normal plant condition output as approximately 1 before the start of the transient. The other outputs are near zero indicating no fault present.

While the time of diagnosis was striking, the above results were not novel as the ANN had been trained on the faults and should therefore have been capable of their accurate diagnosis. A set of further tests were conducted to explore the generality of the network. Combinations of successive transients were modelled on the simulator and the on line ANN used to diagnose the reactor condition in real time. A selection of the results obtained are given in the figures below, Figs 10 and 11.

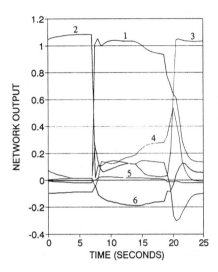

Fig 10: Throttle Opening and Primary Leak

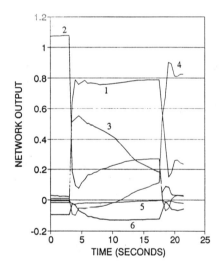

Fig 11: Throttle Opening and Downstream Leak

Most of the combinations were correctly diagnosed. In each case the first transient initiated on the simulator was correctly diagnosed. The second fault condition was introduced and generally correctly identified. In the majority of cases the ANN seemed only able to identify one transient at a time, in each case the ANN results contained a single high valued output. When the second transient was successively diagnosed the output value for the first transient identified usually dropped.

Data for the selected transients were produced at the same power setting. A series of tests were performed to

investigate the effect of varying the power range. A sample of the results obtained is given in the following graphs, Figures 12 and 13.

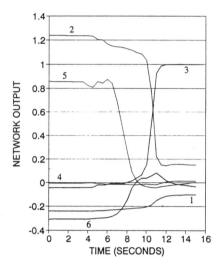

Fig 12: Primary Coolant Leak, 16% Power

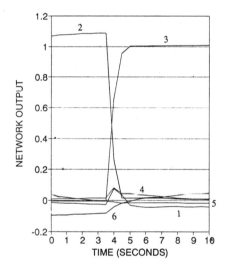

Fig 13: Primary Coolant Leak, 60% Power

The reactor power has a marked effect on the accuracy of the ANN output.

The ANN was trained on a single value of leak size for both the Fast Primary Coolant Leak and the Downstream Steam Leak. Further tests were performed to explore the accuracy and speed of the ANN to diagnose different leak sizes. The leak size selected for the training data allowed a sufficient of number incremental changes for both smaller and larger leaks to be considered. The simulator was then run for a range of incremental leak sizes. A selection of the results obtained are given below in Figures 14 and 15.

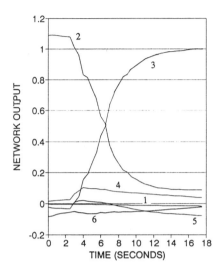

Fig 14: Primary Coolant Leak, 4 kg/s

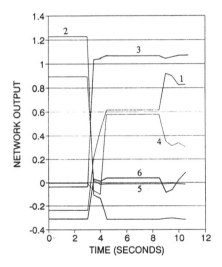

Fig 15: Primary Coolant Leak, 20 kg/s

The results show that the ANN can successively identify both forms of transient for a range of leak sizes. The smaller leaks take longer to be identified.

CONCLUSIONS

An ANN has been developed which successfully identifies five important fault transients from the outputs of a PWR simulator. The identification is performed on-line, in real time and faster than an operator. A range of leak sizes can be input and the leak fault is still identified.

The ANN was trained on only six transients, including normal operating conditions, but the resulting diagnostic tool has been shown to be flexible for a range of leak sizes and conditions. However, the ANN output did become erratic for inputs significantly different from the training conditions. All the positive results have been rapidly obtained and have generally been correctly diagnosed. In the worst case the ANN identified that the plant was in a fault condition and so alert the operator to investigate further. In this situation the ANN output is of questionable benefit as an incorrect diagnosis could persuade the operator to consider an incorrect recovery plan.

There is scope for further work. The accuracy of the ANN could be improved by including more PWR variables in the diagnosis. The input set selected contained only twelve variables out of a possible 108. The results obtained from such a small data set show that the selection was very appropriate. Further work to expand the system to consider a larger number of transients or combinations of conditions would invariably involve considering more plant variables. Alternatively a number of small, simple ANNs could be developed for particular tasks and combined together to form a more complete diagnostic tool [Well95]. The variation of reactor power requires further investigation, possibly the introduction of a scaling factor which would align the inputs to provide a common level for diagnosis. The sensitivity of the ANN also requires further investigation to establish the robustness of the system.

The above features are not only applicable to nuclear reactors but also to many other non-linear systems. The relative success of the embedded ANN module indicates the applicability of Neural Networks to the diagnosis of fault conditions in a complex, critical environment. The widespread use and acceptability of ANNs in such conditions will depend on the safety justification of the technology [Wa95].

REFERENCES

[BB94]   Basu A. and Bartlett E.B., "Detecting Faults in a Nuclear Power Plant by Using Dynamic Node Architecture Neural Networks", Nuclear Science and Engineering 116, April 1994, Pages 313-325

[BU92]   Bartlett E.B. and Uhrig R.E., "Nuclear Power Plant Status using an Artificial Neural Network", Nuclear Technology, Vol 97, Pages 272-281, 1992.

[Hut96]   Hutchison P.G., "Automatic Manoeuvring Room Advisory System", [UNPUBLISHED MOD DOCUMENT]

[Wa95]   Wainwright N., "A Regulators Viewpoint on the use of AI in the Nuclear Industry", Nuclear Energy, Vol 34, No2, Pages 93-97, 1995.

[Well95] Weller P.R., "The Use of Neural Networks for Modelling and Control of Nuclear Reactors", Research Memorandum, Department of System Science, City University, London, UK, 1995.

Neural Network Based Real-Time Fault Location Technique For Transmission Line

Zhihong Chen and Jean-Claude Maun

Department of Electrical Engineering, Free University of Brussels(ULB), CP 165

50 Av. Franklin Roosevelt, 1050, Brussels, BELGIUM

Emails: czhihong, jcmaun@genelec.ulb.ac.be

ABSTRACT

This paper describes the application of an artificial neural network-based algorithm to the single-ended fault location of transmission lines using voltage and current data. From the fault location equations, similar to the conventional approach, this method selects phasors of prefault and superimposed voltages and currents from all phases of the transmission line as inputs of the artificial neural network. The outputs of the neural network are the fault position and the fault resistance. With its function approximation ability, the neural network is trained to map the nonlinear relationship existing in the fault location equations with the distributed parameter line model. It can get both fast speed and high accuracy. The influence of the remote-end infeed on neural network structure is studied. A comparison with the conventional method has been done. It is shown that the neural network-based method can adapt itself to big variation of source impedances at the remote terminal. Finally, when the remote source impedance vary in small ranges, the structure of artificial neural network has been optimized by the pruning method.

I. INTRODUCTION

When a fault occurs on an electrical transmission line, it is very important to find the fault location in order to make necessary repairs and to restore power as soon as possible.

Many algorithms for accurate fault location have been developed. The algorithms based on the steady-state components at the fundamental frequency are still the most popular ones. For long lines, a distributed parameter line model becomes essential for the design of an algorithm. Westlin et.al. [WB76] first introduced a methodology to incorporate this type of model for the determination of the fault distance. The Newton-Raphson method is applied to solve the related nonlinear equations. Unfortunately, due to the heavy computation burden of iterations, the equations are suggested to simplify to the short line model for the application. Moreover, Takagi et al. [TAK⁺82] used the superimposed components with the long line model. The fault current distribution factor is also introduced and assumed to be real. To avoid nonlinear optimization, this algorithm uses the Taylor expansion for hyperbolic functions. Recently, Johns et al. [JMW95] proposed an algorithm directly using the condition that the fault admittance is real, but on-line nonlinear optimization is still necessary.

The modern trend of research, mainly for on-line application, is to get fast estimation and high accuracy simultaneously. The fault location equations based on the fundamental frequency components are strongly nonlinear, especially for the distributed parameter line model and therefore it is not easy, or even impossible, to obtain an analytical solution. Normally, iterative numerical methods should be applied. This is too slow for an on-line application. Moreover, some ill-conditions may occur [WB76]. For most of the existing algorithms, the complexity of the equations and the computation burden have to be reduced. Thus many influence factors can not be considered, such as shunt capacitance, line asymmetry, transposition points [PCM95] , the change of the parameters along the line, and current transformer saturation.

Many successful applications of artificial neural networks to power system have been demonstrated including security assessment, load forecasting, control etc.. Recent applications in protection have covered fault direction discrimination, fault type classification, fault area estimation, autoreclosure technique, and transformer protection [SSS95][DK95][DFK⁺96][AJS⁺94] [KRS94][BMR95]. However, almost all these applications in protection merely use the ANN ability of classification, that is, ANNs only output 1 or 0, and the inputs are usually voltages and currents samples.

In this paper, a new ANN-based algorithm for the single-ended fault locator is proposed for on-line application. This method selects phasors of prefault and superimposed voltages and currents from all phases of the transmission line as the inputs of the ANN, and can output fast and accurate results, avoiding real-time nonlinear optimization. In addition, if necessary, many factors, which are not easy to consider with conventional methods, can be taken into account into the ANN.

II. BACKGROUND OF THE RESEARCH

A. System

The studied system is a two-busbar system shown in Fig. 1.

FIGURE 1 Studied system.

The line is vertical arranged and untransposed. Its series impedance and shunt admittance matrices in phase components are given in Appendix.

B. Assumptions on the operation conditions:

Since the prefault conditions are used for the fault location based on the ANN, we assume :
- The prefault conditions are normal steady-state ones.
- The source impedances at both ends are symmetrical, and remain the same before and during the fault.
- The voltage amplitude at the busbars (V_S and V_R in Fig.1) is in the range [1.1 pu,0.9 pu] and the angle between the busbar voltages is in the range [-20°,20°].

C. Fault type and fault resistance

In the following, only the results for single phase to ground faults are presented. The fault resistance is assumed to be less than 20Ω.

D. Simulation cases:

Table 1 Typical source impedances

Source type	$\dfrac{\lvert Z_1 \rvert}{\lvert Z_{Line} \rvert}$	angle(Z_1)	$\dfrac{\lvert Z_0 \rvert}{\lvert Z_1 \rvert}$	angle(Z_0) $-$angle(Z_1)
strong	0.08	89(°)	1.5	-1(°)
medium	0.3	87(°)	1	-3(°)
weak	1	85(°)	2	-5(°)

For all cases, the source at the sending end is always set weak. This is the worst condition for single-ended fault locators. At the receiving end, possible source impedances are strong, medium or weak.

All above data ranges are typical. The results will show that they do not affect the principle of this method, only the size of the ANN.

III. BASIC EQUATIONS FOR SINGLE-ENDED FAULT LOCATION

To select the inputs of the ANN, we have first to analyze the basic equations for single-ended fault location. Since the distributed parameter line model is only applied to train and test the ANN, and the equations will not be solved, to give a clear and simple description, basic equations are just proposed summarily based on a phase symmetrical short-line model. The short line model can be described by only two parameters: the phase impedance Z_p, and the equivalent earth return impedance Z_e, after eliminating the mutual coupling.

The principle of superimposition allows the faulted system circuit to be decomposed into a prefault and a superimposed circuits, which are shown in Fig. 2 for phase B to ground fault.

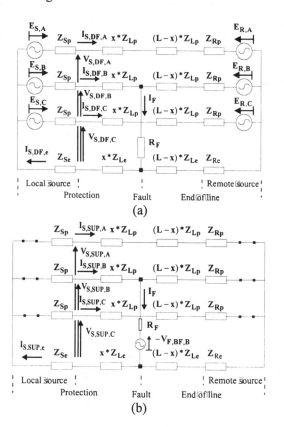

FIGURE 2 The faulted (a) and superimposed (b) circuits for phase B to ground fault.

From Fig. 2, we can develop the basic equations for single-end fault location.

During the fault:

$$V_{S,DF,B} = x * Z_{Lp} * I_{S,DF,B} + R_F * I_F + x * Z_{Le} * I_{S,DF,e} \quad (1)$$

In the superimposed state:

$$V_{S,SUP,B} = \left(L * Z_{Lp} + Z_{Rp}\right) * I_{S,SUP,B} + \left(L * Z_{Le} + Z_{Re}\right) * I_{S,SUP,e}$$
$$- \left[(L-x) * Z_{Lp} + Z_{Rp} + (L-x) * Z_{Le} + Z_{Re}\right] * I_F \quad (2)$$

$$V_{S,SUP,A} = \left(L * Z_{Lp} + Z_{Rp}\right) * I_{S,SUP,A} + \left(L * Z_{Le} + Z_{Re}\right) * I_{S,SUP,e}$$
$$- \left[(L-x) * Z_{Le} + Z_{Re}\right] * I_F \quad (3)$$

$$V_{S,SUP,C} = \left(L * Z_{Lp} + Z_{Rp}\right) * I_{S,SUP,C} + \left(L * Z_{Le} + Z_{Re}\right) * I_{S,SUP,e}$$
$$- \left[(L-x) * Z_{Le} + Z_{Re}\right] * I_F \quad (4)$$

Under symmetrical conditions, eq. (3) and (4) are redundant.

If the remote source impedances Z_{Rp} and Z_{Re} are known, there are 6 independent real equations (3 complex equations based on (1), (2) and (3)) with 4 unknowns (x, R_f, and real and imaginary parts of I_F). The required information to solve equations are the phasors :

$V_{S,DF,B}, I_{S,DF,B}, I_{S,DF,e}, V_{S,SUP,A}, V_{S,SUP,B}, I_{S,SUP,A}, I_{S,SUP,B}, I_{S,SUP,e}$

The equations are nonlinear even for the short line model. Compared with the ones using a distributed parameter line model [WB76], the relationship between the earth return currents of two terminals can not be deleted. This remaining correlation may increase the possibility of ill-conditions if we try to solve the equations iteratively.

IV. SINGLE-ENDED FAULT LOCATION BASED ON ANN

A. Inputs and outputs

To keep the symmetry for all single phase to ground faults, we use following inputs :

$V_{S,BF,A}, I_{S,BF,A}, V_{S,SUP,A}, V_{S,SUP,B}, V_{S,SUP,C}, I_{S,SUP,A}, I_{S,SUP,B}, I_{S,SUP,C}$

This means all information on three phases for the superimposed condition is input to the ANN, while only the information on phase A is used as the reference for the prefault condition.

There are 8 complex variables as input. If the prefault voltage is selected as the vector reference, the number of real variables can be reduced to 15. Normally, the interesting outputs are the fault distance x and the fault resistance R_f.

B. Architecture

The three layer feed-forward neural network is selected to implement the algorithm for single-ended fault location. While the transfer function of the hidden layer neurons is the tanh function, in the output layer we use the linear function. This architecture has been proven to be an universal approximator [BIS95].

Since the desired inputs of ANN for single-ended fault location are the voltage and current phasors, which are complex values, we split them into real and imaginary parts and use a real feed-forward neural network. The complete structure of the ANN for the single-ended fault location is shown in Fig.3.

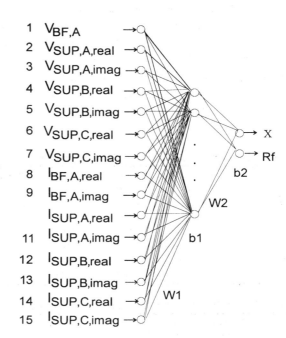

FIGURE 3 ANN architecture for single-end fault location.

C. Learning rule

An improved batch backpropagation learning rule based on Levenberg-Marquardt optimization technique is used to train ANN. It is very powerful, but needs a lot of memory.

V. RESULTS

A. Division an ANN to a set of sub-ANNs according to the measured prefault condition at the local terminal

To get good general performance, for the single-ended fault location, the training and testing data have to cover wide system and fault conditions, e.g. prefault load, fault location and fault resistance. However, the more complex function to fit, the bigger the necessary size of the ANN. To avoid this problem, we can divide an ANN into a set of sub-ANNs according to the measured prefault condition at the local end. However, the bus voltage at the remote end is still considered in the wide range assumed in II.B.

B. Effect of different source impedances

The ANN training and testing for different remote source impedances have been done. Division of an ANN to a set of sub-ANNs based on the measured prefault condition at the local end is used. The results listed in Table 2 and 3 are given only under the conditions that the prefault real power through the line flows from the receiving end to the sending end, and that $1.0pu. \le amplitude(V_s) \le 1.1pu.$. For each of the three remote source impedances, 540 training data and 4319 testing data are generated randomly. The number of iterations for the training is 1000.

Table 2 Results of ANNs for fault distance error(%)

Remote source	Hidden nodes	Data type	Maximum absolute	Mean	Standard deviation
Strong	12	Train	0.376	-0.0007	0.0961
		Test	1.003	0.0151	0.105
Medium	11	Train	0.123	-0.0002	0.036
		Test	0.125	-0.0092	0.031
Wcak	8	Train	0.084	0.0000	0.021
		Test	0.265	0.0677	0.133

Table 3. Results of ANNs for fault resistance error(%)

Remote source	Hidden nodes	Data type	Maximum absolute	Mean	Standard deviation
Strong	12	Train	0.497	-0.0045	0.120
		Test	1.657	-0.1543	0.215
Medium	11	Train	0.251	0.0022	0.055
		Test	0.581	-0.1459	0.140
Weak	8	Train	0.202	0.0003	0.034
		Test	3.009	0.872	1.651

From these results, we can find :
- In all cases, the single-ended fault location based on ANN can estimate the fault position with less than 1% error, and the fault resistance errors are smaller than 3%. By principle, more accurate results may be obtained by increasing the number of training data.
- Generally the more strong the possible influence of the remote source infeed is, the more complex the approximated function, and then the more nodes in the hidden layer are necessary. For this test, the difference is small when the remote source is strong or medium.

- Though the results are the best compared with other neural network structures with different hidden nodes, it seems that, when the remote source is also weak, the ANN still overfits a little, suggesting that, there are a little bit too many weights(hidden nodes) in the network [BIS95]

A comparison with the conventional method can be done under the same conditions. Because normally it is difficult to apply nonlinear optimization for on-line application based on the basic single-ended fault location equations, a simpler and non-iterative method, which considers the fault current flowing the fault resistance to be proportional to the zero-sequence current at the protection for single-phase to ground faults, is selected. The equation for a phase B to ground fault is :

$$V_{S,DF,B} = (Z_{phase}(2,1) * I_{S,DF,A} + Z_{phase}(2,2) * I_{S,DF,B}$$
$$+ Z_{phase}(2,3) * I_{S,DF,C}) * x + R_F * 3I_{S,DF,0} \qquad (5)$$

The results for fault distance are listed in Table 4.

Table 4 Results of the conventional method for fault distance error(%)

Data type	Remote source	Maximum absolute	Mean	Standard deviation
Train	strong	340.310	-10.193	37.613
	medium	33.276	-2.297	3.864
	weak	2.030	-0.076	0.492
Test	strong	753.882	-3.261	20.663
	medium	30.334	-1.500	2.071
	weak	1.994	-0.193	0.337

It can be found that when the remote-end infeed is strong, and the fault is near the remote end with large fault resistance, the conventional method leads to very large errors. Only on-line nonlinear optimization can improve it.

In contrast to the conventional method, the influence of the remote-end infeed on the results of the ANN is not so apparent.

C. Adaptation to the change of source impedances

To check the ability of ANN to adapt itself to large variations of source impedances at the remote end, data for the three different types of remote sources have been used together to train an ANN.

As discussed above, for the phase symmetrical short line model, when the remote end source impedances are unknown, the nonlinear equations have eight unknowns for six equations and become under-determined. Fortunately, under the following conditions, the methods still have opportunities to adapt themselves to the change of the source impedances at the remote end :

- Both Z_{Rp} and Z_{Re} change proportionally;
- Both Z_{Rp} and Z_{Re} are almost reactive;
- One of Z_{Rp} and Z_{Re} is fixed and the other varies.

For practical lines, if the geometry is not completely symmetrical, or the fault does not occur on the central phase of a line with a symmetrical geometry, the equations based on the healthy phases, (3) and (4), are slightly different, and more information can be given.

Though the remote source impedances in Table 1 do not completely fulfill one of the above conditions, an ANN is still trained and tested based on the data from three different types of remote sources together.

Division of an ANN to a set of sub-ANNs is used, too. The following results only cover the prefault conditions that $0° \le angle(V_R - V_S) \le 20°$ and $1.05pu. \le amplitude(V_S) \le 1.1pu.$

The number of training data for each type of source is 1080, testing data are 3000. There are 27 nodes in the hidden layer. The number of training iterations is 300.

Table 5 Results of ANN for fault distance error(%), when remote source impedances change

Data type	Remote source	Maximum absolute	Mean	Standard deviation
Train	strong	1.411	0.0031	0.269
	medium	1.089	0.0007	0.225
	weak	0.773	-0.0009	0.200
	all	1.411	0.0010	0.233
Test	strong	3.545	0.0044	0.359
	medium	0.934	-0.0257	0.218
	weak	0.836	0.0274	0.196
	all	3.545	0.0020	0.269

Table 6 Results of ANN for fault resistance error(%), when the remote source impedances change

Data type	Remote source	Maximum absolute	Mean	Standard deviation
Train	strong	1.746	-0.006	0.387
	medium	1.671	-0.015	0.349
	weak	1.415	-0.004	0.353
	all	1.746	-0.008	0.363
Test	strong	29.844	0.100	2.199
	medium	17.706	0.094	1.254
	weak	2.152	-0.061	0.564
	all	29.844	0.043	1.499

The results show that, when the source impedances at the remote end vary in a large range, the ANN is able to make the internal decision to know which source is practically in operation. Then the fault distance is given based on this decision. The observed maximum absolute error for fault distance is 3.5%.

D. Small range changes of source impedances and ANN structure optimization

A more practical condition is to assume that the source impedances change continually in small ranges. Since the influence of the local source is small, still only the remote source impedances are considered. The remote source is assumed as a medium source in Table 1 with ±5% deviation for the amplitudes of the sequence impedances and ±1° deviation for the angles of sequence impedances.

The structure of the ANN is optimized with the *optimal brain damage* method [CDS90] to get better generalization and a smaller architecture. At first a relatively large initial network is trained with little overfitting for the testing data. Then some unnecessary weights are pruned depending on their sensitivity to the estimation error. After retraining the ANN and repruning the weights, the optimal architecture can be found with the smallest mean square error for the testing data.

To show the results more clearly, the outputs of the ANN are reduced to the fault position only. The prefault system conditions for the following results are $0° \le angle(V_R - V_S) \le 20°$ and $1.0pu. \le amplitude(V_S) \le 1.1pu.$.

The number of training data is 2515 including 512 boundary points. The test data are 2000. The initial network has 25 nodes in hidden layer and is trained by 2000 iterations. Each time 2% of the weights are pruned and the network is retrained with 25 iterations.

Fig. 4 demonstrates the process of finding the optimal architecture and Fig.5 gives the final network structure. Table 7 and Table 8 show the results before and after pruning when the optimal structure has been obtained. The comparison with the conventional method are also included in Table 8.

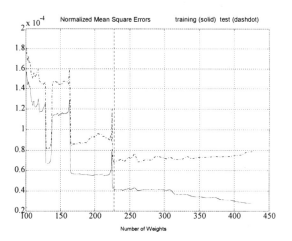

FIGURE 4 The process of finding the optimal architecture

In addition, the pruned network has only 18 nodes in hidden layer with 227 weights. 46.7% of the weights and 7 nodes have been deleted. This leads easier realization and higher execution speed.

For the same network, a more complex pruning method, the *optimal brain surgeon* [HSW93], which theoretically can delete the weights more accurately, was also tested. However, since the necessary network size must remain large and computation burden increases too much, no better result has been obtained.

VI. CONCLUSIONS

This paper has described an ANN-based single-ended fault location method. For on-line application, it can reach fast speed and high accuracy simultaneously, and give designers more opportunities to consider more factors, which are not easy to be treated by conventional methods. This methodology has the ability of adapting itself to large variations of remote source impedances.

The pruning method, *optimal brain damage*, is helpful to get a network with better generalization performance. The designing process of an ANN has proven this when the remote source impedances vary in small ranges. Its ability of automatic input selection will find application for other protection problems.

All selected ranges of values have no influence on the principle of this method, only on the size of the ANN. We can easily extend this method to very high resistive faults(e.g. $200\,\Omega$), more general system configurations, and even two-end fault locator.

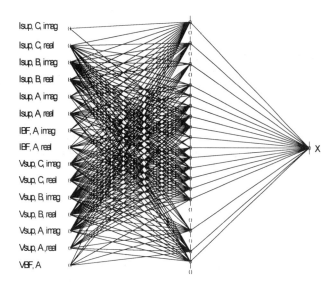

FIGURE 5 The final network structure

Table 7 Comparison of the network structure and performance between before and after pruning.

	Hidden nodes	Number of weights	Data Type	Mean Square Error
Before pruning	25	426	Train	2.794E-5
			Test	7.856E-5
After pruning	18	227	Train	4.297E-5
			Test	6.837E-5

Table 8 Comparison of the fault distance error(%), among conventional method, before and after pruning

	Data Type	Maximum absolute	Mean	Standard deviation
Before pruning	Train	2.391	0.000	0.415
	Test	4.044	0.027	0.696
After pruning	Train	3.232	-0.001	0.515
	Test	3.720	0.022	0.649
Conventional method	Train	45.405	-2.149	4.330
	Test	24.231	-1.946	1.937

Before pruning the results for the testing data are not as good as for the training data. By comparison with other neural network structures, this initial network is known to little overfit [BIS95], though it is the best in our initial try. After finding the optimal architecture by pruning, while the results for the training data become little worse, the results for the testing data are modified. The pruned network gets better generalization performance for the untrained data.

VII. LIST of SYMBOLS

S, F, R : Sending (local) end, **F**ault point, **R**eceiving (remote) end.

A, B, C : phase of the three phase system.

0, 1, 2: Fortescue sequence component.

BF, DF, SUP : Prefault (**B**efore **F**ault) state, **D**uring **F**ault state, and **SUP**erimposed state.

x : distance to fault from the sending end.

L : line **L**ength.

R_f: fault resistance.

$\mathbf{Z}_{phase}, \mathbf{Y}_{phase}$: line impedance and admittance matrices in phase components.

p : phase index.

e : equivalent earth return index.

E, Z: Thévenin's equivalent of the external network.

V, I : complex voltage and current phasors.

VIII. ACKNOWLEDGMENT

We thank Dr. L.Philippot, Mr. D. Wiot and Mr. J.Coemans for valuable discussions. This research is supported by SIEMENS AG.

IX. REFERENCES

[WB76] S.E.Westlin and J.A.Bubenko, "Newton - Raphson technique applied to the fault location problem", IEEE PES Summer Meeting, 1976, Paper No. A 76 334-3.

[TAK+82] T.Takagi et al., "Development of a new type fault locator using the one-terminal voltage and current data", IEEE *Trans.*, Vol. PAS-101, No.8, 1982, pp. 2892-2897.

[JMW95] A.T.Johns, P.J.Moore and B.Whittard, "New technique for the accurate location of earth faults on transmission systems", IEE Proc.C, Vol.142, No.2, 1995, pp. 119-127.

[SSS95] T. S.Sidhu, H. Singh and M. S.Sachdev, "Design, implication and testing of an artificial neural network based fault direction discriminator for protecting transmission lines", IEEE *Trans. Power Delivery*, Vol.10, No.2, 1995, pp. 697-706.

[DK95] T. Dalstein and B. Kulicke, "Neural network approach to fault classification for high speed protective relaying", IEEE *Trans. Power Delivery*, Vol.10, No.2, 1995, pp. 1002-1011.

[DFK+96] T. Dalstein, T. Friedrich, B. Kulicke and D. Sobajic, "Multi neural network based fault area estimation for high speed protective relaying", IEEE *Trans. Power Delivery*, Vol.11, No.2, 1996, pp. 740-747.

[AJS+94] R.K.Aggarwal, A.T.Johns, Y.Song, R.W.Dunn and D.S.Fitton, "Neural-network based adaptive single-pole autoreclosure technique for EHV transmission system", IEE Proc.C, Vol.141, No.2, 1994, pp. 155-160.

[KRS94] M.Kezunovic, I.Rikalo and D.J.Sobajic, "Neural Network applications to real-time and off-line fault analysis", Proc. of Intl. Conf. on Intelligent System Applications to Power Systems, France, Sept., 1994, pp. 29-36.

[BMR95] P.Bastard, M.Meunier, and H.Régal, "Neural network-based algorithm for power transformer differential relays", IEE Proc.C, Vol.142, No.4, 1995, pp. 386-392.

[PCM95] L.Philippot, Z. Chen, and J.-C.Maun, "Transmission system modeling requirements for testing high-accuracy fault locators", Proc. of 1st Intl. Conf. on Digital Power System Simulations, Texas, USA, Apr.,1995, pp. 69-74.

[BIS95] C.M.Bishop, Neural Networks for Pattern Recognition, Oxford University Press Inc. New York,1995.

[CDS90] Y.Le.Cun, J.S.Denker, and S.A. Solla, "Optimal Brain Damage", Advances in Neural Information Processing System II, 1990, pp. 598-605.

[HSW93] B.Hassibi, D.G.Stork, and G.J.Wollf, "Optimal Brain Surgeon and General Network Pruning", Proc. of the 1993 IEEE Intl. Conf. on Neural Network, 1993, pp. 293-299.

[EPR86] EPRI, "Electromagnetic Transients Program (EMTP)", Version 1, Revised Rule Book, Vol.1: Main Program, *EPRI* EL-4541-CCMP, Polo Alto, California, April, 1986.

X. APPENDIX

The series impedance and shunt admittance matrices in phase components obtained by the LINE CONSTANTS support routine of EMTP [EPR86] at 50 Hz, using the "K.C.Lee" untransposed line model.

$$
Z_{phase} = \begin{bmatrix}
4.107778E-02 & & \\
+j*4.46379E-01 & & \\
2.695538E-02 & 4.269062E-02 & \\
+j*1.903961E-01 & +j*4.745678E-01 & \\
2.735199E-02 & 2.855895E-02 & 4.398963E-02 \\
+j*1.617875E-01 & +j*2.215322E-01 & +j*4.853872E-01
\end{bmatrix} \Omega/km
$$

$$
Y_{phase} = j*\begin{bmatrix}
1.058444E-08 & & \\
-2.415754E-09 & 1.131348E-08 & \\
-8.739274E-10 & -2.458611E-09 & 1.182518E-08
\end{bmatrix} S/km
$$

FAST RULE GENERATION AND MEMBERSHIP FUNCTION OPTIMIZATION FOR A FUZZY DIAGNOSIS SYSTEM

Yi Lu[1], and Tie Qi Chen[2]
[1]Department of Electrical and Computer Engineering
The University of Michigan-Dearborn
Dearborn Michigan 48128-1491, U.S.A.
Voice: 313-593-5028, Fax: 313-593-9967
yilu@umich.edu

[2]Department of Physics
Fudan University, Shanghai 200433
The People's Republic of China
tqchen@umich.edu

ABSTRACT

This paper describes a fuzzy rule generation algorithm and a fuzzy membership function optimization algorithm. Both algorithms have been implemented in a fuzzy diagnostic system for the End-of-Line test at automobile assembly plants and the implemented system has been tested extensively and its performance is presented.
Keywords: automotive diagnosis, fuzzy systems, fuzzy rule generation

1. INTRODUCTION

Fault diagnosis has been a classic engineering problem. The techniques for diagnosing faults in dynamic systems range from expert systems to statistic models. In spite of these efforts, automotive engineering diagnosis is continuously being considered as one of the most challenging problems in AI, because the electronic control components in modern vehicles are more like a blackbox; it is difficult for mechanics to diagnose faults accurately unless they are thoroughly familiar with the system specifications and functions. As a result, it is extremely difficult to develop a complete diagnostic model that can fully answer all the questions in terms of faults. The theory of fuzzy logic is aimed at the development of a set of concepts and techniques for dealing with sources of uncertainty or imprecision and incomplete. Fuzzy systems have been successful in many applications including control theory when gradual adjustments are necessary, air and spacecraft control, business and even the stock exchange [Ayo95, NHW91, RhK93, TaS85]. The nature of fuzzy rules and the relationship between fuzzy sets of differing shapes provides a powerful capability for incrementally modeling a system whose complexity makes traditional expert system, mathematical, and statistical approaches very difficult.

This paper describes an effort in applying fuzzy logic to automotive fault diagnosis. Specifically, we describe one algorithm that automatically generates fuzzy rules and one algorithm that optimizes fuzzy membership functions. The importance of automatically generating rules for intelligent systems has been well recognized in the AI community. Typical approaches include neural network techniques[MiP95, Ayo95, TaI94, RhK93], inductive learning algorithms or pseudo-Boolean logic simplification methods[TuZ93], and the fuzzy c-means clustering method[CGH95].

The algorithms described in this paper are different from those existing approaches, are extremely effective and efficient. These algorithms have been implemented in a fuzzy diagnostic system for the End-of-Line test in automobile assembly plants. We have tested the implemented system on large sets of data of different vehicle models acquired directly from various test sites of Ford assembly plants, and the performance of the system is presented.

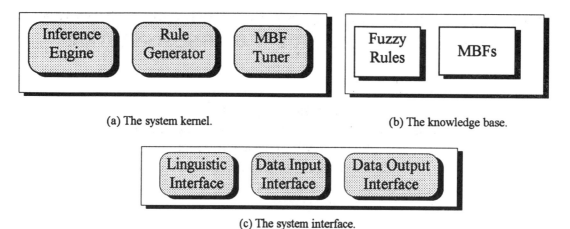

(a) The system kernel.   (b) The knowledge base.

(c) The system interface.

Figure 1 Overview of a Fuzzy System components.

2. OVERVIEW OF A FUZZY DIAGNOSTIC SYSTEM

The major US automotive companies have launched an End-of-Line test system at every North American assembly plant. In order to accomplish this task, a system is designed to collect and analyze Electronic Engine Controller (EEC) data while the vehicle is dynamically tested. Operators drive the vehicles through a preset profile and the vehicles are either passed or failed according to the data collected during the test. The decision is made based on two information sources, the EEC on-board test and off-board test performed by the vehicle test system on EEC generated data. We have developed a fuzzy system to solve the second problem.

Figure 1 gives the overview of the fuzzy diagnostic system. Figure 1(a) shows the system components, an inference engine, a fuzzy rule generator and a membership function (MBF) optimizer. Figure 1. (b) illustrates the knowledge base of the system which is composed of the fuzzy rules and the MBFs. The fuzzy diagnostic knowledge can be generated either from engineer experts or training data through a machine learning algorithm. The system kernel interacts with the knowledge base directly during the fuzzy inference. Figure 1.(c) illustrates the system interface. The system interface provides a linguistic interface between the engineer experts and the knowledge base, and a data input/output interface between the user and the system kernel. The focus of this paper is to present our research results in fuzzy rule generation and membership function optimization.

3. FUZZY RULE GENERATION AND MBF OPTIMIZATION

Fuzzy membership function optimization and fuzzy fule generation are performed during the system learning procedure. The learning procedure first uses an initial set of critical points of membership functions to generate fuzzy rules, then the critical parameters of the membership functions are optimized to give the best performance.

3.1. Fuzzy Rule generation

In a fuzzy diagnostic system, fuzzy rules can be characterized by a set of control variables, $X = \{x_1, x_2, ..., x_n\}$, which are the parameters that reflect the faulty behavior, and a solution variable y that represents the type of fault the system is responsible to diagnose. Each control variable is associated with a set of fuzzy terms $\Sigma_i = \{\alpha_i^1, ..., \alpha_i^{Pi}\}$, and the solution variable is associated with fuzzy terms $\Gamma = \{\tau_1, ..., \tau_q\}$. In this application, we only consider the following rule format:

IF $(x_k^1$ is $\alpha_i^{k1})$ AND $(x_k^2$ is $\alpha_i^{k2})$ AND ...$(x_k^m$ is $\alpha_i^{km})$, THEN y is τ_j.

where $m \leq n$, $\{x_k^1, x_k^2, ..., x_k^m\} \subset X$, $\{\alpha_i^{k1}, \alpha_i^{k2}, ..., \alpha_i^{km}\} \subset \Sigma_i$, and $\tau_j \in \Gamma$.

In general, the rule generator can start from an arbitrary initial point of the critical parameters which will be further optimized during and after the generation of the fuzzy rules. For simplicity, the critical parameters are initialized as follows. Assume there are k_i MBFs for the i-th control variable x_i, then the j-th critical parameter of the MBFs

$$w_{jk} = x_i^{min} + (x_i^{max} - x_i^{min}) \frac{j}{k_i + 1},$$

where x_i^{max} and x_i^{min} are the maximum or minimum value of x_i in the training data set X respectively, and $j = 1, 2, ..., k_i$.

The fuzzy rule generation algorithm can be outlined in the following seven steps:

[Step-1] Initialize the MBFs.

[Step-2] Separate the training data into clusters.

[Step-3] Extract the fuzzy rules.

[Step-4] Re-cluster the training data according to the fuzzy rules.

[Step-5] Calculate the locations of the cluster centers.

[Step-6] If the locations of the cluster centers are close enough to the critical parameters of the MBFs, then go to Step-7; otherwise, update the critical parameters of the MBFs to the cluster centers, then go to Step-2.

[Step-7] Prune the redundant control variables.

cluster #	the largest MBF for each control variable		solution variable		
	x_1	x_2	y LOW	y MEDIUM	y HIGH
0	LOW	LOW	60	100	40
1	MEDIUM	LOW	20	150	30
2	HIGH	LOW	30	140	30
3	LOW	MEDIUM	120	20	20
4	MEDIUM	MEDIUM	100	20	10
5	HIGH	MEDIUM	0	0	0
6	LOW	HIGH	0	0	0
7	MEDIUM	HIGH	0	0	0
8	HIGH	HIGH	10	10	90

Table 1 Statistics of the training data obtained in current iteration.

rule #	antecedent		consequence	# of support samples	support rate
	x_1	x_2	y		
0	LOW	LOW	MEDIUM	100	50%
1	MEDIUM	LOW	MEDIUM	150	75%
2	HIGH	LOW	MEDIUM	140	70%
3	LOW	MEDIUM	LOW	120	75%
4	MEDIUM	MEDIUM	LOW	100	77%
5	HIGH	MEDIUM	*anything*	0	—
6	LOW	HIGH	*anything*	0	—
7	MEDIUM	HIGH	*anything*	0	—
8	HIGH	HIGH	HIGH	90	82%

Table 2 Rules extracted based on the Winner-Take-All mechanism.

rule #	antecedent		consequence	# of support samples
	x_1	x_2	y	
0	*anything*	LOW	MEDIUM	390
1	*anything*	MEDIUM	LOW	220
2	*anything*	HIGH	HIGH	90

Table 3 The final fuzzy rules extracted in current iteration.

Let us use an example to illustrate the algorithm. Let us assume the system have two control variables x_1 and x_2, one solution variable y, and each control and solution variable is associated with three MBFs, namely LOW, MEDIUM and HIGH. According to the values of the control variables, the training data can be separated into $3^2 = 9$ clusters by finding the largest MBF for each control variable. Suppose there are 1000 samples and the statistics of the training data obtained at the current iteration is shown in Table 1. To explain the information listed in this table, we take cluster #2 for example. There are 200 samples belonging to this cluster. For these samples, the largest MBF for x_1 is HIGH and for x_2 is LOW. The antecedent of the rule to be generated from this cluster is "if x_1 is HIGH and x_2 is LOW". The 200 samples are called **the dominated samples** of this will-be-generated rule, and this rule is called **the dominant rule**. So far the consequence of this dominant rule is not determined yet. Among these 200 samples, 30 samples say y is LOW, 140 samples say y is MEDIUM and 30 samples say y is HIGH. Therefore we determine the consequence of rule #2 to be "y is MEDIUM". By using this Winner-Take-All mechanism, we extract 9 rules listed in Table 2. In Table 2, all the generated rules are sorted according to "# of supporting samples" which is used as an effective and reasonable criterion to prune rules. The "support rate" is calculated as the "# of support samples" divided by the dominated samples of the rule. However, the support rate alone cannot be used to prun rules. For example, if one rule has only one dominated sample, then its "support rate" is always 100%; if another rule has 200 dominated samples, but the "support rate" is 75%. Obviously, it's reasonable to keep the latter and discard the former. In Table 2, the consequences of rule #5, #6 and #7 are null because no sample belongs to these clusters. These **null rules** are very useful in the rule reduction process. Without rule #5 ~ #7, only rule #0 ~ #2 in Table 2 can be merged:

Rule #0 ~ #2 → "if x_2 is LOW, then y is MEDIUM"

"# of support samples" of the new rule is 100+150+140 = 390. With the null rules (#5 ~ #7), rule #3 ~ #8 in Table 2 can also be merged:

Rule #3 ~ #5 → "if x_2 is MEDIUM, then y is LOW"
Rule #6 ~ #8 → "if x_2 is HIGH, then y is HIGH"

"# of support samples" of the two new rules are: 120+100+0 = 220 and 0+0+90 = 90 respectively. The final fuzzy rules extracted in this iteration are listed in Table 3. Because the critical parameters of the MBFs will be re-calculated and updated in each iteration, the cluster partition may change and therefore the fuzzy rules extracted in each iteration may be different.

Unlike a fuzzy-neural network in which the fuzzy rules are encoded in a distributed manner among the inter-connection weights, the internal representation of the fuzzy rules in our system is a look-up table. The look-up table representation provides the flexibility, meanwhile it has the least memory requirement. With this representation, the system speed benefits from the situation of less fuzzy rules present because less items in the rule table need to be searched, while the speed of a fuzzy-neural network system doesn't decrease unless the neural network has been reconstructed. We can further improve the system speed by sorting the rule table according to the "# of support samples' of the rules because the table is searched sequentially and the rules fired most frequently are placed at the head of the table.

The above example can be summarized by the following clustering and rule-extraction algorithm which is used in each iterative in the rule generation algorithm:

[Step-1] Initialize the rule table with **the prior rules** which can be preset by the engineer experts, or leave the rule table empty.

[Step-2] Separate the training data into clusters by finding the largest MBF for each control variable.

[Step-3] Choose a cluster sequentially. If the cluster is not empty, extract a rule from the cluster based on the Winner-Take-All mechanism, and label the rule with "# of support samples".

[Step-4] If there are more than one solution variables, repeat Step-3 to extract a consequence for each solution variable. Also a "# of support samples" is obtained for each solution variable. The final "# of support samples" is the minimum one.

[Step-5] If there is no sample in the cluster, a null rule is extracted and the "# of support samples" is 0.

[Step-6] Compare the newly-extracted rule with the prior rules in the rule table. If the newly-extracted rule conflicts with the prior rules, discard it immediately then go to Step-10. If the newly extracted rule is the same as (or contained by) one of the prior rules, go to Step-10.

[Step-7] Try to merge the newly-extracted rule with the rules already in the rule table. If a merge takes place, the redundant control variables are pruned from the antecedent of the newly-extracted rule, and the resulting "# of support samples" for the new rule is the sum of the "# of support samples' of all the merged rules.

[Step-8] In case there are more than one redundant control variables, Step-7 is repeated until the rules cannot be merged any more.

[Step-9] Only after merging, the newly-extracted rule is added into the rule table.

[Step-10] If no more cluster is left, go to Step-11; otherwise, go to Step-3.

[Step-11] Delete all the null rules. If desired, the rules with small "# of support samples' can also be removed.

This clustering and rule-extraction algorithm is based on following two facts. First a sample input may fire more than one rules, since the MBFs for each control variable overlap. However, because of the higher belief values of the MBFs, the dominant rule of this sample largely determines the final classification as long as this dominant rule is included in the current rule set. Secondly, if the knowledge provided by the engineer experts are reliable, most samples from the training data set must conform to (not violate) the knowledge. This guarantees that most rules extracted by this algorithm are consistent with the knowledge of the engineer experts.

The re-clustering and calculation of the locations of the cluster centers in the algorithm are necessary because the training samples belonging to different classes may be included in the same cluster. Thus the rule extracted from such a cluster will not be effective. This is similar to the problem encountered in the fuzzy c-means algorithm. In the fuzzy c-means algorithm, the data points always belong to the nearest clusters, and the data points belonging to different classes are not discriminated while calculating the locations of the cluster centers. Therefore this problem cannot be solved in the fuzzy c-means algorithm. Here, we use **an additional re-clustering procedure** to solve this problem.

The re-clustering procedure uses a measure of distance between a sample and a cluster. Without losing generality, we assume a fuzzy rule is in the following form:

"IF (x_1 is c_1 and x_2 is c_2 and ... and x_n is c_n), THEN (y is c)."

For class c_i of the control variable x_i, there exists a range (x_{icL}, x_{icH}) in which the membership function of class c_i has the largest value. Therefore class c of the solution variable is mapped onto a n-dimensional **hyper-cubed** in the input space by this rule. The n-dimensional hyper-cubed is specified by a set of ranges (x_{icL}, x_{icH}), where $i=1, 2, ..., n$. All the samples inside a hyper-cubed is called a **cluster**. The squared distance between s-th sample and a cluster is defined as follows:

$$SD = \sum_{i=1}^{n} \varepsilon_{xi}(s), \text{ where}$$

$$\varepsilon_{xi}(s) = \begin{cases} (x_{icL} - x_i^{(s)})^2 & if(x_i^{(s)} < x_{icL}) \\ 0 & if(x_{cL} < x_i^{(s)} < x_{icH}) \\ (x_i^{(s)} - x_{icH})^2 & if(x_{icH} < x_i^{(s)}) \end{cases}$$

After all the training samples have been re-clustered, the location of every cluster center is calculated, and the critical parameters of the MBFs are updated to the locations of the cluster centers if they are not close enough to the cluster centers.

3.2 Fuzzy membership Functions and Optimization

In fuzzy logic, each variable is associated with a set of fuzzy membership functions, each of which corresponds to a fuzzy term. In general the MBFs of a fuzzy variable can be defined by a set of **critical parameters** that uniquely describing the characteristics the MBFs. Given the testing data and the fuzzy rules, assigning different values to the critical parameters usually makes the inference engine output different classification results. Similarly, given the training data, assigning different values to the critical parameters usually makes the rule generator generate different fuzzy rules. Fuzzy membership function optimization is to make the control surface (the system response) to react correctly to data by changing either the shape of the underlying fuzzy sets. Assume we have a set of fuzzy rules, FR, and the membership functions:

$MBFx_i = (\alpha_i^1, \alpha_i^2, ..., \alpha_i^{ki})$, for $i = 1, ..., n$ and
$MBFy = (b^1, b^2, ..., b^k)$.

The system response can be written as a function of the fuzzy rules and MBFs:

$FS (FR, MBFx_1, MBFx_2, ... MBFy, \bar{x})$

where $\bar{x}$ is the input sample data to the fuzzy system FS. When fuzzy rule set, FR, is given, we optimize the system response FS by tuning the critical parameters $MBFx_i$, for i = 1, ..., n and $MBFy$.

The MBFs can be optimized by tuning the critical parameters directly [KKN92], for example, using the gradient descent method or tuning the inter-connection weights in the fuzzy-neural network [MiP95]. The gradient descent optimization algorithm requires the objective function to have continuous first-order derivatives. Unfortunately the input-output function of a fuzzy system usually doesn't satisfy this condition, since first the `min-max` implication methods operations are not continuous, and secondly the triangular functions don't have the continuous derivatives. Therefore we choose to use the stochastic annealing method to optimize the membership functions.

A fuzzy system is expected to output the correct status of the solution variable for each input sample. In other words, in the hyper-space of the solution variable, the expected output of a fuzzy system for an input sample is not a point, it is a hyper-cubed. The objective function used in optimization is defined as follows:

$$OF = \phi + \sum_s \varepsilon_y(s), \text{ where}$$

$$\varepsilon_y(s) = \begin{cases} (y_{cL} - y^{(s)})^2 & if\,(y^{(s)} < y_{cL}) \\ 0 & if\,(y_{cL} < y^{(s)} < y_{cH}) \\ (y^{(s)} - y_{cH})^2 & if\,(y_{cH} < y^{(s)}) \end{cases}, \text{ and}$$

$$\phi = \alpha\, N_u / N_t$$

where α is the penalty coefficient which determines how much the penalty affects the objective function, usually a positive real number around 0.01; N_u is the number of undetermined samples; N_t is the number of total samples.

The stochastic annealing method is used in the optimization algorithm. The algorithm evaluates system error E_0 at a point x_0 and then makes a move to a new point x by a random walk. For a fuzzy system, x_0 may represent the critical parameters of the MBFs, $MBFx_i$, i = 1, ..., n, and $MBFy$, and E_0 is the objective function defined earlier. At the new point x, the system error is computed as E. If E is less than E_0, the new point x is accepted. Otherwise, the new point x has P chance of being accepted, where

$$P = \exp\left[-\frac{E - E_0}{T} \right]$$

is a Boltzmann factor, T is called temperature, and error E or E_0 is called energy. The more energy rises, the smaller the probability P is; and the higher the temperature T is, the bigger the probability P is. The new point x is accepted p is greater than a random number generated between 0 and 1.

The optimization process is like annealing. Initially, the temperature is very high, therefore the probability P is very high in spite of the rising of the system. This results the random system move. As the temperature decreases, P is dominated by the system energy.

Therefore the system tends to move to the points where lower energy arises. If the system happens to drop into a local minimum point, it has opportunities to walk out.

When a new point x is accepted by the system, the temperature T is decreased. It has been proved that if the initial temperature is high enough and the temperature decreases slowly enough (for example the logarithmic schedule), the optimization process can always converge to a global optimum [GeG84]. But the logarithmic schedule is too slow to use, therefore we use the exponential schedule,

$$T^{(k+1)} = T^{(k)} \times \beta$$

where $T^{(k)}$ and $T^{(k+1)}$ are temperatures of k-th and $(k+1)$-th iterations respectively; β is the number which determines the temperature decreasing speed, usually a positive real number less than 1.

Initially the temperature is set to the initial system energy E_0. When the system moves to a new point x, if the new energy E is higher than the current temperature T, then T is decreased in the exponential schedule; if E is lower than T, then T is set to E to force the system energy to go down further.

The MBF optimization algorithm can be summarized by the following steps:

[Step-1] Calculation the system error E using the objective function.

[Step-2] If E is less than the preset threshold, then stop.

[Step-3] If this is the first iteration, let T=E and E_0=E, then go to Step-8.

[Step-4] If E>E_0, go to Step-6; otherwise, let E_0=E.

[Step-5] If E<T, let T=E; otherwise, T=T×β. Go to Step-8, where β < 1.

[Step-6] Calculate the probability P, generate a random number r, where $0 \le r \le 1$. If r<P, go to Step-8.

[Step-7] Restore the critical parameters of the MBFs to the old values in the previous iteration, then go to Step-9.

[Step-8] Save the current values of the critical parameters.

[Step-9] Re-cluster the samples which cause errors using the method described in Section 3.5. Here ONLY the samples which cause errors are re-clustered. After the re-clustering, only the critical parameters which are related to the non-empty clusters will be optimized in the next iteration.

[Step-10] If no critical parameters need to be optimized, then stop.

[Step-11] Change the critical parameters by random walk, then go to Step-1. Note ONLY the critical parameters determined by Step-8 are updated.

4. IMPLEMENTATION AND EXPERIMENT

The algorithms described in section 3 has been implemented in a fuzzy system to diagnose vacuum leak in EEC. The fuzzy vacuum leak diagnostic system has five control variables, {throttle position, Lambse_1, Lambse_2, Idle speed DC, Mass air flow}, one solution variable, {vacuum_leak}, and each fuzzy variable is associated with three fuzzy terms for {LOW, MEDIUM, HIGH}. The fuzzy vacuum leak diagnostic system has been implemented on DOS operating system on a PC, has been trained and then tested on two different vehicle models, Ford Thunderbird, and Lincoln Towncar. Both the training and the test data were obtained directly from the test sites of assembly plants of the Ford Motor Company.

For the Thunderbird vehicle model, the engineer experts in the Ford Motor Company provided initialy six fuzzy rules and three membership functions for each system variable. We conducted experiments using the engineering knowledge alone on three data sets, TBIRD_0, containing 22 bad vehicles (vacuum leak HIGH) and 31 good vehicles (vacuum leak LOW); TBIRD_1 containing 20 bad vehicles; and TBIRD_2 containing 1620 good vehicles. The inference accuracy on TBIRD_0 is 50% for bad vehicles and 19.4% for good vehicles. The inference accuracy is only 5%. on TBIRD_1, and 17.5%. on TBIRD_2. These results show that the knowledge base provided by the engineer experts is not sufficient.

We used TBIRD_0 as training data to run our machine learning algorithms described in the previous section. The testing results on TBIRD_0 is 100% in accuracy for bad vehicles and 96.8% for good vehicles, 90% for TBIRD_1 and 98% for TBIRD_2, These results show that our rule generation method is effective.

For the purpose of experiment, we tested the fuzzy knowledge base generated from TBIRD data on three TOWNCAR sets, TOWNCAR_0 containing 22 bad vehicles and 31 good vehicles,. TOWNCAR_1 containing 15 bad vehicles, and TOWNCAR_2 containing 1223 good vehicles. As expected, the results were very poor: the inference accuracy is 22.7% for bad vehicles and 22.9% for good vehicles for TOWNCAR_0, 6.7%. for TOWNCAR_1, and 9.3% for TOWNCAR_2.

When the system was trained on TOWNCAE_0, we generated a new set of fuzzy rules and MBFs. The testing results show that the inference accuracy is 90.9% for bad vehicles and 100% for good vehicles on TOWNCAR_0, 86.7% for TOWNCAR_1 and 99.3% for TOWNCAR_2.

The testing results given above were obtained using squared sinc MBFs. Our system currently supports six types of MBFs, linear, triangular, Sigmoid, β-curve, and Gaussian. We experimented with different types of MBFs and found that the squared sinc functions gave better performance than the other functions.

5. CONCLUSIONS

In this paper, we have introduced an algorithm for fuzzy rule generation and an algorithm for optimizing fuzzy membership functions. These algorithms were integrated into a fuzzy system that detects the vacuum leak in Electronic Engine Controller in automobiles. The fuzzy vacuum leak diagnostic system has been tested on two different vehicle models, Thunderbird and Lincoln Towncar. The testing results show that the system is capable of learning new vehicle model effectively. The algorithms described in this paper generated effective and compact fuzzy rule set and optimal membership functions.

6. ACKNOWLEDGMENT

This work is support in part by a CEEP grant from the School of Engineering at the University of Michigan-Dearborn. The authors would like to thank Mr. Brennan Hamilton from the Ford Motor Company, who provided us with engineering knowledge of automotive diagnosis and all the training and test data.

REFERENCES

[Ayo95]   M. Ayoubi, "Neuro-fuzzy structure for rule generation and application in the fault diagnosis of technical processes," in *Proc. 1995 American Control Conference, Seattle,* (USA), pp. 2757-2761, 1995.

[CGH95]   T. W. Cheng, D. B. Goldgof and L. O. Hall, "Fast clustering with application to fuzzy rule generation," in *Proc. 1995 IEEE Int. Conf.*

Fuzzy Syst., *Yokohama*, (Japan), pp. 2289-2295, 1995.

[GeG84]   S. Geman and D. Geman, "Stochastic relaxation, Gibbs distributions, and the Bayesian restoration of images," *IEEE Trans. Pattern Anal. Machine Intell.*, vol. PAMI-6, pp. 721-741, 1984.

[HaL95]   Brennan T. Hamilton and **Yi Lu,** "Diagnosis of Automobile Failures using Fuzzy Logic," *The eighth International conference on Industrial & Engineering Applications of Artificial Intelligence & Expert Systems*, *Melbourne*, (Australia), June 5 ~ 9, 1995.

[Mip95]   S. Mitra and S. K. Pal, "Fuzzy multi-layer perceptron, inference and rule generation," *IEEE Trans. Neural Networks*, vol. 6, pp. 51-63, 1995.

[NHW91]   H. Nomura, H. Ichihashi, T. Watanabe, "A Self-Tuning Method of Fuzzy Reasoning Control by Descent Method," *Proc. of 4th IFSA Congress, Brussels*, Vol. Eng. pp. 155-158, 1991.

[RhK93]   F. C.-H. Rhee and R. Krishnapuram, " Fuzzy rule generation methods for high-level computer vision," *Fuzzy Sets and Systems*, vol. 60, pp. 245-258, 1993.

[TaI94]   E. Tazaki and N. Inoue, "A generation method for fuzzy rules using neural networks with planar lattice architecture," in *Proc. 1994 IEEE Int. Conf. Neural Networks*, *Piscataway*, (USA), pp. 1743-1748, 1994.

[TaS85]   T. Takagi and M. Sugeno, "Fuzzy Idnetification of Systems and its Applicaitons to Modeling and Control," *IEEE Trans. Syst., Man, and Cybern.*, vol. SMC-15, No.1, pp. 116-132, 1985.

[Tuz93]   I. B. Turksen and H. Zhao, "An equivalence between inductive learning and pseudo-Boolean logic simplification: a rule generation and reduction scheme," *IEEE Trans. Syst., Man, Cybern.*, vol. 23, pp. 907-917, 1993.

FUZZY NEURAL NETWORK MODELS FOR CLASSIFICATION

A. D. Kulkarni and Charles D. Cavanaugh
Computer Science Department
The University of Texas at Tyler, Tyler, TX 75701
Email: akulkarn@mail.uttyl.com

ABSTRACT

During the last few years there has been a large and energetic upswing in research efforts aimed at synthesizing fuzzy logic with neural networks. This combination of neural networks and fuzzy logic seems natural because the two approaches generally attack the design of "intelligent" systems from quite different angles. Neural networks provide algorithms for learning, classification, and optimization whereas fuzzy logic often deals with issues such as reasoning on a high (semantic or linguistic) level. Consequently the two technologies complement each other [Bez93]. In this paper, we combine neural networks with fuzzy logic techniques. We propose an artificial neural network (ANN) model for a fuzzy logic decision system. The model consists of three layers. The first layer is an input layer. The second layer maps input features to the corresponding fuzzy membership values. The last layer implements the decision rules. The learning process consists of two phases. During the first phase the weights between the last two layers are updated using the gradient descent procedure, and during the second phase the membership functions are updated or tuned. As an illustration the model is used to classify pixels from a multispectral satellite image, a data set representing fruits, and *Iris* data.

INTRODUCTION

ANN models have been used for pattern recognition since the 1950 [Ros58]. ANN models are preferred for pattern recognition tasks because of their parallel processing capabilities as well as learning and decision making abilities. ANN models consist of a large number of highly interconnected processing units. ANN models with learning algorithms such as the backpropagation are being used as supervised classifiers, and self organizing networks with learning algorithms such as the competitive learning and Kohonen's feature maps are being used as unsupervised classifiers. Recently many neural network based fuzzy logic decision systems have been used for pattern recognition tasks. ANN models essentially provide algorithms for problems such as optimization, classification, and clustering, whereas fuzzy logic is a tool for representing and utilizing data and information that possesses non-statistical uncertainty. Fuzzy logic methods often deal with issues such as reasoning on a higher (semantic or linguistic) level. Fuzzy sets were introduced by [Zad65] as means of representing and manipulating data that were not precise but rather fuzzy. A fuzzy set is an extension of a crisp set. Crisp sets allow only full membership or no membership at all, whereas fuzzy sets allow partial membership. In a crisp set, the membership or non membership of an element x in set A is described by a characteristic function $\mu_A(x)$, where $\mu_A(x) = 1$ if $x \in A$, and $\mu_A(x) = 0$ if $x \notin A$. Fuzzy logic techniques in the form of approximate reasoning provide decision-support and expert systems with powerful reasoning capabilities bound by the minimum of rules. The permissiveness of fuzziness in human thought processes suggest that much of the logic behind human reasoning in not a traditional two valued or even multi-valued logic, but a logic with fuzzy truths, fuzzy connectives, and fuzzy rules of inference [Zad73]. The exploitation of tolerance for imprecision and uncertainty underlies the remarkable human ability to understand distorted speech, decipher sloppy handwriting, comprehend nuances of natural language, summarize text, recognize and classify images and, more generally, make rational decisions in an environment of uncertainty and imprecision. Soft computing uses the human mind as a role model and, at the same time, aims at a formalization of the cognitive processes humans employ so effectively in the performance of daily tasks [Zad94]. During the past decade fuzzy logic has found a variety of applications in various fields ranging from process control [Lee90] to medical diagnosis [Hal92].

There are many ways to synthesize fuzzy logic and neural networks. The first approach is to use signals and/or weights in a neural network as fuzzy sets along with fuzzy neurons [BH94]. Several authors have proposed models for fuzzy neurons [Gup94]. However, learning algorithms for these models are not yet developed. The second approach to synthesize fuzzy logic with neural networks is to use fuzzy membership functions to pre-process or post-process data with neural networks. The neural network can implement supervised or unsupervised learning [LL91, PM92, MP94, Bez92; NPM92]. No one of the above two approaches is proven to be superior to another. In our paper we have considered the second approach. In the decision system proposed by Lin and Lee [LL91] is an example of the second approach. In their system input/output nodes represent the input states and output control/decision signals, respectively, and in the hidden layers there are nodes functioning as membership functions and rules. The learning algorithm for their network combines unsupervised learning and supervised gradient descent learning. As an illustration they have considered two problems- the scheduling problem as a decision making problem and the fuzzy control of an unmanned vehicle as a control problem. Berenji and Khedkar [BK92] have developed a neural network based fuzzy control system wherein the learning and tuning fuzzy logic controllers is achieved through reinforcement. In reinforcement learning the teacher's response is not as direct, immediate, and informative as in supervised learning and serves more to evaluate the state of the system. Learning in their network is implemented by integrating fuzzy inference into a five-layer feed-forward network. They have used a gradient descent method to improve performance adaptive, and the fuzzy membership functions used in the definition of the labels are modified (tuned) globally to improve performance. They have considered the cart-pole problem. Pal and Mitra [PM92] have developed a fuzzy neural network model using the backpropagation learning algorithm. They have used the model to classify Indian Telugu vowel sounds and to extract inference rules. They have also developed a system which uses AND and OR logical operators to extract inference rules [MP94]. Newton et al. [NPM92] have developed an adaptive fuzzy leader clustering algorithm. They have used the concept of ART-1 [CG87]. This modification of ART-1 type neural network can be used for classification of discrete or analog patterns without prior knowledge of the number of clusters in data sets. Hal et al. [Hal92] have used approximate fuzzy c-means clustering algorithms for segmenting magnetic resonance images (MRIs) of the brain. In this paper we suggest an architecture for a neural network based fuzzy logic decision system. The model consists of four layers. The first layer is an input layer. The second layer is used for fuzzification wherein input feature values are mapped to membership functions. The last two layers implement the fuzzy inference rules. Units in the input and output layers represent input features and output categories, respectively. We have used Gaussian membership functions. The learning algorithm for our network is a supervised gradient descent procedure. The learning process consists of two phases, during the first phase weights for the inference engine are determined so as to minimize the mean squared error between the desired and actual output. During the second phase parameters for the fuzzy membership functions are updated or the membership functions are tuned so as reduce the mean squared error further. As an illustration the model is used to classify multispectral images, a fruit data set, and *iris* data.

FUZZY LOGIC DECISION SYSTEM

Fuzzy sets allow partial membership. In a crisp set membership or non-membership of an element x in a set is described by a characteristics function $\mu_A(x)$, where $\mu_A(x) = 1$ if $x \in A$ and $\mu_A(x) = 0$ if $x \notin A$. Fuzzy set theory extends this concept by defining partial membership. A fuzzy set A in the universe of discourse U is characterized by a membership function μ_A which takes values in the interval [0,1]. A fuzzy set A in U may be represented as a set of ordered pairs. Each pair consists of a generic element x and its grade of membership function; that is, A = $\{x, \mu_A(x) \mid x \in U\}$, x is called a support value if $\mu_A(x) > 0$. A linguistic variable x in the universe of discourse U is characterized by $T(x) = \{T_x^1, T_x^2, ...T_x^k\}$ and $\mu(x) = \{\mu_x^1, \mu_x^2 .. \mu_x^k\}$, where $T(x)$ is the term set of x; that is the set of names of linguistic values of x with each T_x^i being a fuzzy number with membership function μ_{xi} defined on U. For example if x indicates a reflectance value the $T(x)$ may be *low*, *medium*, or *high*. A general model of a fuzzy decision system is shown in Figure 1. The input vector x which

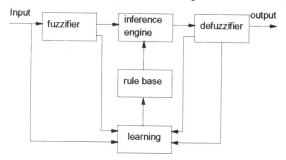

Figure 1. Fuzzy logic decision system

includes the input state linguistic variables x_i and the

output state vector y which includes the output linguistic variables y_i can be defined as

$$x = \{(x_i, U_i, \{ T_{xi}^{1}, T_{xi}^{2}, ..., T_{xi}^{ki} \}, \{ \mu_{xi}^{1}, \mu_{xi}^{2}, ..., \mu_{xi}^{ki} \}) /$$
$$i = 1 ...n\}$$
$$y = \{(y_i, U_i', \{ T_{yi}^{1}, T_{yi}^{2}, ..., T_{yi}^{li} \}, \{ \mu_{yi}1, \mu_{yi}^{2}, ..., \mu_{y}^{li} \}) /$$
$$i = 1 .. m\} \qquad (1)$$

The fuzzifier in Figure 1 performs the mapping from an observed feature space to fuzzy sets in certain input universe of discourse. A specific value x_i is mapped to the fuzzy set T_{xi}^{1} with degree μ_{xi}^{1} and the fuzzy set T_{xi}^{2} with degree μ_{xi}^{2} and so on. The fuzzy rule base contains a set of fuzzy logic rules R. For a multi-input and multi-output system,

$$R = \{R_1, R_2, ... R_n\}$$

where the ith fuzzy logic rule is

$$R_i = \text{ if } (x_1 \text{ is } T_{x1}, \text{ and } ... \text{ and } x_p \text{ is } T_{xp}),$$
$$\text{then } (y_1 \text{ is } T_{y1} \text{ and } ...\text{and } y_q \text{ is } T_{yq}) \qquad (2)$$

The p pre conditions of R_i form a fuzzy set T_{x1} x...x T_{x2} and the consequence of R_i is the union of q independent outputs. The inference engine is to match the preconditions of rules and perform implication. For example if there are two rules

$$R_1: \text{ if } x_1 \text{ is } T_{x1}^{1} \text{ and } x_2 \text{ is } T_{x2}^{1} \text{ then } y \text{ is } T_{y}^{1}$$
$$R_2: \text{ if } x_1 \text{ is } T_{x1}^{2} \text{ and } x_2 \text{ is } T_{x2}^{2} \text{ then } y \text{ is } T_{y}^{2}$$
$$(3)$$

Then the firing strengths of rules R_1 and R_2 are defined as α_1 and α_2 and are given by

$$\alpha_i = \mu_{x1}^{i} \wedge \mu_{x2}^{i} \qquad (4)$$

where $\wedge$ is the fuzzy *and* operation and is defined as

$$\alpha_i = min (\mu_{x1}^{i}, \mu_{x2}^{i}) \qquad (5)$$

The above two rules, R_1 and R_2, lead to the corresponding decision with membership function μ_y^{i}, i = 1, 2, which is defined as

$$\mu_y^{i} = \alpha_i \wedge \mu_y^{i} \qquad (6)$$

The output decision can be obtained by combining the two decisions

$$\mu_y = \mu_y^{1} \vee \mu_y^{2} \qquad (7)$$

where $\vee$ is the fuzzy or operation, which is defined as

$$\mu_y = max (\mu y^{1}, \mu y^{2}) \qquad (8)$$

Equations (5) and (8) describe the commonly used fuzzy AND and OR functions. However, these functions can be defined in many other alternative ways. The defuzzification block is required only for control systems. The most commonly used defuzzification method is the center of area method.

ARTIFICIAL NEURAL NETWORK

The ANN model for a fuzzy logic decision system is shown in Figure 2. The model consists of three layers. Each layer consists of a number of simple processing units. Layer L_1 is the input layer and layer L_2 performs the functions of the fuzzifier block shown in Figure 1. Units in L_2 may be compound units so as to implement a desired membership function. We have chosen Gaussian membership functions. However, membership functions of other shapes such as triangular or π-shaped functions can be used. Initially, membership functions are determined using the mean and standard deviation values of input variables. Subsequently, during learning these functions are updated. Layers L_2 and L_3 represent a two-layer feed-forward network. The connection strengths connecting these layers encode fuzzy rules used in decision making. In order to encode decision rules, we have used a gradient descent search technique [Pao89]. The algorithm minimizes the mean squared error obtained by comparing the desired output with the actual output. The model learns in two phases. During the first phase of learning the weights between layers L_2 and L_3 are updated so as to minimize the mean squared error. Once the learning is completed the model can be used to classify any unknown input sample. Layers in the model are described below.

Layer L_1. The number of units in this layer is equal to the number of input features. Units in this layer correspond to input features, and they just transmit the input vector to the next layer. The net-input and activation function for this layer are given by

$$net_i = x_i$$
$$out_i = net_i \qquad (9)$$

where net_i indicates the net-input, and out_i indicates the output of unit i.

Layer L_2. This layer implements membership functions. In the present case we have used five term variables {*very_low, low, medium, high, very_high*} for each input feature value. The number of units in layer L_2 is five times the number of units in L_1. The net-input and activation function for units are chosen so as to implement Gaussian membership functions which are given by

$$f(x; \sigma, m) = \exp [-\{(x-m)^2/2\sigma^2\}] \qquad (10)$$

where m represents the mean value and σ represents the standard deviation for a given membership function. The net-input and output for units in L_2 are given by

$$net_i = x_i$$
$$out_i = f(x; \sigma, m) \qquad (11)$$

Layers L_2 and L_3. These layers implement the inference engine. Layers L_2 and L_3 represent a simple two-layer feed-forward network. Layer L_2 serves as the input layer and L_3 represents the output layer. The number of units in the output layer is equal to the number of output classes. The net-input and out-put for units in L3 is given by

$$net_i = \sum_j out_j \, w_{ij} \qquad (12)$$

$$out_i = 1 / \{ 1 + \exp [-(net_i + \phi)] \} \qquad (13)$$

where out_i is the output of unit i and ϕ is a constant. Initially weights between layers L_2 and L_3 are chosen randomly, and subsequently updated during learning. The membership are initially determined based on the minimum and maximum for input features. The algorithm minimizes the mean squared between the desired and the actual outputs. The learning algorithm is described below.

Step 1: Present a continues values input vector $x = (x_1, x_2, \ldots x_n)^T$ to layer L_1, and obtain the output vector $o = (o_1, o_2, \ldots o_m)^T$ at layer L_3. In order to obtain the output vector o, calculations are done layer by layer from L_1 to L_3.

Step 2: calculate change in weights. In order to calculate change in weights the output vector o is compared with desired output vector or target vector d, and the mean squared error is then propagated backward. The change in weight Δw_{ij} is given by

$$\Delta w_{ij} = -\alpha \, \partial E / \partial w_{ij} \qquad \mathbf{(14)}$$

where α is a training rate coefficient (typically 0.01 to 1.0) and E represents the mean squared error at L_3. Equation (14) can be represented as

$$\partial E / \partial w_{ij} = \partial E / \partial o_j \, . \, \partial o_j / \partial net_i \, . \, \partial net_i / \partial w_{ij} \qquad (15)$$

where

$$\partial E / \partial o_j = -(d_i - o_i) \qquad (16)$$

$$\partial o_j / \partial net_i = \partial f(net_i) / \partial net_i$$
$$= o_i (1 - o_i) \qquad (17)$$

$$\partial net_i / \partial w_{ij} = o_j \qquad (18)$$

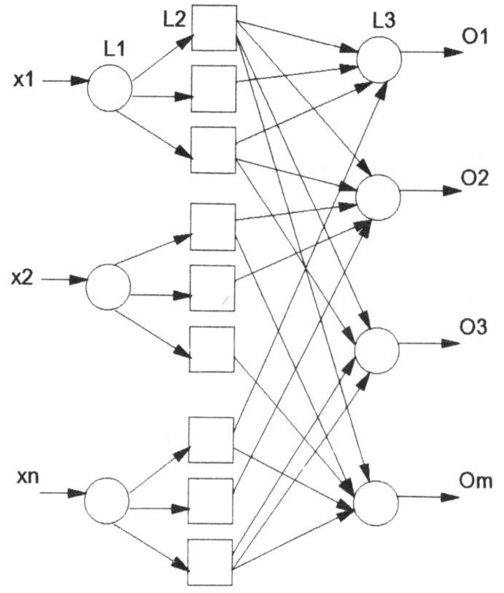

Figure 2. Fuzzy neural network model

In Equation (18) o_j represents output of unit j in L_2. Equation (14) can be rewritten as

$$\Delta w_{ij} = -\alpha \, \delta_i \, o_j \qquad (19)$$

where

$$\delta_i = o_i (1 - o_i) (d_i - o_i) \qquad (20)$$

In order to update membership function, we need to find change in parameters i. e. mean values and standard deviations that define membership functions. Again using gradient decent we get

$$\Delta m = -\beta \, \partial E / \partial m \qquad (21)$$

where

$$\partial E / \partial m - \partial E / \partial o_j \quad \partial o_j / \partial m$$

$$-\partial E / \partial o_j = \sum_{i=1}^{n} \delta_i \, w_{ij}$$

and

$$\delta_i = f'(net_i) (d_i - y_i)$$
$$= o_i (1 - o_i) (d_i - y_i) \qquad (22)$$

For Gaussian functions $\partial o_j / \partial m$ is given by

$$\partial o_j / \partial m = \exp [\{(x_k - m)^2 / 2\sigma^2 \}\{(x_k - m) / \sigma^2 \}] \qquad (23)$$

From Equations (21), (22), and (23) we get

$$\Delta m = \beta \, \exp [\{(x_k - m)^2 / 2\sigma^2 \}\{(x_k - m) / \sigma^2 \}] \sum_{i=1}^{n} \delta_i \, w_{ij} \qquad (24)$$

Similarly for changes in standard deviation values we get

$$\Delta\sigma = -\gamma \, \partial E/\partial\sigma \quad (25)$$

where $\quad \partial E/\partial\sigma = \partial E/\partial o_j \; \partial o_j/\partial\sigma$

From Equation (25) we get

$$\Delta\sigma = \gamma \exp[\{(x_k - m)^2/2\sigma^2\}\{(x_k - m)/\sigma^3\}] \sum_{i=1}^{n} \delta_i \, w_{ij}$$

$$(26)$$

Step 3: Update weights and membership functions

$$\begin{aligned}
w_{ij}(k+1) &= w_{ij}(k) + \Delta w_{ij} \\
m_i(k+1) &= m_i(k) + \Delta m_i \\
\sigma_i(k+1) &= \sigma_i(k) + \Delta\sigma_i \quad (27)
\end{aligned}$$

The update procedure can be implemented in two phases. During the first phase we update weights and consider membership functions as constants, where as during the second phase we update membership functions and keep the updated weights unchanged.

Step 4: Obtain the mean squared error at L_3.

$$\varepsilon = \sum_{i=1}^{m} (o_i - d_i)^2 \quad (28)$$

If the error is greater than some minimum value ε_{min} then repeat Steps 2 through Step 4.

COMPUTER SIMULATION

We have developed software to simulate the fuzzy-neural network decision system. As an illustration the system was used in three illustrative examples: a) recognition of pixels in remotely sensed multispectral images, b) classification of fruits, and c) classification of Iris data samples. In our first illustration we have used the fuzzy-neural network model as a supervised classifier to classify pixels in a remotely sensed image. The technique of remote sensing to a great extent relies on the interaction of electro-magnetic radiation with the matter. The remotely measured signal expressed as a function of the wave length is referred to as the "spectral signature" of the object on which measurements have been made. In principle, spectral signatures are unique; that is; different objects have different spectral signatures [Kul94]. It is therefore possible to identify an object from its spectral signature. The pixels are classified based on their spectral signatures. We have used data that are obtained from a sensor called Thematic Mapper

(TM), which is a multispectral scanner that captures data in seven spectral bands. The sample feature space, with only two feature values, is shown in Figure 3. The original image for spectral band 5 is shown in Figure 4a, and the classified output is shown in Figure 4b. The scene represents the Mississippi river bottom land area, and is of the size 512x512. We used five linguistic term values {*very low, low, medium, high, very high*} to represent a gray value of a pixel in the scene. Each pixel was represented by a vector of five gray values. We used only five bands (bands 2, 3, 4, 5, and 7), because these bands showed the maximum variance and contained information needed to identify various classes. During the training phase the network was trained using training set data. We selected four training set areas, each of the size 10 scans x 10 pixels. These represent four different classes. Each class is represented by a small homogeneous area. Only a small fraction of samples (400 pixels) of the entire data set (256134 pixels) were used as training samples. We used five units in layer L_1. These units correspond to the five gray values that represent a pixel. Layer L_2 contained twenty-five units. The twenty-five units in layer L_2 correspond to the twenty-five term values, five for each band. The four units in the output layer represent four output categories. The four units in the output layer correspond to the four categories. The target output vectors for four classes were defined as (1,0,0,0), (0,1,0,0), (0,0,1,0), (0,0,0,1). These vectors represented classes 1 through 4, respectively. The network was trained with training set data and training set data was reclassified to check the accuracy. The training set data was reclassified with 100 percent accuracy. In the decision making phase the entire scene was classified. In this examples we have chosen four classes which are non-overlapping in the feature space. The learning was accomplished in two phases. During the first phase the weights between layers L_2 and L_3 were updated using the gradient descent process. These layers implement the inference engine. The weights were adjusted so as to minimized the mean squared error between the actual and the desired output values. During the second phase, the mean squared error was further minimized by tuning the membership functions. The curve representing the mean squared error versus the number of iterations is shown in Figure 5.

In our second example, we considered fruit data. We consider three types of fruits: grapefruits, lemons, and apples. The weight and volume were used as input features for classification. The classes were well separated in the feature space. The feature space showing fruit samples is shown in Figure 6. We had two input features and three output classes. We had two units in L_1, ten units in L_2, and three units in L_3. The model was trained using training samples. The fuzzy membership functions before and after learning are

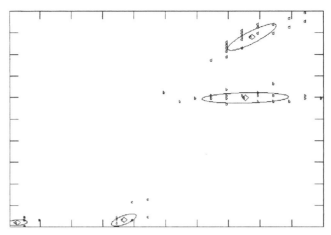

Figure 3. Feature space, Mississippi Scene, Bands 5 and 7

Figure 4a. Raw Data, Mississippi Scene

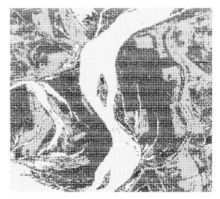

Figure 4b. Classified output, Mississippi Scene

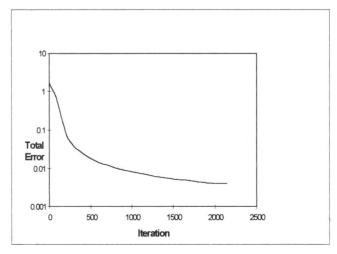

Figure 5. Mean squared error versus iterations, Mississippi Scene

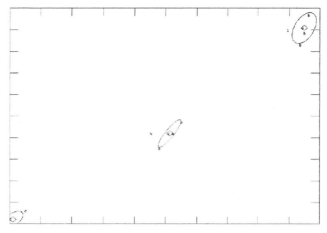

Figure 6. Feature Space, Fruit Data

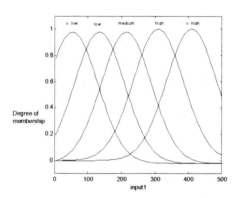

Figure 7a. Fuzzy membership functions,
fruit data, (before Learning)

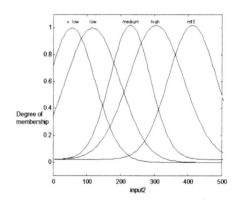

Figure 7b. Fuzzy Membership functions,
fruit data, (after learning)

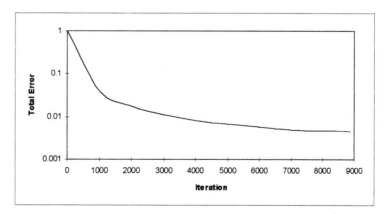

Figure 8. Mean squared error verses iterations, fruit data

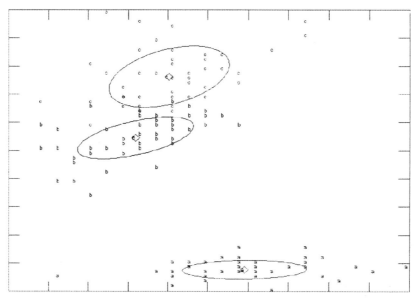

Figure 9. Feature Space (Iris data)

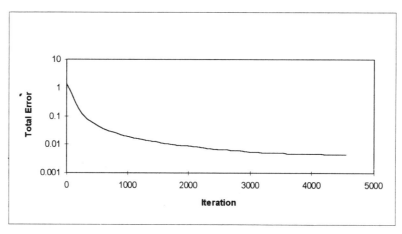

Figure 10. Mean squared error verses iterations, Iris data

shown in Figure 7a and 7b, respectively. The training set data was reclassified with one hundred percent accuracy. The curve showing the mean squared error versus the number of iterations is shown in Figure 8.

In the third example, we considered the well known Iris data set. The feature space for data samples for iris data is shown in Figure 9. We have chosen this data, because it contains over-lapping clusters in the feature space. This data set contains one hundred fifty samples. Each sample is represented by four features, and there are three output classes. Layer L_1 contained four units representing four features, layer L_2 contained twenty units which represent twenty term values, and layer L_3 contained three units that correspond to three output classes. Mean vectors for the three classes were used to train the classifier. In order to test the classifier all one hundred fifty samples were classified, the classification error was found to be 7.33 percent. Since, this data set represents overlapping clusters, one hundred percent correct classification in this case is not possible. The curve showing the mean squared error versus the number of iterations is shown in Figure 10.

DISCUSSION AND CONCLUSIONS

The satellite image in the above examples was classified also with a simple three layer BP network. We found that fuzzy-neural network decision system could converge faster than a simple backpropagation network. However, the fuzzy logic decision system has increased dimensions. Also with a fuzzy logic decision system we can interpret decision rules in terms of linguistic variables. This can be achieved by displaying the fuzzy membership values at layer L_4 and the output decision vector. In the present case we used Gaussian membership functions with twenty-five percent overlap. However, other functions such as triangular and π-functions with different overlaps can be used for fuzzification. We have chosen the minimum and maximum values for membership functions from histograms obtained from training set data. We also found that in the case of non-overlapping classes, often it is possible to minimize the mean squared error below any desired valued ε_{min} by just adjusting weights that encode fuzzy inference rules, and fine tuning of membership functions may not be always necessary. In conclusion we can say that neural networks represent a powerful and reasonable alternative to conventional classification methods. Our experiment suggest that by combining fuzzy logic with neural networks we can develop more efficient decision systems.

REFERENCES

[BK92] H. R. Berenji, and P. Khedkar. Learning and tuning fuzzy logic controllers through reinforcements. *IEEE Transactions on Neural Networks*, 3: 724-740, 1992.

[Bez92] J. C. Bezdek. Computing with uncertainty. *IEEE Communications Magazine*, 30:24-36. 1992.

[Bez93] J. C. Bezdek. Editorial-fuzzy models-what are they and why. *IEEE Transactions on fuzzy systems*, 1: 1-5, 1993.

[BH94] J. J. Buckley, and Y. Hayashi. Fuzzy neural networks. In *Fuzzy Sets, Neural Networks, and Soft Computing*. Editors: Yager, R. R. and Zadeh, A., Van Nostrand, New York, 233-249, 1994.

[CG87] G. A. Carpenter, and S. Grossberg, S. A massively parallel architecture for a self organizing neural pattern recognition machine. *Computer Vision Graphics and Image Processing*, 37: 54-115, 1987.

[Gup94] M. M. Gupta. Fuzzy neural networks: theory and applications. *Proceedings of SPIE*, 2353: 303-325, 1994.

[Hal92] L. O. Hall, et al. A comparison of neural network and fuzzy clustering techniques in segmenting magnetic resonance images of the brain. *IEEE Transactions on Neural networks*, 3: 672-682, 1992.

[Kul94] A. D. Kulkarni. *Artificial neural networks for image understanding*. Van Nostrand Reinhold. New York, NY, 1994.

[Lee90] C. C. Lee. Fuzzy logic in control systems: Fuzzy logic controller Part I. *IEEE Transactions on Systems, Man, Cybernetics*, 1:404-418.

[LL91] Chin-Teng Lin, and George C. S. Lee. Neural network based fuzzy logic control and decision system. *IEEE Transactions on Computers*, 40: 1320-1336, 1991.

[NPM92] S. C. Newton, S. Pemmaraju, and S. Mitra. Adaptive fuzzy leader clustering of complex data sets in pattern recogniticn. *IEEE*

Transactions on Neural Networks, 3: 794-800, 1992.

[PM92] S. K. Pal and S. Mitra. Multilayer perceptron, fuzzy sets, and classification. *IEEE Transactions on Neural Networks*, 3: 683-697, 1992.

[MP94] S. Mitra, and S. K. Pal, S. K. Logical operation based fuzzy logic MLP for classification and rule generation. *Neural Networks*, 7: 353-373, 1994.

[Pao89] Y. H. Pao. *Adaptive pattern recognition and neural networks*. Addison-Wesley, Reading, MA, 1989.

[Ros58] F. Rosenblatt. The perceptron: A probabilistic model for information storage and organization in brain. *Psychology Review*, 65: 368-408, 1958.

[Van91] M. V. Vannier, et al. Validation of magnetic resonance imaging (MRI) multispectral tissue classification, *Computer Medical Imaging and Graphics*, 15: 217-223, 1991.

[Zad65] L. A. Zadeh. Fuzzy sets. *Information and Control*, 8: 338-352, 1965.

[Zad73] L. A. Zadeh,. Outline of a new approach to analysis of complex systems and decision processes. *IEEE transactions on Systems, Man, and Cybernetics*, 3: 28-44, 1973.

[Zad94] L. A. Zadeh. Fuzzy logic, neural networks, and soft computing. *Communications of the ACM*, 37 :77-84, 1994..

CAUSAL GRAPHS AND RULE GENERATION: APPLICATION TO FAULT DIAGNOSIS OF DYNAMIC PROCESSES

M. Ouassir, C. Melin

Heudiasyc Laboratory URA CNRS 817
Université de Technologie de Compiègne
Centre de Recherches de Royallieu
BP 529 COMPIEGNE Cédex 60205 FRANCE
Tél.: (00) 3 44 23 44 23
Fax : (00) 3 44 23 44 77
e-mail : majid.ouassir@hds.utc.fr
e-mail : christian.melin@hds.utc.fr

Abstract. This paper describes a systematic off-line algorithm for constructing a rule-based fault diagnostic system using the signed digraph (SDG) model. In order to overcome certain limitations related to the SDG (Iri and al. 1979), (Kramer and Palowitch 1987) and (Oyeleye and Kramer 1988), we have developped an alternative strategy for eliminating spurious interpretations attributed to system compensations and inverse responses from backward loops and forward paths in the process when fault exist. Moreover, this method apply fuzzy logic to represent the rule base, in order to explain fault propagation and to ascertain fault origins. The algorithm was then successfully applied to a level controlled tank process to extract rules and these rules were evaluated using on-line data to discover the fault origins in an early stage and perform the diagnosis.

Keywords. Causal Modelling ; Qualitative propagation ; Fault detection ; Fault Diagnosis ; signed digraph ; Fuzzy Sets

1 INTRODUCTION

Signed Directed Graph (SDG) uses concept from graph theory to represent the cause-effect relationships between variables in process systems. The nodes of the SDG correspond to variables and the branches represent the causal influences between nodes. The influences are represented by $\{-, +\}$ on the arcs, indicating that the cause and effect variables tend to change in the same or opposite direction. It is a pure qualitative approach, since the permissible values assigned to the node variables are restricted only to the set $\{-, 0, +\}$, with - for low, 0 for normal and + for high deviation. It is a well recognised method but has several limitations:

a)- Iri and al. (1979): The currently availible SDG algorithme of Iri used the initial direction of deviation as the basis for diagnosis, but did not include complex dynamic system variable activities such as non single transitions and ultimate responses caused by compensatory and inverse responses.

b)- Kramer and Palowitch (1987): In their treatment of certain loops and forward paths, preserved the dominant path of fault propagation and eliminated non dominant ones to simplify SDG and improve resolution. Their diagnosis method establishes a rule base through SDG, but does not suggest a way to handle negative feedback loops and negative forward paths on the uncontrolled loops.

a)- Oyeleye and Kramer (1988): In order to account for ultimate response phenomena, they applied the DeKleer and J.S.Brown (1984) qualitative physical approch and the property that the system will eventually achieve a steady state. They added a non physical forward path to SDG to form an extended SDG for explaining the propagation paths of feedback loops attributed to their compensatory and inverse responses.

However, as with Kramer and Palowitch (1987), experience, skill, and understanding are required to judge which path is the dominant of fault propagation.

c)- Ulerich N. H.; Powers (1988): developed a causal fault tree for on-line hazard calculation and fault detection from digraphs. They expanded verification gates with dynamic cause of delay to illustrate fast and slow responses; they did not describe how those verification gates were derived merely from digraphs.

Some researchers make an effort by using certain quantitative data to improve the diagnostic resolution of SDG. However, these numerical data must be obtained through expert experience, and experts themselves are often at a loss to clearly state the magnitude of these quantities.

The purpose of this paper is to show how diagnostic rules can be obtained from SDG and how fuzzy logic concept can be integrated into diagnostic rules so as to provide a qualitative interpretation and eliminate spurious interpretations attributed to system compensations and inverse responses from backward loops and forward paths in the process. The algorithm and rules are then tested by developing a fault diagnosis procedure for a tank level controller process .

2   A SYSTEMATIC KNOW-LEDGE EXTRACTION

The goal of this research is to discover the fault origins in an early stage and to reduce the number of spurious interpretations. Because it is difficult to obtain the expert experience needed to compile a rule base for a fault diagnosis system. In this research, we propose to use a SDG which serves as a basic structure to narrow the search space. To reduce spurious interpretations, we try to replace the original SDG by a set of equation of constraint (Fig. 1). The paths in de SDG and constraint equations are used in ordre to generate the diagnosis rules. In the algorithm, we permet to the variables to change their values from normal to low, from normal to high , from low to normal and from high to normal with the use of Oyeleye and Kramer (1988) rules with two constraints will be used in the algorithm: addition and multiplication constraintes (DeKleer and J.S.Brown

1984) (Forbus 1986) as follow:

Rule 1. Any node can change its sign provided that the new sign equals the net influence on the node. (Note that the ambiguous influence, ?, matches any node sign.)

Rule 2. Direct transitions from (+) to (−) and (−) to (+) are forbidden.

Rule 3. A node is permitted to change its sign only if there is a simple causal path between the new even and the primary deviation.

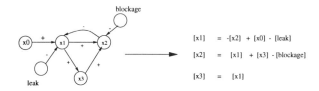

Fig. 1.   Signed directed graph (SDG) and Constraint Equations

Algorithm of rule generation

The algorithm of derivation is brievly described as follow: The   set of observed variables is $\{E_s^1, E_s^2, ...E_s^n\}$ and the set of non-observed variables is $\{E^1, E^2, ...E^m\}, with (n \leq m)$,

(0) Initially : nodes E^k et E_s^k $(k = 1..;m, n)$ are
     set to 0
     While Fault list is Not empty do
(1) Assume fault P_i to + (or −)
     Propagat this fault by constraint equations to E^k
     with the use of Kramer rules
     If E^k is attached to E_s^k in the SDG then
     $E_s^k = E^k$ and write the rule:
     If (E_s^1) And ...And (E_s^n) then P_i
(2) For each output variables E^j attached to E^k in the SDG
     Calcul the influence E^j (Kramer rules)
     If E^j is attached to E_s^j in the SDG then
     $E_s^j = E^j$ and write the rule :
     If (E_s^1) And...And (E_s^n) Then P_i
(3) If the initial state is reached, then stop the algorithm

3 APPLICATION

To demonstrate our approch for diagnosis analysis, we consider a system and its SDG in figure 2. The process is a tank level controlled with a PI controller.

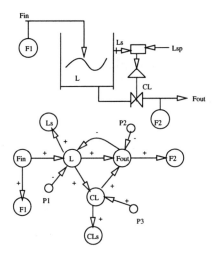

Fig. 2. A level controlled tank process and its associated SDG

F_1	Water in-flow for the tank
F_2	Water out-flow through the valve
L_s	Level at the output of level sensor
L_{sp}	Level set point
CL_s	Valve opening

TABLE 1 Observed variables for On-Line Diagnosis

P_1	leak in the tank
P_2	blockage in the tank outlet line
P_3	failure in the control valve or Level controller

TABLE 2 Fault Origins

In order to test our method, we use a quantitative simulation (mathematical equations) of the process. The measured variables and the control signals available for fault diagnosis system are shown in Table 1. Table 2 shows the fault origins in the system and Figure ??

3.1 Generation of rules

The root failure chosen for purposes of exemple is $P_2 = +$, blockage in the tank outlet line, $P_1 = -$ and leak in the tank.

The constraint equations obtained from the SDG of the process are shown in table 3

(1) $[L] = [F_{in}] - [F_{out}] - [P_1]$
(2) $[F_{out}] = [L] + [CL] - [P_2]$
(3) $[CL] = [L]$
(4) $[L_s] = [L]$
(5) $[CL_s] = [CL] - [P_3]$
(6) $[F_2] = [F_{out}]$
(7) $[F_1] = [F_{in}]$

TABLE 3 qualitative constraints for the example

Example: rules for P_2 .

Initial State: $[L_s] = 0, [CL_s] = 0, [F_2] = 0, [F_1] = 0,$ $[P_1] = 0, [P_2] = 0$ *and* $[P_3] = 0$
If $[P_2] = +$ blockage in the tank outlet line
by eq.2 $[F_{out}] = [L] + [CL] - [P_2] = 0 + 0 - (+) = -$
F_{out} attached to F_2 so
by eq.6 $[F_2] = [F_{out}] = +$
We write the rule
If $[Ls] = 0$ and $[CLs] = 0$ and $[F2] = -$ and $[F1] = 0$
then P_2
by eq.1 $[L] = [F_{in}] - [F_{out}] - [P_1] = (0) - (-) - (0) = +$
L is attached to à L_s so:
$[L_s] = [L] = +$
We write the rule
If $[Ls] = +$ and $[CLs] = 0$ and $[F2] = -$ and $[F1] = 0$
Then P_2
by eq.4 $[CL] = [L_s] = +$
CL is attached to CL_s so:
$[CL_s] = [CL] = +$
We write the rule
If $[Ls] = +$ and $[CLs] = +$ and $[F2] = -$ and $[F1] = 0$
Then P_2
by eq.2 $[F_{out}] = (+) + (+) - (+) = (+) - (+) = ?$
The use of Kramer Rules then $[F_{out}] = 0$
F_{out} is attached to F_2 so:
$[F_2] = [F_{out}] = 0$
We write the rule
If $[Ls] = +$ and $[CLs] = +$ and $[F2] = 0$ and $[F1] = 0$
then P_2
by eq.1 $[L] = [F_{in}] - [F_{out}] - [P_1] = (0) - (0) - (0) = 0$
L is attached to L_s so:
$[L_s] = [L] = 0$
We write the rule
If $[Ls] = 0$ and $[CLs] = +$ and $[F2] = 0$ and $[F1] = 0$
then P_2
by eq.3 $[CL] = [L_s] = 0$
CL is attached to CL_s so:
$[CL_s] = [CL] = 0$
We write the rule
If $[Ls] = 0$ and $[CLs] = 0$ and $[F2] = 0$ and $[F1] = 0$
then P_2

Here we reach the initial state (loop)
Algorithme terminated for this fault (P_2)

Then, we have obtened a set of 4 rules for P_2 and this set describes different states of process variables (observed) for a fault origin F_2 as follow:

Rule.1 If $[L_s] = +$ And $[CL_s] = 0$ And $[F_2] = -$ And $[F_1] = 0$ Then P_2

Rule.2 IF $[L_s] = +$ And $[CL_s] = +$ And $[F_2] = -$ And $[F_1] = 0$ Then P_2

Rule.3 IF $[L_s] = 0$ And $[CL_s] = +$ And $[F_2] = 0$ And $[F_1] = 0$ Then P_2

Rule.4 IF $[L_s] = 0$ And $[CL_s] = +$ And $[F_2] = 0$ And $[F_1] = 0$ Then P_2

The set of rules for the failures P_1 and P_3 are derived in the same way

Comparation and contribution:
Diagnostic Algorithm of Iri and al. (1979): Following the algorithm of Iri and al. (1979), we would find the unique rule 2. We can not find the fault origin P_2 if we observe the pattern rule 1, rule 3 and rule 4. None of these rules (1,3,4) meet the criteria of Iri's diagnostic algorithm. The pattern observed is $\{[L_s] = +, [CL_s] = +, [F_2] = -, [F_1] = 0\}$, and the composant CFCM (Iri and al. 1979) corresponding is shown in Figure 3

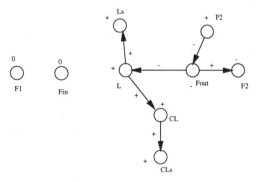

Fig. 3.   the CFC the SDG for the pattern
$\{[L_s] = +, [CL_s] = +, [F_2] = -, [F_1] = 0\}$

Oyeleye and Kramer (1988): could not explain the system's compensation or inverse responses if they have not skill and experiences in order to judge which is the dominant or nondominant path of fault propagation,

and they require the condition branch to represent different process states (Oyeleye and Kramer 1988). Figure 4, shows the condition branch in order to handle system's compensation or inverse responses .

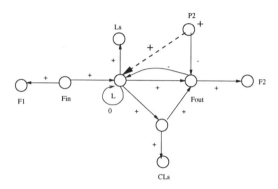

Fig. 4.   the extended SDG with condition branch

We have confidence that the proposed approach can improve the resolution of the previous SDG-based diagnosis methods.

3.2   Fuzzy sets representation of rules

Fuzzy logic presents a measure theory for handling linguistic terms, and thus, both analytical and heuristic symptoms can be embedded as fuzzy rules within the rule base diagnosis system. In this way, Each rule set associated to fault origine can be sinply translated into a fuzzy rule set in the following form:

$$\text{IF}(s_1 \text{ is } TS_{11})...And...(s_n \text{ is } TS_{n1}) \text{ THEN } (f_k \text{ is } TF_1)\}$$

......

$$\text{IF}(s_1 \text{ is } TS_{1m})...And...(s_n \text{ is } TS_{nm}) \text{ THEN } (f_k \text{ is } TF_m)\}$$

where s_i and f_k are linguistic variables (LV) (Dubois and H.Prade 1980), defined on universes S_i and F, respectively; TS_{ij} is the jth predicate of the ith input LV, where TS_{ij} can be defined as:

$$TS_{ij} = \{(s_i, \mu_{TS_{ij}}(s_i)) | s_i \in S_i\}$$

$\mu_{TS_{ij}}(s_i) \in [0, 1]$ is the degree of membership of LV to the fuzzy subset TS.

Zadeh (Zadeh 1965) also interprets $\mu_{TS_{ij}}(s_i)$ to be the possibility that the proposition is true for the value s_i. The logical operator which describes the relations among the input LV is the *min* opérator (Zadeh 1965).

inference system. Now, given real values for the antecedents variables $(s_1, s_2, ..., s_n)$, the forward inference task has to determine the possibility of the conclusion propositions assuming the rules are true. The membership grade for the antecedent variables are first calculated and the degrees of rules fulfilment $(\mu_{R1}, ..., \mu_{Rm})$ are then aggregated by *max* operator.

Finally, the possibility that the conclusion propositions are true, is calculated through cutting or scaling the membership functions $(\mu_{TF_1(f_k)}, ..., \mu_{TF_m(f_k)})$ by the values $(\mu_{R1}, ..., \mu_{Rm})$.

In our work, the linguistic variables LV are Negative, Zero and Positive, with Negative = - , Zero = 0, and Positive = +. Figure 5 shows the fuzzy representation of observed variables $\{L_s, CL_s, F_2, F_1\}$

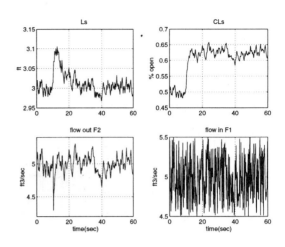

Fig. 6. System with a small blockage (P_2) at t=10s

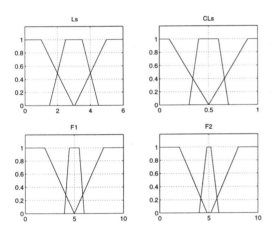

Fig. 5. Fuzzy Representation of Rules of the example of figure 2

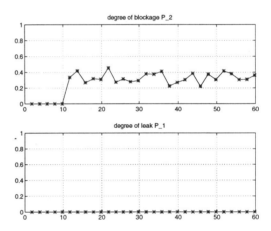

Fig. 7. fuzzy inference of P_1 and P_2 (small blockage)

3.3 *Simulation results*

Two fault cases are studied to test the proposed method in this paper. The fault origins are P_1 and P_2 2. The diagnosis algorithm is programmmed in Matlab and C.

Fault case F_2: we inject the failure P_2 (a small Blockage in the outlet) at $t = 10s$. Figure 6 shows the failure mode of P_2 at t=10s and the dynamic changes of process variables. the detailed diagnosis analysis procedure after process subjected to this fault is described as follows (Note that the procedure is executed in parallel process form and samples on-line measurement of (L_s, CL_s, F_1, F_2) every 2 s and makes possible fault interpretation):

Cause identification process : The inference process corresponding to the failure P_2 and P_1 is shown in figure 7. The inference process gives the degree of membership of each fault. The proposed diagnosis system analysed the situation at 10s after the process alerted to a fault as follows:

- $SL = \{P_1, P_2\}$ a set of fault origin
- the degree of P_1 is 0 all times
- Evolution of degree of P_2 in the time

This means P_2 is the most possible fault origin obtained by the proposed algorithm after the inference.

Fault case F_1: we inject the failure P_1 (a small Leak in the bac) at $t = 10s$. Figure 8 shows the failure mode

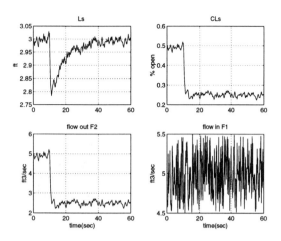

Fig. 8. System with a small leak (P_1) at t=10s

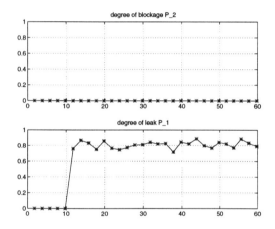

Fig. 9. fuzzy inference of P_1 and P_2.(small leak)

of P_1 at t=10s and the dynamic changes of process variables.

Cause identification process : The inference process corresponding to the failure P_1 and P_2 is shown in figure 9. The inference process gives the degree of membership of each fault.

The proposed diagnosis system analysed the situation at 10s after the process alerted to a fault as follows:

- $SL = \{P_1, P_2\}$ a set of fault origin

- the degree of P_2 is 0 all times

- Evolution of degree of P_1

This means P_1 is the most possible fault origin obtained by the proposed algorithm after the inference.

In this two situation of faults P_1 and P_2, the diagnostic système detect their deviations and decide that they cause failures in the system (see Figure 7 and Figure 9). But the control loop (see Figure 6 and Figure 8) reduce the influence of deviations of the process with regard to the desired input/output behaviour.

Fault case F_2: Now in Figure 10 and Figure 11, when we have injected a increasing blockage in the time.

Fault case F_1: the same in Figure 10 and Figure 11, we have injected a increasing leak in the time.

From this simulation results, using the inference process could explain the system's compensation or inverse responses that do not need to judge which is the dominant or nondominant path of fault propagation (Kramer and Palowitch 1987), and do not require the condition branch to represent different process states (Oyeleye and Kramer 1988). We have confidence that we can generate the set of rules from the SDG describing the system's behavior, though some inaccuracy on the deduced values exists which cannot be avoided because of the nature of the fuzzy representation.

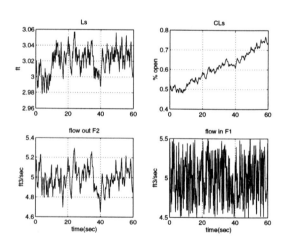

Fig. 10. System with a incresing blockage (P_2) at t=10s

4   DISCUSSION AND CONCLUSION

The proposed approach has merits when compared to Iri and al. (1979) and Kramer and Palowitch (1987). It offers a solution to some known problems associated

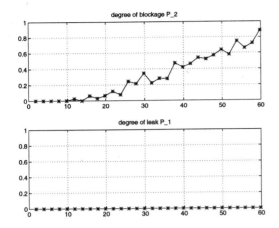

Fig. 11. fuzzy inference of P_2 (incresing blockage)

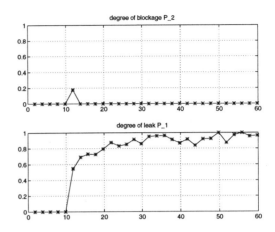

Fig. 13. fuzzy inference of P_1 (increasing leak)

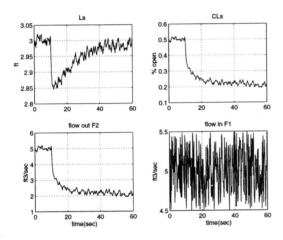

Fig. 12. System with a increasing leak (P_1) at t=10s

integral controller, is used as an example to illustrate the effectiveness of the proposed method. Simulation results show that the proposed method gives significant improvement over the conventional approach in reducing spurious or erroneous interpretations.

REFERENCES

DeKleer, J. and J.S.Brown (1984). A qualitative physics based on confluences. In: , *Artif. Intell.*. Vol. 24.

Dubois, D. and H.Prade (1980). Fuzzy sets and systems. In: *Academic Press: New York*. Vol. Part 3.

Forbus, K.D. (1986). Qualitative process theory. In: *Artificial Intelligence*. Vol. 29.

Iri and al. (1979). An algorithm for diagnosis of system failures in the chemical process. In: *Computers Chem. Eng.*. Vol. 3. Printed in Great Britain. pp. 489–493.

Kramer and Palowitch (1987). A rule-based approach to fault diagnosis using the signed direct graph. In: *AICHE J.*. Vol. 33. pp. 1067–1078.

Oyeleye and Kramer (1988). Qualitative simulation of chemical process systems: Steady-state analysis. In: *AIChE Journal*. Vol. 34, No.9. pp. 1441–1454.

Ulerich N. H.; Powers, G.J. (1988). On-line hazard aversion and fault diagnosis in chemical processes: The digraph + fault-tree method. In: *IEEE Trans. Reliab.*. Vol. 37.

Zadeh, L. A. (1965). Fuzzy sets. In: *Inf. Control*. Vol. 8.

with the SDG: (1) it produces spurious (multiple) interpretations as a consequence of the qualitative nature of the SDG and (2) it may produce an erroneous (incorrect) interpretation when the node variables go through nonsingle transition.Our approch is capable of handling the nonsingle-transition problem in the controlled variable without modification of the origin SDG, eg., ESDG or conditional branch approach (Kramer and Palowitch 1987) ,(Oyeleye and Kramer 1988) or simplification according to states approch (Iri and al. 1979).

We have confidence that the proposed approach can improve the resolution of the previous SDG-based diagnosis methods, through some inaccuracy exists in the deduced value, which cannot be avoided because of caracteristics of the fuzzy representation.

A level controlled tank process with a proportional-

DEFEASIBLE LOGIC ON AN EMBEDDED MICROCONTROLLER

Michael A. Covington

Artificial Intelligence Center, The University of Georgia
Athens, GA 30602-7415, U.S.A.
E-mail: mcovingt@ai.uga.edu

ABSTRACT

Defeasible logic is a system of reasoning in which rules have exceptions, and when rules conflict, the one that applies most specifically to the situation wins out. This paper reports a successful application of defeasible logic to the implementation of an embedded control system. The system was programmed in d-Prolog (a defeasible extension of Prolog), and the inferences were compiled into a truth table that was encoded on a low-end PIC microcontroller.[1]

Advantages of defeasible logic include conciseness and correct handling of temporal succession. It is distinct from fuzzy logic and probabilistic logic, addressing a different set of problems.

DEFEASIBLE LOGIC

Consider the following rules for controlling an air conditioner:

(1) Run the air conditioner when the temperature is over 78°.

(2) Do not run the air conditioner when the temperature is above 78° but below 81° and the AC line voltage is low (to reduce demand when the power company is heavily loaded).

(3) Never run the air conditioner if it was turned off less than 4 minutes ago (to protect the compressor).

To a human observer, these seem like perfectly reasonable rules, but according to classical logic, they are contradictory. Suppose the temperature is 79° and the line voltage is low. Rule (1) says to run the air conditioner and rule (2) says not to. Likewise, rule (3) conflicts with rules (1) and (2) whenever the temperature is high but the air conditioner has just shut off.

---

[1]I want to thank Donald Nute, David Billington, and Don Potter for encouragement and assistance with this project.

The reason we humans do not notice the contradiction, or at least do not object to it, is that these rules conform to a familiar pattern of human reasoning known as DEFEASIBLE LOGIC [N92] — logic that can change its mind, logic in which rules can have exceptions. We find it intuitively reasonable for the more specific rules to override the more general ones.

Defeasible logic is what we use when reasoning with incomplete information, so there is always the possibility of a more specific rule being invoked as we learn more about the situation. Classical logic is a model of how we reason when we are sure we know all the relevant facts — which, in real life, is not often. Classical conclusions are guaranteed true, but defeasible conclusions are only apparently true, based on the available evidence. To put this another way, defeasible logic is NON-MONOTONIC; further information can cause it to abandon a conclusion that it would have reached from fewer premises.

As stated, the rules are not quite a full set. They do not say what to do when the temperature is below 78°, nor what to do when the temperature seems to be above 81° but below 78° (an impossible situation). We can remedy this by adding two more rules:

(4) Do not run the air conditioner unless the other rules say to do so.

(5) If the temperature is over 81°, then it is necessarily also over 78°.

Rules (3) and (5) are ABSOLUTE rules, as indicated by the words "never" and "necessarily" — no other information can override them. Rules (1), (2), and (4) are DEFEASIBLE rules, which means that they can be overridden by absolute rules and by defeasible rules that are more specific.

Defeasible logic provides a natural, concise way for human beings to describe the conditions under which things should happen. The remainder of this paper deals with how to translate defeasible rules into a program for an embedded controller.

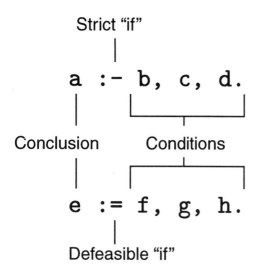

Figure 1   Strict and defeasible rules.

D-PROLOG

Defeasible inference engine

An extension of Prolog called d-Prolog that performs defeasible inference has been developed by Donald Nute [N96, N97]. d-Prolog is implemented in Quintus Prolog as an extension to the inference engine.

As in Prolog, d-Prolog rules are of the form "A if B and C and D..." with any number of conjoined conditions (Fig. 1). Strict rules denote "if" by the symbol :- and cannot be overridden; if the conditions are true, so is the conclusion. Defeasible rules denote "if" by := and are overridden by strict rules and by defeasible rules that take more information into account. For the complete mechanism see [N96, N97]; not all of it is used in this paper.

Unlike Prolog, d-Prolog expresses negation explicitly by the prefixed operator neg. Thus, given a query X, the d-Prolog inference engine can conclude any of four different things:

- "Yes" (X can be inferred and neg X cannot).

- "No" (neg X can be inferred but X cannot).

- "I don't know" (neither X nor neg X can be inferred).

- "Contradiction" (both X and neg X can be inferred).

In addition, a "yes" or "no" answer can be absolute or defeasible depending on whether defeasible rules were used in deducing it.

Figure 2 shows how the air conditioner control rules are expressed in d-Prolog. The inputs are four bits, just_off (compressor was turned off less than 4 minutes ago),

volt_low (line voltage is low), over_81, and over_78 (temperature over 81° and 78° respectively). The conclusion is run_ac (run the air conditioner) or neg run_ac (don't run the air conditioner).

In Rule 4, true is a dummy condition that is always true. Thus, Rule 4 establishes a DEFAULT: "Run the air conditioner unless some other rule specifies otherwise." The d-Prolog inference engine knows that true takes no information into account and is therefore overruled by a defeasible rule with any other set of conditions.

Intermediate steps of reasoning are of course permitted, although there happen to be none in this example. A chain such as

```
a := b.
b := c.
c := d.
d := e.
```

(where b, c, and d are neither inputs nor outputs) is just as legitimate in d-Prolog as in ordinary Prolog.

MICROCONTROLLER IMPLEMENTATION

Hardware

Although this rule set is hardly large enough to demonstrate the full power of defeasible logic, a prototype air conditioner controller based on it has been implemented on a PIC16F84 microcontroller. The prototype controls an LED rather than an air conditioner, has switches rather than thermostats, and uses 4 seconds rather than 4 minutes as its timeout cycle. Nonetheless, it demonstrates that the control logic has been implemented correctly.

The PIC16F84 has 1K words of program memory, 64 bytes of RAM, and, in this application, runs at 0.1 MHz, and costs $6.48 in single quantities. It was chosen for convenience in programming and testing, since even cheaper processors are adequate for this application, such as the PIC12C508 ($1.50).

Truth table

Because of its small size, the PIC does not run the d-Prolog inference engine directly. Instead, it stores a truth table that was generated by running d-Prolog on a Sun workstation.

The truth table is generated as follows:

1. Take the list of inputs (inputs in Fig. 2) and negate some or all of its elements.

2. Assert the resulting set of premises (some affirmative and some negative).

```
run_ac := over_78.                        % Rule 1

neg run_ac := volt_low, over_78, neg over_81.    % Rule 2

neg run_ac :- just_off.                   % Rule 3

neg run_ac := true.                       % Rule 4

over_78 :- over_81.                       % Rule 5

inputs([just_off,volt_low,over_81,over_78]).    % declarations needed by
outputs([run_ac]).                        % truth table generator
```

Figure 2   Air conditioner control rules in d-Prolog.

3. Determine whether each of the outputs (also declared in Fig. 2) is true or false. If a contradiction or a "don't know" state is discovered, issue a warning message. (The need for rules (4) and (5) in the air conditioner system was in fact pointed out by the inference engine in this manner.)

4. Retract the temporarily asserted premises.

5. Backtrack to step 1 and try a different combination.

The combinations of premises are tried in binary-number order (visible in Fig. 3), first 00000000 (all negated), then 00000001, 00000010, 00000011, and so on, ending with all premises non-negated. This makes it possible to use the input bit pattern as an offset into the table.

PIC software

On the PIC, the truth table is implemented as a subroutine which, given an input bit pattern, returns an output bit pattern. In our situation there are 4 significant bits of input, just_off, volt_low, over_81, and over_78. These are stored in the four least significant bits of the input byte, and the truth table has $2^4 = 16$ entries. The output has only one significant bit, run_ac, which is stored in the lowest bit of the output byte.

The PIC has a particularly strict form of Harvard architecture, with separate memories for program and data. There is no way to read data from the program memory into data registers. Instead, data tables are implemented as successive RETLW instructions ("return with literal in W"), each of which contains a byte in the bottom 8 bits of the 14-bit instruction word. Accordingly, a table lookup subroutine adds the offset to the program counter, jumps to the appropriate RETLW, and returns with the appropriate table entry. Figure 3 shows how it is done; as a precaution, the irrelevant input bits are masked off to guarantee that execution does not jump outside the table.

The largest possible truth table, using this technique, has 8 input bits and up to 8 output bits, and occupies 256 bytes. If that's not enough, minor changes in implementation can accommodate much larger tables. If a truth table were all that was needed, it could reside in a programmable logic array rather than a microcontroller, but in typical applications, the services of a CPU are needed to gather input data and keep time.

FURTHER ISSUES

Power of defeasible logic

Can defeasible logic do anything that classical logic cannot do? Yes and no. By definition, since its output is a binary truth table, defeasible logic does nothing that other logics cannot do. The difference is in how it does it. Defeasible rules correspond to a natural kind of human thinking and provide a concise representation for complex sets of conditions. The whole reason for having programming languages, after all, is to accommodate the thinking patterns of human beings; the computer itself would be content with ones and zeroes.

Temporal persistence

To show how defeasibility pervades human thinking, consider a simple but notorious problem in temporal reasoning that is highly relevant to embedded control. It is the rather macabre Yale Shooting Problem and consists of the rules:

(1) Normally, a gun that is loaded at time t will still be loaded at time $t + 1$.

(2) Normally, a person who is alive at time t will still be alive at time $t + 1$.

```
; Generated by TRUTAB.PL
;   Input bits:  [0,0,0,0,just_off,volt_low,over_81,over_78]
;   Output bits: [0,0,0,0,0,0,0,run_ac]
;
; CPU: PIC16C84/F84
;
LOOKUP   ANDLW B'00001111'   ; eliminate unused bits
         ADDWF PCL,F         ; add offset to program counter
         RETLW B'00000000'   ; premises=[neg just_off,neg volt_low,neg over_81,neg over_78]
         RETLW B'00000001'   ; premises=[neg just_off,neg volt_low,neg over_81,over_78]
         RETLW B'00000001'   ; premises=[neg just_off,neg volt_low,over_81,neg over_78]
         RETLW B'00000001'   ; premises=[neg just_off,neg volt_low,over_81,over_78]
         RETLW B'00000000'   ; premises=[neg just_off,volt_low,neg over_81,neg over_78]
         RETLW B'00000000'   ; premises=[neg just_off,volt_low,neg over_81,over_78]
         RETLW B'00000001'   ; premises=[neg just_off,volt_low,over_81,neg over_78]
         RETLW B'00000001'   ; premises=[neg just_off,volt_low,over_81,over_78]
         RETLW B'00000000'   ; premises=[just_off,neg volt_low,neg over_81,neg over_78]
         RETLW B'00000000'   ; premises=[just_off,neg volt_low,neg over_81,over_78]
         RETLW B'00000000'   ; premises=[just_off,neg volt_low,over_81,neg over_78]
         RETLW B'00000000'   ; premises=[just_off,neg volt_low,over_81,over_78]
         RETLW B'00000000'   ; premises=[just_off,volt_low,neg over_81,neg over_78]
         RETLW B'00000000'   ; premises=[just_off,volt_low,neg over_81,over_78]
         RETLW B'00000000'   ; premises=[just_off,volt_low,over_81,neg over_78]
         RETLW B'00000000'   ; premises=[just_off,volt_low,over_81,over_78]
;
; End of generated code.
```

Figure 3   PIC microcontroller code generated by the d-Prolog program.

(3) Normally, a person who is alive and is shot with a loaded gun at time t will be dead at time $t + 1$.

All three rules are defeasible: sometimes an adversary sneaks in and secretly unloads a gun, sometimes a person dies spontaneously, and some people survive gunshots. Ordinarily, though, we expect people to die when shot with a previously loaded gun, and to remain alive otherwise.

Now suppose the gun was loaded at t_0 and a person who is alive gets shot with that gun at $t_0 + 1$. Then is that person dead at $t_0 + 2$? Defeasible logic correctly infers "presumably, yes," because rule (3) is more specific than rule (2) and therefore overrides it.

Crucially, *defeasible logic is not just a logic of defaults; it is a logic that chooses intelligently between one default and another.* A logic based purely on defaults would be stymied in the same situation because it cannot tell which default should be overridden [HM87, SS95].

The Yale Shooting Problem typifies a kind of reasoning that pervades embedded control. Things normally stay the same unless acted upon, and things that are acted upon normally change in the specified way.

Embedded controllers perceive time in discrete units — indeed, one of the common ways to use a microcontroller is to let its watchdog timer reboot it several times a second,

whereupon it wakes up, makes a decision, and goes back to sleep. Thus the sequence $t_0, t_0 + 1, t_0 + 2, \ldots$, ontologically troublesome in the real world, is exactly right for a microcontroller.

Defeasible versus fuzzy logic

A frequently asked question about defeasible logic is whether it is anything like fuzzy logic or probabilistic logic. The answer is, "Not really." Defeasible logic attacks a quite different set of problems and solves them in a different way.

There are two ways to use defeasible reasoning. We can use it simply as a more concise and human-friendly notation for formulas that could be expressed in classical logic. In that case, it contributes nothing to the power of an embedded system, but potentially a great deal to the ease of programming it. Or we can use defeasible logic to represent uncertain information, as in the Yale Shooting Problem. In the latter case, defeasible logic represents uncertainty in a quite different way than other technologies.

The purpose of fuzzy logic is to compute compromises numerically between conflicting conditions that are both true to a degree. This is not necessarily a matter of uncertainty; the premises may be perfectly certain but express judgments that are not binary. In practice, fuzzy logic programs are numerical models that are adjusted empirically, like other kinds of numerical models, to give the desired results [B92, SUK+94].

Probabilistic reasoning deals with premises that may or may not be true, but whose probabilities are known or estimated.

Neither of these is like defeasible reasoning. To say that a conclusion is defeasible is not to say that it is true only to a degree (as in fuzzy logic) nor that it has only a certain probability of being true. Defeasible logic makes no claims about likelihood; it only claims that if a conclusion is defeasible, further information can cause it to be overridden. To characterize the defeasibility further, one enumerates the kinds of premises that would cause the conclusion to be withdrawn. One need not know the likelihood of these premises actually turning out to be true.

Used in this way, defeasible logic models human reasoning from incomplete information. Instead of dealing precisely with every situation (as in classical logic), human beings make generalizations that cover most situations, then enumerate the exceptions by means of more specific rules. Defeasible logic allows the designer of an embedded control system to describe a complex truth table this way, using generalities and exceptions, rather than requiring exceptionless classical rules or numerical parameters.

CONCLUSIONS

Defeasible logic should take its place alongside fuzzy logic, probabilistic reasoning, and conventional computer programming in the embedded system designer's toolkit. It isn't the solution to every problem, but in the right cases, it provides a convenient design methodology that leads to rapid implementation. In particular, the literature on fuzzy logic sometimes expresses a wish for other non-classical reasoning techniques, and defeasible logic is one of them.[2]

REFERENCES

[B92]   Jim Bezdek. Editorial: fuzzy models — what are they, and why? In *Fuzzy logic technology and applications,* ed. Robert J. Marks II, pp. 3–7. New York: IEEE, 1992.

[HM87]   S. Hanks and D. McDermott. Nonmonotonic logic and temporal projection. *Artificial Intel-*

ligence 33:379–412, 1987.

[J94]   R. Jager, H. B. Verbruggen, and P. M. Bruijn. Demystification of fuzzy control. In *Fuzzy reasoning in information, decision and control systems,* ed. S. G. Tzafestas and A. N. Venetsanopoulos, p. 165–197. Dordrecht: Kluwer, 1994.

[N92]   Donald Nute Basic defeasible logic. In *Intensional logics for programming,* ed. L. Fariñas del Cerro and M. Penttonen, pp. 125–154. Oxford: Oxford University Press, 1992.

[N96]   Donald Nute. d-Prolog: an implementation of defeasible logic in Prolog. In *Non-monotonic extensions of logic programming: theory, implementation, and applications,* ed. J. Dix, L. M. Pereira, and T. Przymusinski, pp. 161–182. Research report 17/96, Institut für Informatik, University of Koblenz-Landau, 1996.

[N97]   Donald Nute Defeasible Prolog. In M. Covington, D. Nute, and A. Vellino, *Prolog programming in depth,* 2nd ed., pp. 345–405. Upper Saddle River, N.J.: Prentice-Hall, 1997.

[SS95]   Erik Sandewall and Yoav Shoham. Nonmonotonic temporal reasoning. In *Handbook of logic in artificial intelligence and logic programming,* ed. D. M. Gabbay, C. J. Hogger, and J. A. Robinson, vol. 4, pp. 439–498. Oxford: Clarendon Press, 1995.

[SUK+94] Hartmut Surmann, Ansgar P. Ungering, Thorsten Kettner,and Karl Goser. 1994. What kind of hardware is necessary for a fuzzy rule based system? *Proceedings, Third IEEE Conference on Fuzzy Systems,* vol. 1, pp. 274–278. New York: IEEE, 1994.

---

[2] See for example [J94].

A graphical representation for defeasible rules has been developed [N97] and a system for entering rules into the computer graphically is being developed (Nute, personal communication).

FACILITATION OF IMMUNOASSAY THROUGH OBJECT IDENTIFICATION

Susan L. Collins[*], Kai H. Chang[+], James H. Cross[+], and W. Homer Carlisle[+]
[*]Westinghouse Savannah River Company
Aiken, SC 29080, USA
and
[+]Department of Computer Science and Engineering
Auburn University, Alabama 36849-5374, USA
Email: kchang@eng.auburn.edu

ABSTRACT

Immunoassay is a biological technique that is used for detecting and measuring target compounds. Recently, the use of immunoassay techniques has moved into the environmental arena. With the increasing use and need for environmental assay techniques, quick and efficient methods of automatic detection and measurement are needed. A positive immunoreaction results in a measurable colorimetric reaction when an antigen binds to the antibody under appropriate thin-film material configuration. By applying the antibody to a test plate surface in a specific pattern formation, an automatic pattern recognition software routine along with specialized electronics can be utilized to automatically analyze the results from an assay exposure. The use of automated equipment eliminates the need for highly trained technicians to analyze the test results. This paper reports a system that automatically reads a test plate and converts the data into a meaningful image representation. This image can then be analyzed and the positive reactions are automatically identified as a particular object. The object identification routine utilizes an intelligent probability weighting function based upon feature parameters extracted from the object.

Keywords: Practical applications, image processing, rule-based systems, immunoassay

1. INTRODUCTION

Immunoassay is a biological technique that is used for detecting and measuring target compounds. An antibody is specifically formulated for each individual target compound, the antigen. The antibody is bound to a test plate in such a manner that its receptors are physically arranged to provide maximal area for the binding of the antigen. The antibody is a clear liquid and is invisible to the human eye once it has dried on a thin-film substrate. The high specificity and affinity to the antibody results in a measurable colorimetric reaction when an antigen binds to the antibody under appropriate thin-film material configuration [3]. The test plate containing the antibody is exposed to the sample, either a vapor or liquid stream to determine if the antigen is present in the sample. The quantity of antigen present in the vapor or liquid test stream is determined by the magnitude of the colorimetric reaction when it is exposed to the test plate. The optical density of the test plate can be measured to determine if the antigen is present even at low part per billion concentrations.

Immunoassay techniques have been utilized as a source of medical detection for many years. Recently, the use of immunoassay techniques has moved into the environmental arena [5, 6, 7]. With the increasing use and need for environmental assay techniques, quick and efficient methods of automatic detection and measurement are required. Research is in progress to develop a portable system that can be taken into the field to sample vapor streams and provide on-site analysis of the observed reactions. The trend toward more stringent regulations requiring environmental remediation and monitoring has resulted in a greater need for such sensing and measuring devices. The equipment currently in use to measure immunoassay responses is laboratory type equipment that can be very costly. Such equipment also requires trained technicians to operate and maintain. There is a need for cost-effective measurement devices that are field ready and do not require highly trained personnel to operate.

Many different types of antibodies will be applied to the test plate. The antibodies are produced by live animals that are inoculated with the antigen to create the antibody, thus the time it takes to produce antibodies can be lengthy and is dependent upon the biological processes. The cost of producing the antibodies can be greater than $100 per milligram of antibody. Since only a small amount of antibody is necessary to produce a response with the required resolution, the antibody area will be approximately 2 mm^2 in size. Techniques have already been developed that enable the technician to manually apply the antibodies to the test plate in patterns unique to each antibody type. Standard patterns will be defined for each type of antibody. The antibodies that have reacted with target antigens will be identified by their doping patterns. This will aid in preventing the possibility of mislabeling, since the antibody is difficult to see once it has dried.

The scope of this paper is the data acquisition, graphical representation, image analysis and pattern recognition of the above described test plate. The data is acquired using a fiber optic probe which measures the optical densities of areas on the test plate. The data from the fiber optic probe is digitized and stored in a data file. The data file is then processed to produce a graphical image of the test plate. The image is also enhanced and smoothed for display and analysis purposes. The image analysis software is used to determine which, if any, of the patterns indicate positive after an exposure to the sample to be tested.

2. SYSTEM DESCRIPTION

Test plates will be developed for detection of several classes of environmental and health hazards. Broad spectrum test plates will be developed and used to determine if any of many types of common antigens are present in the sample stream. If a broad spectrum test plate identifies an antigen, then a more specific test plate for that class will be tested. Each type of antibody is only specific to one antigen and will only bind with that antigen. For example, there are 32 different type of Legionella bacteria that can be monitored by immunoassay techniques and a specific antibody is required to detect each type of the bacteria.

The more specific test plates can also be used to determine an estimate of the concentration of antigen in the sample. This system is intended to be an immediate response system that will provide preliminary results that will be followed up by more precise, time consuming laboratory procedures. Since the system will be used to identify possible hazards, immediate results are necessary so responsive actions can be taken quickly to prevent exposure of materials that are harmful to the environment and humans. The system is considered a first detection instrument.

The test plate is circularly shaped and made of flat plastic coated with an appropriate thin-film surface material. The test plate must have exacting flatness so as to not affect the optical density through deformities. As a result of the biological nature of the data and other ultimate uses planned for this technology, the data will be scanned in a spiral helical nature. The measurement device uses a spindle servo motor that yields a constant circumferential velocity. The test plate is placed on the spindle and a fiber optic probe is used to measure the optical density of the test plate as the plate rotates. Since the servo mechanism operates at a constant circumferential velocity, all data is scanned at the same rate no matter what its radial position on the test plate. The scanning rate of the probe is 1.2 meters per second circumferential velocity and increments 10 micrometers in the radial direction for each rotation. This translates to a resolution of 150 micrometers per sample in the linear direction and 10 micrometers in the radial direction. This resolution is required for the success of other future planned work with this system. A sample test plate with the sampling direction and darkened test patterns is shown in Figure 1. The purpose of the timing marks will be discussed later.

For proof of principle of the imaging technique, a system has been developed that can digitize and display the test plate data, perform digital filtering, and analyze up to ten antigen patterns. The system is based on a 486 PC running the Windows operating system. The software is written in C++ with a Windows interface. There are four main phases for the software development: the digitization, image reconstruction, digital filtering and smoothing, and analysis of the test plate for patterns.

The digitization phase encompasses reading the optical density from the fiber optic probe. The output of the fiber optic probe is fed into an electronic current amplifier with a -10 to 10 VDC output. A 16 bit analog to digital converter running at 8000 samples per second is used to digitize the data. This sample rate has been determined based upon the size of the patterns and the required resolution.

The image data must also be displayed on a computer monitor to verify the digitization, translation and filtering of the data. An immediate problem is to convert the helical scan data into a Cartesian coordinate system. In order for the results to be meaningful, image representation must accommodate the unequal resolution. Since the system maintains a constant linear velocity, the number of revolutions per second is, therefore, not constant. In order for the software to identify when it has completed a revolution, timing marks with a preset pattern are manufactured on each test plate. These timing marks are used to assist in reconstructing the image from the digital data. Since a single antibody pattern may span several hundred revolutions, it is important that the data location reference be maintained so as not to skew the image.

For testing, ten simulated patterns have been used to evaluate the performance of the system. The patterns are simulated since the biological aspects of this project are still under way and are not expected to be completed until a later date. The patterns may be placed at any location on

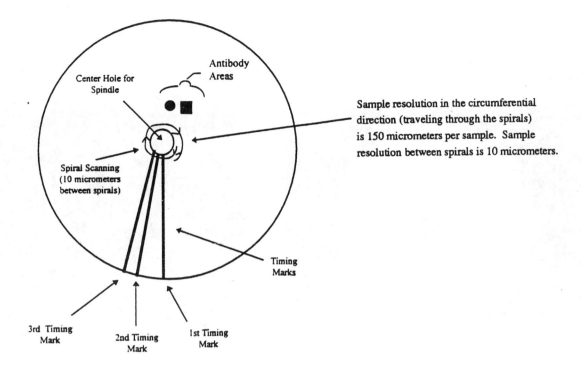

Figure 1 Sample Test Plate

the test plate with any orientation, so the pattern recognition algorithm has to consider object rotation. There may be slight variations in the scale of the patterns as well. Although actual antibody areas will be 2 mm^2 in total area, the simulated patterns will be made with a 5 mm^2 square area. This lessens the difficulty in making the simulated patterns.

3. SOFTWARE DEVELOPMENT

The software development can be described by five components: digitization, image reconstruction, filtering and smoothing, feature extraction, and object identification.

3.1 DIGITIZATION

The software for the digitization of the fiber optic probe data was written using the LabVIEW software development environment from the National Instruments, Inc.[4] The sampling rate of 8000 samples per second was chosen after trying a few different rates. Since the biological aspects of the project have not been completed, it was determined that for proof of principle the raw data be converted into a binary image instead of a gray scale image. To create the template data file, the test plate was sampled continuously and the raw binary data stored in a data file. Two memory buffers are configured and as the converter

digitizes the data, it places that data into one memory buffer while the other buffer it written to the hard disk. This method allows for long term continuous read and store capability.

3.2 IMAGE RECONSTRUCTION

Since the raw data is taken in a helical scanning pattern that does not correspond to the Cartesian coordinates, it cannot be displayed directly on a screen. Functions were written that sequentially read the input data file to determine the boundaries of each complete 360° circular scan. Since the sampling rate for the data is high, it is possible to have more data points per scan than the actual resolution of the screen display which is configured as 1024x768, so only the number of points in the actual screen resolution can be displayed and the software must take this into account.

The position of the first timing mark indicates the start of each circular scan cycle (see Figure 1). The timing marks are three continuous straight lines spaced at fixed distances from each other extending outward from the center of the test plate. Identification of the timing marks is based upon locating the known width pattern of the three marks. The first and second timing marks are exactly twice the distance apart as the second and third marks.

A complete circular scan can be identified from the first

timing mark to the next first timing mark. This data is then converted into the Cartesian coordinate image. Bresenham's circle drawing algorithm is used to convert the scanned circular data into an image [2]. A sample reconstructed image is shown in Figure 2.

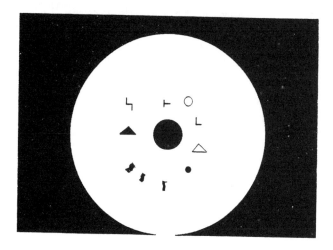

Figure 2 A sample reconstructed test plate image

3.3 FILTERING, SMOOTHING, AND EDGE RECOGNITION

The first step in the filtering and smoothing process is to perform a Laplacian omnidirectional edge enhancement algorithm. This edge enhancement algorithm generates an output image that is dark at all points except for edge pixels which are bright. Since the original image may contain noise that causes either rough edges or gaps in the edges, erosion and dilation operations are performed to smooth the noise out of the edge enhanced image. The dilation operation uses a 3x3 mask in a manner similar to the Laplacian Omnidirectional edge enhancement algorithm, except it looks at all 8 neighboring pixels to determine if any one of these neighbors are edge pixels, and if so, sets the pixel in question to be bright. If none of the neighbors are edge pixels, then the pixel in question is set to black. It creates a smoother, larger object image with any small gaps filled in.

An erosion operation is then applied to sharpen the image and remove small noise pixels from the image. The erosion algorithm operates by applying a 3x3 template mask to its 8 neighboring pixels and sets the pixel in question to dark if any of its neighboring pixels are not bright pixels. Application of the dilation and then the erosion algorithms, also called closing, smooths the edges and fills in the small gaps. The closing operation does not affect the shape or size of the object. The operation makes the edge pixels bright and everything else dark. A sample result of the closing operation is shown in Figure 3.

Figure 3 A sample processed test plate image

Application of the edge enhancement algorithm and the closing operation to the input image creates an image that has the small noise pixels removed and has crisp, continuous edges in the resultant image. With this sharper image, it is now possible to extract image features.

3.4 FEATURE EXTRACTION

The following image features are used in this application: object boundary points, direction changes in boundary points, perimeter, major axis and minor axis lengths, major axis and minor axis ratio, bounding figure points, number of boundary slope changes, object roundness, bounding box area, object area, and x and y centroids.

The *boundary points* consist of the outer-most or perimeter pixels of an object. They are stored in a dynamic linked list as they are identified. An edge following algorithm was developed. It follows the perimeter points and appends them to a object boundary list. The edge following algorithm uses an ordered nearest neighbor approach to find the next boundary point on the perimeter. The algorithm is based upon the eight neighboring pixels and uses the direction of last move to determine the next search direction. It starts at the top, leftmost edge pixel of an object and moves in a clockwise direction around the object's perimeter. Each move is assigned, from the pixel p in question, a direction number as follows:

7 8 1
6 p 2
5 4 3

The direction of last move is used to determine the order that the neighboring pixels should be traversed to select the next boundary pixel. For example, if the last move was 1, the neighbors will be traversed in the following order: 1, 8, 2, 7, 3, 6, 4, and the first edge pixel encountered will be selected as the next pixel in the boundary. A case statement is used for this purpose. The algorithm is also defined so the search will not backtrack along the path already taken.

Of course, even with the filtering and smoothing operations, gaps in the edge boundary may still exist. The algorithm will first search its neighbors for the next edge pixel, and if none are found, it will then search the nearest neighbor two pixels away. It will continue for up to neighbors three pixels away. These fixed points were selected since the maximum number of pixels per object is not very large and the gaps should not be larger than this distance.

The *perimeter length* of the object is also calculated during the edge following algorithm. Each straight direc-

tional move will add 1.0 to the perimeter length and each diagonal move will add 1.414 to the perimeter length. *Directional changes* is also calculated by the edge following algorithm. It counts the number of directional changes that are made during the edge following. This number gives a rough approximation of the number of directional changes for each object, and is useful in object identification.

The *bounding figure points* (i.e., X_{min}, Y_{min}, X_{max}, and Y_{max}) for the object are also determined in the edge following algorithm. Since the objects will not be in overlapping space, the object bounding box is used to eliminate that area from any further searching. The values are also used to calculate the *bounding box area.*

The next feature is the *number of slope changes.* This feature is valuable since it provides a method for identifying the overall shape and complexity of the object. It is calculated by comparing the slope at two points along the boundary pixel list. The first slope uses the first and the third points and the second slope uses the third and fifth points (Figure 4). If the difference of the slopes is greater than a threshold, then a slope change is assumed.

The *major and minor axes* for the object are also useful features. The major axis is the longest distance line that can be drawn between two edge boundary points. The mi-

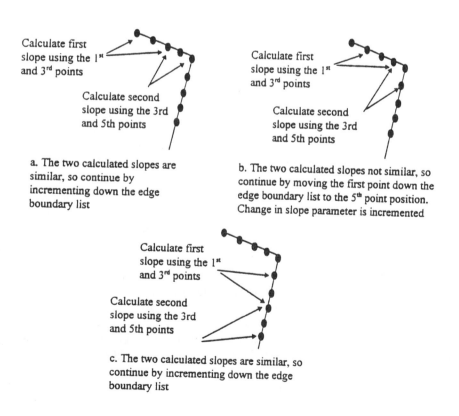

Calculate first slope using the 1st and 3rd points

Calculate second slope using the 3rd and 5th points

a. The two calculated slopes are similar, so continue by incrementing down the edge boundary list

Calculate first slope using the 1st and 3rd points

Calculate second slope using the 3rd and 5th points

b. The two calculated slopes not similar, so continue by moving the first point down the edge boundary list to the 5th point position. Change in slope parameter is incremented

Calculate first slope using the 1st and 3rd points

Calculate second slope using the 3rd and 5th points

c. The two calculated slopes are similar, so continue by incrementing down the edge boundary list

Figure 4 Caculation of Slope Change

nor axis is the longest distance line between two edge boundary points that is also perpendicular to the major axis. The major axis and minor axis determination algorithm returns the end points for each axis and also the length of each line. Since the minor axis depends upon the major axis, determination of the major axis must be performed first. This is done by traversing through the boundary points and calculating the distance from each point to every other point. The two points that provide the maximum distance form the major axis. Although this approach may seem inefficient, i.e., $O(N^2)$, the number of boundary point (N) is very small (less than 100 and will be even smaller for the actual implementation) and does not present a problem for this application. Once all points have been checked, the major axis is assigned the maximum points and the slope of the major axis is calculated. The minor axis, which is perpendicular to the major axis can now be calculated. To facilitate calculation of the minor axis, the entire object is translated and rotated so the major axis is positioned at the x axis. The minor axis is now the longest line between two points that have the same x coordinate. It was found during testing that sometimes the resolution of the object may not be great enough so the longest distance points are not exactly perpendicular to the x-axis. To accommodate this situation, the software allows a variance of several x coordinates when calculating the minor axis.

Another important and useful feature is the *total area of the object*. It is useful in the sense that an object is to be identified as either empty (no immunoreaction) or solid (positive immunoreaction.) To calculate the area, the original unenhanced image is used to count all of the bright pixels that reside within the object boundary. For every bright pixel inside the boundary, the area is incremented by 1.0. At the same time, the sums of the x-coordinate and y-coordinate of these pixels are also calculated. These sums are used to calculate the *object centroid*. The x- and y- coordinates for the centroid are found by taking the respective sums and dividing by the total object area.

Once the above parameters have been found, they can be used to calculate additional features. The object area and perimeter values can be used to calculate the *object roundness*, i.e., roundness = $(4\pi \times \text{Area})/\text{Perimeter}^2$. This function returns a value from 0 to 1 with a rounder object giving a value closer to 1. The major axis and minor axis lengths can be multiplied to obtain the *bounding box area* for the object. The *major axis to minor axis ratio* is also a useful feature for object identification. After all of these feature parameters are calculated, they can be used in the object identification algorithm. The features for one of the object images shown in Figure 2 are given in the following.

Starting Edge Pixel   (574, 259)

Bounding Box Coordinates   LX 563, RX 593, TY 259, BY 291

Change Dir 7   Slope Changes 5

Perimeter 102   Area 80

Centroid Sums, XC 46240   YC 22000

Major Axis (571,290) (585, 260) Length 33.105892

Minor Axis (566,267) (590, 284) Length 30.413813

Roundness 0.094889

Bound Box Area 1006.876404

Maj/Min 1.088515

Centroid 578.000000 275.000000

Object ID is 4

Probability Weighting Array

7   27   5   50   -30   15   27   30   17   0

3.5 OBJECT IDENTIFICATION

For testing purpose, ten simulated object classes were used (similar to the patterns in Figure 2). Perfect model patterns were first used to develop the rule base for object identification. The identification approach uses the concept of probability weighting functions. An array of n elements is created for each object pattern detected on the test plate, where n is the number of possible object classifications. Each element of the array corresponds to a specific object class. After the identification algorithm is executed, the value of each array element indicates the likelihood of the target pattern being a particular object class. The array element with the largest value is the most probable object class. For example, in the feature list of the previous section, the object is classified as Class 4.

Using the features obtained from the models, rules are developed and weights are assigned to the appropriate array elements. Adding (or substracting) weight to the m-th array element based upon a feature implies the feature supports (or does not support) the classification of object m. The amount that is added or subtracted from the array value depends upon the certainty of the feature implying a specific object class. The certainty values used in this system vary from -10 to 10. Each rule may be applied to more than one object class and they can have varying weights. The rules are implemented in C++. A sample rule follows.

Example Rule 1
// This rule utilizes the major axis to minor axis ratio for
//object identification. If the ratio is less than 1.15, then the
//probability of it being object 2 or object 3 is greater than
//the probability of it being object 7. Also, in this case, it
//is not likely to be objects 6, 8, and 9.

```
if (majtomin <= 1.15 )
    {
    objs[2] += 8;
    objs[3] += 8;
    objs[6] += -5;
    objs[7] += 5;
    objs[8] += -5;
    objs[9] += -5;
    }
    else
```

// If the ratio is not less than 1.15, then the probability of it
//being objects 2, 3, and 7 are not likely, so weight is
//subtracted from the array value for these objects. Also, in
//this case, it is likely to be object 9, so weight is added to
its array value.

```
    {
    objs[2] += -5;
    objs[3] += -5;
    objs[7] += -5;
    objs[9] += 5;
    }
```

Once all of the rules have been fired, the element with the highest value is the most likely object class. In order to correctly identify all of the ten object classes, 14 rules have been developed. The rules were developed based upon the features of the perfect models [1]. Six to ten objects of varying size, position, and rotation for each of the ten object classes were used as the models. To assist the rule development, a tool was built to analyze the association and disassociation between a feature and each object class. Since many of the object classes have similar features, many rules are required to achieve correct identification.

The selection of weights for the rules are based upon the consistency of the feature for each object class. For instance if the major axis to minor axis ratio for each object of class n, is always 0.99 to 1.01, then a measured value of 1.00 indicates high likelihood that the object belongs to class n. On the other hand, if the ratio ranges from 0.55 to 2.55, a measured value of 1.0 would not suggest an as-strong support. In this case, it may be appropriate to assign a smaller weight.

Each of these object classes has been tested using patterns of various sizes, rotations, and positions. Each object class was tested with 24-36 test patterns. A 100% correct identification has been achieved for each object class.

Although only ten object classes were used in this project, more classes can be added as desired. Models of the new classes could be run through the software and the feature parameters could be analyzed. Weighting values would be assigned as appropriate to the existing rule base.

4. CONCLUSION

The system development effort has focused on developing the recognition software. Ten sample classes were used in testing the software. These classes have been appropriately identified in all of the test cases. By using the probability weighting function to assist in identifying the objects, new objects can easily be added to the system. Models of the new object could be run through the software and the feature parameters could be analyzed. This identification algorithm for immuno-optical testing has been shown to work well and will be used as soon as the rest of the areas of the total project are ready. Since this effort has shown to be successful, further work is planned to enhance the resolution of the input image for identification of smaller target size objects. Upon completion of the biological aspects, testing with actual immunoassay test plates will be performed. This work will help to make the identification of the immuno responses a quick and automatic process and remove human decision from the testing. This method can provide quick results that will allow immediate responsive actions to be taken.

5. REFERENCES

[1] Collins, S. L. Facilitation of Immunoassay through Object Identification, Technical Report 96-07, Department of Computer Science and Engineering, 1996.

[2] Foley, J. D., & Van Dam A. (1982). Fundamentals of Interactive Computer Graphics. Massachusetts: Addison-Wesley Publishing Company.

[3] Giaver, L. (1973). The Antibody-Antigen Reaction--A Visual Observation. J. Immunology, Vol 110, pp. 1424 - 1426.

[4] LabVIEW for Windows, (1995). Version 3.01, National Instruments, Inc..

[5] Lukens, H., & Williams, C., (1977) A Solid Substrate Immunological Assay for Monitoring Organic Environmental Contaminants. EPA Report EPA-600/1-77-018.

[6] Lukens, H., (1993) Detection of Narcotics with an Immunoassay Film Badge. Proceedings of the INNM

Security Conference. Scottsdale, Arizona.

[7] Lukens, H., (1993) Detection of CBW and Explosives
     with an Immunoassay Film Badge. Proceedings of the
     INNM Security Conference. Scottsdale, Arizona.

THE AUTOMATED CATALOG: AN EXPERT DATABASE SYSTEM

Q. Dong   W.D. Potter   D.E. Nute   L.W. Hill
Artificial Intelligence Center
The University of Georgia
Athens, Georgia 30602–7415, U.S.A.
E-mail: potter@cs.uga.edu

ABSTRACT

In this paper we discuss the development and implementation of a loosely coupled expert database system supporting catalog sales for a large industrial supplier, Richmond Supply Company. The system is called an Automated Catalog System (ACS) because it automates most of the sales, purchasing, and technical support processes. The ACS enables the user to search a database of products using any combination of the seller-defined parameters and select items for full-page graphic and text display. It helps the user build purchase orders and send them via the push of a button through his or her modem. Since the average customer is not very familiar with the technical aspects of the products they are going to purchase, expert product selectors have been incorporated. The product selectors, which are small expert systems on different product categories such as casters, paint, and gloves, aim to help users make better, more informed purchase decisions.

INTRODUCTION

As an important medium for interaction between customers and retailers, Catalog Marketing is one of the fastest growing and hottest businesses existing today. Thousands of companies have become involved in cataloging as a cost-effective, profitable, and widely-accepted marketing medium. According to facts compiled by the Direct Marketing Association (DMA), catalogs now account for twenty-two percent of the postal mail stream [Muld96]. For the individual consumer, the number of catalogs he or she receives in a given month can be overwhelming.

Nonetheless, with shifts in information media and technology, the nature of the catalog is rapidly changing. As the catalog industry grows bigger, competition becomes more intense, and customers grow more demanding, traditional print catalogs are losing their attraction. Interactive, electronic catalogs are emerging with the benefit of advanced computer technology in a variety of areas. CD-ROMs, with their compact size yet tremendous storage capacity and impressive multimedia capabilities, are a medium far superior to paper. Database technology enables efficient storage, retrieval, update, and analysis of large amounts of catalog-related information. Finally, the incorporation of Artificial Intelligence (AI) techniques, especially expert systems techniques, makes cataloging an intelligent and helpful marketing tool. The automated catalogs which result from the combination of these three techniques, though clearly related to their paper-bound counterparts, are a new medium of different dimensions.

This paper is the result of an effort at designing and implementing one such catalog for Richmond Supply Company in Augusta, Georgia. It is called an Automated Catalog System (ACS) because it automates most of the sales, purchasing, and technical support processes. The ACS enables the user to search a database of products using any combination of the seller-defined parameters and select items for full-page graphic and text display. It helps the user build purchase orders and send them via the push of a button through his or her modem. The average customers are not very familiar with the technical aspects of the products they are going to purchase. The product selectors, which are small expert systems on different product categories such as casters, paint, and gloves, aim to help the users make better, more informed purchase decisions. The system incorporates a Database Management System (DBMS) with expert systems. Visual Basic is used to develop the system interface and the electronic catalog. Microsoft Access is our DBMS, and LPA Prolog for Windows is used to build the product selectors.

Advantages of the Electronic Catalog

Compared to the traditional print catalog, the electronic catalog has a number of advantages:

- Human-Computer Interaction

For all its quality and colorfulness, paper is still a limited medium for transferring information. It is totally passive, for the information printed in the catalog cannot be changed, reordered, or searched. It just sits there waiting to be manipulated by the reader.

The electronic catalog completely changes the way we interact with its information. Search parameters are classified in scientific categories and supplied by the application. The User only needs to make one or more selections, and with the click of a button, the system displays its search results. The user can narrow the search or select one particular item from the result list for more details. As we can see, the system removes most of the searching burden from the user and performs the search itself. The user only has to know what information is required and not where it is located or how to find it. The user and the system interact in a way which very much resembles the interaction between the customer and sales person. This shift makes the tedious old job of searching more natural and enjoyable.

- Efficient and Powerful Information Retrieval

Different customers use different information to locate their particular items. For example, to find Ansell Edmont Redmont Heavy-Duty Coated Neoprene Gloves, some customers may use the company's name *Ansell Edmont*, some may use the trade mark *Edmont*, while others may use the keywords *Heavy-Duty*, *Coated* or *Neoprene*. It is always the print publisher's challenge to link the printed information with sufficient ancillary information, also printed, to enable the reader to successfully locate all the pertinent information. It is not easy to cross-reference thousands of terms in a clear and meaningful manner.

However, in most computerized database systems, it is a trivial task to associate a product with all its relevant information. Modern query languages, such as SQL, are able to perform very complex queries very efficiently. In our automated catalog system, the user can search the products by any combination of the search parameters.

- Reasonable Presentation of Information

In our automated catalog, the customer will never see information that he or she does not want to see. If you are searching for gloves, the system will not show you belts. A list of search results is displayed after you have specified one or more search parameters. The results include the products' item number, a brief description, price information, and a bit of sales information like "50,000 Sold Last Year" or "New Product".

The presentation of initial search results enables the customer to compare similar products at a glance and choose the products of interest to view. When displaying products, the catalog has one item per page which allows more detailed information to be displayed, and the user is able to focus on the selected product. This is a more natural and reasonable way of displaying information. Eliminating unnecessary information and distraction greatly reduces the stress of searching the traditional print catalog.

Motivation to Incorporate Expert Systems Techniques

Automated catalogs provide a better way of doing business. Most of the automated catalogs are typical database applications. However, we take one step further by incorporating AI techniques with database techniques. Our catalog includes product selectors on different product categories to assist customers in making purchasing decisions.

Product selection can be very difficult as the product variety and options increase. Incorrect product selection not only results in waste of time and resources, but could cause serious personal injury and property damage. This is especially true in the case of safety product selection. Making a smart purchasing decision is far more complex than browsing pictures and text in a catalog. It usually requires careful consideration of several technical parameters, and help from experienced professionals. For example, to select casters, the customer has to bear in mind factors such as loading capacities, floor conditions, motive power, floor saving, noise, temperature, and chemical resistance. The typical customers are people who are broadly familiar with the product's area, but lack specific knowledge and experience to make the correct selection.

Mail order companies try to help their customers by providing technical support lines, attaching selection guidelines in the catalog or publishing separate manuals. It may be thought that technical lines and written documents are sufficient at providing customer support. Why go through the trouble of developing expert systems? Technical lines are normally very busy and people feel uncomfortable talking to a company representative because of sales pressure. Expert systems aim to imitate the reasoning ability of a human expert. They work even better than human experts in the aspect of providing tireless, consistent, objective and impartial support. Written instruction, like a printed catalog, lacks the necessary interaction with users, and people respond much more strongly to the interactive nature of an expert system. It is well known that reading a manual is usually considered as a course of last resort, as witnessed by the adage [Barr88]: "when all else fails then read the instructions".

The external benefits of incorporating expert systems

technology lie in enhanced customer satisfaction and competitiveness:

- Increased sales effectiveness

- Reduction in human errors and omissions; consistency and standardization in customer support; raised quality of service.

- Specialized expertise available at all times and in all locations to the customers.

Internal benefits come from enhanced efficiency and effectiveness within the company:

- Expert systems free experienced professionals from their more routine activities, so that they have more time to devote to other more specialized, novel, and demanding tasks. The number of professionals needed for technical support can thus be reduced.

- Since the customers can get support right at their desktops, cost of customer support calls are greatly cut down.

- Developing expert systems provides experts an opportunity to examine and refine their own expertise. Expert systems also improve training facilities.

In our automated catalog system, the customers can select a product category under the **Expert Systems** menu option and start their consultation. When they search for products, there is also an icon indicating whether a selector is available for this product; and they can click the icon to access the corresponding selector. The expert system typically asks the user a sequence of relevant questions and uses the responses to reach the appropriate recommendation. The user can then use the **Catalog** button to go back to the database to display a product or to place an order.

EXPERT DATABASE SYSTEMS

The idea of integrating AI and Database technology has been of interest since the beginning of the AI and DB fields 40 years ago. Both AI and database technology have seen substantial development and evolved into two mature research fields since the pioneering days, but, strangely enough, there has been little communication or technical exchange between them. Only in recent years has the importance and need for AI-DB integration begun to gain new attention. The merging of the two fields creates the new term *Expert Database Systems* (EDS) [Kers86, Kers88], which is both a thriving research area and an important and practical application development tool.

Expert systems are computer programs which aim to replicate the reasoning of an expert in solving fairly complex real-world problems [Barr88]. Expert systems lend themselves to supporting decisions which are not well-structured, and rely largely on experience and acquired knowledge. From applications in medical diagnosis and mineral exploration to fault analysis and product selection, expert systems are spreading into almost all business areas.

The core of an expert system is its knowledge base. A knowledge base contains facts and rules. The size of the knowledge base varies according to the size of the application and the complexity of the problem. Even though an expert system is suitable for complex knowledge representation and inferencing, it lacks facilities to store and manipulate large amounts of data. Expert systems tend to be memory-based: all information is required to be in primary memory during program execution. Without a sophisticated indexing method, rules and facts are searched mainly in a sequential order. These factors limit the amount of knowledge and data which can be stored in an expert system.

A *Database* is an organized collection of related data. It is capable of storing large volumes of data vital to operating a business. Databases help decision-making by providing meaningful information in a convenient format. Business applications such as inventory management, order tracking, and record keeping are commonly handled by database systems quickly, accurately and efficiently. A *DBMS* is a commercial software program serving as the interface between the user and the database. It facilitates operations such as entering information into a database, updating, deleting, manipulating, storing and retrieving. The DBMS hides much of database's internal complexity from the user and makes it possible to access information quickly and easily. Databases are disk-based. With the decreased price and increased capacity of the computer's secondary storage, a DBMS is able to manage very large databases.

Understanding the strengths and limitations of expert systems and DBMS's helps to explain the need for their integration. Through the use of an existing commercial DBMS, expert systems can access thousands of rules and facts . If enhanced with expert systems features, a DBMS will possess advanced reasoning capabilities . Their integration is based on the need to manage, access and reason about large amounts of information intelligently—a task shared by expert systems and DBMS's. The result of this integration is an Expert Database System.

It is most logical to define an Expert Database System as a tool for developing applications requiring both a DBMS and one or more expert systems [Kers86]. There are four main approaches to develop such a system [Brod86, Kers86, Kers88, Mylo88, Rupa91]:

1. Extend a DBMS by enhancing it with powerful inference facilities.

2. Extend an expert system by adding database facilities.

3. Loosely couple existing expert systems with a DBMS.

4. Tightly integrate database and expert systems functionalities.

The first three approaches are considered evolutionary approaches, while the fourth is revolutionary. The first approach involves the addition of an inference engine and additional data constructs to a DBMS.

The second approach attempts to augment expert systems with a DBMS facility. For example, many researchers, noting the close relationship between Prolog and relational databases, seek to enhance expert systems with database facilities by using Prolog as a relational database language.

When loosely coupled, as in the third approach, both the expert system and the DBMS will maintain their own functionality and communicate through a well-defined interface. The communication between the two systems is limited and can only take place through the interface. The major advantage of such an approach is the large degree of system independence. Existing systems can still be used with some minor modification. However, loose coupling fails to make use of full DBMS functionality, and there is duplicate data in both systems which introduces the usual problems with data redundancy.

As a final option, tight coupling supports dynamic communication between the two systems. This coupling tightly incorporates functionalities of both expert systems and DBMS's. The goal of tight coupling can be achieved either by developing a third-party program which understands and manages both systems, or by designing a new intelligent database language which takes advantage of the strengths of both systems. Tight coupling results in more constructive improvements, but also requires expensive restructuring and redesigning of the system.

THE LOOSE COUPLING APPROACH ADOPTED IN ACS

The approach we adopt in our current system is the third approach discussed in the previous section: loose coupling of expert systems and the DBMS. We use different software to develop different components of the project. Microsoft Access is used to store and update our product database; Visual Basic is used to implement the user interface and the electronic catalog system which includes the search, display and order modules; LPA Prolog is still the language to develop expert systems. Each component is developed

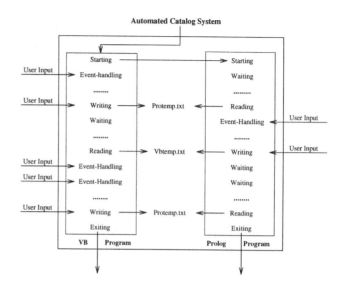

Figure 1 Communication Channel between Expert Systems and Electronic Catalog System

separately and is independent of the others. Finally, a communication link is set up to bring them together.

After we implemented the different components in our system, it became necessary to set up a communication channel to bring these components together. The communication channel ensures the correct operation of the whole system as well as accurate information exchange between expert systems and the database.

We use two temporary files for this purpose. One file is written by the VB program and read by the Prolog program. We name it *protemp.txt*. The other file is written by the Prolog program and read by the VB program. We name it *vbtemp.txt*. During the execution of the system, both the VB and the Prolog programs are running at the same time. One is running in the foreground handling user input, the other is running in the background waiting for messages through the communication channel. At some point, the user might want to perform a task which requires executing the background program. The foreground program informs the background process through the communication channel by writing to the appropriate file, and then hides itself in the background to wait for messages. The background program reads and deletes the file, and then becomes the foreground program to handle user input. The execution goes on like this until the user chooses to exit the system.

The VB program is the main program and it starts the Prolog program. The user first sees the VB interface, where he or she can search, display and order products. The selectors are accessed from the VB interface. When exiting from the selectors, the user is returned to the VB interface. The VB program exits when the user chooses to quit. It sends an ending signal to the Prolog program, and the Prolog pro-

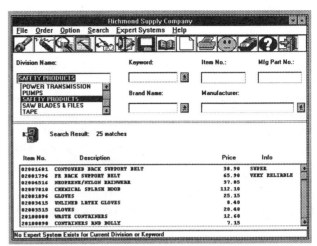

Figure 3   Search Screen

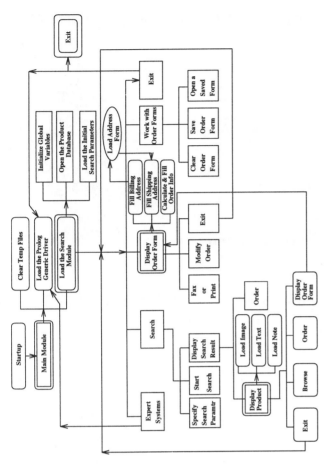

Figure 2   Logical Control of the Electronic Catalog

gram exits also. Figure 1 is an illustration of the communication channel between the two systems.

THE ELECTRONIC CATALOG SYSTEM

We call the part of the program developed using Visual Basic the *Electronic Catalog System* because it automates the search, display and ordering process. It is the front-end of the product database and is central to the ACS system. It contains four basic modules: the **Main Module**, the **Search Module**, the **Display Module**, and the **Order Module.**

The **Main Module** is the startup module of the VB program. It first clears the temporary files which were used as the communication channel and might be left over from a previous session. It then calls the Prolog program to load the generic driver for the expert product selectors and, at the same time, it loads the **Search Module**. The loading of the **Search Module** includes initializing global variables, opening the product database, and loading the initial options of search parameters such as division names and key-

words. Then the **Search Module** is ready for user input. The user can conduct a search, access the selectors, display an order form, work with order forms, or exit. When the user specifies the search parameters and starts a search, a list of search results will be displayed. The user has the option of displaying the catalog page of a product, placing an order for a product, or displaying the order form. Displaying a product calls the **Display Module**, which loads the image, text, and notes associated with the product. Displaying the order form loads the **Order Module**, which fills both billing and shipping addresses, and calculates and loads the order information. The user can perform different tasks at the **Display** or the **Order Module**. When the user exits, he or she is returned to the **Search Module**. He or she can then perform more tasks or exit from the whole system.

The Search Module

When starting up ACS, after displaying its logo screen, the system displays the **Search Screen** supported by the **Search Module**. This screen is the first screen the user sees. The product selectors and other modules are all accessed through this screen. When exiting from the selectors or other modules, the user is returned to the **Search Screen** and can only exit the system from there.

Figure 3 captures the Search Screen. The user selected *Safety Products* from the **Division Name Combobox** as the search parameter, and the search results are displayed in the **Search Result Listbox**.

Besides providing access to other modules and the selectors, the most important function implemented by the **Search Module** is the search engine. It allows the user to search products by specifying one or any combination of the six search parameters: division name, keyword, brand name, manufacturer, item number, and manufacturer's part number.

During loading up, the **Search Module** first initiates the global variables and opens the product database. It then opens the **Division Table** in the database and fills the combo boxes with initial selections. The **Division Name Combobox** is filled with the first-level divisions, and the other combo boxes contain the complete index of keywords, manufacturers, and brand names respectively. The user can start with any of the search parameters.

The hierarchical division system is implemented like this. Double-clicking a division name expands the division and fills the combo box with subdivisions and a go-back sign. The user can either double-click on one of the subdivisions to expand further or click on the go-back sign to go to an upper level. If a division name can not be further expanded, double-clicking on it does nothing.

To effectively guide the user through the large search space, the selections in the combo boxes adjust themselves according to user's needs. For example, when a user selects *Safety Products* in the **Division Name Combobox**, the **Keyword Combobox** lists only keywords associated with that specific division, as do the other combo boxes. If a user selects the subdivision *Personal Protective Equipment* under *Safety Products*, the list of options in the other combo boxes is further restricted. The user will only see what is relevant to the search, which makes selection easier. On the other hand, the user is forced to choose meaningfully and wisely. It is not possible for the user to specify *Caster* as division name, *Hard Hats* as keyword, and *Rubbermaid* as brand name to perform a futile search.

The Display Module

The **Display Module** can be called from the **Search Module** by clicking on the **Product Display** icon or double-clicking on a particular item in the **Search Result Listbox**. It can also be accessed from the product selectors by clicking on the **Catalog** button. The **Display Screen** is the interface of the **Display Module**. Figure 4 illustrates the **Display Screen** which displays a Rainfair^{TM} Neoprene/Nylon Rain Suit.

Order Module

The **Order Module** can be accessed from both the **Search Module** and the **Display Module**. Figure 5 shows the **Order Screen** with a sample order.

When loading up, the **Order Module** first looks for the address files. We use *address.txt* to store the customer's billing address, and *saddress.txt* for the shipping address. If the system cannot find the first file, the **Address Module** will be called and the user can input his or her address.

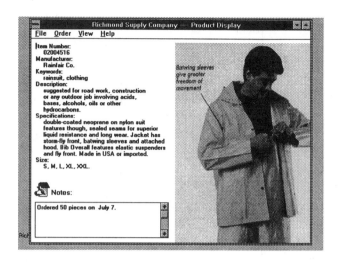

Figure 4   The Display Screen

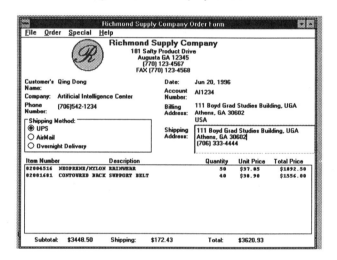

Figure 5   The Order Screen

This address is saved to the *address.txt* file, so the system can refer to it later. If the shipping address file is not found, the system assumes that it is the same as the billing address. When a user places an order for a specific product in either the **Search Module** or the **Display Module**, the order information is stored in an array. The **Order Module** retrieves the information, and displays it on the **Order Screen** after some calculation and formatting.

PRODUCT SELECTORS

Expert systems consist of three main components: a user interface, an inference engine, and a knowledge base. In our application, it is necessary to separate the knowledge base from the user interface and the inference engine so the inference engine can work with knowledge bases in different

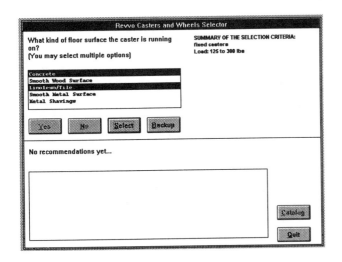

Figure 6   The Product Selector Interface

product domains and the user can see a uniform interface. Based on lessons we learned developing the **Paint Selection Expert System** and the **Sling Selector Expert System** [Frie89, Pott88], we implemented a generic driver including the user interface and the inference engine modeled after the Paint system components.

The interface is responsible for managing the program's communication with the user. The interface presents the questions in a logical sequence, accepts user replies, informs the user of the system's conclusions, allows the user to correct wrong answers, and displays the result of the consultation. Figure 6 is an illustration of the interface.

The inference engine carries out the reasoning of an expert system. The inference engine of our generic driver employs the backward-chaining method. It starts by finding the first rule describing a product, and tries to satisfy its condition. It examines such rules one by one in a depth-first manner until it exhausts them all.

In our knowledge base, the predicate *item* is used to describe a product. The following is an example of such a clause in our Caster knowledge base.

```
item('TFN 3003 P4J', [ctype(fixed),load(445),
    wtype('HR'), btype('ball')]).
```

The first argument of the *item* clause is a string used to identify the product. The second argument is a list of conditions. The condition in the example means that the caster type of that specific item is fixed, its load capacity is up to 445 pounds, its wheel type is hard rubber, and its wheel bearing type is ball bearing.

DIRECTIONS FOR FUTURE DEVELOPMENT

Our future development of the ACS system includes the following goals:

- Provide facilities for building knowledge bases. The ACS system intends to incorporate selectors on all major product domains to provide maximum support for the customer. Our generic driver is capable of handling different knowledge bases. However, the rules in the knowledge bases should use the predefined predicates and follow certain syntax. A facility to automate the construction of the knowledge bases is under development.

- Implement an editor for updating the database. Our database is in Microsoft Access format. Access has its own interface for adding, deleting and modifying items in a table, but this general interface may not be very convenient and friendly for users who are not familiar with Access. An editor should be implemented to provide easy modification of the database. The editor would only be used by the company's data processing personnel.

- Combine the product database and knowledge bases. The catalog database stores the catalog-related information of a product, such as its manufacturer, brand name, and price. The knowledge bases contain knowledge on the selection of the product—its unique features, its advantages and disadvantages, etc. The two types of information are stored in two places, which results in data redundancy and creates problems in updating. It is possible to combine knowledge bases with the database by creating more tables and relations. The system can extract a snapshot of the appropriate information from the database, and provide the knowledge bases to the selectors at the run time.

BIBLIOGRAPHY

[Barr88] Barrett, M. L and Beerel, A. C. (1988) *Expert Systems in Business: A Practical Approach.* Chichester: Ellis Horwood.

[Brod86] Brodie, M. L. and Mylopoulos, J. (1986) (ED.) *On Knowledge Base Management Systems.* Springer-Verlag.

[Frie89] Fried, L. M. (1989) *The Paint Expert System and The Paint Knowledge Base Development Tool.* Master's Thesis. University of Georgia, Athens, Georgia.

[Kers86] Kerschberg, L. (1986) (Ed.) *Expert Database Systems: Proceedings From the First International*

Workshop. The Benjamin Cummings Publishing Company, Inc.

[Kers88] Kerschberg, L. (1988) (Ed.) *Proceedings From the Second International Conference on Expert Database Systems.* The Benjamin Cummings Publishing Company, Inc.

[Muld96] Muldoon, K. (1996) *How to Profit Through Catalog Marketing.* Third Edition. Lincolnwood, Ill.: NTC Business Books.

[Mylo88] Mylopolous, J. and Brodie, M. (1988) (Ed.) *Readings in Artificial Intelligence and Databases.* San Mateo, CA: Morgan Kaufmann Publishers, Inc.

[Pott90] Potter, W.D., T.H. Tidrick, T.M. Hansen, L.W. Hill, D.E. Nute, L.M. Fried, and T.W. Leigh, (1990) SLING: A Knowledge-Based Product Selector. *Expert Systems With Applications: An International Journal.* 1: 161–169.

[Rupa91] Ruparel, B. (1991) Designing and Implementing an Intelligent Database Application: A Case Study. *Expert Systems with Applications.* 3: 411–430.

CyberTax™: Add Intelligence to Tax Forms with Prolog

Tony F. Chang
VerTec Solutions, Inc.
P. O. Box 49811
Greensboro, NC 27419
Email: tchang@earthlink.net

ABSTRACT

CyberTax™ is a commercial product which allows user to fill-in their personal tax forms just like on papers; plus, it minimizes the number of fields that must be filled in by filling out fields based on rules and perform the calculation when needed. The architecture of CyberTax™ are explained. The methodology and implementation of the system is discussed. Results showed that it is very easy to integrate rule-based system(with a Prolog engine) into an existing application written in high-level computer languages.

INTRODUCTION

CyberTax™ is a commercial product which allows user to fill-in their personal tax forms just like on papers; plus, it minimizes the number of fields that must be filled in by filling out 'same fields' on different form automatically; transfers values automatically from one form to another and perform the calculation when needed. Once the tax return has been completed, the system sends the data to a verification module to ensure that there are no obvious mistakes. After passing the verification process, the tax information may be sent electronically, or by mail.

CyberTax™ is the combination of VerTec Solution's TranZform™ product and intelligent Prolog rules to handle the complicated tax forms. TranZform™ is implemented in C++, and all the rules are written in Prolog! This is because the fact that C/C++ doesn't provide the power of Prolog's assertional database, symbolic pattern matching and search algorithms, and automatic memory management features. On the other hand, Prolog is quite difficult and inefficient when dealing with Graphic User Interface. With Amzi!'s logic server, we brought C/C++ and Prolog together. We used C/C++ to present data to the user while we used Prolog to get the right data.

Prolog was chosen to represent the tax rules because of the nature of the tax rules--they change every year and are made up of rules for relating data. Also, since tax forms contain so many duplicate values, Prolog provided the best means to handle the redundant values on all the forms. Due to Prolog, the user has no need to enter the same information more than once.

The workflow for CyberTax™ is simple due to Prolog handling most of the logic for each form. Users enter their tax information by tabbing to and entering data into simple edit boxes. After the data is entered, it is passed to the built-in Knowledge Base which analyzes the information. This approach provides a quick and effective means to perform all of the related computations in all of the supported tax forms.

The second and the third section of this paper described the architecture of CyberTax™ which uses a logic-base to represent the relationships between fields on income tax forms. The fourth and the fifth section are focused on the implementation of the intelligent-rules into a C++-based commercial application and the representation of the tax rules in Prolog.

PROJECT DEFINITION

Adobe's Portable Document Format (PDF) is a standard means to communicate sophisticated documents across heterogeneous networks, computers and printers. Many organizations are encoding their documents in PDF format including the Internal Revenue Service (I.R.S.) which has made available all the federal tax forms for personal filers on the Internet. Each year more taxpayers are downloading these forms, and many are filing their tax returns electronically.

Our vision was to build a new product that assists in filling out the I.R.S. provided forms directly in the Acrobat Reader. Our goals were to:

- *Provide an Intuitive User Interface*—The interface is the actual tax forms themselves with the fields that must be filled in highlighted, and help and instructions available with a single mouse click.

- *Calculate Fields and Validations*—We want to minimize the number of fields that must be filled in, transferring values automatically from one form to another and recalculating when needed. This provides the tax filer with an instant assessment of their tax situation, and valuable tool for trying various 'what-if' scenarios.
- *File Printed or Electronic Forms*—Once filled in, we wanted the filer to be able to print out the forms and either mail them in, or send the return electronically.
- *Easy Maintain and update the Rules in Future Tax Years*—As the tax code and tax forms change each year, we want to be able to quickly and easily update the rules for calculations, so we needed an intuitive and easy-to-read means of representing the forms, fields and calculation rules.

SOLUTION: CYBERTAX™

CyberTax is not a "stand-alone" software product, but a software enhancement to the Adobe Reader v2.1 that allows the user to fill in the IRS forms in Acrobat. It contains these components:

- The Adobe® Acrobat® Reader and Portable Document Format (PDF),

- VerTec's TranZform™ plug-in software that allows Acrobat forms to be filled in,

- An Amzi! logic-base for representing the tax rules, and

- Third party module for electronic tax filing.

The following diagram shows how all the pieces are tied together:

PDF Tax Forms → Acrobat Reader + IRS
Instructions Database
↓
CyberTax™ Plug-In
(with TranZform)
↓
Amzi! Logic Server
↑
↓
Tax Rules Logic-Base

Figure 1: CyberTax™ Architecture

The Acrobat Reader provides the user interface. It displays and prints PDF forms. The CyberTax™ plug-in

allows the user to type values into the forms and links forms with instructions and help. Every time a value is entered, it is saved in the form, and passed to the Amzi! Logic Server which executes the tax rules in our logic-base. These rules perform calculations and check for errors. The results are displayed to the user via CyberTax™ and Acrobat.

I.R.S. Forms in Portable Document Format (PDF)

Adobe has designed a format for describing complex documents. PDF format supports color, graphics, text and document navigation all in one portable, electronic format. The I.R.S. provides all its personal tax forms and instructions in this format, and Adobe makes its Acrobat Reader freely available.

The TranZform™ Acrobat Plug-In

TranZform™ consists of two Adobe Acrobat plug-ins. The Definer plug-in allows a developer to take any Acrobat document and define portions of the document to be input and/or output fields. The Fillin plug-in, when distributed with the newly defined PDF Form file, lets the end-user tab or click on a defined area and enter data.

An Amzi! Logic-Base

The Amzi! Logic Server allows declarative logic rules, expressed in the logic programming language Prolog, to be easily integrated into an application. Declarative logic was chosen to represent the relationships on the tax forms because the natural mapping between Prolog and the tax rules, it is a lot easier to maintain and make changes of the rules in Prolog than any other programming language.

CyberTax™

The workflow for CyberTax™ is quite simple. Users enter their tax information by tabbing to and entering data into edit boxes. After the data is entered, it is passed to the logic-base which analyzes the information and calculates the values for as many fields as possible on the form. This approach provides a quick and effective means to perform all of the related computations in all of the supported tax forms. In additions, a user could enter guess values into the edit boxes. The program would then handle the computations, thus allowing a user to get a feel for what the tax situation would be under certain circumstances.

Instead of building an inference engine in our current product, we used the Amzi!® Logic Server® to provide an inference engine to execute the tax rules. By using an embeddable logic-base component, we avoided the need to either design and implement our own rule language for tax forms, or code the tax form relationships in a difficult to read and maintain procedural manner.

Finally when the user has filled in all the forms, and the logic-base has checked their accuracy and done all the necessary calculations, the forms can either be printed directly from the Acrobat Reader, or they can be filed electronically.

TAX RULES AND PROLOG

In this part we will examine the implementation of the logic-base and the interaction between the main application(in C++) and the logic rules.

CyberTax™ contains approximately 800 logic rules and supports 22 of the most commonly used federal tax forms, including multiple W-2s, W-2Gs, and 1099Rs. The rules are used to define how calculations are made on the forms. Currently, CyberTax™ does not use the logic-base to provide tax advice.

Representing knowledge with rules

A Prolog logic-base is a collection of data, facts and rules. In CyberTax™, the data is entered into the forms by the user. The rules are a permanent part of the program, and are changed each year to accommodate changes in the tax code. The CyberTax™ logic-base contains the following:

*   Facts to represent the values of each field on each form

*   Facts that specify what values have to be carried forward to other forms

*   Facts that indicate what fields are related to other fields

*   Facts that indicate what fields are totals fields that need to be calculated

*   Rules that specify how to calculate the fields on each of the forms

The facts for related and total fields help CyberTax™ efficiently recalculate the values on the forms.

A typical way to map data in Prolog is:

value(formName, fieldName, fieldValue).

For CyberTax™, we gave all the supported tax forms a unique name in our database, for example: fm1040, scheduleA, and scheduleB. All the fields on a form are represented by a unique number. To avoid the memory overhead, the user-entered values are stored in the following format:

formName(fieldName, fieldValue).

For example:

fm1040(fd0100, 1039). // field number 100 on form 1040
schedulea(fd0100, 2253). // field number 100 on Schedule A
scheduleb(fd0200, 2821). // field number 200 on Schedule B
schedulec(fd0300, 250). // field number 300 on Schedule C

Copying Values from One Form to Another

In order to simplify the data entry, CyberTax™ is able to copy values from one field to another field in the same form or different forms based on the rules in the database. To represent this, we used another fact called 'carryforward' which has the following format:

carryforward(source_form, source_field, dest_form, dest_field).

For example:

carryforward(fmw2, fd0120, fm1040, fd0375).

This represents the fact that the value in Form W-2, field #0120 will need to be carried forward to Form 1040, field #0375.

A more complex rule is used to handle conditional carry-forward cases. For example, in form 1040, line 8a(taxable interest) will only to copy to Schedule B if the amount is over US$400. In Prolog, we can represent it as:

carryforward(fm1040, fd8a, scheduleb, fd0010):-
  value(fm1040, fd0010, X),
  X > 400.

So this clause will only 'succeed' if the value is over 400. Otherwise it fails, and the calling function does not carry forward the value. After the user enter data into a field and all the data validations are done, the application will query the rule database to determine if there is any need to duplicate specific values to other fields and forms.

Maintaining relations between fields

As the user types values into the tax forms, we need to know what fields need to be recalculated. To do this, the logic-base has facts that indicate what fields are dependent upon what other fields.

CyberTax™ stores this information in a fact named 'related' which takes three arguments: form name, field name, and related field names (a Prolog list in square brackets). It means that if any field in the third argument changes, the field (2nd argument) need to be re-calculated. For example:

```
related(fm1040, fd1130, [fd1050, fd1060, fd1080, fd1100, fd1105, fd1107, fd1125]).
```

This says for Form 1040, field fd1130 is related to the values for the fields in the list [fd1050, fd1060, . . .]. With this fact, a lot of unnecessary calculation can be avoid – in the previous example, if the value of fd1130 changed, we only need to update all the fields that are related to it rather than all the fields in the form(that's all the fields in the square brackets).

Streamlining Total Fields Calculations

Again, to avoid the unnecessary calculations, CyberTax™ uses a fact 'totalfield' to identify the list of total fields that need to be computed. The 'related' facts (described above) determine if there is any relationship between the newly entered or changed fields and the total fields. Thus, Prolog will only perform the calculation if the new field is related to ANY total fields in the same form, instead of performing ALL the computations for that form. For example:

```
totalfield(fm1040, fd0697).
totalfield(fm1040, fd1125).
totalfield(fm1040, fd1170).
```

These give the total fields for Form 1040.

Representing tax rules with Prolog

Each field on a form that is calculated is represented by a Prolog rule.

A typical tax rule:
The value of Form 1040, Field 0167 is equal to Field 0160+ Field 0167.
Can be translate into:

```
fm1040(fd0167, X):-
    value(fm1040, fd0160, A),
    value(fm1040, fd0163, B),
    X is A+B,
    retractall(value(fm1040, fd0167, Z)),
    asserta(value(fm1040, fd0167, X)).
```

This rule reads as follows:

- Get the value from field fd0160 on form fm1040

- Get the value from field fd0163 on form fm1040

- Add those two values together

- Replace the value of the field fd0167 in the logic-base with the newly calculated value.

So Form 1040 has a bunch of rules named 'fm1040', with the first argument being the field and the second is a variable which is the value returned from the rule.

Here is another rule. This one has two parts. Notice the first part checks if field fd0420 in form fm2441 has a value greater than 0. If it does then fd0371 is set to 'DCB'. If not, the second part is executed and fd0371 is set to blank.

```
fm1040(fd0371, X):-
    value(fm2441, fd0420, A),
    A > 0,
    X = 'DCB',
    retractall(value(fm1040, fd0371, Z)),
    asserta(value(fm1040, fd0371, X)), !.
fm1040(fd0371, X):-
    X = ' ',
    retractall(value(fm1040, fd0371, Z)),
    asserta(value(fm1040, fd0371, X)).
```

Most of the 800 or so rules in the logic-base are similar to those shown above.

Executing the Rules

Instead of having to call the predicates above, CyberTax™ uses the symbolic power of Prolog to call them using a single predicate:

compute(formName, fieldName, fieldValue).

The 'compute' predicate takes three arguments: form name, field name, and the value (which is returned). The code for compute is simply:

```
compute(Form, Field, Value):-
    X =.. [Form, Field, Value],
```

call(X).

The Prolog operator =.. is called univ and it takes the list containing the form, field and value and converts into a Prolog term that is form(field, value), which you will note is the same style as is used above for expressing the rules.

With the backtracking ability in Prolog, any changes in any fields in the formula will cause recalculation to ensure the accuracy of each fields. The result is a logic-base that captures the rules of the tax forms in a readable manner that is easy to maintain and leaves the order of execution to Prolog.

The data for each fields are saved in Prolog's dynamic database (in memory) via the asserta in the rules. The C/C++ interface is used to retrieve the values for display on the forms themselves in Acrobat.

INTEGRATED PROLOG TAX RULES WITH C/C++ APPLICATION

As we mentioned above, we added Amzi!'s logic server into our application in order to use the tax rules wrote in Prolog. In order to use a Prolog logic-base in C++ we needed to write the rules and compile them into an executable module. Then within CyberTax™ we do the following:

1. Initialize the Prolog engine

2. Load the pre-compiled logic-base file

3. Wait for the user to enter a value into a tax form field (in Acrobat)

4. Post new values to and query the logic-base

5. Repeat 3 and 4 until the user is done

6. Shut down the Prolog engine.

Steps 1, 2 and 6 are accomplished by single calls to the Logic Server API. The more interesting interactions are the code for step 4.

The logic-base in CyberTax™ implements the rules to perform the computations for all of the tax forms. It is invoked every time the user enters a numeric value into a tax form field. The routine that interfaces with all the Prolog rules described above is called 'Computation'. It performs the following tasks:

1. Carries forward the newly entered form field value to other form fields (if needed)

2. Calculates all the total fields for the current form

By doing this, all the values on all the forms are always properly calculated and keep up-to-date.

'Computation' uses the Logic Server API to interface with the Prolog engine. The API provides the C/C++ programmer with the ability to easily query Prolog rules. The beauty of the Logic Server is the communication between the Prolog engine and the application program can be accomplished using character strings. Thus, performing a Prolog query from an application is simply a matter of constructing a query string in Prolog syntax (just like you are using a Prolog interpreter) and call the appropriate API function to execute it.

As we mentioned above, the first task in Computation is to carryforward the value just entered to other form fields (if any). This is done by querying whether or not the current field has any carryforward rules defined for it. For example, to query whether Form 1040, Field #0100 is a carry forward field, we do:

```
sprintf(text, "carryforward(fm1040, fd0100, schedulea, fd0200)");
pEng->CallStr(t, text);
```

If there is any rule matches it, CallStr will succeed and return 'true,' else it will fail and return 'false.' Computation continues to query the logic-base looking for carryforward definitions. For each one it changes the value in Prolog's dynamic database and saves it in the Acrobat form.

CyberTax™ has a re-calculate function which is used to perform calculations for a form. In our main application, we used the following logic to perform the re-calculations.

```
// perform re-calculation for Form 1040
If (totalfield(fm1040, X) == TRUE)
        compute the value of field X on fm1040.
```

In the case of performing computations, we use a similar interface:

```
sprintf(text, "compute(fm1040, fd0100, X)");
pEng->CallStr(t, text);
pEng->GetArg(3, cLONG, &longBuf);
```

Here we used the GetArg function retrieve the arguments from the Prolog predicate. The first argument of this function indicates the argument number; the second argument is the type of the variable; and the last argument is the pointer to where to put the value. Amzi Prolog supports several common used data types included: string, integer, short, long, float, double, atom and term. The Logic Server API also provides functions for building and decomposing complex Prolog terms like lists and structures.

CONCLUSION

Adobe Acrobat provides visual display of IRS-provided tax forms, TranZForm™ allows the form to be used for input and output by the user, and an Amzi! logic-base provides all of the rules that embody the tax code. CyberTax™ is a combination of the above components. Prolog was chosen to represent the tax rules because of the nature of the tax rules--they change every year and are made up of rules for relating data. Also, since tax forms contain so many duplicate values, Prolog provided the best means to handle the redundant values on all the forms.

The strengths of Prolog for CyberTax™ are two-fold. First we have shown you how Prolog's declarative rules provide a natural mapping with the logic of the tax forms. Second, Prolog's specialized execution behavior (backtracking) automatically recalculates needed values. In addition, we have shown how easy it is to integrate prolog rules into your application written in high-level languages (like: C, C++) without performance penalty using the appropriate tools. It is true that the same tax rules can be implement in C or C++; however, implementation in Prolog is a lot easier in certain cases (like tax rules).

Finally, I would like to take a chance to thank Brian Sobus, the developer of TranZform™ and Dennis Merritt, the developer of Amzi! Prolog Logic Server™ for their invaluable advice and effort during the development period.

REFERENCES

[AMZI95]   Amzi! Prolog + Logic Server User Manual, 1995.

[BRAT90]   I. Bratko, PROLOG: Programming for Artificial Intelligence, *Addison Wesley Publishing Company*, 2nd Edition, 1990.

[COVI88]   Covington, Nute, Vellino, Prolog Programming in Depth, *Scott, Foresman and Company*, 1988.

AUTOMATIC SOFTWARE VERIFICATION BASED ON REVERSE ENGINEERING AND DEDUCTION: A BRIEF DESCRIPTION OF THE ALICE SYSTEM

Vicent R. Palasí and Francisco Toledo
Departament d'Informàtica, Universitat Jaume I,
Campus Penyeta Roja, 12071 Castelló, Spain
e-mails: {palasi, toledo}@inf.uji.es

ABSTRACT

A fully automatic method for verification of programs and its implementation (named "the ALICE system") are introduced. The method is discussed and the problems that it entails are solved. Furthermore, the paper contains a description of the different parts of the ALICE system as well as of its theoretical foundations.

Keywords: Automatic verification, Logic Programming, Deduction, Reverse Engineering.

1 INTRODUCTION

Since 1969, when software's chronic crisis was detected, it has been clear that the lack of an automatic procedure of verification (that is, an automatic procedure for checking the correctness or incorrectness of a program at compile-time) is a great drawback (see, for instance, [Gib94]). In absence of such a procedure, the technique used for determining if a program is correct or not is "program testing" at run-time. But, as Dijkstra said, "Program testing can be used to show the presence of bugs, but never to show their absence!". So it is generally accepted that, while there be no automatic procedure of verification, the conception of Computer Science as a Engineering will not be well established.

We say that a program is correct if the behaviour expected from it is the same as its actual behaviour. Therefore, it is obvious that we can refer to the correctness of a program only if we have a description of the behaviour expected from it, which is known as specification. This specification must be written in a formal language if we want to study the problem of correctness from an automatic point of view.

Although a lot of research has been employed in order to find an automatic procedure which serves to obtain correct programs, the most interesting results can be divided into two groups. Firstly, there are the procedures derived from Hoare Logic (see, for example, [Hoa71], [Coh90]). Secondly, there are the techniques known as transformational development (see, for example, [Bau82], [PaS83]). It must be remarked that none of them is fully automatic. (I.e, in Hoare Logic, the user must supply the invariant suitable for each loop, and, in transformational development, he must choose the transformation rule which will be applied in each moment). Consequently, these theories allow us to do "semiautomatic verification" or "computer-aided verification", but never "automatic verification".

This paper aims to introduce a fully automatic method for verification of programs, based on a logical formalism, which implementation is the ALICE[1] system. Given a program and an algebraic specification, this system determines whether the program is correct w.r.t. the algebraic specification.

2 FUNCTIONAL ARCHITECTURE OF THE ALICE SYSTEM

If we want to check the correctness of a program P w.r.t.

---

[1] It is the acronym of "ALgebraic Inference of the Correctness of Environments". The formal description of the ALICE system and the proof of its correctness appear in the following research reports: [Pal96a], [Pal96b], [Pal96c], [Pal96d]. However, the brevity of this paper makes it impossible to describe the system with full particulars. Consequently, we only explain the intuitive ideas and we leave the formal description for the former reports.

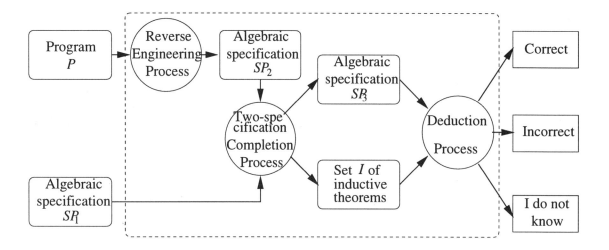

Figure 1: Functional architecture of the ALICE system

an algebraic specification SP_1[2], we must see if the program and the specification are "equivalent" according to a reasonable notion of equivalence (which is formally defined in [Pal96a]). This checking implies comparing two objects (program and specification) of different nature (almost every verification technique is based on this idea), although it seems more suitable to compare two objects of the same nature. The *Reverse Engineering Process* (a part of the ALICE system, studied in Section 4) transforms program P into an algebraic specification SP_2 that is "equivalent" to it. In this way, our approach to verification implies comparing two specifications.

When the Reverse Engineering Process is finished, we will obtain two specifications SP_1 and SP_2. We will have to check if these specifications are equivalent (according to the meaning of equivalence defined in Section 3). The *Two-specification Completion Process* creates an algebraic specification SP_3 and a set I of inductive theorems such that proving the equivalence between SP_1 and SP_2 is the same as proving I in the initial algebra of SP_3.

Therefore, the problem of determining whether P is correct w.r.t. algebraic specification SP_1 has been reduced in our approach to proving several inductive theorems in a given initial algebra. Therefore, in order to solve this problem, we can use the *Deduction Process* of the ALICE system (which is an inductive theorem prover based on techniques of Automatic Deduction). Since bibliography about proof of inductive theorems exists (see [KoR90], [Red90] for Horn clauses), this process will not be explained here.

To sum up, the functional architecture of the ALICE sys-

tem appears at Figure 1[3](everything within the discontinuous line is transparent to the user).

3   VISIBLE SEMANTICS

Though there have been several proposals of initial ([GTW78]), final ([Kam83]) and behavioral ([Niv87], [BBK94]) algebraic semantics, none of them is completely suitable for dealing with the problem of proving the correctness of a program w.r.t. an algebraic specification. The following example makes this clear and shows the intuitive ideas of the semantics which we are going to use: "visible semantics".

Let us suppose that there is an imperative programming language in which the operation of multiplication is not built-in, and also that a function to compute this operation is programmed. The result can be seen at Figure 2a[4].

If we wish to study the correctness of this function, we must have to write a specification of it. The result (using initial semantics) can be seen at Figure 2b[5]. Following the general idea explained in Section 2, we transform function "*" into the specification SP_2 showed in Figure 2c[6] (via the

---

[2]It can be proved ([Pal96b]) that this problem is undecidable (even if only partial correctness is required). However this theoretical result does not imply that the method has no practical utility (as in PROLOG happens).

[3]It could be argued that, since the problem is undecidable, the system could not finish. However, in practice, we cannot wait an infinite time for a response. If ALICE has not finished after a fixed period of time, we will have to stop the proof and answer "I do not know". Similar approach was taken in practical PROLOG implementations by some of the unification algorithms for solving the "occur-check" step.

[4]We use the programming language AL, which is described in [Pal96c].

[5]We assume that operations *suc*, *zero* and $+$ (with prefix notation) have been defined. Their signatures and equations have been omitted for reasons of brevity.

[6]Operation *eval_function* is part of the algebraic semantics of the language. It is assumed that all the operations of this alge-

Figure 2a	Figure 2b	Figure 2c

obs *(a,b: **nat**) **ret** c: **nat**
  c:=0;
    while b > 0 **do**
    c:=c+a;
    b:=b-1
    endwhile
endfunction
	spec SP_1 **is**
	  sorts
	  nat
	  signature
	  zero: $\longrightarrow$ nat
	  suc: nat $\longrightarrow$ nat
	  +, *: nat nat $\longrightarrow$ nat
	  equations
	  $\forall$ a,b:nat
	  +(zero,b)=b
	  +(suc(a),b)=suc(+(a,b))
	  *(zero,b)=zero
	  (suc(a),b)=+((a,b),b)
	endspec
		spec SP_2 **is**
		  sorts
		  nat
		  signature
		  *: nat nat $\longrightarrow$ nat
		  equations
		  $\forall$ a,b:nat
		  *(a,b)=eval_function(
		    " **obs** *(a,b: **nat**) ...
		    ... **endfunction** ",a,b)
		endspec

Figure 2: Example of formal verification via algebraic specification

Reverse Engineering Process). In our approach, we can say that function "*" is correct w.r.t. specification SP_1 if specifications SP_1 and SP_2 (which is the algebraic semantics of function "*") are "equivalent".

It is easy to see some properties that this equivalence must fulfil:

1. Two specifications can be equivalent despite having different signatures (in our example, SP_2 has at least one operation (eval_function) which does not belong to SP_1).

2. In fact, we can divide operations of a specification into two classes: those that we wish to specify (such as "*", in our example) and those that we describe only because they are required to define the former (such as "eval_function" or "+"). We call them "observable" and "hidden", respectively. It is easy to see that two equivalent specifications must have the same observable operations.

3. Finally, it is well known that a term represents a "computation" of the specified software system. Terms that have some hidden operation represent computation states that are not visible in the external behaviour of the system. Therefore, only terms that only have observable operations (hence, "totally observable terms") should be considered to define the equivalence.

Therefore, we can suppose that two specifications are equivalent if their totally observable terms "behave" in the same way. But this approach is not completely correct. The problem becomes clear when we wish to specify data types like stacks, lists, sets, etc. For these data structures, it is advisable to obtain an abstraction of the internal implementation similar to that of the conventional behavioural semantics. To reach this goal, we will divide the observable sorts[7] into two classes. The "nonvisible" ones are the sorts that have abstraction of the internal implementation (they will normally represent data structures). The remaining sorts are "visible" sorts (they will normally represent atomic types). Therefore, we will consider that two specifications are equivalent if their visible results are the same, that is, if their totally observable terms of visible sort "behave" in the same way.

To sum up, the semantics which we have developed has two levels of "observability": that which distinguishes between observable and hidden operations, and that which distinguishes between visible observable and nonvisible observable sorts. Furthermore, this semantics is based on the initial one (since this latter is one of the simplest, most intuitive and most widespread semantics) and, for this reason, it is named "initial visible semantics"[8].

Although, because the sake of brevity, the formal description of "initial visible semantics" cannot be explained here (see [Pal96a]), the intuitive ideas of this theory have been given in the present section.

---

braic semantics have been defined. Their signatures and equations have been omitted for reasons of brevity. (For a more detailed description of the problem of the algebraic semantics of a language, see Section 4).

---

[7] A sort s is observable if there is an observable operation of sort s.

[8] Given a specification SP, we will use the notation $Vis - I[SP]$ to refer to its initial visible semantics.

4 THE REVERSE ENGINEERING PROCESS

As has been said in Section 2, the Reverse Engineering Process is the process which transforms an imperative program into an equivalent algebraic specification (see Figure 1). It is remarkable that this process is different for each programming language used to write the program, although the underlying ideas are always the same. In this section, we will explain these ideas. The reader must find a complete and formal description of the Reverse Engineering Process for a particular language[9] in [Pal96c].

4.1 REQUIREMENTS FOR A SEMANTICS DEFINITION OF A PROGRAMMING LANGUAGE

We wish to define a process which transforms a program into a visible algebraic specification that is "equivalent" to it (see Figure 1). But before describing this process, we must explain what the word "equivalent" means for us. We say that a program P and a visible algebraic specification SP_1 are equivalent if the semantics of P is the same as that of SP_1. Since the semantics of an algebraic specification has already been defined[10], we must only define that of a program. Moreover, since we have to compare the two semantics, it would also be useful to define the semantics of a program in an algebraic way.

In the literature, there have been several attempts at defining the semantics of an imperative language via algebraic specifications (see for instance, [GoP81]). But these works cannot be applied here for the following reasons:

- They are oriented to the definition of the semantics of a language and not to that of a particular program.

- The way they define data types is not useful for our purposes.

The example of Section 3 can help us to understand the last reason. Let us focus on the signature of operation "eval_function". Since function "*" has two parameters of natural type and one result of the same type, the signature of "eval_function" should be "function nat nat $\longrightarrow$ nat".

However, we should be able to translate any imperative function into algebraic notation. Therefore, there must be one operation such as "eval_function" for each combination of parameters and result types[11]. In fact, the number of operations such as "eval_function" could

be infinite. Analogously, the number of operations such as "eval_assignment", "eval_expression", "eval_statement" (and, as a general rule, all the operations which deal with different types) could also be infinite.

This problem is solved in [GoP81] by creating the generic sorts (that is, types) *value* (which means any value regardless of its type) and *lval* (which means any list of *values*). A program [12] is translated into the algebraic notation as an operation with the signature *lvalue* $\longrightarrow$ *value*. In this way, there is one single "eval_function", one single "eval_assignment" and so on. But the example in Section 3 shows that this solution is not acceptable here. If we have a function declared as "obs *(a,b:**nat**) **ret** c: **nat**", its translation into algebraic notation must be an operation with the signature "nat nat $\longrightarrow$ nat", since it must be compared to an operation with the same signature (that defined by the user in SP_1).

Consequently, our solution must be type-dependent enough to translate each imperative function into an algebraic operation with the same types. But it must be type-independent enough to avoid having infinite "eval_function", "eval_assignment", etc.

4.2 TRANSFORMING PROGRAMS INTO SPECIFICATIONS

The solution proposed here is dual. On the one hand, the programming language will be specified from the sorts *value, lval*, which allows us to avoid the existence of infinite operations. On the other hand, we will add to each particular program a set of sorts and operations which cause each imperative function to be translated into an operation with the suitable signature.

In this way, the semantics of any program P will be a visible specification $Trad(P)$ composed of two parts:

1. A type-independent and program-independent part. Functions such as eval_function belong to this part and work only with the generic types[13]

2. A part that is different for each program and is composed of the following operations and equations (which specify the management of the data types of a program):

   - For each type T in the program, there are two type conversion operations gen_T (which converts from this specific type to the generic type *value*) and esp_T (which converts from *value* to the specific type T)

---

[9]To be exact, the AL language described in [Pal96c].

[10]The intuitive definition is in Section 3 an the formal definition is in [Pal96a].

[11]The problem is greater in a programming language which allows the user to define an indefinite number of new types.

[12]A function in object-oriented programming languages or languages based on abstract data types.

[13]It is not included here because it is similar to that of [GoP81]. Furthermore, it is fully described in [Pal96c].

- For each function with a heading as **obs** f(a_1 : T'_1,...,a_n : T'_n) **return** a_{n+1} : T'_{n+1} there is one operation with the same name, defined by one equation $f(a_1, ..., a_n) = esp_{T_{n+1}}(eval - function(\textbf{obs } \textbf{f}(a_1 : T'_1,...,a_n : T'_n) \textbf{ return } a_{n+1} : T'_{n+1}, gen_{T_1}(a_1) :: gen_{T_2}(a_2) :: ... :: gen_{T_n}(a_n)))$[14].

(The meaning of this equation is the following: In order to define a particular function over several given values, we convert these values to the generic type *value*, by applying gen_T. Then, "eval_function" defines the result of the function working only with generic types[15]. Finally, we convert back the result from type *value* to its specific type, by applying esp_T.)

To sum up, the semantics of P can be defined as:

$$Semantics(P) = Semantics(Trad(P))$$

Therefore, the Reverse Engineering Process is the algorithm which transforms P into $Trad(P)$. Moreover, since we are dealing with initial visible semantics, the above statement is equivalent to:

$$Semantics(P) = Vis - I[Trad(P)]$$

And this equation is the definition of the semantics of any program P.

5 THE TWO-SPECIFICATION COMPLETION PROCESS

The ALICE system aims to deduce whether program P is correct w.r.t visible specification SP_1. In other words, whether P and SP_1 are equivalent or whether P and SP_1 have the same semantics, that is:

$$Semantics(SP_1) = Semantics(P)$$

Since we are dealing with initial visible semantics, we have that:

$$Semantics(SP_1) = Vis - I[SP_1]$$

Furthermore, as we have seen in the above section:

$$Semantics(P) = Vis - I[Trad(P)]$$

Consequently, statement $Semantics(S) = Semantics(P)$ is equivalent to:

$$Vis - I[SP_1] = Vis - I[Trad(P)]$$

This expression means that the initial visible semantics of SP_1 and $Trad(P)$ are the same. In other words:

$$SP_1 \text{ and } Trad(P) \text{ are initially visibly equivalent.}$$

Therefore, the proof of the correctness of a program can be reduced to the proof of the initial visible equivalence between two algebraic specifications. As can be seen in Figure 1[16], the Two-Specification Process reduces this equivalence to the proof of a set of inductive theorems and this proof is done by the Deduction Process.

If we were dealing with initial semantics, the reduction performed by the Two-Specification Completion Process would be easy. In order to prove that SP_1 and SP_2 are equivalent, it would be enough to deduce satisfaction of equations belonging to SP_1 in the initial algebra of SP_2 and vice versa. The same method can be applied to the initial behavioural semantics, though it would be necessary to work with behavioural satisfaction instead of conventional satisfaction. But this is no problem, since the former can be reduced to the latter (see [BiH94]). However, this method is not valid in initial visible semantics[17]. The reason is that SP_1 and SP_2 may have different signatures, and consequently the equations of SP_1 may not make sense in the initial algebra of SP_2 or vice versa. Therefore, our method is different. We create a specification SP_3 (named, the V-reunion[18] of SP_1 and SP_2) whose initial algebra contains all the information contained in the initial algebras of SP_1 and SP_2 (and some information which relates all two specifications). Then, it can be proved (see [Pal96d]) that proving the initial visible equivalence between two specifications can be reduced to proving several inductive theorems[19] over their V-reunion.

Therefore, the Two-specification Completion Process is the one that, starting from two specifications, builds their

---

[14] a::b::c means the list of values [a,b,c].

[15] Since the definition is done inside the type-independent part of the specification.

[16] In this figure, specification $Trad(P)$ is named SP_2.

[17] Although the concept of visible satisfaction can be defined (see [Pal96a]).

[18] The formal definition of this concept and its discussion appears at [Pal96d].

[19] The choice of these inductive theorems is discussed at [Pal96d].

Figure 3a	Figure 3b
spec SP_2 **is**	**spec** SP_3 **is**
sorts	**sorts**
nat	nat, nat'
signature	**signature**
zero: $\longrightarrow$ nat	$<$ *All the operations of* SP_2 $>$
suc: nat $\longrightarrow$ nat	zero' : $\longrightarrow$ nat'
$*$: nat nat $\longrightarrow$ nat	suc' : nat' $\longrightarrow$ nat'
$<$ *Other operations of the algebraics*	$+'$,$*'$: nat' nat' $\longrightarrow$ nat'
semantics$>$[20]	yes: $\longrightarrow \gamma$
equations	plus: $\gamma \; \gamma \longrightarrow \gamma$
$\forall$ a,b:nat	trans: nat nat' $\longrightarrow \gamma$
$*$(a,b)=esp_{nat}(eval_function(**obs**	**equations**
$*$(a,b ... **endfunction** ,	$\forall$ a,b:nat; a' ,b' :nat'
gen_{nat}(a) :: gen_{nat}(b)))	$<$ *All the equations of* SP_2 $>$
$<$ *Other equations of the algebraics*	$+'$ (zero' ,b')=b'
semantics$>$[20]	$+'$ (suc' (a'),b')=suc' ($+'$ (a' ,b'))
endspec	$*'$ (zero' ,b')=zero'
	$*'$ (suc' (a'),b')=$+'$ ($*'$ (a' ,b'),b')
	plus(yes,yes)=yes
	trans(zero,zero')=yes
	trans(suc(a),suc' (a'))=trans(a,a')
	trans($*$(a,b),$*'$ (a' ,b'))=plus(trans(a,a'),trans(b,b'))
	endspec

Figure 3: An example of automatic verification using the ALICE system

V-reunion and the inductive theorems stated in [Pal96d][21].

6   USING THE ALICE SYSTEM: A TOY EXAMPLE

In order to illustrate our method, we introduce an example[22] of automatic verification using the ALICE system. Let us suppose (as in Section 3) that we have programmed the imperative function "$*$" (see Figure 2a) and we wish to know whether it is correct w.r.t to specification SP_1 (see Figure 2b). Firstly, as it has been discussed in Section 4, function " $*$" is transformed into the specification SP_2[23] (which appears at Figure 3a) by the Reverse Engineering Process. Then, the Two-specification Completion Process returns the specification SP_3 (see Figure 3b)[24] and the following set I of inductive theorems:

$$trans(x,x') = yes \; \& \; trans(y,y') = yes \; \& \; x = y \Rightarrow x' = y'$$
$$trans(x,x') = yes \; \& \; trans(y,y') = yes \; \& \; x' = y' \Rightarrow x = y$$

As it has been seen, checking the correctness of "$*$" w.r.t SP_1 is equivalent to proving I over the initial algebra of SP_3. So we can use the Deduction Process (just an inductive theorem prover) to prove the former inductive theorems and, therefore, to solve our problem[25].

7   CONCLUSIONS AND FUTURE WORK

A system (named ALICE) which automatically deduces whether a program is correct w.r.t. an algebraic specification has been introduced. In fact, the system reduces the proof of correctness to the proof of some inductive theorems. This latter can be implemented by an inductive theorem prover, such as those described in the bibliography.

---

[20]These operations and equations are described in [Pal96c].

[21]The complexity of this process is linear.

[22]It is recommended to read this example by consulting Figure 1.

[23]In this specification, "zero", "suc" and "$*$" are observable and the other operations are hidden. Moreover, "nat" is a visible sort.

[24]SP_3 is named the V-reunion of SP_1 and SP_2.

[25]Since there is a lot of bibliography about inductive theorem provers, this final part of our system is not explained here (see, for example, [KoR90], [Red90]).

It is advisable to use the ALICE by individually applying it class by class. If the result is "correct", the class is correct. If the result is "incorrect", there is an error in the class and we will have to debug it. On the other hand, if the system cannot give a conclusive answer, we will have to mark the class as "doubtful"[26].

There are two future research lines. On the theoretical side, we wish to extend this method of verification to object-oriented programming. The aim is to describe a system such as ALICE for object-orientation and to prove its correctness. To do this, we will probably have to use order sorted algebra ([GoM92]) instead of many-sorted algebra. On the practical side, we wish to testing the correctness of different programs w.r.t. their algebraic specifications using the ALICE system.

8 REFERENCES

[Bau82] BAUER, F.L. *From Specifications to Machine Code: Program Construction Through Formal Reasoning.* Proceedings of Sixth International Conference on Software Engineering, Washington, 1982, pp. 84-91.

[BBK94] BERNOT, G. BIDOIT, M. KNAPIK, T. *Behavioural approaches to algebraic specifications.* Acta Informatica **31** (1994), pp. 651-671.

[BiH94] BIDOIT, M. HENNICKER, R. *Proving Behavioural Theorems with Standard First-Order Logic.* Research Report LIENS 94-11. Laboratoire d'Informatique. Ecole Normal Superieure (1994).

[Coh90] COHEN, E. *Programming in the 1990's. An Introduction to the Calculation of Programs.* Springer-Verlag, 1990. ISBN: 3-540-97382-6.

[Gib94] GIBBS, W.W. *Software's chronic crisis.* Scientific American **271**, num. 3 (September 1994), pp. 72-81.

[GoP81] GOGUEN, J.A. PARSAYE-GHOMI, K. *Algebraic Denotational Semantics Using Parameterized Abstract Modules.* Formalization of Programming Concepts, LNCS **107** (1981), pp. 292-309.

[GoM92] GOGUEN, J.A. MESEGUER,J. *Order-Sorted Algebra I: Equational Deduction for Multiple Inheritance, Polymorphism and Partial Operations.* Theoretical Computer Science **105**, num. 2 (November 1992), pp. 217-273.

[GTW78] GOGUEN, J.A. THATCHER, J.W. WAGNER, E.G. *An initial algebra approach to the specification, correctness and implementations of abstract data types.*, in: R.T.Yeh, ed., Current Trends in Programming Methodology; IV Data Structuring (Prentice-Hall, Englewood Clifts, NJ, 1978), pp. 80-149.

[Hoa71] HOARE, C.A.R. *Proof of a program: Find.* Communications of the ACM **14**, num. 1 (1971), pp. 39-45.

[Kam83] KAMIN, S. *Final data types and their specification.* ACM Transactions on Programming Languages and Systems **5** (1983), pp. 97-123.

[KoR90] KOUNALIS, E. RUSSINOWITCH,M. *Mechanizing Inductive Reasoning.* Proceedings of the AAAI Conference, Boston, 1990, pp. 240-245.

[Niv87] NIVELA, P. *Semántica de Comportamiento en Lenguajes de Especificación.* PhD thesis, directed by Fernando Orejas Valdes. Facultat d'Informatica. Universitat Politecnica de Catalunya (1987).

[Pal96a] PALASI, V.R *Visible Semantics: An Algebraic Semantics for Automatic Verification of Algorithms.* Research Report LSI-96-26-R. Departament de Llenguatges i Sistemes Informatics. Universitat Politecnica de Catalunya (1996).

[Pal96b] PALASI, V.R. *Automatic Verification of Programs: the algorithm ALICE.* Research Report LSI-96-31-R. Departament de Llenguatges i Sistemes Informatics (1996).

[Pal96c] PALASI, V.R. *Semàntica Algebraica del Llenguatge AL.* Research Report LSI-96-47-R. Departament de Llenguatges i Sistemes Informatics (1996).

[Pal96d] PALASI, V.R. *Deducció Automàtica de l'Equivalència Inicial Visible.* Research Report LSI-96-64-R. Departament de Llenguatges i Sistemes Informatics (1996).

[PaS83] PARTSCH, H. STEINBRUGGER, R. *Program Transformation Systems.* ACM Computing Surveys **15**, num. 3 (September 1983), pp. 199-236.

[Red90] REDDY, U. *Term Rewriting Induction.* Proceedings of the 10th International Conference on Automated Deduction, Kaiserslautern, 1990, pp. 162-177.

---

[26]This fact is consequence of the undecidability of the first-order logic.

MODEL-BASED DIAGNOSIS FOR REACTOR COOLANT PUMPS OF EDF NUCLEAR POWER PLANTS.

Marc Porcheron, Benoit Ricard
Electricité de France, Research and Development Division
Advanced Information Processing Section
6, quai Watier, 78401 Chatou Cedex, FRANCE
E-mail: Marc.Porcheron@der.edfgdf.fr

ABSTRACT

DIAPO is an AI diagnostic system for *reactor coolant pump sets* in EDF nuclear power plants. About 200 pump and plant state parameters can be used to describe the phenomenon to be diagnosed. Basically, diagnosis relies on the cooperation of a set of *fault models* used by an abduction procedure in order to explain observations which are elaborated from monitoring data or provided by plant staff. Diagnosis of a dysfunction thus consists in the conjunction of identified initial causes, recognized faults, recognized abnormal situations and locations of faulty components. In addition, prediction of future states of the pump can be provided. Heuristic knowledge is also used in order to focus diagnosis. A full scale prototype of DIAPO has been jointly developed by EDF and Jeumont-Industrie and validated. An industrial version is currently under development.

Key words: model-based reasoning; diagnosis; practical applications

INTRODUCTION

Major components of EDF nuclear power plants are continuously monitored in order to detect anomalies in their behavior. In addition, dedicated AI expert systems assist plant operators in identifying the underlying faults when such components behave abnormally. Among those systems, DIAPO diagnoses *reactor coolant pump sets* (RCPS). Diagnosis performed by DIAPO presents the following major features:

- it is elaborated from monitoring data while the RCPS operates;
- it can be based on an incomplete limited subset of RCPS observed parameters;
- it consists in an explanation of observations in terms of hypotheses drawn from different fault models of the RCPS; explanations are built by applying abductive reasoning techniques to these fault models;
- fault models integrate temporal knowledge in order to take into account the dynamics of the RCPS;
- it may account for multiple faults when their manifestations are independent.
- it provides predictions about the possible evolution of the abnormal RCPS behavior.

These features enable DIAPO to satisfactorily operate in a context typical of industrial system diagnosis, as synthesized in [Dvorak, Kuipers 89]: such systems are generally dynamic (the system exhibits time-varying behavior); few system parameters are observable (sensors are scarce, sometimes unreliable, plant operation restricts observability); diagnosis must be performed while the system operates (such systems are designed to operate with multiple minor faults, but incipient failure detection is mandatory).

A prototype of DIAPO has been developed in cooperation with the RCPS manufacturer, Jeumont-Industrie. An industrial version of the system is currently under development, and will eventually be integrated into EDF new plant monitoring architecture designed for the fifty French nuclear power units [Joussellin et al. 94].

This paper is organized as follows: section 2 describes the RCPS diagnostic problem; section 3 outlines DIAPO functional architecture; section 4 deals with diagnostic techniques involved; section 5 presents the way these techniques are put into operation in DIAPO; section 6

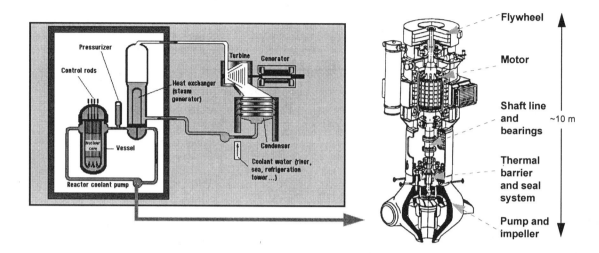

FIGURE 1 Operating principle of a Pressurized Water Reactor plant and RCPS exploded view.

outlines major results and perspectives; section 7 considers the relationship with related works.

THE PROBLEM

The *reactor coolant pump set* (RCPS) is one of the major components in the primary cooling system. It plays a key role in nuclear plant operation, as it is responsible for ensuring circulation of the coolant in the primary system, between the reactor core and the steam generator (Figure 1). Good functioning of the RCPS is therefore decisive in maintaining plant availability; stopping the RCPS results in plant shutdown and a consequent serious loss of production capacity. It should be added that in a pressurized water nuclear plant, this functional role is compounded by a safety role, as primary coolant also serves to cool the reactor core.

The RCPS is an imposing machine some ten meters in height and 80 tons in weight with a flow rate of up to 7 m3/s. The nominal rotation speed is of about 1500 rpm. Three functioning phases are considered: start-up, nominal operating conditions and rundown before standstill.

Experts in the design, operation and maintenance generally distinguish the following main functional assemblies, from bottom to top (Figure 1):
- the *pump section* per se, in which the primary coolant water is set in motion by an *impeller*;
- the *thermal barrier*, and the *seal system*, preventing the hot, slightly radioactive primary coolant from rising above the shaft line;
- the *shaft line*, consisting in three shafts coupled one to the other, crossing the machine from top to bottom; their stability is ensured by *bearings*;
- the *motor*, located on the top, which ensures rotation of the shaft line; a *flywheel* allows to increase the slow-down time in case of electrical power loss.

The RCPS are continuously monitored by an automated system which uses some 30 sensors to provide measurements characterizing the condition of the machine and alarms. Main ones are thermo-hydraulic (bearing temperatures, seal flow...) and vibratory measurements (shaft and housing vibrations...). Data provided by the RCPS monitoring system and stored in a data-base and can be retrieved for "real-time" or delayed consultation. Together with these monitoring data, plant state parameters and additional information concerning the pump (e.g. maintenance and technological information) are available. This leads to about 200 parameters which may be used to describe a dysfunction.

Some 150 RCPS exist in the different nuclear power plants in France. Their expected lifetime is over 30 years. A vast RCPS diagnostic expertise exists, built up over twenty years of designing and operating these machines. RCPS are therefore familiar machines, both in terms of their design and in their operation, and experts are readily available both at EDF and Jeumont-Industrie.

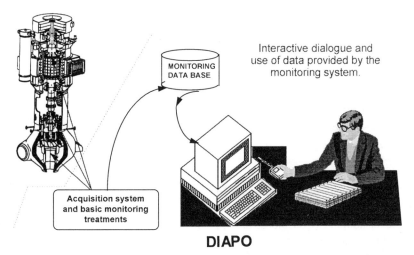

DIAPO

Figure 2. Diagnostic process.

At the present time, diagnosis of these machines is most often performed by these experts when an alarm is given or a measurement evolves in abnormal fashion. One of the main objectives of the diagnosis is to determine as quickly as possible whether or not it is necessary to shut down the machine (which leads, as we have said, to plant shutdown).

DIAPO aims at assisting experts and at providing plants with local diagnostic capacities.

SYSTEM OVERVIEW.

Typically, the user initiates a diagnostic session when an alarm is given or a measurement evolves in an abnormal fashion. This constitutes the initial observation set provided to DIAPO. In order to establish the complementary relevant observations during the diagnostic process, DIAPO makes use of both data stored in the monitoring data-base and of complementary information interactively provided by the user (cf. Figure 2).

At the end of the session, DIAPO provides the user with an explanation of observations in terms of a combination of diagnostic hypotheses. Each diagnostic hypothesis is a couple *(H, D)*, where *H* may be a *component failure*, an *established abnormal situation*, an *incriminated location*, or a *primary cause*, and *D* is the suspected temporal extend of *H*. DIAPO can also provide predictions about the possible evolution of the abnormal RCPS behavior encountered.

Diagnoses can be computed from incomplete observation sets. Namely, a data-base query may fail or the user may answer *"I don't know"* to some query

during diagnosis, typically because some parameter is unavailable or some observation is not practicable.

Queries to the user remain necessary because the monitoring system cannot automatically provide *all* information needed to make an "intelligent" diagnosis. Information delivered by the monitoring system is limited to the characterization of certain variations in measurements provided by sensors. However, other information potentially important to the diagnosis is obviously not known to the monitoring system (e.g. the fact of having performed repairs on a bearing, or a boron injection in the primary system).

Finally, it must be noticed that, although diagnosis might use "on-line" monitoring data, it is not automatically triggered by the monitoring system and has no feedback action on the RCPS.

DIAGNOSTIC PRINCIPLES

RCPS expert knowledge analysis pointed out that diagnosis essentially consists in an *explanation of observations,* based on *causal knowledge of RCPS faults.* Although knowledge of the correct behavior of the RCPS and associated consistency-based reasoning methods (cf. [Reiter 87], [de Kleer, Williams 87]...) could also be involved, RCPS fault models are sufficiently complete to justify using an abductive approach [Console, Torasso 91]. Basically, diagnosis in DIAPO thus relies on the cooperation of a set of fault models which contribute to explain observations describing an abnormal behavior of a RCPS.

Fault models and abduction.

A fault model as used in DIAPO is a logical theory constituted by a set of relations linking back the observations expected when the RCPS behaves abnormally to the "explanations" thereof. The generic representation of such a relation is an inference of the form: $C_1 \wedge C_2 \wedge ... \wedge C_n \xrightarrow{\Delta} E$ which can be interpreted as: *"the conjunction of $C_1, C_2, ... , C_n$ causes E after a delay Δ."*

Because any fault model accounts for a certain *description level* of the physical processes involved, some causal relationships may be *incomplete* with regards to the detail level adopted [Console et al. 89]. For example, the conditions under which a "primary water rise damages the pump bearing" may be too complex to be specified within the fault model. Nevertheless DIAPO fault models allow representation of such relations wherein some conditions or processes have been abstracted or ignored. For that purpose distinguished terms are included in the left hand side of the relation to model the actual condition (process) abstracted (ignored) in the model In such cases, the inference takes the following form, α representing the abstracted condition :

$$C_1 \wedge C_2 \wedge ... \wedge C_n \wedge \alpha \xrightarrow{\Delta} E .$$

Keeping up with the example, the knowledge "a primary water rise *may* damage the pump bearing " will be represented by the following term:

$$Primary\_water\_rise \wedge \alpha \xrightarrow{\Delta}$$

$$loosening\_of\_pump\_bearing$$

A delay Δ in a causal relation consists in a set of temporal constraints. These constraints relate to the respective beginnings and ends of the elements involved in the relation; they may be approximate, viz. bear only an interval into which the actual delay might take place. Keeping up with the example, the delay Δ in the above term may represent the knowledge that the loosening of the bearing may occur *between zero and six months* after the primary water rise.

Abductive diagnosis procedure used by DIAPO relies on [Console et al 89]. The set of observations consist in *positive* observations which enable satisfying symptoms, and *negative* ones which enable rejecting them. The *explanations* of the observations are the most specific formulae of the language of abductible terms (i.e. terms without any cause in the models) which imply the positive observations and are consistent with the negative ones. (An explanation F_1 is called more specific than an explanation F_2 iff $\vdash F_1 \rightarrow F_2$). If an explanation F is put in normal disjunctive form, each term of the disjunction

represents an *elementary explanation*. An elementary explanation E_i is a conjunction of abductible terms and negations of abductible terms of the form :

$$E_i \equiv C_{i1} \wedge \cdots \wedge C_{in} \wedge \neg C'_{i1} \wedge \cdots \wedge \neg C'_{ik}$$
$$[1] \quad F \equiv \vee E_i$$

Negative terms are derived from "negative" observations by use of the inference scheme ($C \rightarrow E$; $\neg E \vdash \neg C$).

In formula [1] each term C_{ij} is a couple associating one term of the model and a temporal extend computed by applying propagation mechanisms to causal relationship delays and observation dates. Because of the approximate nature of certain delays and observation dates, each temporal extend consists in four parameters, of which the first two (resp. last two) define an interval of uncertainty with respect to their beginning (resp. end): *(earliest date of beginning, latest date of beginning, earliest date of end, latest date of end)*.

Let us consider this very simple fault model (taken from the RCP fault model and omitting temporal delays):

Shaft vibration $\wedge$ α $\rightarrow$ Bearing knee-joint blocked
Primary water rise $\wedge$ β $\rightarrow$ Bearing knee-joint broken
Bearing knee-joint blocked $\rightarrow$ High ratio D1/D2
Bearing knee-joint broken $\rightarrow$ High ratio D1/D2
Seal#1 blocked $\rightarrow$ Low QFJ1 flow

High ratio D1/D2 and *Low QFJ1 flow* are symptoms; *Bearing knee-joint blocked, Bearing knee-joint broken* and *Seal#1 blocked* are diagnostic hypotheses; α and β are abstract conditions.

Let Ψ ={*Low QFJ1 flow, High ratio D1/D2*} be the set of observations.

The explanation formula is :
F= (*Seal#1 blocked $\wedge$ Primary water rise $\wedge$ β*) $\vee$
(*Seal#1 blocked $\wedge$ Shaft vibration $\wedge$ α*)
The diagnosis is:
D= (*Seal#1 blocked $\wedge$ Bearing knee-joint blocked*) $\vee$
(*Seal#1 blocked $\wedge$ Bearing knee-joint broken*)

It is worth noting that specific effects of combinations of independent faults may be explicitly represented in the model, and thus naturally diagnosed by the method. Moreover, an elementary explanation can account for a multiple fault insofar as its effects consist in a strict conjunction of the manifestations of each independent constituent.

Finally, when several fault models are available, the complete diagnosis is the conjunction of the diagnoses obtained with each fault model. It is worth noting that a given concept may be represented by different "viewpoints" in different models. For example, *a failure mode of a component* part of a model based on the materiel representation of the device, may also be seen as a *primary cause* in a model representing the causal relationships involved in malfunction propagation,... An hypothesis may thus be assessed in a given model, and rejected in a more complete one.

Application to RCPS diagnosis.

Outline of the main fault models used by DIAPO

Four main fault models are used by different tasks in DIAPO. They are displayed in table 1. Subsequent paragraphs present the main characteristics of each model.

"Associative" fault model

The associative fault model includes a representation of direct relations between the hundredth RCPS faults and their symptoms together with the temporal constraints bearing on their respective occurrences. It can be seen as a causal model of limited depth, as the fault is considered as the cause of its symptoms but not detail is held on the precise mechanism through which one entails the other. Figure 3 shows an excerpt from this model.

The explanation of the observations obtained by exploiting this model is a formula of form [1] above, where positive terms are RCPS faults which are diagnosed as being present, and negative ones are RCPS faults which are eliminated.

Causal fault model

The causal fault model gives a detailed description of the causal relationships involved in RCPS dysfunctions. This causal knowledge is formalized in a *network* whose nodes represent the causal objects (internal RCPS events and conditions, observable manifestations, maintenance and control procedures on the RCPS, etc.) and whose arcs represent causal relations between them. The arcs hold temporal constraints describing the delay between the occurrence of a cause and that of the effect. Figure 4 shows an excerpt from the network used by DIAPO, which holds some one thousand nodes.

The explanation of the observations obtained by exploiting this model is a formula of form [1] above, where positive terms are initial causes of the RCPS dysfunction which are diagnosed as being present, and negative ones are eliminated initial causes.

This model allows to keep a trace of the causal paths that have been established, from initial causes to ultimate consequences, including intermediate events. This provides a precise reconstruction of the degradation mechanism that might have occurred. Moreover, this model provides predictions about future consequences of established events. Figure 5a-b shows a typical diagnostic result obtained with this model. Computed temporal extends are indicated below each node.

Prototypical fault model

This model allows to establish hypotheses bearing on the abnormal *situation* encountered. It is constituted by a hierarchy of about 20 "prototypes" of abnormal situations. A situation causes a set of symptoms characteristic of several faults, and can be refined into a set of more precise situations (see figure 6). The abductive reasoning in this type of model is similar to a *classification* process.

The explanation of the observations obtained by exploiting this model is a formula of form [1] above, where positive terms are abnormal situations of the RCPS which are diagnosed as being present, and negative ones are eliminated situations.

Failure localization model

With this model, it is possible to establish hypotheses as to the *localization of faulty components in the RCPS structure*. The structure is represented by an arborescence whose 125 nodes are parts of the RCPS and whose arcs describe the "geographical" inclusion relation between these parts (sub-parts). The extremities of the arborescence are components.

Table 1 Description of fault models

Fault model	Contents	Function	Inference method
Associative fault model	Direct relations between RCPS faults and symptoms	Assess possible faults from the presence of symptoms	Abduction
Causal fault model	Causal representation of RCPS malfunction propagation	Assess possible causes of observations	Abduction
		Propose possible consequences of current RCPS state	Abduction/deduction
Failure localization model	Material decomposition of the RCPS + component failure modes	Identify faulty parts of the RCPS	Abduction
Prototypical model	Hierarchy of possible abnormal situations of the RCPS	Establish the situation of the RCPS from observations	Abduction

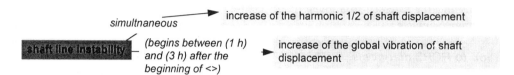

Figure 3: Excerpt of the associative fault model used by DIAPO

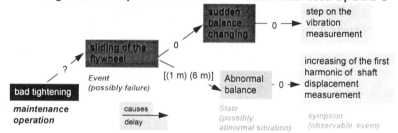

Figure 4: Excerpt from the causal model used by DIAPO

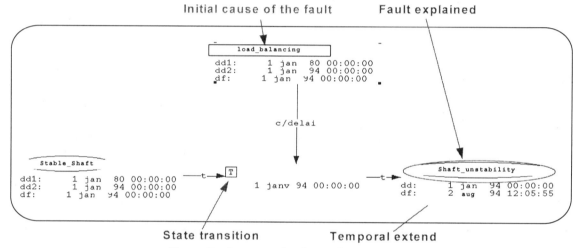

Figure 5a: An example of causal explanation

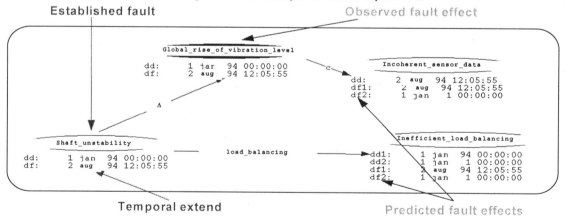

Figure 5b: An example of causal prediction

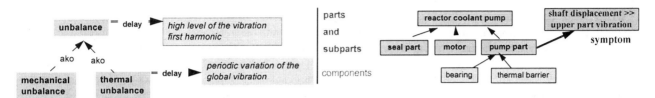

Figure 6: Excerpt from the hierarchy of situations used by DIAPO

Figure 7: Excerpt from the localization model used by DIAPO.

Symptoms are associated to nodes, characterizing the fact that a faulty component is situated in the corresponding parts of the RCPS.

The localization is established by determining the nodes whose symptoms are satisfied and progressively refining the sub-parts. As in the case of the prototypical model, the abductive reasoning in this type of model is similar to *classification*.

Figure 7 shows an excerpt of the physical structure of the RCPS as represented in DIAPO.

The explanation of the observations obtained by exploiting this model is a formula of form [1] above, where positive terms are *locations* in the RCPS structure (viz. faulty parts, subparts or components), which are diagnosed as being established, and negative ones are eliminated locations.

Final diagnosis

The final explanation is obtained by making the conjunction between the explanations obtained with each of the four models above. Once expressed in normal disjunctive form, its appears as a disjunction of explanations using the abductible terms *of the different models* such as that of the form below :

$$C_i \wedge D_j \wedge S_k \wedge ... \wedge \neg C'_i \wedge \neg L'_j \wedge ...$$

Such an explanation can be interpreted in the following way: *"The observations are explained by cause C_i in conjunction with fault D_j and situation S_k; cause C'_i did not occur and the problem is not situated in location L'_j".*

It is worth noting that:
- some hypotheses might be entailed in one model and rejected in a more complete model;
- in a given model, all hypotheses may be eliminated, thus providing only negative terms in the final diagnosis.

OPERATIONAL ISSUES

Completion of observations

Initial observations usually do not allow to determine the truth value of all symptoms. This makes it necessary to complete the observations during the diagnosis:
- to obtain an accurate description of the actual phenomenon, and associated diagnosis when obtaining positive information (e.g. "there is a step on the vibration parameter at a given date"),
- to rule out some possible hypotheses by invalidating symptoms they entail when obtaining negative information (e.g. "there is *no* step on the vibration parameter at a given date")

In order to complete observations DIAPO exploits knowledge specific to the domain of RCPS diagnosis, and more generally that of rotating machines. This "first principles" knowledge of the field is derived from *laws of dependency* between observations based on physics, mathematics and other laws relative to system operation, maintenance, etc. Here are a few samples of this knowledge:

L1: If a scalar descriptor has a mean value higher than X in one time frame, then the mean value is greater than Y in this time frame for every Y smaller than X.

L2: If there is a simultaneous step in the amplitude of the harmonic of a vibration and in its phase, then there is a simultaneous (vectorial) step on the harmonic.

L3: If there is a rise in a scalar descriptor, then there is a variation in this descriptor in the same time frame.

This knowledge is represented either by production rules or in the hierarchies of types of objects used to represent the observations (e.g. L3, which states that a rise is a specific type of variation), or is implicitly put into application by using arithmetic functions to describe part of observations (e.g. L1).

When the "completion knowledge" above cannot establish the missing information, external requests may be addressed to the user or to the RCPS monitoring data base.

Focusing diagnosis.

Complexity of abductive procedures may be critical when dealing with large fault models. Two means may be used in order to reduce this complexity :

- to exploit in priority less complex fault models and use the results to focus abduction in more complex ones. Fault model complexity may be estimated from structural parameters such as : the average number of causes of a given term in the model and the average length of the paths linking primary causes to final effects. Once an hypothesis has been assessed in a given fault model, abduction can be limited to the "connected component" of this hypothesis, for the relationship represented in a more complex model.
- to use heuristic domain knowledge to evoke hypotheses in the fault models and then to focus abduction on their connected compone ts, as suggested above. Here is an example of such focusing knowledge, in the domain of RCPS diagnosis: *If a high level is observed in the ratio between the global levels of the two sensors d1 and d2, then evoke the "Pump bearing with broken attachment" fault, beginning between 0 and 1 year before the start of the high level*

Global Diagnostic procedure

Diagnosis is initiated by computing an explanation of the initial observations provided by the user. Focusing techniques introduced above are put into operation. As long as the result obtained is not satisfactory, the diagnostic procedure progressively completes the observations using the type of completion knowledge described in paragraph 5.1. Diagnosis is considered unsatisfactory as long as some observations remain unexplained while some fault models have not been entirely exploited.

MAJOR RESULTS AND PERSPECTIVES

A full scale prototype of DIAPO has been jointly developed by EDF and the RCPS manufacturer Jeumont-Industrie between 1992 and 1994 [Porcheron, Ricard 93]

[Porcheron et al. 94]. It has been developed using ILOG's SMECI™ Lisp-based expert system shell. Validation of the prototype by future end-users confirmed the relevance of the implemented methods: about 80% of proposed cases have been correctly diagnosed, failures being due to minor lacks in the knowledge base. The decision to industrialize the system, towards fitting out the fifty EDF nuclear power plants has been consequently taken. This industrial version is currently under development.

The diagnostic methods used in DIAPO displays significant advantages:

- Diagnosis is elaborated from monitoring data during pump operation and can be based on an incomplete limited subset of RCPS observed parameters;
- Diagnosis provides an explanation of observed events by their possible "causes" as described in different fault models : faults, abnormal situations, primary causes, and locations. Solutions are dated, according to temporal knowledge which takes into account the dynamics of the RCPS. Multiple faults are handled when explicitly described, or when their manifestations are independent.
- DIAPO can issue predictions based on expected consequences of the diagnosis.

The main limitations of this diagnostic method are:

- It strongly relies on the hypothesis that fault model are complete. It can therefore only be applied to systems the malfunctions of which are sufficiently well known. An important effort must also be dedicated to knowledge acquisition and validation
- Dynamics of the system is difficult to take completely into account, as it would dramatically increase fault model complexity.
- Completing initial observations during diagnosis can be difficult.

Besides industrializing DIAPO, current works include the development of a generic architecture in order to reuse these diagnostic methods to others applications. Complementary methods, such as *consistency based diagnosis*, *case based reasoning*, etc., will eventually be included in this architecture, in order to tackle the limitations mentioned above.

RELATED WORKS

Model-based diagnosis has been widely used in order to deal with more or less complex artifacts [Hamscher 91], [Dague et al. 90], [Milne et al. 96]... "Pure" consistency-based methods fit well when a model of the correct behavior of the device can be elaborated. Conversely, abductive methods based on fault-models are well adapted to devices the faulty behavior of which is sufficiently well-known. Both approaches have been integrated in recent works [de Kleer, Williams 89] [Console, Torasso 91] [Poole 89] [Struss, Dressler 89]. As noted before, methods used by DIAPO heavily rely on abductive approaches dealing with faults models, although a limited amount of knowledge on the correct behavior of the RCPS is used especially for completion of observations during diagnosis.

DIAPO fault models describe straightforward causal relationships between *symbolic representations* of concepts such as *states, events, symptoms,* etc. rather than influence between behavioral parameters derived from numerical or qualitative modeling of the equipment, as in for example [Kuipers 86][Milne et al 96].

Concerning the issues related to taking into account the dynamic of the RCPS, DIAPO models may be viewed as a weak combination of far-from-complete *behavioral models* and *state transition graphs* as described in [Console et al 92].

ACKNOWLEDGMENTS

The development of DIAPO included knowledge engineering participation by P. Parent at Jeumont-Industrie. Expertise was brought among others by R. Chevalier, C. Bore, L. Mougey, H. Martinal, P. Voinis, at EDF and, under the coordination of E. Lejeune, MM. Avet, Mazuy, Monjean, Toubault, at Jeumont-Industrie. Participation to DIAPO at EDF R&D division also included B. Monnier and J.L. Doutre. Coordination with the Power Generation Division was ensured by J.L. Busquet, J.E. Donnette and B. Monnier.

REFERENCES

[Console et al. 89]
 L. Console, D. Theseider Dupre, P.Torasso. "A theory of diagnosis for incomplete causal models". In *Proc. International Joint Conference on Artificial Intelligence*; Detroit, 1989.

[Console et al. 92].
 L. Console, L. Portinale, D. Theseider Dupre, P.Torasso. "Diagnostic reasoning across different time points". In *Proc. 10th European Conference on Artificial Intelligence,*pp 369-373. Vienna, Austria;1992.

[Console, Torasso 91]
 L.Console and P.Torasso. "A spectrum of logical definitions of model-based diagnosis". In [Hamscher et al.-92].

[Dague et al. 90]
 P. Dague, P Devès, P. Luciani, P. Taillibert. "Analog systems diagnosis". In [Hamscher et al.-92].

[de Kleer, Williams 87]
 J. de Kleer, B.C. Williams. "Diagnosing multiple faults". In [Hamscher et al.-92].

[de Kleer, Williams 89]
 J. de Kleer, B.C. Williams. "Diagnosis with behavioral modes". In [Hamscher et al.-92].

[Dvorak, Kuipers 89]
 D. Dvorak and B.Kuipers. "Model-based monitoring of dynamic systems". In [Hamscher et al.-92].

[Hamscher 91]
 W.C. Hamscher. "Modeling digital circuits for troubleshooting". In [Hamscher et al.-92].

[Hamscher et al. 92]
 W. Hamsher, L. Console, J. De Kleer (Eds). "Readings in Model-Based Diagnosis". Morgan Kaufman; 1992.

[Joussellin et al. 94]
 A. Joussellin, B. Ricard, J. Morel, R. Chevalier, B. Monnier "PSAD: An Integrated Architecture for Intelligent Monitoring and Diagnosis of EDF Power Plants"; Proc. of ISAP'94, Montpellier, France 1994.

[Kuipers 86]
 B.J. Kuipers. "Qualitative simulation". Artificial Intelligence, 29(3): 289-338, Sept. 1986.

[Milne et al. 96]
 R. Milne, C. Nicol, L. Travé-Massuyès, J. Quevedo. "TIGER: knowledge-based gas-turbine condition monitoring". AI communications 9(3) 89-154 (1996).

[Poole 89]
 D. Poole. "Normality and faults in logic-based diagnosis". In [Hamscher et al.-92].

[Porcheron et al. 94]
 M.Porcheron, B. Ricard, J.L.Busquet, P.Parent "DIAPO, a case study in applying advanced AI techniques to the diagnosis of a complex system"; *11th European Conference on Artificial Intelligence (ECAI'94);* Amsterdam, Netherlands; August 1994.

[Porcheron, Ricard 93]

M.Porcheron, B. Ricard "DIAPO, an industrial application of advanced AI methods for the diagnosis of a complex system"; *TOOLDIAG'93*; Toulouse, France; March 1993.

[Reiter 87]

R. Reiter. "A theory of diagnosis from first principles". In [Hamscher et al.-92].

[Struss, Dressler 89].

P. Struss, O. Dressler. "Physical negation: integrating fault models into the general diagnostic engine". In [Hamscher et al.-92].

DESIGN OPTIMIZATION WITH UNCERTAIN APPLICATION KNOWLEDGE

Ravi Kapadia
Dept. of CS, Vanderbilt University
Box 1679 Station B
Nashville, TN 37235

ravi@vuse.vanderbilt.edu

Markus P.J. Fromherz
Xerox PARC
3333 Coyote Hill Road
Palo Alto, CA 94304

fromherz@parc.xerox.com

Abstract

Designing modern electro-mechanical devices is a non-trivial task, due to their versatility, their increasing performance requirements, and the integration of software and hardware. Competition in the market place motivates additional requirements, such as high performance and low cost, and an important part of design is often to optimize conflicting objectives.

In this paper, we present *model-based techniques* for design optimization. We develop a parameterized machine model in the form of design, scheduling and cost constraints and present a tool that determines values for design variables that optimize the performance and cost of the machine.

Furthermore, we present a method for analyzing design solutions in the absence of precise workload knowledge. A qualitative method for classifying optimal sets of machine parameter values with respect to workload distribution, helps the designer to understand the interaction between potential application contexts and optimal solutions, and may be used to generate configurable designs.

1 Introduction

Designing modern electro-mechanical devices is a non-trivial task, due to ever increasing functionality and performance requirements. To manage this complexity, modern systems are equipped with increasingly sophisticated, specialized control software. However, the interactions between system hardware, control software and system workload exacerbates the analysis of these systems.

The primary goal of design is to propose a concrete device that delivers a desired functionality. The design process requires a specification of the functionality, and a technology for realizing the artifact (i.e., a set of primitive components or building blocks and their relations). Competition in the market place motivates additional requirements, such as high performance and reliability for typical workload, and low manufacturing and maintenance cost. Typically, market studies are used to identify the "average" user's workload distribution, and the system is designed to optimize performance with respect to this average distribution. But individual users are likely to subject the system to workload distributions that vary from the average. Thus, the design process must be sensitive to this *uncertainty* about the application context.

We are investigating the application of *model-based computing* [6] techniques to system design. The central idea of model-based computing is to develop declarative, multi-use models of machines. Together with suitable generic algorithms, such models facilitate automatic derivation of software for tasks such as simulation, control, scheduling and diagnosis. This approach has been successfully applied to the development of generic scheduling software for reprographic machines (e.g., printers and photocopiers) [6].

In the next section, we discuss our view of the design task in the context of reprographic machines, and review related work. To address the design task, we develop a formal description (model) of the relevant design variables, the constraints imposed by the machine on these variables, the costs of realizing different values for them, and present a computational architecture for design in Section 3. We analyze the design solutions generated by our approach in the face of incomplete workload distribution knowledge to provide the designer with some confidence of their applicability, and discuss the use of this analysis in generating configurable designs in Section 4. We close with our conclusions.

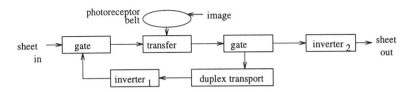

Figure 1: Schema of a simple reprographic machine

2 Designing Reprographic Machines

We use examples from the domain of modular, networked reprographic systems to illustrate our design methods. In this section, we outline the configuration, behavior and control of reprographic machines, and define and discuss the design task.

2.1 Reprographic Machines

Figure 1 shows the schematic of a simple reprographic machine. In this form, the machine prints simplex (one-sided) or duplex (two-sided) sheets. A sheet enters the reprographic machine through the input port. An image is laid down on the revolving photoreceptor belt and transferred to the sheet (*transfer*). A simplex sheet bypasses inversion (i.e., passes through $inverter_2$ without inversion) on its way to the output port, while a duplex sheet is routed to the *duplex transport* and inverted ($inverter_1$), another image is transferred on the back side of the sheet (*transfer*), and the sheet is inverted again ($inverter_2$) before it is sent to the output port. The belt is divided into an integral number of slots, each of which can accommodate one image. An image can be placed in any slot, except in the region where the two ends of the belt meet. In our examples we assume this region, called the *seam gap*, to be one slot wide.

In order to generate the output in minimum time, a reprographic machine is equipped with software that controls the actions of its components. The heart of this software, the *scheduler*, receives a stream of job descriptions. A job description includes an ordered sequence of single and double-sided sheets. The scheduler plans the actions of each component, and determines the times at which they must be initiated, to produce the desired output. For a detailed description of the scheduling process, refer to [6].

The transportation and printing of sheets is constrained in various ways by the physics of the machine. For example, sheets cannot overlap in the paper path, sheets and images must be synchronized, images can be placed on the photoreceptor belt only at certain places, and inversion takes longer than by-

passing, which implies that an inverted sheet cannot be followed immediately by a non-inverted sheet at the entrance of the inverter. An optimal schedule for a job description is one which produces the output in the shortest time, i.e., the *schedule length* is minimal. An optimal design for a given workload is one which optimizes the objective function, typically maximizing performance (i.e., generating the documents in the workload in the shortest time) and minimizing cost.

2.2 The Design Task

Designing an electro-mechanical system such as a reprographic machine involves *problem formulation*, for example, specification of the composition of the machine together with a set of design variables; and *design generation*, which attempts to find satisfactory design variable values. Tong and Sriram [9] call this *exploratory* design. Feedback from design generation is used to refine problem formulation (refer Figure 2). This separation is based on the notion that a designer often solves ill-defined design problems by formulating well-defined routine sub-tasks (which can be solved using automated tools), and then integrating their solutions.

In Section 3, we address routine design generation. Given the structure of the machine with a set of design variables with unknown values, a solution to the design problem is the assignment of values to these variables such that various objectives (e.g., performance, cost) are optimized. The routine design step must not simply generate "the optimal design", but present it in the context of alternative design solutions.

The performance of a reprographic machine depends not only on the machine design, but also on the workload (i.e., jobs submitted to the machine). The routine design step uses estimates of workload distribution to identify optimal designs. As mentioned earlier, such estimates are uncertain at best. We apply the idea of exploratory design to deal with this uncertainty. By analyzing routine design solutions (Section 4), we can provide feedback about their applicability in the face of uncertain workload distribution knowledge.

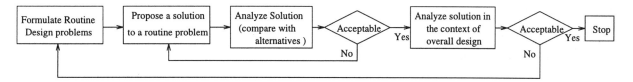

Figure 2: Problem formulation and routine design

2.3 Related Work

The objectives that a design must optimize can be classified as *static* or *dynamic*. Optimizing a design for static objectives (e.g., number of components, area or volume occupied by device, and single, isolated use scenarios) can be done independently of workload. On the other hand, dynamic design objectives include the performance of the system which depends on the workload. To the best of our knowledge, existing AI-based approaches to engineering design focus on optimization for static objectives.

Heatley and Spear [7], and Mittal and Araya [8] address the problem of geometric design in the reprographic machine domain (e.g., the exact placement of rollers). The focus is on static optimization such as minimizing the number of components. Bakker et al. [1] use model-based diagnosis techniques for *redesign*. Starting from an existing design (i.e., model) and specifications for the target design (i.e., observations), their diagnostic candidates form design solutions. The candidate generation process is applicable only to static systems like arithmetic function circuits. Beldiceanu and Contajean [2] describe the design of elliptic filters but their approach is akin to optimizing a reprographic machine for the production of a single sheet, and their techniques cannot be extended easily to optimize for a sequence of sheets.

3 Design Optimization Using Model-based Computing

As motivated in the previous section, given a machine model with a set of design variables whose values are unknown, the routine design task consists of determining consistent values for these variables such that performance and cost of the resulting design are optimized. In this section, we show how this can be achieved by using machine models that combine design, scheduling and cost constraints.

Model-based computing makes use of constraint logic programming as the language for specifying models and implementing reasoning algorithms. The models are *veridical* and *declarative*. A veridical model (i.e., one that is true to "first-principles") captures

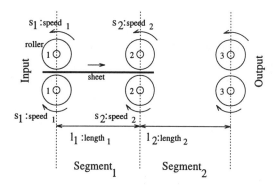

Figure 3: Duplex transport

the physics of the underlying components and the construction of systems from these components. A declarative model is a conjunction of logical formulas that can be described in the vocabulary of the domain, and can be produced by engineers familiar with the domain. Since models are independent of task-specific algorithms, they can be re-used in multiple applications.

3.1 Constraint-based Design Models

Consider a simple, but realistic design in which the duplex transport in Figure 1 consists of two transport components (see Figure 3), whose lengths and speeds are design variables. Let l_i and s_i be the length and speed of component i, respectively. In the following discussion, we present a typical and complete constraint-based description of a machine, starting from design constraints, scheduling constraints [4], and a cost model (see [5] for a detailed discussion). Design constraints are used to generate design solutions, while scheduling and cost constraints are used for maximizing performance and minimizing cost, respectively.

Design constraints. Consider constraints on design variables s_i (millimeters per millisecond) and l_i (millimeters).

Consistency. Consistency constraints define allowable designs by constraining the design variables.

This includes domain constraints, such as:
$$l_1, l_2 \in [150..175], s_1, s_2 \in [25..35].$$
Note that in our application l_i and s_i assume discrete values in their respective domains. As another example for consistency constraints, the speed of the second loop component must be no less than the speed of the first one to prevent a sheet from buckling (see Figure 3):
$$s_1 \le s_2.$$

Abstraction. The time spent by a sheet traveling along the duplex loop is a compilation of the transit times of the loop's components. Let k be the time difference between the printing of the front side and the back side of a duplex sheet. We assume that k is a discrete variable. Also, assuming that the transit times for transfer and inversion are t_1 and t_2 milliseconds, respectively, we get:
$$k = l_1/s_1 + l_2/s_2 + t_1 + t_2 \text{ milliseconds.}$$
Different combinations of l_i and s_i may result in the same k and, therefore, in the same productivity. We use k (called the length of the duplex loop) in the scheduling constraints presented below. While we could, in principle, use l_i and s_i instead, the abstraction allows us to simplify the design problem to generating and evaluating values of k before mapping them to detailed designs in terms of l_i and s_i at different costs.

Scheduling constraints. The scheduler accepts a job description and — using an *instantiated* machine model (i.e., all design variables have been assigned values) — determines an optimal schedule. In our example a schedule is determined by the sequence of input sheets; all other operations can be derived from this sequence.

Let the output sequence (job) $O = o_1, \ldots, o_n$ be an ordered sequence of n single- and double-sided sheets ("s" and "d" for short) specified by the user. A schedule is an explicitly timed sequence $S^e = x_1, \ldots, x_n$, where x_i denotes the input time of sheet o_i and ranges over discrete points in time.

As a simplification, we assume that all sheets are aligned to a period of p milliseconds (corresponding to the sheet length), i.e., all x_i are multiples of p, and that the transit times of the transfer and inverter components, t_1 and t_2, respectively, are both p milliseconds. Without loss of generality, we can choose $p = 1$. As an example, suppose that the design variable for duplex loop length, $k = 5$ and $O = \text{sdd}$, a correct schedule would be $S^e = (0, 1, 2)$ (i.e., the page of o_1 is printed at time 0, the pages of o_2 are printed at times 1 and 6, and the pages of o_3 are printed at times 2 and 7). Given O, the scheduler's task is to find an S^e such that x_n is minimal (when x_n is minimal, the schedule length is minimal, thus, the schedule is optimal) and the following constraints are satisfied.

Order constraint. The order of O has to be preserved by S^e:
$$\forall i = 1, \ldots, n \cdot y_i < y_{i+1},$$
where $y_i = x_i$ if $o_i = \text{s}$, and $y_i = x_i + k$ if $o_i = \text{d}$ (i.e., y_i is the time when the last page of a sheet is printed).

Loop constraint. A sheet returning through the loop (k units of time after it entered the machine) must not collide with an incoming sheet:
$$\forall i = 1, \ldots, n \cdot \forall j(\max(i - k, 1) \le j < i) \cdot o_j = \text{d} \to x_i \ne x_j + k.$$
Note that for $O = \text{sdd}$ and $k = 5$, both $S^e = (0, 1, 2)$ and $S^e = (3, 0, 1)$ are valid schedules (the output order is still sdd), but only the latter is optimal.

Inversion constraint. The inversion of a sheet takes p milliseconds, while bypassing the inverter takes no time. As a consequence, inverted sheets followed by non-inverted sheets have to be at least p milliseconds apart. In our example, duplex sheets are inverted at the output, while simplex sheets bypass the inverter:
$$\forall i = 1, \ldots, n - 1 \cdot (o_i = \text{d} \land o_{i+1} = \text{s}) \to x_i + k + 1 \ne x_{i+1}.$$
So, for $O = \text{dds}$ and $k = 5$, $S^e = (0, 1, 8)$ is a valid schedule, while $S^e = (0, 1, 7)$ is not.

Seam gap constraint. No image can be laid out on the seam of the photoreceptor belt. The seam occurs every $g * p$ milliseconds (starting at $-p$) and — for a fixed speed of revolution — is p milliseconds long.
$$\forall i = 1, \ldots, n \cdot (x_i + 1) \bmod g \ne 0 \quad \land \quad (o_i = \text{d} \to (x_i + k + 1) \bmod g \ne 0).$$

Cost model. Associated with each design decision is a cost. We define the cost of a design as the weighted sum of the costs of individual decisions. The discussion and development of appropriate cost models is beyond the scope of this paper. It is up to the design engineer to decide which design decisions are important to measure. In our examples, we use a simple model that defines the cost of transport components as directly proportional to their speeds and lengths, i.e., the overall cost[1]
$$C = \sum_{i=1}^{2} w_{i1} * l_i + w_{i2} * s_i,$$
where w_{ij} are constants. Finally, a limit L on the overall cost of producing the artifact may be imposed by asserting $C \le L$. The cost model is used to prune

---

[1]This refers to the cost for realizing components with the chosen design variables, not the cost for designing the system.

expensive designs and assist in ranking design solutions.

3.2 Design Optimization

We have built a tool, Design Optimizer, that uses the above models to generate and evaluate design solutions. The task of this tool is to determine a legal set of design variable values that is consistent with the design constraints, and optimizes cost and productivity. The tool is developed in CLP(fd) — a constraint logic programming language over finite domains.

3.2.1 Architecture

Design Optimizer consists of the following elements (see Figure 4).

Machine model. Design constraints, scheduling constraints, and a cost model, as described in the previous section, are used to generate abstract and detailed designs.

Job descriptions. To verify the productivity of a proposed design d_j, it is necessary to simulate the performance of d_j on expected jobs. We assume the availability of classes of jobs $C_1, \ldots C_k$, (where all jobs in a given class occur with equal probability). $J = \{J_1, J_2, \ldots, J_n\}$ is a *training set* for the design of the machine where a job description J_i, is a tuple $\langle O_i, p_i \rangle$:

- O_i is J_i's output sequence (e.g., **sddds**).
- p_i, the priority of the job, which ranges from 0 (low priority) to a user-defined maximum value (high priority). The priority of J_i is proportional to its probability (which depends on the class C_j to which J_i belongs).

Design Generator. The generator accepts an uninstantiated machine model. Design variables in this model are instantiated by applying of design constraints. Cost constraints are applied next to prune the search space. Finally, the performance of the resulting set of instantiated designs ($\{d_1, \ldots, d_m\}$) is evaluated by the schedule simulator.

Schedule Simulator. For each instantiated design d_j and job description J_i, the simulator generates a schedule with length S_{ij}. By examining the schedules generated for all jobs, the simulator assigns $SCORE_j$, a measure of the "closeness" to optimality to design d_j, which is a weighted sum of deviations of the lengths of schedules for jobs in J from the best known value,

i.e.,

$$SCORE_j = \sum_{i=1}^n p_i * (S_{ij} - S_i), \qquad (1)$$

which is a measure of the overall closeness to the ideal design. Note that if d_j is optimal, $S_{ij} = S_i$. If, however, d_j cannot generate a schedule for J_i, $S_{ij} = \infty$. Thus, $SCORE_j = 0$ if design d_j is optimal for all jobs in J, while $SCORE_j = \infty$ if d_j cannot generate a schedule for one or more jobs in J.

Design Analyzer. The simulator's results are analyzed to present the design engineer with a range of the best solutions. If comprehensive knowledge of the training set J is available at design time, the analyzer simply returns the results of the simulator. Otherwise, as explained in Section 4, it combines the results of multiple executions of the simulator to present the user with a range of solutions and the explanations for their applicability.

Mapping. The final stage in the algorithm performs a one-to-many mapping from abstract designs to component parameter values, using the abstraction constraints in the design model. Along with instantiated component parameters, this stage also computes a cost for each solution.

Suggested designs. Design Optimizer computes a collection of designs ranked by the extent to which they satisfy all constraints over all jobs, including a measure of the cost of the design.

3.2.2 Algorithm

The current framework for design optimization is a generate-and-test algorithm. Let d_j be a set of design variables, J_i a job, and $S_{ij} = f(J_i, d_j)$ the objective function (the schedule length to be optimized) for design instance d_j and job J_i. Then, the abstract algorithm is:

```
repeat:
    generate d_j;
    for each J_i:
        score d_j based on S_ij;
    rank d_j according to its score.
```

3.2.3 Example with exact workload distribution knowledge

The sample design task is to determine optimal values for length and speed of the duplex transport components (see Figure 3). With no loss of generality, we

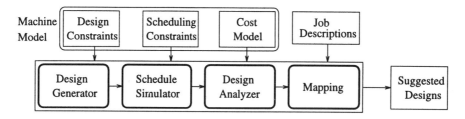

Figure 4: Architecture of Design Optimizer

refer to a design with a duplex transport length of k units as d_k. We assume that (see Section 3.1):

- The seam gap occurs every 16 slots (i.e., $g = 16$).
- Parameter domains are $l_i \in [150..175]$, $s_i \in [25..35]$, and $t_1 = t_2 = 1$. Thus, $k \in [12..16]$.
- The weights in the cost model are set to $w_{11} = w_{21} = 0$ and $w_{12} = w_{22} = 1$, resulting in the cost constraint $C = s_1 + s_2$.
- A user-specified cost limit $L = 70$ is imposed, i.e., $s_1 + s_2 \leq 70$.
- The training set J contains $O_1 = 10$ simplex (10s), $O_2 = 10$ duplex (10d), $O_3 =$ one simplex, 12 duplex (1s12d), $O_4 =$ one simplex, 12 duplex, one simplex (1s12d1s), $O_5 = 15$ duplex (15d), and $O_6 =$ one simplex, 15 duplex (1s15d).

Discovering counter-intuitive solutions. First, consider that the probabilities for all J_i are known exactly. Let them be $p(J_1) = 40\%$, $p(J_2) = 40\%$, $p(J_3) = 5\%$, $p(J_4) = 5\%$, $p(J_5) = 5\%$, and $p(J_6) = 5\%$. Expressing priorities as integers proportional to the probabilites, we have $p_1 = 8$, $p_2 = 8$, $p_3 = 1$, $p_4 = 1$, $p_5 = 1$, and $p_6 = 1$.

The resulting designs are described in Table 1, where d_{13} and d_{16} are the best designs. Optimal $S_{ik} = S_i$ for a job J_i are shown in bold face. These results, where "slower" designs (i.e., designs with larger loop times) perform better than "faster" ones, may at first seem counter-intuitive. However, the slower designs, in particular d_{16}, are better synchronized with the seam gap, which occurs at time $= \{15,31\}$, i.e., on every 16th slot (starting from 0). Making the duplex loop length k a multiple of the belt revolution time synchronizes the front and back sides of a sheet with respect to the recurring seam gap, so that there is never a need to skip slots except at the seam. As belt utilization is increased, productivity increases. Similar observations have been made in designing pipeline processor units, where adding delay can improve overall performance [3]. Note that this result would not be found if the design was optimized for single-sheet jobs only.

Mapping to component parameter values. In the final stage, Design Optimizer maps abstract designs to component parameter values (transport speeds and lengths). The program enumerates all consistent combinations of component parameter values, with costs, that realize the abstract designs of interest. An abbreviated version of the actual output of the program for designs d_{13} and d_{16} is presented below[2].

```
Selecting Design Parameters
  to realize ranked durations:
Duration [13],
design parameters: [[150,25],[175,25]],
                   cost: 50.
design parameters: [[156,26],[175,25]],
                   cost: 51.

...

Duration [16],
design parameters: [[175,25],[175,25]],
                   cost: 50.
```

Note that
```
    design parameters:  [[150,25],[175,25]],
                    cost:  50
```
means that $l_1 = 150, s_1 = 25, l_2 = 175, s_2 = 25$, and that C, the cost to realize these parameter values, is 50 units.

4   Design Analysis for Uncertain Application Knowledge

The design of an artifact usually results in a compromise, in that we try to find a solution that satisfies most users. More precisely, the training set chosen during design is a best guess at a characterization of the market requirements. Thus, while the requirements for different users may be radically different, they all get a common "average" solution.

As explained above, it is rare that the workload distribution is known exactly at design time. What can be expected from market studies is a qualitative "feel"

---

[2]There are five sets of design parameters that realize d_{13}. The complete set is presented in [5].

Duplex loop (k)	Design (d_j)	Schedule length (S_{ik})						$SCORE_k$
		10s	10d	1s12d	1s12d1s	15d	1s15d	
13	d_{13}	9	23	**25**	**27**	41	41	28
16	d_{16}	9	25	27	29	**30**	**32**	28
12	d_{12}	9	**22**	36	38	39	39	38
14	d_{14}	9	24	26	28	43	43	42
15	d_{15}	9	25	27	29	45	45	56

Table 1: Discovering counter-intuitive solutions

for workload distribution, resulting in coarse classifications such as "frequent" (high probability) and "infrequent" (low probability). The analysis stage described in this section identifies a set of distinct, quantitative ranges for workload distributions (given a qualitative classification), and the corresponding designs that are optimal in each range.

4.1 Using Qualitative Classification

We assume that the (potentially large) training set contains accurate output sequences, but that the probability distribution of the jobs is not known precisely at design time. Following this assumption, we present the analysis for a two level qualitative classification; see [5] for a generalization to an arbitrary number of classes.

Input. The analysis algorithm assumes the availability of the following:

- Let H be the set of frequent (high-probability) jobs, i.e., $H = \{H_1, \ldots, H_h\} \subseteq J, |H| = h$, and L the set of infrequent (low-probability) jobs, i.e., $L = \{L_1, \ldots, L_l\} = J \backslash H, |L| = l$. We assume that the probability of each element within a set is uniform, i.e., $\forall i, j, p(H_i) = p(H_j), p(L_i) = p(L_j)$, and $\sum_{i=1}^{h} p(H_i) = X, \sum_{i=1}^{l} p(L_i) = 1 - X$.

- Let $T = [\langle d_1, SCORE_1^H, SCORE_1^L \rangle, \ldots, \langle d_m, SCORE_m^H, SCORE_m^L \rangle]$, be a design scoring list, where d_j is the jth design, $SCORE_j^H = \sum_{i=1}^{h}(S_{ij} - S_i)$ over H and $SCORE_j^L = \sum_{i=1}^{l}(S_{ij} - S_i)$ over L. Note that T can be obtained by two separate executions of Design Optimizer over H and L respectively. Since exact p_i values are unavailable, the scoring function is simply a summation of the deviations, (i.e., for all i, p_i is set to 1 in Equation 1).

Algorithm. We want to discover those designs d_j that perform best over all J, i.e., over both L and H. The overall performance of a d_j depends on the probability distribution (X) of the jobs in H and L.

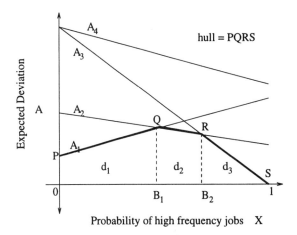

Figure 5: Boundary points

The relation that governs *expected deviation* (A_j) from optimal schedule lengths for each design (d_j) is:
$A_j = SCORE_j^H/h * X + SCORE_j^L/l * (1 - X)$.
The lower the value of A_j, the better is the design d_j. Note that $SCORE_j^H, SCORE_j^L, h,$ and l are all non-negative, so for $0 \leq X \leq 1, A_j \geq 0$.

In Figure 5, assuming four designs, we show the plot of four expected deviation functions versus probability distribution X. From this figure, it is evident that over different ranges of X, different designs are optimal. Expected deviation is minimized along the "convex hull" $PQRS$. $0, B_1, B_2, 1$ correspond to the X-coordinates of $PQRS$. Thus, for $0 \leq X \leq B_1$, design d_1 with expected deviation function A_1 (corresponding to segment PQ of $PQRS$) is optimal; for $B_1 \leq X \leq B_2$, design d_2 with expected deviation function A_2 (segment QR) is optimal; and for $B_2 \leq X \leq 1$, design d_3 with expected deviation function A_3 (segment RS) is optimal.

The resulting information enables the designer to understand how different workload distributions influence the generation of optimal designs. This helps the designer relate the important boundaries when estimating workload distribution. Second, this analysis may be used to avoid commitment to a specific design in the absence of accurate workload distribution until a time that such information becomes available. In particular, one can predetermine a set of designs with associated distribution boundaries, and later, given a user-defined value for X, look up the design d_j for which $B_{i-1} \leq X \leq B_i$, and adapt machine parameters accordingly[3].

4.2 Example: Design with Qualitative Classification

In this extension to the example in Section 3.2.3, we relax the assumption that we know the probability distribution of expected jobs at design time. Assume that jobs have been categorized explicitly into two classes: frequent jobs (J_1 and J_2) and infrequent jobs (J_3 through J_6), where J_1 and J_2 constitute fraction X of the jobs, while J_3 through J_6 constitute the remaining $1 - X$. From Tables 2 and 3, which show the results of design generation and simulation stages on H and L, respectively, $T = [\langle d_{12}, 0, 38 \rangle, \langle d_{13}, 1, 20 \rangle, \langle d_{14}, 2, 26 \rangle, \langle d_{15}, 3, 32 \rangle, \langle d_{16}, 3, 4 \rangle]$.

For each d_j, the expected deviations are:
- $A_{12} = 0/2 * X + 38/4 * (1 - X)$
- $A_{13} = 1/2 * X + 20/4 * (1 - X)$
- $A_{14} = 2/2 * X + 26/4 * (1 - X)$
- $A_{15} = 3/2 * X + 32/4 * (1 - X)$
- $A_{16} = 3/2 * X + 4/4 * (1 - X)$

Output. Figure 6 shows the plot of three deviations A_{12}, A_{13} and A_{16} (the rest are omitted for clarity), where X is the variable on the x-axis and A_j is plotted on the y-axis. The hull $PQST$ identifies the designs and their corresponding probability distributions. The x-coordinates of $PQST$ are $B_0 = 0$, $B_1 = 0.8$, $B_2 = 0.9$, and $B_3 = 1$. Thus, d_{16} is best for $0 \leq X \leq 0.8$, d_{13} is best for $0.8 \leq X \leq 0.9$, and d_{12} is best for $0.9 \leq X \leq 1$.

5 Conclusions

The aim of this work is to investigate the applicability of model-based computing to the problem of making design decisions, analyzing them in terms of cost and productivity for real-world design problems, and generating configurable designs to tackle the problem of incomplete knowledge about the application context.

The benefit of an unbiased formal representation of all constraints for the evaluation of productivity and cost is most evident if the resulting designs

---

are counter-intuitive (as in the example in Section 3.2.3). Our design analysis allows for automated classification of designs that can be used at configuration time to look up appropriate parameter values for user-specific workload information. The 2-level qualitative classification of workload distribution shown in this paper can be extended to handle an arbitrary number n of classifications, by considering the convex hyper-surfaces formed by hyper-planes in n-dimensional space [5].

In the future, we will investigate other forms of uncertainty in the application context, such as incomplete training sets. To overcome the problem of not being able to consider all possible jobs in the training set, we will focus on job classification methods such that all jobs within a class are optimized by the same design. The design analysis and compilation approach may be extended to dynamic adaptation to deterioration: if the system senses that some component parameters are deviating beyond the specified boundaries (e.g., slower paper transport through worn rollers), the entire configuration could be re-optimized taking this deviant component behavior into account. By adjusting appropriate parameters dynamically, the system could maintain higher productivity levels for longer periods of time.

We believe that model-based computing — developing software that captures real-world objects in declarative executable descriptions — is an approach suitable for various application domains. As more and more applications profit from explicit system models, modeling will become a core activity in hardware and software development.

Acknowledgements. We would like to thank Danny Bobrow, Gautam Biswas, Vijay Saraswat, and Marc Webster for their assistance in developing and presenting the ideas in this paper.

References

[1] R. Bakker, S. van Eldonk, P. Wognum, and N. Mars. The Use of Model-based Diagnosis in Redesign. In *Proc. 11th European Conference on Artificial Intelligence*, 1994.

[2] N. Beldiceanu and E. Contajean. Introducing Global Constraints in CHIP. In *Math. Comput. Modeling*, (20) 12, pages 97–123, 1994.

[3] A. El-Amawy and Y. Chang Tseng. Maximum Performance Pipelines with Switchable Reservation Tables. In *IEEE Trans. on Computers*, Vol 44, No. 8, August, 1995.

---

[3]Of course, not all parameters can be adapted dynamically. In the example of Section 3.2.3, the speeds of rollers may be adapted dynamically, but the distances between roller pairs is likely to remain fixed.

Duplex loop (k)	Design (d_k)	Schedule length (S_{ik})		$SCORE_k^H$
		10s	10d	
12	d_{12}	9	**22**	0
13	d_{13}	9	23	1
14	d_{14}	9	24	2
15	d_{15}	9	25	3
16	d_{16}	9	25	3

Table 2: Design for typical (high probability) jobs

Duplex loop (k)	Design (d_k)	Schedule length (S_{ik})				$SCORE_k^L$
		1s15d	1s12d	1s12d1s	15d	
16	d_{16}	**32**	27	29	**30**	4
13	d_{13}	41	**25**	**27**	41	20
14	d_{14}	43	26	28	43	26
15	d_{15}	45	27	29	45	32
12	d_{12}	39	36	38	39	38

Table 3: Design for low probability jobs

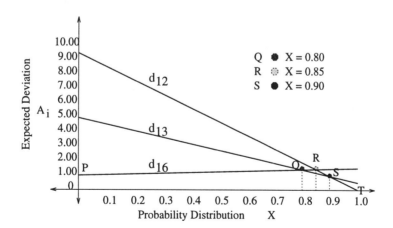

Figure 6: Comparing designs

[4] M. Fromherz and B. Carlson. Optimal Incremental and Anytime Scheduling. In P. Lim and J. Jourdan, editors, *Proc. Workshop on Constraint Languages/Systems and their Use in Problem Modeling at ILPS'94*, pages 45–59. ECRC, TR 94-38, Nov. 1994.

[5] M. Fromherz and R. Kapadia. Towards Model-based Design Optimization of Electro-mechanical Systems. Xerox PARC, TR, May 1996.

[6] M. Fromherz and V. Saraswat. Model-based Computing: Using Concurrent Constraint Programming for Modeling and Model Compilation. In U. Montanari and F. Rossi, editors, *Proc. Constraint Programming*, pages 629–635, Cassis, France, September 1995. Springer-Verlag, LNCS 976.

[7] L. Heatley and W. Spear. Knowledge-based Engineering at Xerox. In *Artificial Intelligence in Engineering Design*, Vol. III, pages 179–195, Academic Press Inc., 1992.

[8] S. Mittal and A. Araya. A Knowledge-based Framework for Design. In *Artificial Intelligence in Engineering Design*, Vol. I, pages 273–293, Academic Press Inc., 1992.

[9] C. Tong and D. Sriram. In *Artificial Intelligence in Engineering Design*, Vol. I, pages 1–53, Academic Press Inc., 1992.

IMPROVING DECISION MAKING BY COMPARING EXPLANATIONS OF MULTIPLE POINTS OF VIEW

Ahmed Almonayyes
Mathematics and Computer Science
Kuwait University,
P.O.Box: 5969 Safat - 13060 - Kuwait
e-mail: sami@mcc.sci.kuniv.edu.kw
phone: (+965) 4811188 ext. 5305

ABSTRACT

Even though today case-based reasoning is applied in a wide range of different areas, there are only few systems which make use of case based techniques for decision making in complex domains such as international conflicts. In this paper, we present a case-based prototype of an intelligent decision-support system for the task of understanding international conflicts. After this, we describe a computational approach for indexing and retrieving conflict cases by using several pieces of explanatory knowledge whereby each describes a particular point of that accounts for the emergence of a conflict.

MOTIVATION

In the domain of international conflict, policy makers use old cases to assist them in decision making and help them in selecting appropriate policy options. However, effective decision making requires good understanding of a conflict situation. Better understanding of a conflict should lead to greater confidence in decision, and higher quality of the decision. Therefore, policy makers often use old cases to understand and evaluate situational dynamics by comparing and contrasting them with a new conflict situation. Use of such cases allows policy makers to generate policy options, alerts them to the causal factors operating during a conflict situation, and enables prediction of what might happen if a course of action is chosen.

However, policy makers may lack the ability to do good analogical reasoning in conflict analysis for several reasons. Firstly, they may have trouble remembering appropriate cases (at the appropriate time) which are needed to make good analogical decisions. Secondly, in a highly complex domain, such as international conflicts, decision makers may overlook important information that tells them which aspects of a conflict situation are the crucial ones to focus on. That is, they may use analogies regarding a present conflict situation without really understanding the degree of similarities between cases and, as a result, the limitations of such analogies. Thirdly, policy makers are often predisposed to formulate explanations about a conflict situation according to their own point of view. This predisposition may lead them to overlook other explanations which may be crucial in the formation of more appropriate decisions. Finally, policy makers tend to be guided by cases that predict desired outcomes. That is, they prefer historical cases which mostly suggest successful outcomes, and fail to notice the ones which suggest failure.

The aim of this work is to provide computational methods that address the above shortcomings. These computational methods are implemented as part of a CBR system that helps policy makers to do better analogical reasoning, and as a result, improve their understanding of a conflict situation. The system will augment policy makers memories by providing them with the relevant experiences which help them with such tasks as coming up with new explanations, identifying alternative courses of action, critiquing and evaluating their decisions, and warning of potential problems associated with these decisions.

THE POLITICA SYSTEM: AN OVERVIEW

POLITICA is a prototype of a Case-Based Decision Support System that has a memory of past conflict cases. The problem understanding model in POLITICA is implemented by means of the following steps: 1) generate a set of explanatory hypotheses to account for the initial events of the conflict; 2) use the Explanation-Based Learning (EBL) [DM86] approach to generate additional

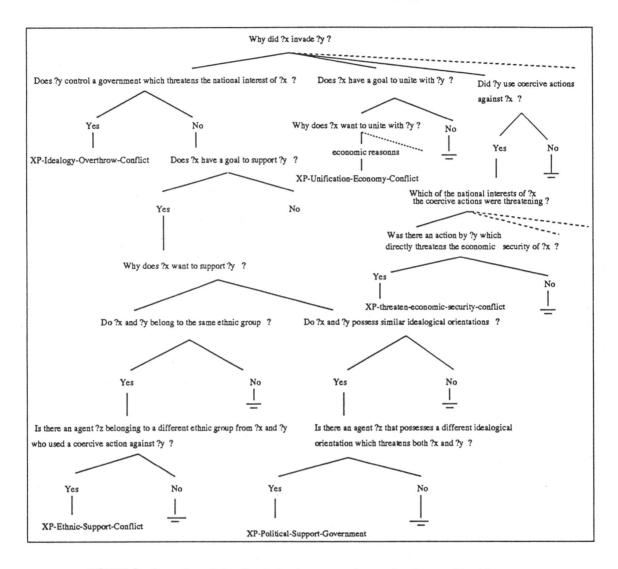

FIGURE 2   A portion of the discrimination network organization used in this work

system, about a conflict. It is also used to direct the system towards exploring other sources of a conflict that may otherwise be overlooked by policy makers. The third step is to retrieve the XPs indexed under the terminal node reached by the traversal process given in the previous step. The XPs are then used to retrieve several cases, each of which is explained according to several points of view. Each point of view gives a particular explanation of *how* and *why* a conflict takes place. The basic assumption here is that cases which share the largest number of similar explanations are more likely to give an accurate prediction of events since the underlying factors, which are crucial to decision making, are taken into account.

For example, the Iraq-Kuwait 1990 conflict can be explained by several XPs. Each XP explains a particular viewpoint of the conflict. For instance, one retrieved XP called *XP-economic-assets-conflict* explains Iraq's

invasion of Kuwait as a result of economic reasons, where Iraq's sudden interest in Kuwait occurred after its chronic economic problems which led to its attempt to seize Kuwait's oil revenues to solve those problems. Another XP called *XP-military-security-conflict* explains Iraq's military action in view of its pressing need to gains access to the Persian Gulf for military security reasons. Given these to explanatory hypotheses, the system retrieves two sets of cases where each set contains a number of cases indexed under each of the two XPs. For instance, one of the retrieved cases is the **Japan-Manchuria-1930** case. This case describes the situation where Japan faced a severe incompatibility between its growing population and the availability of the resources. The predicament created a deep economic and social depression which forced Japan to seek vital raw materials and energy resources by invading Manchuria, Japan continued its aggression against its neighbors (e.g. China) to fortify

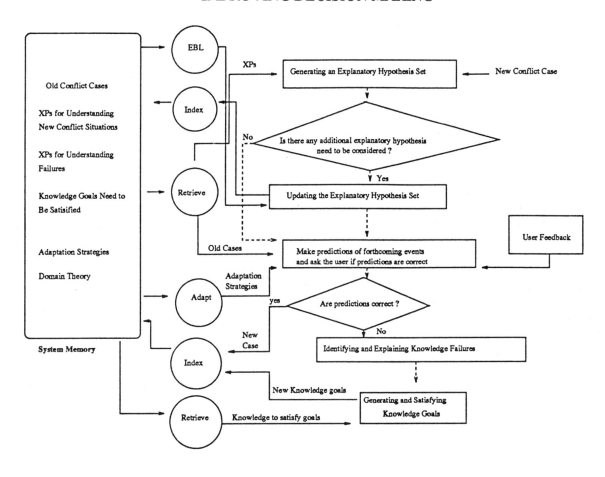

FIGURE 1 The basic problem understanding model

explanatory hypotheses to improve the understanding of the conflict. These explanations are then added to the Explanatory Hypothesis Set (EHS) which represents a short list of candidate hypotheses that explain the new conflict situation from several points of view; 3) locate similar previous experiences (i.e. conflict cases) in memory on the basis of the information generated by the previous two steps; 4) retrieve those experiences from memory; 5) select the most appropriate experiences from those retrieved; 6) transfer the information from the old case to the new one to make a possible prediction of forthcoming events; 7) test the predictions (i.e. expectations) against the actual results by getting feedback from the user; 8) If success is found, update memory and return success; and 9) otherwise, identify and explain the failure, generate knowledge goals to resolve this failure, satisfy the knowledge goals, and update memory. Figure 1 illustrates the basic problem understanding model.

A conflict case integrates several pieces of explanatory knowledge whereby each describes a particular point of view that accounts for the emergence of a conflict. The explanatory knowledge used in this work is based on the

notion of Explanation Pattern (XP) [Sch86]. An XP embodies a causal network to lead from a set of premises toward a conclusion. The conclusion is the event that the XP explains. It links actions of the disputants to their goals and beliefs yielding an understanding of their involvement in the conflict.

After the system explains the new conflict case by using several XPs, it uses the XPs to retrieve several sets of cases where each set contains a number of cases indexed under each XP. Old cases which share the largest number of XPs with the new case are preferred over cases with less XPs (given the fact that most of the other features are similar).

A three step process is used to generate the explanatory hypothesis set. The first step is to make an initial set of facts by using a hardwired selection procedure which singles out all those facts that are normally examined before understanding any conflict situation. The second step is to use the results of a hardwired inference process which is based on a discrimination network of multiple-choice questions [RH92] shown in figure 2. The role of the discrimination network is to select only those relevant facts from the large amount of information, given to the its victory and to establish a regional power in that part of the world.

INDEXING AND RETRIEVING A CONFLICT CASE

In POLITICA, conflict cases are organized at two levels. At the first level, cases are indexed on the basis of several XPs, and then discriminated at the second level on the basis of other types of features [Kol93]. These features include the disputants actions, their goals, and the state of the world at the time of processing a new conflict. The core idea is that rather than discriminating on the full case description (i.e. the set of all known facts), the system discriminates only on a small subset of features which are crucial to the usefulness of the retrieved case. A case is useful when it potentially allows good predictions to be made about a new conflict case. Thus, in order to retrieve the best case, a new conflict case must be assessed in terms of the disputants goals, their actions, and the general state of each disputant (e.g. the fact that South Vietnam is controlled by a pro-western government). In this work, it is considered more important to discriminate between cases on the basis of features related to issues such as idealogies, security (political, economic, and military), or ethnic matters, rather than on the basis of surface features such as continents, or disputants names. However, it is worth pointing out that the set of all known facts will only be used during the ranking phase of the problem understanding process, and after the retrieval phase has been accomplished, to determine the best matching case to new conflict.

POLITICA uses an indexing scheme based on three labels to construct indices that differentiate groups of cases from others. The *primary* label indexed disputants at a higher conceptual level in terms of their roles (e.g. invader, invadee) in the conflict rather than their actual names. The other two are called *secondary* labels and contain the type name of the attribute (e.g. the structure of the government) and its value (e.g. pro-western). Such indexing scheme provides two strengths to the retrieval process; First, several conflicts which involve the same disputants are indexed in the appropriate places in the memory according to their roles in the conflict, rather than their names. For example, in the Korean conflict, North Korea invaded South Korea. Here, North Korea is assigned the role of being the aggressor. However, in the Vietnamese conflict, the USA played a different role in the conflict in spite of the fact that is used the same type of coercive action (i.e. invasion). The role of the USA in this conflict was to support the government of North Vietnam against the Vietcong. It is worth pointing out here that defining a role for each party in the conflict is a

highly subjective view. In order to ensure consistency in the analysis of conflicts, the western point of view has been adopted in describing all cases of conflicts.

The second strength of the indexing scheme is that the same conflict can be indexed in different places in memory according to which question the understanding model is currently addressing (e.g. in the Korean conflict, the system retrieves different sets of cases depending on whether the question being processed is *Why did USA invade South Korea?* or *Why did North Korea invade South Korea?*.

The case memory of POLITICA includes case conflicts with different descriptions each of which can be indexed and retrieved from several discrimination networks of organizational structures called Generalized Episodes (GE) [Kol93]. Retrieving a case indexed under a generalized episode is primarily a partial-matching process. The system determines the correctness of a match, along some feature slot (or dimension), of two cases using an abstraction hierarchy. The matching procedure assesses the similarity of the slot fillers of the feature in two cases by climbing a hierarchy tree of objects. If there is a common node in the tree for both fillers, the system will determine that they are similar in terms of a higher level concept.

Retrieving the best case requires going through three phases. In the first phase, a number of XPs are retrieved each of which explains a different point of view on the emergence of a conflict. The retrieved XPs are all passed on by the system to the next phase. In the second phase, the system traverses each of the discrimination networks, where each network is associated with a particular XP, in an attempt to find the generalized episode that best matches the current case. Only the features given in the initial set of facts are used as dimensions of a new case to be matched against the values labeling the arcs (i.e. links) in the network. At each level of the network that is being searched, a number of GEs are retrieved and passed on to a matching function that matches the new conflict case against the norms of each GE. The GE with the best match score is selected, and its descendants are matched similarly. The search stops when the system reaches the most specific GE which either has no other GEs indexed below it, or has a better matching score than its children. In the third phase, the system collects all the cases indexed under the retrieved GEs and passes them to a matching function which selects the best case from that collection. The matching procedure determines the degree of match between the new case and an old case by using a numeric matching scheme. This scheme assesses the similarity between cases on the basis of all the features given in the cases descriptions. Old cases are compared to the new case feature by feature. A score representing the total degree of match between the old and new cases is

```
(define-xp XP-Political-Support-Government
            (({explain}
                  (*invade*
                       (actor (? x)
                       target (? y))))
            ({premises}
                  ((G-*support*-state (agent-1 (? x)
                       agent-2 (? y)
                       value (yes)))
                  (*control*-state
                       (agent (? y)
                       object (government) of (? y) value (pro (P-orientation (? p)))))
                       (*idealogy*-object
                       (procedure (check-idealogical-orientation (agent (? x) object (? p)))))))
            ({links}
                  ((((*control*-state
                       (agent (? y) object (government) of (? y) value (pro (P-orientation (? p))))
                  initiate
                       ((G-*support*-state
                       (agent-1 (? x) agent-2 (? y) value (yes)))))
                  ((((*coercive-action* (actor (? z) target (? x))))
                  threaten
                       ((G-*preservation*-state
                       (agent (? x) object (*national-interest*) value (yes)))))

                       .
                       .
                       .

            )
            ); define
```

FIGURE 3   **Partial representation of the XP** *XP-Political-Support-Government*

computed by combining the match scores resulted from comparing their corresponding features. However, the matching procedure is also used to rank cases by comparing their explanations with the explanation associated with a new conflict. The causal structure given in the new case is created by merging all the explanations (i.e. the retrieved XPs and the explanations created by EBL constructed in the earlier stages of the problem understanding). The system determines the degree of similarity between the explanations of two cases by checking the relevance of features in both explanations. Cases with higher scores of relevant features suggest better explanations than those with lower scores, and as a result, should be ranked higher by the matching procedure. The evaluation function used to assess similarities between two cases is defined as:

$$similarity(O,N) = \alpha \sum_{f' \in N.SAKF} sim(f', O.SAKF) +$$

$$\beta \sum_{f' \in N.SAKF \cap O.SAKF} [sim(f', O.clinks) + sim(f', N.clinks)]$$

Where f' is a fact in a new conflict; $N.SAKF$ and $O.SAKF$ are the set of all known facts given for the new and old cases respectively. The α parameter represents the weight of a match (i.e. the importance of the match) between a fact in the old case and the same fact in the new case. The β represents the additional weight given for a match between a relevant fact in the old case and the same fact in the new case, and also given for a match between a relevant fact in the new case and the same fact in the old case. In this work, α and β are given the values 1 and 1.5 respectively. This means that a match between two relevant facts is more important than a match between two irrelevant facts. The terms $O.clinks$ and $N.clinks$ represent the causal structures given for old and new conflicts. The term sim represents the similarity function which computes the value of match between two features.

The strength of the retrieval approach taken here lies in the fact that the system can use initial set of facts as retrieval cues to concentrate only on a small subset of network labels that index potentially useful cases. Also, the similarity of two cases can be judged not only by the

```
(NAME: ?-HAVE-COMMON-POLITICAL-THREAT-NODE
 PARENT:  ?-GOAL-IDEALOGY-SUPPORT-NODE
 QUESTION: (*COERCIVE-ACTION*
             (ACTOR(? Z) TARGET (USA)))
        LINK:THREATEN
            (G-*PRESERVATION*-STATE
            (AGENT (USA) OBJECT(*NATIONAL-INTEREST*) VALUE(YES)))
      (*COERCIVE-ACTION*
            (ACTOR(? Z) TARGET(SOUTH-VIETNAM)))
       LINK:THREATEN
           (G-*PRESERVATION*-STATE
           (AGENT(SOUTH-VIETNAM) OBJECT(*NATIONAL-INTEREST*) VALUE (YES)))
 LIST-OF-OPTIONS: NIL
 LEFT:  XP-POLITICAL-SUPPORT-GOVERNMENT
 RIGHT:  NIL
 )
```

FIGURE 4   A conceptual representation of the question

```
(*BELIEF*-STATE
      (AGENT(VIETCONG) OBJECT(COMMUNISM)
      VALUE(YES)))
   (*CONTROL*-STATE
      (AGENT(SOUTH-VIETNAM) OBJECT(GOVERNMENT)
      OF(SOUTH-VIETNAM)
      VALUE (PRO (P-ORIENTATION(WESTERN-BLOCK))))
   (*OVERTHROW-AUTHORITY*
      (ACTOR(VIETCONG) TARGET(SOUTH-VIETNAM)))

LINK:THREATEN

  (G-*PRESERVATION*-STATE
      (AGENT(USA) OBJECT(*NATIONAL-INTEREST*) VALUE (YES)))
```

FIGURE 5   A conceptual representation of the causal link that answers the first sub-question

number of features that are similar in the descriptions of two cases, but also by the total number of features which are included in the causal structures of both cases.

EXPLAINING CONFLICTS: AN EXAMPLE

To illustrate how this process works suppose the system is instructed to explain and predict the outcome of the USA military intervention in the South Vietnam-North Vietnam 1963 conflict. The system is given the following events [Kr93]:

In 1960, the Vietcong led by Pathet Lao attempted to overthrow the pro-western government of South Vietnam. The Vietcong are supported by the communist state of North Vietnam. The USA decided to intervene in South Vietnam with military force to support the pro-western government against the communist Vietcong.

The first step is to collect the initial set of facts (ISF). The set includes the parties involved in the conflict, namely, the pro-western government of South Vietnam, the Vietcong, North Vietnam, and the USA. It also includes the actions taken by the adversaries in the conflict. These actions are the facts that Vietcong wanted to overthrow the government of South Vietnam, the supporting action of the North Vietnamese to the Vietcong, and finally the action taken by the USA to intervene militarily on behalf of the pro-western government of South Vietnam. It is worth mentioning here that the understander (i.e. the system) at this point does not attempt to build an explanation of the events in the conflict, or explain the

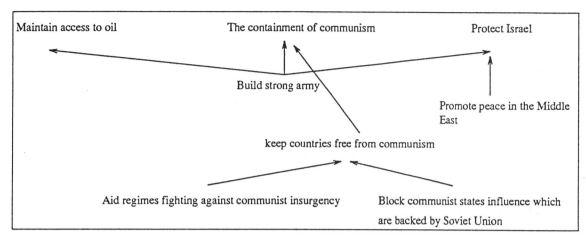

FIGURE 6   The iDEALOGY tree of USA during the Vietnamese conflict

initial conduct of the adversary (i.e. USA) which led to its involvement in the conflict. The main purpose here is to collect all those facts which are intuitively relevant to the conflict before continuing the understanding process in the next step.

The second step is to apply a built-in inference process that is based on the discrimination network organization shown in figure 2. The traversal process of the discrimination net is determined by what questions are asked and which answers are inferred from the set of facts given about the conflict. Each fact inferred is added to the initial set of facts. The traversal process continues until the system reaches a node where an appropriate XP is found that explains the new situation. The initial set of facts inferred so far is used to bind the variables given in the *premises* of the retrieved XP. Figure 3 shows the XP *XP-Political-Support-Government* which is used to generate one of the explanatory hypotheses of the Vietnamese conflict.

For example, in figure 2, the root node "Why did USA invade South Vietnam ?" unifies with "USA invaded South Vietnam" by binding the variables ?X and ?Y to the USA and South Vietnam respectively. The set of variable bindings is used to instantiate the next question "Does South Vietnam control a government which threatens the national interest of the USA ?". Since no fact directly matches this question is found, the link "no" is traversed and the third question "Does USA have a goal to support South Vietnam ?" is attempted.

Let us assume that the system reaches the last question "Is there an agent ?Z that possesses a different ideological orientation which threatens the national interests of the USA and South Vietnam ?. A representation of this question is shown in figure 4. Notice that the question basically consists of two sub-questions for which the system needs to find answer. Also notice that the concept *coercive-action* is an abstract concept of a more specific action taken by ?Z against both the USA and South

Vietnam. At this point, the system uses a deductive inference process to elaborate on the question by building additional causal links to find more specific answer to the question. Here the Vietcong is unified with ?Z and the link "yes" is traversed to retrieve the XP *XP-Political-Support-Government*. However, if the traversal process reaches an empty child, it returns a null value as its output. The null value indicates that neither an XP nor a question can be found below this point in this portion of the tree. Figure 5 shows a conceptual representation of the causal link which answers the first sub-question.

In the third step, the system applies the retrieved *XP-Political-Support-Government* (shown in figure 3) to the conflict. The variables ?X, ?Y, and ?Z are bound to USA, South Vietnam, and Vietcong respectively. The *premises* are verified and the causal links are justified to form a hypothetical explanation for the USA military intervention.

However, additional information beyond what the XP can provide is needed to understand other aspects of the USA's military involvement. For example, the fact that North Vietnam is a communist state that is probably backed by the communist Soviet Union and China does threaten the USA's national security in the region. This additional information is inferred by using domain-specific knowledge structures called *ideology trees* (shown in figure 6) which represents the state's long range political goals that can explain their actions. The *ideology trees* presented in this work are different from *goal trees* as given in [Car79]. In Carbonell's work, *goal trees* consist of relative importance links in addition to the subgoal links which connect the nodes in the tree. Relative importance links are not included in this work because the aim of this research is not to understand the different reasons that make the various political parties (within a single country) take different political decisions about certain issue (this is because of the different subjective interpretations given for this issue). Rather, the aim here is to explain a conflict situation in terms of the long term goals of a single political party that is in power when the conflict occurs.

Suppose the system determines that the old case **North Korea-South Korea-1950**, indexed under *XP-Political-Support-Government*, is the most similar case to the current conflict. The system adapts the events of the old case to suit the current situation, and accordingly can make predictions about forthcoming events of the current conflict[1] . Once the conflict has been resolved, the system asks the user about the decision made by the American policy makers during the **South Vietnam-North Vietnam-1963 conflict**. The decision was to intervene militarily in South Vietnam (as was the case in the Korean conflict) to assist the government of South Vietnam in defeating the Vietcong and take pre-emptive steps in obstructing any attempt by the North Vietnamese to invade South Vietnam. The system then asks the user about the final outcome of that decision. The outcomes are: 1) the failure of USA to achieve its goals (e.g. the preservation of the South Vietnam's government and the defeat of the Vietcong), and, 2) the communist Vietcong successful control of the government of South Vietnam. Based on this decision and its outcomes, the system executes a comprehensive failure analysis of possible causes of failure to the USA policy in the **South Vietnam-North Vietnam-1963** conflict. The analysis is based on defining the major factors of international and domestic contexts which led to the successful outcomes of the American policy in the Korean conflict and contributed to their failures in the Vietnamese conflict. Two XPs for problem understanding failures can be identified here:

• **XP-IC-UN-Military-Mandate**: This kind of XP indicates that there is a failure to obtain a clear United Nation mandate in the Vietnamese conflict, while there was a clear mandate in the Korean case. This XP indicates that by the absence of a clear UN mandate to support a military intervention by ?X against ?Y removes any moral legitimacy of ?X to take an aggressive action against ?Y. This weakens ?X's international position which produces a degradation effect on ?X's morale to fight against ?Y. The result is a possible military defeat for ?X. To remedy such failure, a knowledge goal is generated to find a method to obtain international support for ?X's military intervention against ?Y.

• **XP-DC-NO-Public-Support**: This XP indicates a situation where ?X's failure to get public backing for military action against ?Y lowers its morale which could

lead to its military defeat. As a result, a knowledge goal of obtaining domestic public support must be generated.

Each of the above XPs indicates a lack of knowledge that caused a failure (e.g. missing indices for failure XPs). Once blame assignment has been made for a particular failure, it is a simple process to specify the required knowledge to address this failure. That is, the explanation of a failure generates a knowledge goal which specifies what knowledge needs to be acquired to resolve that failure.

CONCLUSIONS AND FUTURE WORK

In this paper, a computational model of problem understanding was presented. Four main tasks for the understanding model were identified: generating an explanatory hypothesis set, updating the explanatory hypothesis set, identifying and explaining knowledge failures, and generating and satisfying knowledge goals. The purpose of -the first two tasks (i.e. generating and updating the explanatory hypothesis set) is to explain a conflict situation from multiple points of view. Each point of view suggests an explanation of how and why a conflict occurs. The explanatory hypotheses in the set are used to retrieve a set of several conflict cases from memory. The best candidate case is selected from the set to generate predictions about forthcoming events in the current conflict. Followed by feedback from the user, the system attempts to identify and explain expectation failures that arise when it fails to make correct predictions. The system uses a library of explanation patterns (for problem understanding failures) to explain the failures. Each XP is used to generate knowledge goals which represent the need to bridge a gap in the system knowledge base. These goals can then be satisfied by employing available domain-specific knowledge.

The similarity metric discussed in the present paper is similar in some aspects to the CBR+EBL similarity metric [Bento93], however, the major difference here lies in the fact that the purpose of retrieving explanations in our work is to *predict*, **not** *to explain*, the outcome of a new case. In other words, the explanations needed to understand a new case have already been constructed by the system and used as a basis for the retrieval of old cases (remember that old cases are indexed under XPs). Also, the similarity metric discussed here takes advantage of the availability of an explanation already given to a new case. It uses the available explanation to select the best case from a set of retrieved cases. An interesting extension to this work is to use introspective reasoning [SL95] to refine the retrieval criteria by learning new features to be used in indexing.

---

[1]The adaptation process is beyond the scope of this paper. For the purpose of illustrating the theory, it is sufficient here to assume that this process has already taken place.

There are several advantages of using the multi-level indexing scheme discussed in this paper. Firstly, several conflicts which involve the same disputants are indexed in the appropriate places in memory according to their roles in the conflict. Secondly, the same conflict can be indexed in different places in memory according to which question the understanding model is currently addressing (e.g. in the Korean conflict, the system retrieves different sets of cases depending on whether the question being processed is *Why did USA invade South Korea?* or *Why did North Korea invade South Korea?*). Thirdly, the system can use the facts given in the initial set of facts as retrieval cues to concentrate only on a small subset of network labels that index potentially useful cases. Finally, the similarity of two cases can be judged not only by the number of features that are similar in the descriptions of two cases, but also by the total number of features which are included in the causal structures of both cases.

The current model incorporates 25 conflict cases, 12 XPs, and a simplistic from of domain knowledge. Future developments of this model may take into account more complex knowledge (e.g. military, cognitive and emotional propositions) to widen the scope of analysis. Moreover, the understanding model can be improved to accommodate the analysis of the internal as well as international conflicts.

The paper reports an ongoing research; we are currently experimenting with several similarity metrics on data set of international conflicts to see whether our similarity metric would actually provide more accurate predictions. Also, we are looking into the possibility of using the methodology and system architecture discussed in this paper to improve problem understanding in other domains.

REFERENCES

[Bento93] C. Bento and E. Costa (1993). A Similarity Metric for Retrieval of Cases Imperfectly Explained. *In Proceedings of the First European Workshop*, EWCBR-93, Kaiserslautern, Germany (pp. 92-105).

[Car77] J. Carbonell. Subjective Understanding: Computer Models of Belief Systems. *Ph.D.*, Yale University, USA.

[DM86] G. Dejongand and R. Mooney. Explanation -Based Learning: An Alternative View. *Machine Learning* (pp. 145-176).

[Kol93] J. Kolodner. Case-Based Reasoning Morgan Kaufmann Publishers, Inc.

[Kri93] J. Krieger. The Oxford Companion to Politics of the World; Oxford University Press, 1993.

[RH92] A. Ram and L. Hunter. The Use of Explicit Goals for Knowledge to Guide Inference and Learning. *Technical Report GIT-CC-92/04*, College of Computing, Georgia Institute of Technology, Atlanta, Georgia, 1992.

[Sch86] R. Schank. Explanation Patterns: Understanding Mechanically and Creatively. Lawrence Erlbaum Associates, Hillsdale, New Jersey, 1986.

[SL95] F. Susan. and B. Leake. Using Introspective Reasoning to Refine Indexing. *In Proceedings of the Fourteenth International Joint Conference on Artificial Intelligence*, Montreal, Canada, 1995.

TRAIN TIMETABLE AND ROUTE GENERATION USING A CONSTRAINT-BASED APPROACH

Hon Wai Chun
Department of Electronic Engineering, City University of Hong Kong
Tat Chee Avenue, Kowloon, Hong Kong
Email: eehwchun@cityu.edu.hk

ABSTRACT

This paper describes research in modelling train timetable and route generation as a constraint-satisfaction problem (CSP). The key objective of this research is to design a constraint-based scheduling algorithm that can be used to generate a train timetable given headway requirements at different times of the day. The key constraint is to avoid track circuit or route contentions while maximising train utilisation. The objective of the scheduling algorithm is to determine how service levels can be increased without jeopardising passenger safety. This research investigated traffic at a train terminus where two types of trains are competing for the use of the same tracks; trains that are reversing and trains that are being dispatched from the depot. The contention problem is particularly serious during the rush hour train build-up. The current timetable and train routing are generated using two separate rule-based systems. However, due to the complexity of constraints involved, the current systems cannot generate a plan that can meet the desired service levels.

INTRODUCTION

This paper describes a constraint-based model and algorithm for a train timetable and route generation program. This program determines how trains can be efficiently dispatched from a depot to meet rush hour service requirements while satisfying resource constraints. Our research focused mainly on generating a timetable and routing for the building up of train services.

However, it can also be applied to the reverse case of breaking down the service after rush hour ends. The research was performed using data from one of the world's busiest subway systems. The constraint-based algorithm described in this paper was tested on one of the busiest lines within this subway system.

The subway authority currently uses a rule-based expert system to generate the train timetable based on headway requirements, i.e., the time between the arrival of two consecutive trains. The routing is then generated using a separate semi-automatic system which is based on heuristics. This routing program determines how each train should travel to get to its destination. Subway trains normally travel in a cyclic manner from one final terminus to other and back. The key problem is what happens when the train reaches the final terminus and needs to reverse while at the same time trains are being dispatched from the depot to the terminus. There is a tremendous amount of contention for the same set of tracks during the rush hours. The timetable and route generation is currently performed by separate systems. The timetable generation system must therefore ensure there is enough slack or buffer for the routing system to be able to generate the routes. This extra time buffer causes inefficiency. The route generation program, on the other hand, is confined to work within the timetable that is given. The subway authority cannot see how service levels can be increased without increasing the efficiency of the scheduling algorithm. This paper proposes a constraint-based algorithm that combines timetable and route generation into one process. Headway constraints for timetable generation are considered at the same time as resource constraints for route generation.

Our research focuses on designing and developing a constraint-based resource allocation algorithm [DUNC94, PUGE94b] that can assist the human scheduler by generating different "what-if" scenarios or timetables based on different sets of criteria, such as headway requirements and train speeds. The resources to be allocated in this study are the track circuits and the routes.

Our constraint-based scheduling algorithm performs three main scheduling tasks - (1) generate a train timetable, (2) generate the route sequence each train must travel, and (3) determine the travel time within turnaround and headshunt interlocking area. The most important concern for the subway authority is the ability to increase service levels during rush hours. The scheduling algorithm must be able to generate a timetable and route sequence that can maximise the utilisation of the track circuit resources in order to meet the service level demands.

PHYSICAL CONSTRAINTS

Figure 1 is a simplified diagram of the track circuits and signals within the turnaround and headshunt interlocking area that are used within the study. There are two platforms at the final terminus -- an "Arrival Platform" for trains arriving from the

other terminus and the "Departure Platform" for trains travelling to the other terminus. The study only involved this terminus since there is no depot at the other terminus and hence no track-circuit resource contention.

The track circuits within the turnaround and headshunt interlocking area allow trains to reverse from the Arrival Platform to the Departure Platform. Subway trains can travel in both directions and have a driver compartment at each ends of the train. As part of the reversing process, the train driver must walk from one end of the train to the other. In some cases, there may be an additional driver at the other end to reduce the amount of time needed to reverse a train. While trains are using the tracks to reverse, other trains may be dispatched from the depot through "Depot Track 1" and "Depot Track 2." Furthermore, all tracks within the turnaround and headshunt interlocking area can be used for train travel in either directions.

The tracks in the turnaround and headshunt interlocking area are divided into routes. Routes are further divided into track circuits. Figure 1 shows the names of the track circuits. A route is defined to be a set of track circuits between two signals. Since trains can travel in either direction on the same tracks, there are signals for both direction of travel. The relevant signal is on the right hand side of the direction of train travel.

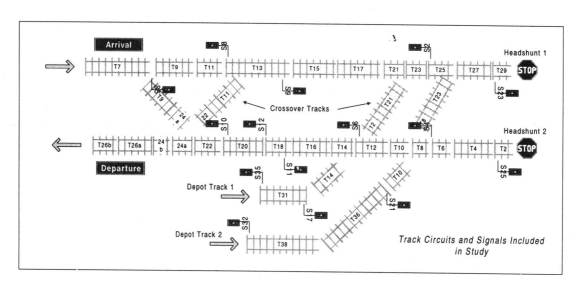

Fig. 1 Track circuits and signals within the turnaround and headshunt interlocking area.

- **Track Circuits.** For the purpose of our research, a "track circuit" is the smallest piece of railway track that can be uniquely identified. However, the length of each track circuit may be different. The *state* of a track circuit may either be "down" indicating that a train is currently over this piece of track circuit or "up" indicating that the track circuit is free. The track circuits in Figure 1 are drawn roughly to proportions. A total of 33 track circuits are included into the study. The same track numbers are used to identify tracks branching out from track circuits, such as the crossover tracks, and their parent branches.

- **Signals.** Figure 1 illustrates all the signals that were used in our research. There are a total of 16 signals that will define the starting point for 42 possible routes that a train may take. However, several routes end at the Departure Platform, which does not have a signal after track circuit T26b. For uniformity, the Departure Platform will be considered as a "pseudo-signal."

- **Routes.** For this study, a "route" is defined as a sequence of track circuits that starts and ends at a signal (including the pseudo-signal). Before a train enters a route, it must first "call" and "set" the route. Setting a route reserves a section of track so that no other train will use the same track resource. A set of Boolean equations or constraints defines when a route can be set. The equations ensure that the route is safe to enter.

Our scheduling system differentiates routes that span two immediate signals as "basic routes." Routes that are composed of more than one "basic routes" are called "composite routes." The scheduling algorithm considers a total of 22 basic routes and 20 composite routes. To travel from a start signal to the target destination signal may require the train to traverse a sequence of basic routes. This sequence is referred to as a "route sequence."

- **Train Lengths.** To determine track circuit occupancy, the length of the train is needed. The length of the trains used in this research is 177 metres from axle to axle, while the end to end length is 182.5 metres.

TIMING CONSTRAINTS

Several different types of timing data were used by our scheduling algorithm. This includes the run times of the train, the signal timing, the time to change the crew, and the transmission time.

- **Run Times.** Only the nominal run times are used in this study. The scheduling algorithm used the run times to determine when routes will be freed up for other trains to use. The run times are defined as the travel time needed between any two signals without stopping.

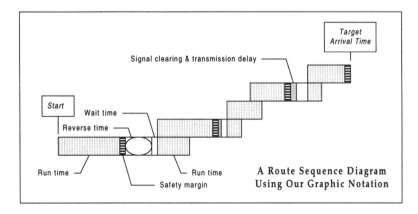

Fig. 2 The representation of a route sequence and its timing.

- **Signal Timing.** It takes about 7 seconds to set a route if point movement is required. It takes about 3 seconds to set a route if point movement is not required.

- **Double-ended Timing.** Double-ended crew is the case where there is a driver at both ends of a train. This technique is used to reduce the time needed to reverse a train. Normally, to reverse a train, the driver should stop the train and walk from the one end to the other to restart the train. The time required for the whole process is around 4 minutes. With double-ended crew, the additional driver will restart the train after the current driver stops it. Reversing, in this case, only takes about 15 seconds to finish but requires a larger crew size to operate.

- **Transmission Delay.** A 2-second transmission delay from issuing the route setting command (by computer or operator) to actuation of the route setting activity is required.

Figure 2 is a Gantt chart that shows the type of timing information that is represented and used by our scheduling algorithm. Each row, along the vertical axis, represents one particular route and the different types of timing involved when a train travels in that route. The complete sequence of rows represents the sequence of routes a train takes from the start signal, e.g. the depot, to the end signal, e.g. the Departure Platform.

THE CONSTRAINT-BASED MODEL AND ALGORITHM

To solve this problem, we represented the timetable and route generation problem as an object-oriented [LEPA93] constraint-satisfaction problem (CSP) [KUMA92]. Although CSP or constraint-programming has a relatively long history [STEE80], with constraint language extensions found in Prolog [COLM90, VANH89], and Lisp [SISK93], it is only recently that constraint-programming became more popular with the availability of the ILOG's C++ class libraries [PUGE94a]. This library provided a very efficient and clean implementation of constraint-based programming features in a

conventional language. Our scheduling algorithm was implemented using the ILOG C++ class libraries.

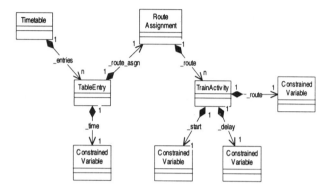

Fig. 3 The UML class diagram of the constrained variables.

In general, any scheduling and resource allocation problems can be formulated as a constraint-satisfaction problem (CSP) which involves the assignment of values to variables subjected to a set of constraints. CSP can be defined as consisting of a finite set of n variables $v_1, v_2, ..., v_n$, a set of domains $d_1, d_2, ..., d_n$, and a set of constraint relations $c_1, c_2, ..., c_m$. Each d_i defines a finite set of values (or solutions) that variable v_i may be assigned. A constraint c_j specifies the consistent or inconsistent choices among variables and is defined as a subset of the Cartesian product: $c_j \subseteq d_1 \times d_2 \times ... \times d_n$. The goal of a CSP algorithm is to find one tuple from $d_1 \times d_2 \times ... \times d_n$ such that n assignments of values to variables satisfy all constraints simultaneously.

When the train timetable and route generation problem is formulated as a CSP, the problem is represented as a set of constrained variables. The first type of constrained variable represents entries in a train timetable. There are two timetables in our problem – a timetable for the Arrival Platform (T1) and a timetable for the Departure Platform (T2). Each timetable is a list of constrained variables. The domain of the variables will depend on the desired headway values requested by the user.

Each timetable entry also contains a "route assignment." Each "route assignment" represents a sequence of "train activity." Each train activity represents the selection of a particular route, at a particular start time, with a particular delay. The route, start time, and delay are all represented as constrained variables. The domain of the route variable is all routes that can start from the current location of a train. The domain of the start time is related to the previous route's end time. The domain of the delay represents the amount of time a train might need to wait at a signal. Figure 3 is a simplified Unified Modelling Language (UML) class diagram that shows the essential classes in our design.

This constraint-based formulation was designed from the requirement that the input to the scheduling algorithm will be a table of desired headway values at the Departure Platform. Therefore the schedule algorithm will schedule a time for the arrival of a train at the Departure Platform. However, we will need to work backward to figure when the train departs from the depot and how long the train waits at each intermediate signal. On the other hand, this train might be a reversing train from the Arrival Platform. In this case, the scheduling algorithm works forward from when the train leaves the Arrival Platform until it reaches the Departure Platform. This combination of searching backwards and forwards at the same time complicates the algorithm, but is a necessity given the subway authority's input requirement. Figure 4 illustrates this combined search.

Although the search is a combined backward and forward search, the scheduling of the timetable entries is a forward process; timetable entries are instantiated from the earliest entries first. In other words, the timetable is generated from the first train

to arrive at the T2 Departure Platform to the last train of the desired scheduling period.

During the constraint-based search, the algorithm predicts when a train leaving the Departure Platform will return back to the Arrival Platform from the other terminus using nominal turnaround times. In addition, the algorithm merges the time a train leaves the Arrival Platform with the timetable entry for the arrival of that same train at the Departure Platform. This merging operation is the major source of backtracking for the scheduling algorithm. The time a train arrives at the Departure Platform from the Arrival Platform might not match exactly the headway required for the Departure Platform and the train must wait inside the turnaround area for the next departure time slot.

For each time value that the algorithm assigns to the timetable, the algorithm also selects a route sequence that will lead the train to arrive/depart at the desired time. The route sequence with the shortest total runtime that does not interfere with any other previously made route assignments will be selected first.

For each potential route sequence that is selected, the algorithm also determines how much time the train should wait at each signal for signal clearance. The minimum amount of waiting time will be used. This is equivalent to "stretching" the route sequence as shown in Figure 5 (the arrows highlight the "stretching" action and not the temporal evolution).

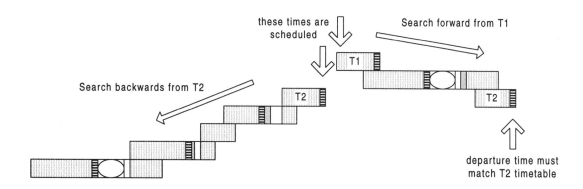

Fig. 4 The combined backward and forward search.

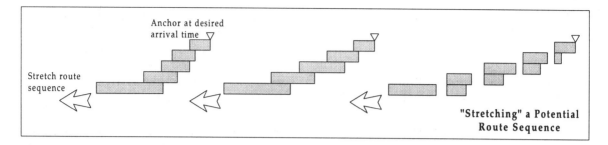

Fig. 5 "Stretching" a route sequence by adjusting the waiting time.

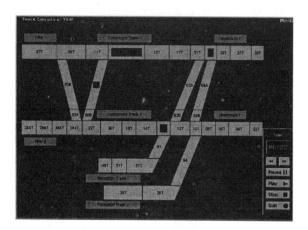

Fig. 6 The graphic simulator used to display the scheduling results.

If the proposed route sequence generated by the scheduling algorithm does not conflict with any other previously assigned routes, then that route will be assigned to the train. On the other hand, if no feasible routes can be found, the algorithm will try to adjust the waiting times at the signals. If no solution can be found, the algorithm then adjusts the arrival/departure times at the station. If still no solution can be found, then the algorithm tries to assign a different route sequence to the previous timetable entry. This "undoing" of previously made assignments is performed automatically by the backtracking mechanism provided by constraint programming. This backtracking will continue until a solution is found or when all possible solutions have been tried. Once the timetables and route sequences have been generated, graphic simulation software is used to display and visualise the resulting schedule (see Figure 6).

ADVANTAGES OF CONSTRAINT PROPAGATION

To illustrate the problem complexity, for the 3 hours of morning train dispatch there are around 70 choice points for just the timetable generation - each with potentially over 100 possible values. For route selection, there are a total of 130 route sequence combinations within the interlocking and headshunt area, each with an average of 4 routes, each route with possibly 60 timing variations. The total complexity of just the morning dispatching is far too much for any rule-based approach.

The constraint-based approach, proposed in this paper, was able to generate timetables and route sequences within a reasonable time because of constraint propagation. Constraint propagation eliminated invalid choices before they can be selected for search. In the case of timetable generation, routes that conflict with assigned routes will be eliminated and timetable entries that violated headway requirements will not be selected. The constraint-based approach makes use of arc consistency [MACK77] and constraint propagation [WALT72] to reduce the domain size of each constrained variable before search. Smaller domain size means smaller and more focused search space.

RESULTS FROM TEST CASES

The constraint-based scheduling algorithm was tested on many test cases. The algorithm was designed to minimise changes in the desired headway value as much as possible. However, minimising headway changes in early morning forces the headway at the final target to fluctuate slightly.

The following table lists test cases that achieved the desired 105-second headway values, i.e., roughly 34 trains per hour. Currently, the trains operate at roughly 112 or 113-second headway values. Also listed is the number of trains that were dispatched from 6am to 8:40am, the number of trains that had a headway value of exactly 105 or 106 seconds, and the average deviation from 105 seconds within the 7:40am to 8:40am peak.

Table 1. Summary of Testing Results

Test Case No	Total trains since 6am	No. of 105 or 106 trains	Avg. Deviation from 105 sec.
2	66	29	1.4
5	72	22	4.7
10	73	21	4.8
13	73	22	4.7
14	73	23	4.7
15	73	23	4.7

As an example, Test Case 15 was produced with the following input headway table.

Table 2. The Headway Table Used for Test Case 15

Station	Start	End	Headway	No Trains
TSW2	6:00	6:30	240	8
TSW2	6:30	7:00	180	10
TSW2	7:00	7:25	120	13
TSW2	7:25	7:40	112	8
TSW2	7:40	8:40	105	34

Out of these test cases, the best result was obtained in Test Case 2 where trains were dispatched for the 105-second headway with an average deviation of only 1.4 seconds. The current rule-based systems can at most dispatch up to 112-second headway values.

SYSTEM IMPLEMENTATION

The object-oriented constraint-based [LEPA93] allocation system was implemented in C++ using the ILOG Solver class library [PUGE94a] and the RTL Scheduling Framework developed by Resource Technologies Limited [CHUN96a, CHUN96b]. The graphic user interface that simulates the generated schedule was developed using C++ graphic components provided by ILOG Views. The system was developed using platform independent coding and can execute within Windows 95/NT or Unix environment.

CONCLUSIONS

This paper documents our research in modelling train timetable and route generation as a constraint-satisfaction problem. The constraint-based scheduling algorithm was tested using data from one of the busiest subway systems in the world. The results showed that the scheduling algorithm was able to generate timetables and routes that had a higher service level than that was previously possible with a rule-based approach.

ACKNOWLEDGEMENTS

The author would like to thank the Hong Kong Mass Transit Railway Corporation for the cooperation received and for making actual data available for this research. Part of this research was performed in cooperation with Resource Technologies Limited (http://www.rtl.com.hk/~rtl) in Hong Kong and with assistance from ILOG (http://www.ilog.com) in Singapore.

REFERENCES

[CHUN96a] H.W. Chun, K.H. Pang, and N. Lam, "Container Vessel Berth Allocation with ILOG SOLVER," The Second International ILOG SOLVER User Conference, Paris, July, 1996.

[CHUN96b] H.W. Chun, M.P. Ng, and N. Lam, "Rostering of Equipment Operators in a Container Yard," The Second International ILOG SOLVER User Conference, Paris, July, 1996.

[COLM90] A. Colmerauer, An Introduction to Prolog III, Communications of the ACM, 33(7), pp.69-90, 1990.

[DUNC94] T. Duncan, "Intelligent Vehicle Scheduling: Experiences with a Constraint-based Approach," ILOG Technical Report 94-04.

[KUMA92]   V. Kumar, "Algorithms for Constraint Satisfaction Problems: A Survey," In *AI Magazine*, 13(1), pp.32-44, 1992.

[LEPA93]   C. Le Pape, "Using Object-Oriented Constraint Programming Tools to Implement Flexible "Easy-to-use" Scheduling Systems," In *Proceedings of the NSF Workshop on Intelligent, Dynamic Scheduling for Manufacturing*, Cocoa Beach, Florida, 1993.

[MACK77]   A.K. Mackworth, "Consistency in Networks of Relations," In *Artificial Intelligence*, 8, pp.99-118, 1977.

[PUGE94a]   J.-F. Puget, "A C++ Implementation of CLP," In *ILOG Solver Collected Papers*, ILOG SA, France, 1994.

[PUGE94b]   J.-F. Puget, "Object-Oriented Constraint Programming for Transportation Problems," In *ILOG Solver Collected Papers*, ILOG SA, France, 1994.

[SISK93]   J.M. Siskind and D.A. McAllester, "Nondeterministic Lisp as a Substrate for Constraint Logic Programming," In *Proceedings of the Eleventh National Conference on Artificial Intelligence*, Washington, DC, pp.133-138, July, 1993.

[STEE80]   G.L. Steele Jr., *The Definition and Implementation of a Computer Programming Language Based on Constraints*, Ph.D. Thesis, MIT, 1980.

[VANH89]   P. Van Hentenryck, *Constraint Satisfaction in Logic Programming*, MIT Press, 1989.

[WALT72]   D.L. Waltz, "Understanding Line Drawings of Scenes with Shadows," In *The Psychology of Computer Vision*, McGraw-Hill, pp.19-91, 1975.

APPLICATION OF NEURAL NETWORK TO DISTURBANCE ESTIMATION IN ROBOTIC ASSEMBLY

S. P. Chan

School of Electrical and Electronic Engineering
Nanyang Technological University
Block S1, Nanyang Avenue
Singapore 639798
Republic of Singapore
e-mail: espchan@ntu.edu.sg

ABSTRACT

The success of robotic assembly for odd form electronic components depends heavily on the ability to monitor and control the insertion force. For this purpose, a joint disturbance observer can be applied. However the observed disturbance signal includes the effects of model uncertainties. An approach of using a neural network to learn the parametric and unstructured uncertainties in robot manipulators is proposed. Furthermore a true teaching signal for learning the uncertainties is obtained. After learning, the neural network is embedded in the structure of the joint torque disturbance observer to compensate for the uncertainties in the robot dynamic model. As the result, accurate estimate of the external disturbance force can be deduced.

INTRODUCTION

The unstructured and non-standard shape of odd form components prevented the design of insertion machines to fully automate the assembly of printed circuit board (PCB). In a flexible manufacturing system (FMS) environment, assembly robots can be used in place of human operators. However, the protruded leads of the components could easily be jammed during insertion due to factors such as misalignment, bending, position error, gripper orientation error, or location error in the PCB's hole. Hence it is imperative to be able to monitor and control the insertion force in order to prevent damaging the components, PCB or robot.

Adding compliance unit or force/torque transducer to the end-effector [Lee87] will necessary increase the overall system cost. Alternatively, torque sensing techniques which require additional sensed signal such as current [NLH88] or acceleration [Tai89, YMO92]

have been proposed. In [Cha95], a joint torque perturbation observer is introduced to estimate the external reaction force during component insertion. Only the joint displacement signal and its first time derivative, i.e. velocity signal, are needed. Since joint position and velocity signals are normally available, this disturbance observer is suitable for implementation. However the uncertainties in friction torque limited the accuracy of the observed reaction torque. Friction models have been widely studied [Arm88, HM88, CNA+91]. It has been well established that friction depends on the direction of rotation, but the character of the function, in particular for low velocities, gives rise to some disagreements. In general the friction torque is a nonlinear function of joint velocity. However, it is difficult to represent the characteristic accurately in terms of a simple mathematical model. Frequently, a simple model which is linear in parameters is assumed. Parameter linearity is suitable for on-line identification but it may omit the negative velocity dependence at low velocity.

In this paper, a new approach of using a neural network to learn the uncertainties in the dynamic model of robot manipulators is proposed. The neural network is then embedded within the structure of the joint torque disturbance observer. As the result, improved estimation of the external disturbance due to insertion force is obtained.

JOINT TORQUE DISTURBANCE OBSERVER

For PCB assembly using SCARA robots, only the motion of the third joint which is prismatic is involved during the component insertion stroke. The dynamic equation of motion for this axis is [Sch90]

$$m_3\ddot{q}_3 + T_{f3}(\dot{q}_3) + m_3 g + T_{r3} - T_3 \qquad (1)$$

where q_3 is the joint displacement, T_3 is the applied joint torque, m_3 is the joint mass, $T_{f3}(\dot{q}_3)$ is the friction torque, g is the gravitation constant and T_{r3} is the reaction torque due to external disturbance. The subscript 3 denotes the third (z) axis. The corresponding internal model $\hat{G}_{p3}$ for the z axis is

$$\hat{m}_3\ddot{q}_3 + \hat{b}_3\dot{q}_3 + \hat{m}_3 g - \hat{T}_3 \qquad (2)$$

where $\hat{m}_3$ is the estimated value of m_3 and $\hat{b}_3\dot{q}_3$ is the estimated friction torque. The structure of the joint torque perturbation observer [Cha95] is shown in Figure 1. The observed disturbance torque is

$$T_{per3} - \frac{1}{\tau_3}\int(\hat{T}_3 - T_{ref3})dt \qquad (3)$$

where τ_3 is a positive gain constant and T_{ref3} is the reference joint torque. Substituting (2) for $\hat{T}_3$

$$T_{per3} - \frac{1}{\tau_3}\left[\hat{m}_3\dot{q}_3 + \hat{b}_3 q_3 + \int(\hat{m}_3 g - T_{ref3})dt\right] \qquad (4)$$

From (4), it is clear that only information of the joint displacement and its first time derivative are required.

The applied joint input torque is

$$T_3(t) - T_{ref3}(t) - T_{per3}(t) \qquad (5)$$

Substitute (5) into (4) obtains

$$T_{per3} - \frac{1}{\tau_3}\left[\hat{m}_3\dot{q}_3 + \hat{b}_3 q_3 + \int(\hat{m}_3 g - T_3 - T_{per3})dt\right] \qquad (6)$$

Using the dynamic equation (1) and rearrange the terms

$$T_{per3} + \frac{1}{\tau_3}\int T_{per3}dt - \frac{1}{\tau_3}\int[(\hat{m}_3 - m_3)\ddot{q}_3 + (\hat{b}_3\dot{q}_3 - T_{f3}) \\ + (\hat{m}_3 - m_3)g - T_{r3}]dt \qquad (7)$$

In the transform domain

$$T_{per3}(s) - (1 + \tau_3 s)^{-1}\left[\Delta m_3\big(s^2 q_3(s) + g\big) + \Delta T_{f3}(s) - T_{r3}(s)\right] \qquad (8)$$

where s is the Laplace transform variable,

$$\Delta m_3 \triangleq \hat{m}_3 - m_3 \qquad (9)$$

denotes parametric uncertainty in the joint mass, and in the time domain

$$\Delta T_{f3} \triangleq \hat{b}_3\dot{q}_3 - T_{f3}(\dot{q}_3) \qquad (10)$$

denotes the unstructured uncertainty in friction torque. The disturbance signal of (8) represents a measurement of the combined effects due to variation in joint mass, variation in friction torque, and reaction torque. A low pass filter is also included to enhance robustness of the measurement from system noise.

In most applications, the joint mass can only be estimated to within a certain accuracy. The friction torque $T_{f3}(\dot{q}_3)$ is also difficult to be modeled exactly. Sensitivity analysis performed on (8) shows that

$$\frac{\partial T_{per3}}{\partial \Delta m_3} - (1 + \tau_3 s)^{-1}\big(s^2 q_3(s) + g\big)$$
$$\frac{\partial T_{per3}}{\partial \Delta T_{f3}} - (1 + \tau_3 s)^{-1} \qquad (11)$$

Uncertainties in the joint mass affect the measured disturbance signal through the joint acceleration signal, whereas those of the friction torque affect the measurement directly.

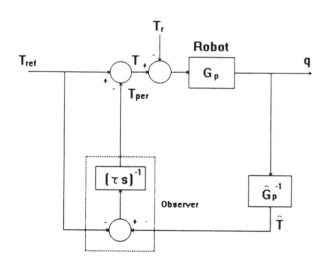

FIGURE 1 The torque perturbation observer

NEURAL NETWORK UNCERTAINTIES COMPENSATOR

The disturbance signal of (7) consists of two major components, one corresponding to the model uncertainties and the other due to the reaction torque. Under ideal condition when the model parameters exactly match those of the robot manipulator, T_{per3} provides an accurate measurement of T_{r3}. Due to model uncertainties, this condition is not always possible to achieve. The proposed approach uses a simple model (2) in conjunction with an adaptive enhancement to account for the modelling error as a disturbance torque ΔT_3. The key feature is to train a neural network to learn the uncertainties in joint mass and friction torque. With the internal model of (2) and using the definition of (9) and (10)

$$\Delta T_3(\dot{q},\ddot{q}) = \hat{m}_3\ddot{q}_3 + \hat{b}_3\dot{q}_3 + \hat{m}_3 g - [m_3\ddot{q}_3 + T_{f3}(\dot{q}_3) + m_3 g]$$
$$= \Delta m_3(\ddot{q}_3 + g) + \Delta T_{f3}$$

$$(12)$$

The disturbance torque is a complicated function of joint acceleration and velocity. Since a neural network is capable of performing nonlinear mapping, it can be adopted here. When properly trained, the neural network provides an approximation to the nonlinear map. Hence it can be employed to compensate for the model uncertainties.

With supervised learning, the neural net is first trained off-line using pre-recorded data set. In order for it to compensate the model uncertainties, a correct teaching signal must be available. The problem is simplified by assuming the unknown joint mass m_3 to remain constant. The learning scheme is shown in Figure 2. Compare with the time domain equation (12), it is clear that in the absence of external disturbance, T_{per3} of (7) provides the measurement of ΔT_3 through a low pass filter. Hence it can be used as a teaching signal for training the neural network to realize the nonlinear mapping from $\dot{q}$ and $\ddot{q}$ to ΔT_3. Through learning

$$T_n(\dot{q},\ddot{q}) \sim T_{per3}(t) \qquad (13)$$

where $\dot{q}$, $\ddot{q}$ are the inputs and T_n is the output of the neural network. In the transform domain

$$T_n(s) \sim (1+\tau_3 s)^{-1}\left[\Delta m_3(s^2 q_3(s) + g) + \Delta T_{f3}(s)\right] \quad (14)$$
$$= (1+\tau_3 s)^{-1}\Delta T_3(s)$$

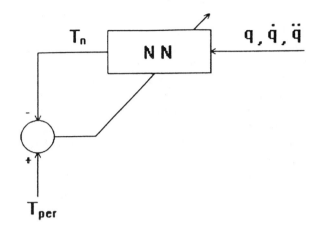

FIGURE 2 Training of the neural network

After training the neural network is embedded within the structure of the joint torque disturbance observer as shown in Figure 3 to estimate the external disturbance due to insertion force. The output of the neural network compensator T_n is compared with the observed disturbance T_{per3} and attempts to cancel the effect due to model uncertainties. From (14) and (8) the disturbance estimator output

$$T_o(s) = T_n(s) - T_{per3}(s) \qquad (15)$$
$$\sim (1+\tau_3 s)^{-1} T_{r3}(s)$$

which provides an estimate of the external disturbance torque.

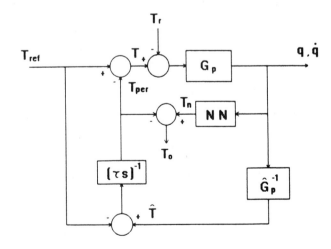

FIGURE 3 Structure of the compensator

SIMULATION STUDY

Computer simulations are performed to verify the proposed approach of applying a neural network to estimate external disturbance. Numerical integration is conducted by means of the fourth-order Runge-Kutta method with constant step size of 0.1 ms. A 2 ms sampling interval is used to simulate a digital implementation of the control method. The actual joint mass is 3.0 kg while its assumed value is 2 kg. A value of 0.005 is chosen for the gain constant τ_3. The initial condition of the integrator which constitutes the joint torque disturbance observer is set to zero. Similar to Canudas de Wit *et al.* [CNA+91], the friction torque is given by a nonlinear function of joint velocity

$$T_{f3}(\dot{q}_3) = \left[\alpha_0 + \alpha_1|\dot{q}_3|^{1/2} + \alpha_2|\dot{q}_3|\right]\text{sgn}(\dot{q}_3) \quad (16)$$

where $\alpha_0 = 5.0$, $\alpha_1 = -20.0$, and $\alpha_2 = 80.0$. It is modelled by a viscous term with $\hat{b}_3 = 80.0$.

The neural network uncertainties compensator implemented is a three-layered network consisting of input, hidden, and output layer. According to (12), the input signals to the network are the joint velocity $\dot{q}$ and the joint acceleration. However in most industrial robot, joint acceleration signal is usually not available. In this paper, a dynamic network that has time delay elements in the input layer is utilized which enables the neural net to learn without the information of joint acceleration signal. A network configuration with five input nodes is adopted with the input signals being $q(k)$, $q(k-1)$, $q(k-2)$, $\dot{q}(k)$, $\dot{q}(k-1)$ where k denotes the time index of the sampling period. The desired output is the compensation torque corresponding to model uncertainties. The number of units in the hidden layer is chosen to be 5. Units with nonlinear sigmoid transfer characteristic are used for both the output and hidden layers. Learning proceeds with the error propagation algorithm of the generalized delta rule according to Rumelhart *et al.* [RHW86]. Training of the network is conducted off-line using data obtained from a specially designed motion test in which the joint is driven by a triangular acceleration wave having long period. In this manner, the network is able to learn the model uncertainties for the ranges of velocity and acceleration which are within the normal operating condition of the robot. Since network learning is conducted off-line, there is no practical constraint on the training time.

During the first 2.5 s of the assembly cycle, the robot picks up a component at the pick up point and moves the end effector to a position which is vertical above the insertion point. The z axis then travels downward to perform the actual insertion operation. For the downward stroke, a trajectory consisting of three segments of quintic polynomials is selected to maintain continuity in velocity and acceleration.

$$q_{d3}(t) = \begin{cases} 0.72(t-2.5)^3 - 1.06(t-2.5)^4 + 0.42(t-2.5)^5 & 2.5 \le t < 3.5s \\ 0.08 + 0.02(t-3.5) & 3.5 \le t < 3.8s \\ 0.086 + 0.02(t-3.8) + 2.0(t-3.8)^3 \\ \quad -17.5(t-3.8)^4 + 37.5(t-3.8)^5 & 3.8 \le t \le 4.0s \end{cases}$$

$$(17)$$

where q_{d3} is in unit of m. From the initial position (0 mm), the end-effector rapidly approaches an assembly location (80 mm) which is just about the insertion point. During the insert phase (80 to 86 mm), the end-effector travels with reduced constant velocity. This will provide enough response time for the robot to stop the arm motion if the sensed reaction torque exceeds a prescribed threshold. Finally it comes to a complete stop at the terminal position (90 mm) in the set-down phase. The joint reference torque T_{ref3} is determined using the computed torque method [Mar73]

$$T_{ref3} = \hat{m}_3(\ddot{q}_{d3} + K_p e + K_v \dot{e}) + \hat{b}_3\dot{q}_3 + \hat{m}_3 g \quad (18)$$

where $e = q_{d3} - q_d$, and $\hat{m}_3$, $\hat{b}_3$ are the assumed values of m_3, b_3. The gain constants K_p and K_v are both set equal to 100.

It is interesting to note that the neural network has never been trained with the desired trajectory (17). However, this trajectory also consists of velocities and accelerations which are within the ranges of value in the special training data set. If the network learnt the mapping properly, its generalizing ability should enable it to give the correct output. Hence it is reasonable to believe the network is capable of providing the appropriate model uncertainties information during the actual insertion process. This generalization is demonstrated in Figure 4 which shows the network output T_n coincides well with signal T_{per3}. In the absence of mis-insertion, the deduced reaction torque T_o is near zero. The initial transient waveform of T_o comes from the observed disturbance signal. The assumed initial condition for the integrator of the disturbance observer may be different from the actual state of the robot. Anyway, the observer output converges quickly to provide useful information. This side effect can be reduced by choosing a larger value for the constant τ_3.

In most applications, it is only necessary to deduce the reaction torque during the insert phase. By choosing a constant velocity trajectory, the effect due to uncertainty in joint mass can be eliminated. For time between 3.5 s and 3.8 s, ΔT_3 and hence T_{per3} both remain constant under normal condition. The output of the neural network T_n also remains at a constant value

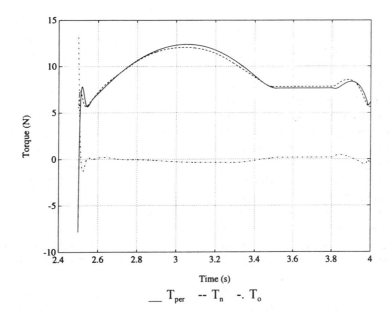

FIGURE 4 Performance of the disturbance estimator under normal condition

during the insert phase and T_o practically equals to zero.

A mis-insertion condition is simulated by blocking the downward movement of the end-effector on purpose. In this study the leads are characterized as linear elastic spring with stiffness of 10,000 N/m. The interaction of the component leads with the PCB generates a reaction torque which affects the dynamics of the robot according to (1). Due to modelling errors, the changes in joint velocity and acceleration give rise to a disturbance torque ΔT_3. Nevertheless, the neural network output T_n is capable of compensating for the effect of model uncertainties. As the result, T_o provides an accurate estimation of the reaction torque during occurrence of jamming. Figure 5 clearly indicates the estimated disturbance output T_o closely follows the actual reaction torque T_{r3}. The small time delay between the two signals is due to the time constant of the low pass filter. This lag time can be reduced by decreasing the value of τ_3.

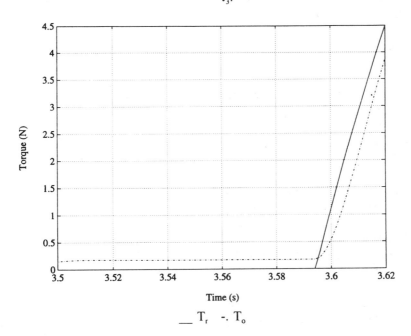

FIGURE 5 Performance of the disturbance estimator under abnormal condition

CONCLUSIONS

A neural network compensator has been applied to compensate for the parametric and unstructured uncertainties in the dynamic model of robot manipulators. By embedding the compensator within the structure of a joint torque disturbance observer, a true teaching signal is obtained to train the network. The compensated disturbance observer is capable of deducing measurement of the joint reaction torque due to external disturbance. This approach offers distinct advantages over the conventional method of only using a structured model. To demonstrate its usefulness, the assembly of odd form electronic components using a SCARA is considered. Accurate estimation of the reaction force during the downward component insertion stroke is made available from the compensated disturbance observer. The computer simulation study being conducted clearly demonstrates the performance of the proposed technique. Further work is still required to deal with the case when the joint mass does not remain constant but deviates from the training condition due to payload variations.

REFERENCES

[Arm88]   B. Armstrong. Friction: Experimental determination, modelling and compensation. In *Proc. IEEE Int. Conf. on Robot. and Autom.*, Philadelphia, 1988, pp. 1422-1427.

[Cha95]   S. P. Chan. A disturbance observer for robot manipulators with application to electronic components assembly. *IEEE Trans. Industrial Electronics*, vol. 42, no. 5, pp. 487-493, 1995.

[CNA$^+$91]   Canudas de Wit, P. Noel, A. Aubin, and Brogliato. Adaptive friction compensation in robot manipulators: Low velocities. *Int. J. of Robotics Research*, vol. 10, no. 3, pp. 189-199, June 1991.

[HM88]   V. Held and C. Maron. Estimation of friction characteristics, inertial and coupling coefficients in robotic joints based on current and speed measurements. In *Proc. IFAC Robot Control*, SYROCO'88, Karlsruhe, 1988, pp. 208-212.

[Lee87]   J. Lee. Apply force/torque sensors to robotic applications. *Robotics*, vol. 3, pp. 189-194, 1987.

[Mar73]   B. Markiewicz. Analysis of the computed torque drive method and comparison with conventional position servo for a computer-controlled manipulator. *Technical Memo 33-601*, Jet Propulsion Laboratory, Pasadena, CA, 1973.

[NLH88]   F. Naghdy, B. Luk, and K. Hoade. Stochastic force control in a robotic arm. In *Proc. IFAC Robot Control*, SYROCO'88, Karlsruhe, 1988, pp. 145-150.

[RHW86]   D. Rumelhart, G. E. Hinton, and R. L. Williams. Learning internal representations by error propagation. In D.E. Rumelhart and J.L. McClelland (eds), *Parallel Distributed Processing: Explorations in the Microstructure of Cognition. Vol. 1: Foundations*, MIT Press, Mass. 1986.

[Sch90]   R.J. Schilling. *Fundamentals of Robotics: Analysis and Control*. Prentice-Hall, 1990.

[Tsi89]   T.C. Tsia. A new technique for robust control of servo systems. *IEEE Trans. Ind. Elect.*, vol. 36, no. 1, pp. 1-7, 1989.

[YMO92]   F. Yu, T. Murakami, and K. Ohnishi. Sensorless force control of direct drive manipulator. In *Proc. IEEE Int. Symposium on Ind. Elect.*, Xian, May 1992, pp. 311-315.

ADAPTIVE FUZZY DIAGNOSIS FOR THE DETECTION OF AUTOMOTIVE FAILURES — A NEURAL NETWORKS APPROACH

Yi Lu Murphey and Thomas F. Jakubowski
Department of Electrical and Computer Engineering
The University of Michigan-Dearborn
Dearborn, MI 48128-1491
yilu@umich.edu
Phone: 313-593-5420, fax: 313-593-9967

ABSTRACT

This paper presents the application of an adaptive neural network architecture in the implementation of a fuzzy diagnostics system. The fuzzy diagnostic system attempts to diagnose automotive failures for the End-of-Line testing procedures. The neural network architecture consists of three neural network subsystems, an **Action Selection Network (ASN)** which is the implementation of the fuzzy diagnosis, an **Action Evaluation Network (AEN)** which plays the role of *critic* in evaluating the results of ASN, and a **Stochastic Action Modifier** which combines the results of **ASN** and **AEN** to finalize the output. The fuzzy diagnostic system implemented on the neural network architecture has been tested on a large set of data directly downloaded from automobile assembly plants.

Keywords: automotive diagnosis, neural network, fuzzy diagnostic systems

1. INTRODUCTION

Testing of a product is one of the essential parts of any manufacturing process. The strength at which this statement holds increases as the complexity of the product increases. Perhaps one of the most technologically advanced and complex products being manufactured in mass quantities today is the automobile. With every new model produced each year, the number of microcontroller and microprocessor based control units on each vehicle increases. Over the past decade, the increase in the amount of on-vehicle electronics is astonishing. Monitoring the quality of the finished product is one of the challenges facing engineers. As vehicles become more complex, the stringency of the testing must also increase; however, in an ever increasingly competitive marketplace, the amount of time and expenses in the testing process must be minimized. The solution to this complex problem is through the automation of the testing procedure.

One of the standard testing regimens which a newly manufactured vehicle is subject to, upon completion of the production process, is known as End of Line, EOL, Testing. During the EOL tests, a vehicle is tested in procedures designed to invoke problematic scenarios and real-world conditions. Test data is generated by such devices as the Electronic Engine Controller, EEC. This data can be analyzed using real-time computational methods to determine whether or not a vehicle is operating up to federal and corporate quality levels.

Recent efforts to improve the effectiveness and efficiency of Automotive Diagnostics has included the use of fuzzy logic for diagnostics purposes[HaL95, Luy96]. In this paper, we present our research in the implementation of the fuzzy diagnostic system in a neural network architecture. The neural network implementation makes the system trainable, so that the fuzzy membership functions do not need to be manually defined. This is advantageous in manufacturing environment when the system is required to test different vehicle models. The following sections describe the system architecture, implementation and test results.

2. SYSTEM ARCHITECTURE

The neural network architecture implemented in the system is similar to the one used by Berenji and Khedkar [BeK92]. The system consists of three neural network subsystems, an **Action Selection Network (ASN)** which is the implementation of the fuzzy diagnosis, an **Action Evaluation Network (AEN)** which plays the role of *critic* in evaluating the results of ASN, and the **Stochastic Action Modifier** which combines the results of **ASN** and **AEN** to finalize the output.

2.1 The Action Selection Network

Fuzzy Rules Set

Rule 1:
If **Lambse1** *is* **LOW** *AND*
If **Lambse2** *is* **LOW** *AND*
If **MAF** *is* **MED** *AND*
If **Idle DC** *is* **LOW** *AND*
If **TPS** *is* **NOT LOW**
THEN Vacuum Leak is <u>**HIGH**</u>

Rule 2:
If **Lambse1** *is* **LOW** *AND*
If **Lambse2** *is* **NOT LOW**
THEN Vacuum Leak is <u>*LOW*</u>

Rule 3:
If **Lambse1** *is* **NOT LOW** *AND*
If **Lambse2** *is* **LOW**
THEN Vacuum Leak is <u>*LOW*</u>.

Figure 1 A set of fuzzy rules

The Action Selection Network(ASN) implements a fuzzy diagnostic system. We use an example of fuzzy rule set shown in Figure 1 to illustrate the ASN. The ASN that implements the fuzzy rules is illustrated in Figure 2. The ASN consists of five layers. The input nodes consist of the state variables of the fuzzy system, namely, Lambse1, Lambse2, MAF, TPS, and Idle DC. The values of these parameters are generated by the EEC sensors throughout the EOL testing: the Lambse parameters are fuel to air ratios, TPS refers to throttle position and MAF to mass air flow. The second layer consists of the antecedent fuzzy terms for each of the input variables. In our case, we have three fuzzy terms, low, medium, and high. Each node in the input layer has a corresponding node in the antecedent layer for the possible states which the particular parameter may fall, and each node in the antecedent layer has exactly one input. The weights connecting the antecedent layer to the input layer can be viewed as the centers of the

membership functions. The spreads of these fuzzy memberships are available via the bias to the rule layer. Together, these weights and bias, when incorporated in a radial basis approach produce a membership value for each of the state variable labels in the system. Each antecedent node is connected only to the rule nodes which they impact. An antecedent node may have more than one output. The rule nodes perform a minimization operation on the membership values. The forth layer of the network implementing the consequence labels serves as a defuzzification process, which is a mass centroid method in this implementation. The defuzzification involves the process of summing the products of each of the centers of the consequence label memberships and multiplying by the membership each of these consequence labels is determined to fire with. To finalize the defuzzification, the output is divided by the sum of the consequence labels firing strength. The result of the defuzzification is the force which is represented by the variable *F*. The force does not represent the end result of the system's action as the AEN's influence must be considered. These consequence labels also determine the result of the fuzzy systems rule base and provide one of the two possible means in which the system can be tuned: firstly the parameters of the fuzzy membership can be tuned, secondly the consequence parameters can be tuned. An implementation of one or each of these allow a fuzzy control system to learn.

As previously indicated, the ASN uses the radial basis method to compute the membership functions. Triangular membership functions are the form chosen for this diagnostic control system. The radial basis method computes the membership by taking the difference between the value of the antecedent node, which is the system state variable in the input layer, and the rules. A difference of zero produces a membership of 1.0 or 100%. The greater the magnitude of this difference, the lower the membership for the parameter. The rate at which the membership falls off, the width of the membership function can either be controlled by a bias element or by implementing the weight as a multi-dimensional vector. When using a bias to determine the width of the membership function, a larger bias value indicates a wider curve, or slower fall off.

2.2 Action Evaluation Network(AEN)

The (AEN) serves as a critic to the decision made by the ASN. The technical name for this role is Adaptive Critic Element, ACE. The AEN(see Figure 3) is a three level Neural Network with one hidden layer. The network is maximally connected and uses a form of back propagation for training. The inputs to the network are the same state variables as the ASN as well as a failure signal. The signal can be binary, i.e. yes or no, or can be more complex. Each node in the hidden layer has $n+1$ inputs when the bias is included in this count and n is the number of nodes in the input layer. The resulting output of the AEN is a score of the ASN decision based on the state of the system. The score is combined with the failure signal to produce an internal reinforcement value.

In essence, the AEN looks at the state variables of the system. If the last time these state variables were introduced, the decision of the ASN was correct, the AEN will add very little to the ASN's decision. The current failure state of the system is also considered. However, if the system failed, or had a significant amount of error at the state variables occurred as inputs to the system, the AEN evaluates the decision of the ASN, as if saying, "The decision made last time by the

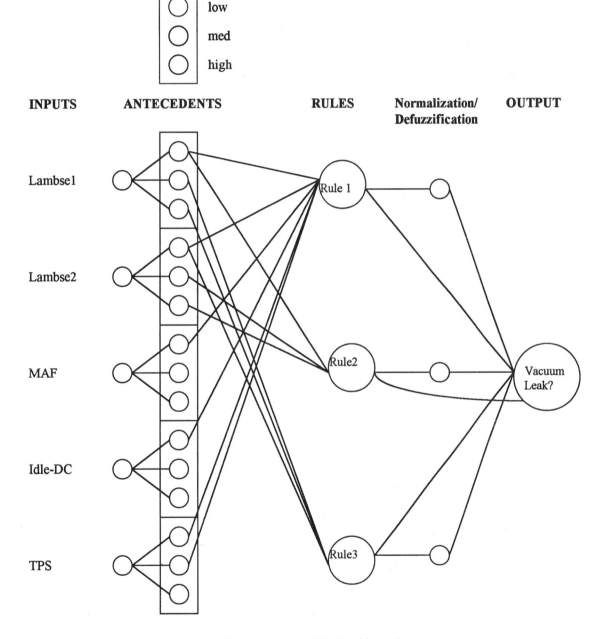

Figure 2: Action Selection Network

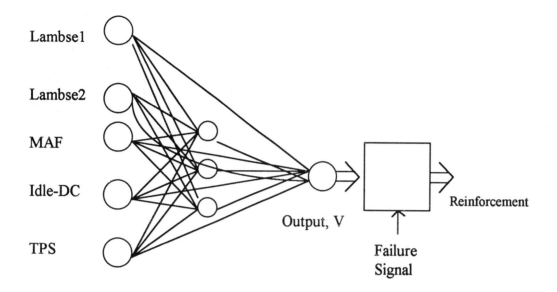

Figure 3: Action Evaluation Network

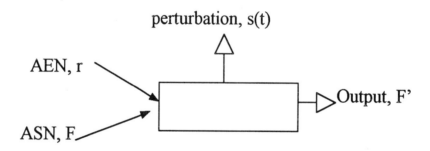

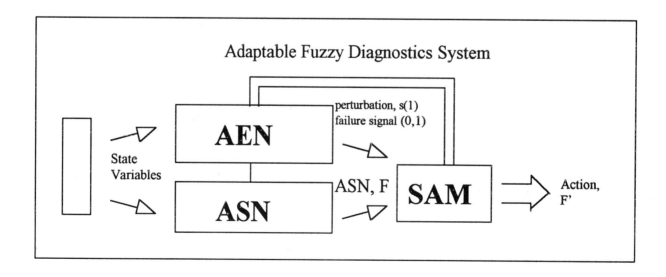

Figure 4: Combination of system components into a cohesive unit.

system was wrong when it was faced with these circumstances, so why would this one be correct?" The score, or reinforcement value, of the ASN's decision is sent to the Stochastic Action Modifier for further analysis and manipulation so that a final system output may be determined. The score sent to the SAM is termed r_{hat} for discussion. In this application, the AEN contains five input layer nodes: the five state variables, the three hidden layer nodes corresponds to the three fuzzy rules, and one output node which is the score of the decision. The rule nodes perform a minimum operator: each node receives several input values corresponding to the appropriate linguistic variables of the rule. A rule only fires at the strength of the lowest of its input requirements.

2.3 Stochastic Modifier(SAM)

The SAM combines the results of the AEN and the ASN to produce an ultimate decision on which action to take. The result of the SAM's processing is the resulting value of the entire Adaptive Fuzzy System shown in Figure 4. The SAM uses a Gaussian random variable whose mean is that of the ASN output and standard deviation is the reinforcement value generated by the AEN.

The Stochastic Action Modifier uses the reinforcement, r_{hat}, generated by the AEN and the recommended action, F, generated by the ASN to compute the final action. As indicated, the SAM utilizes a Gaussian random variable to obtain the final result, F'. The Gaussian random variable has mean F and standard deviation of $\sigma(r_{hat}(t-1))$. The model indicates that σ can be a non-negative monotonically reducing function such as $\exp(-r_{hat})$. The results of this computation is such that for a high reinforcement, r_{hat}, value, the magnitude of the difference between F and F' is low. As the reinforcement value decreases, the magnitude of this difference will increase. The SAM also computes a perturbation value, $s(t)$, for each step. The perturbation, the difference between $F'(t)$ and $F(t)$ divided by the Gaussian computation is used to scale the ASN's learning rate.

3. NETWORK LEARNING

Learning in the ASN is in essence a tuning of the fuzzy rules parameters. Only two layers of the network have weights which are subject to change, the weights representing the fuzzy rules and the weights representing the consequence labels. The algorithms used in training and updating the weights of the ASN consider both the change in the rate of failure the system, $\delta Fail$, and the

change in the width of each of the membership parameters. The purpose of training the ASN is to minimize $\delta Fail$ which also maximizes the success of the system. During the training, a failure signal allows the system to adjust the weights of the membership function by an amount $\delta LRate$. The amount by which the membership functions are tuned is a dynamic variable in that it is a function of the membership parameters. The implementation allows the consequence tuning algorithms be bypassed if desired. If a training data set has a population occupied by a vast majority of only one particular outcome, i.e. all vehicles with vacuum leaks, the tuning of the consequence labels may prove ineffective or even detrimental to the system's performance.

After the ASN Training, a table is generated for purposes of training the AEN. This table is critical in allowing the AEN to be trained in an efficient manner. The table stores the cases for which a system fails during the training of the ASN. The training algorithm also compute a reinforcement force for which the AEN will learn to produce an appropriate reinforcement value.

As indicated, the AEN is a back propagation network by nature. For purposes of training the evaluation network, the weights of the network are initially trained for an output value of 1.00 for all cases. Once the ASN is trained the table of reinforcement values is introduced to the network for purposes of training for the cases of the specific system. The system is then trained for a with an assortment of values having high reinforcement and the assortment of parameters in the table generated during the training of the ASN. The error of the network is minimized, so that the patterns of high reinforcement are learned as well as those of low reinforcement.

The training algorithm used for the weight updates in the AEN involves back propagation with momentum, and a stochastic update strategy as opposed to a batch update strategy. For purposes of design and testing the system's performance the local error function uses Samad's coefficient. This coefficient allows a network to perform using either "classic" back propagation "fast" back propagation.

The final step in the network training is the analysis of the system with the test data set. This test data set is processed and a prediction of system performance is generated.

4. IMPLEMENTATION

A Graphical User Interface (GUI) has been implemented to facilitate the use of the diagnostic system. The program is written for a PC platform running MS-DOS with a mouse input device. The system's menu-tree consists of a multi-layered structure which allows for such options as training the networks, running the diagnostics system, changing the files used for training or running the diagnostics, and viewing the fuzzy memberships in their before and after training forms.

The first layer of the menu structure has four options: training, diagnostics, utilities, and viewing the fuzzy memberships. Selecting the training option brings the user to the neural network training menu screen. This menu screen allows the user to train the ASN or the AEN. The user must train the ASN before the AEN because of the role the ASN performs in the AEN's training. Clicking the appropriate button begins the network training which will continue until an threshold is reached for the overall change in output error, δFail or until the keyboard is pressed to stop the training. The graphical progress of the training allows the user to see the progress of the network's learning. The output shows the user a "Good Count" and a "Bad Count." These values indicate the limits of the ASN's diagnostic abilities. Without the addition of the AEN, the ASN will miss-diagnose as many cases as are present in the "Bad Count" column for the training data set.

The AEN training can be started by activating the appropriate menu choice. The user is presented with graphical output as the network is presented with the data for training, and then as the network learns the reinforcement patterns. The final phase of the AEN training is the computation of the prediction value for the network's diagnostic abilities.

The diagnostic option on the main menu screen allows the user to execute system diagnostics. The system cycles through the active data file and presents the results of the diagnostics to the user in text output form. A data file is also generated so the results of processing can be more thoroughly considered.

The utilities option on the main menu brings the user to a utilities menu. The options in the utilities menu include options to change the data file for processing. Another option allows the user to change the threshold for the ASN training. Other options allow the user to load in a new fuzzy data set or to save the current optimized fuzzy set to a data file for further analysis.

The final option on the main menu screen allows the user to display the fuzzy memberships graphically. A major portion of this fuzzy diagnostics system is the ability of the system to adjust the fuzzy membership parameters for a given training set. The option allows the user to scroll through the fuzzy memberships for each

of the state variables. Numerical data is also displayed for the user.

5. TEST RESULTS AND CONCLUSION

In testing the system, there are two areas of particular interest to note. The first is in the tuning of the fuzzy membership functions. After the training, there is a fairly significant movement in several of the fuzzy membership parameters, and the system's performance is enhanced by the fuzzy tuning training with an improvement of over 50% over the original fuzzy diagnostics results. The addition of the AEN further minimizes the diagnostic error, in which errors can only be generated in the cases where there is a direct conflict within two data samples: one set of state variable inputs specifies a vehicle vacuum leak as HIGH, and subsequently an identical state variable set specifies a data set as LOW, the opposite also may hold. Obviously this problem cannot be overcome without additional knowledge. On a test set of ten thousand Thunderbird data samples, the original fuzzy diagnostics system produced 38 incorrect diagnosis with the fuzzy memberships at their default values. After training the fuzzy parameters, we reduce the number of diagnostics errors to 10. Finally the addition of the AEN further reduces the number of diagnostic errors to 1. Figure 5 highlights these results.

The experimental results show that the neural networks architecture discussed throughout this paper can serve to provide a basis for implementing a reliable and effective fuzzy diagnostic system. This optimum performance is achieved through the optimization of the membership functions enhanced by the addition of the evaluation unit which further improves the system performance. However, we must understand that the performance of a fuzzy system is limited to the abilities of its fuzzy set, and the optimization of fuzzy rules is not our next step of research.

6. ACKNOWLEDGMENTS

The authors wish to thank Mr. Brennan Hamilton of The Ford Motor Company for providing engineering knowledge and the test data needed for the design and test of the system.

7. REFERENCES

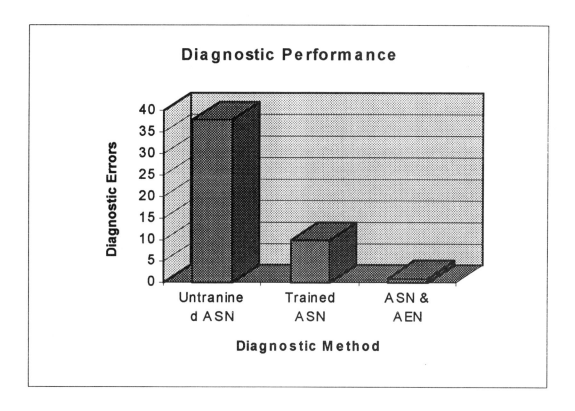

Figure 5. Performance chart of three different approaches.

[BeP92] Hamid R Berenji, and Pratap Khedkar, "Learning and Tuning Fuzzy Logic Controllers Through Reinforcements", *IEEE Transactions on Neural Networks*, Vol. 3 No. 5, September 1992.

[Day90] Dayhoff, Judith E. *Neural Network Architectures*, Van Nostrand Reinhold: New York, 1990.

[GuY88] Gupta, Madan M. and Takeshi Yamakawa, (Editors). *Fuzzy Logic in Knowledge-Based Systems, Decision and Control*, Elsevier Science Publishers: New York, 1988.

[HaL95] Brennan T. Hamilton and Yi Lu, "Diagnosis of Automobile Failures using Fuzzy Logic," *The eighth International conference on Industrial & Engineering Applications of Artificial Intelligence & Expert Systems*, Melbourne, Australia, June 5 ~ 9, 1995

[Kha90] Khanna, Tarun. *Foundations of Neural Networks*, Addison-Wesley Publishing Company: Reading, Massachusetts, 1990.

[Kos92] Kosko, Bart. *Neural Networks and Fuzzy Systems*, Prentice Hall: Englewood Cliffs, NJ, 1992.

[Luy96] Yi Lu, "A Self-Learning Fuzzy System for Automotive Fault Diagnosis," *The Ninth International conference on Industrial & Engineering Applications of Artificial Intelligence & Expert Systems*, Japan, pp. 167 - 172, June, 1996.

[Swa95] Swan, Thomas. *Mastering Borland C++ 4.5*, Sams Publishing: Indianapolis, IN, 1995.

[Was93] Wasserman Philip, D. *Advanced Concepts in Neural Computing*, Van Nostrand Reinhold: New York, 1993.

DIAGNOSTIC REASONING OF MECHANICAL FAULT VIA FUZZY NEURAL NETWORKS

Li Yue, *Wen Xisen, Qiu Jin
Department of Mechatronics Engineering and Instrument,
National University of Defense Technology,Changsha,Hunan, P.R.China.410073
Email: S801@ nudt.edu.cn

*National University of Defense Technology,Changsha,Hunan, P.R.China.410073
Email: WXS@ nudt.edu.cn

ABSTRACT

The method and strategy of combining traditional neural networks with fuzzy analysis technique, which discussed deeply in this paper, are valuable for application of solving the problems of diagnostic reasoning for a variety of mechanical faults. The fuzzy fault diagnosis put into practice by fuzzy processing of input-output of the conventional structure of neural networks. The network weights are trained and optimized by using genetic algorithm.The results of experiment performed on gear couples of NF125 type motor engine demonstrate that the hybrid system of fuzzy neural networks has a good ability of recognizing the states of samples and validity of diagnostic reasoning for border fuzzy situation of fault feature data.

KEY WORDS
fuzzy analysis, neural network, genetic algorithm

INTRODUCTION

The essence of fault diagnostic reasoning process is to recognize and analyze the fault state [XY95]. In order to form the mapping relation between the fault state and fault symptom and recognize the fault state accurately, we often divide fault of being investigated object into some categories during the process of fault diagnosis in a mechanical system or mechanical component. Usually, the relationship is not parallelism one by one between fault category set and fault state set, it is actually a kind of full mapping between two[Guo95]. Because of the characteristic of fuzzy and random in a mechanical system, it is very difficult to distinguish perfectly all kinds of faults by using single decided function. Artificial neural networks are capable of learning, self adapting to environment changes and handling fuzzy input information. This technique has many advantages and useful properties making it attractive for the automation of mechanical fault detectors [SD93].Neural networks do not require the complex mathematical modeling due to their ability to learn the desired mapping from examples . Once appropriately trained, the links or interconnects within the neural network itself will contain the non linearity of desired mappings. Indeed, neural networks can be viewed as nonlinear adaptive system identification units, which rely on pattern recognition for the identification procedure, Artificial neural networks have been widely used partly because of their multi-input parallel processing capabilities, which are most suitable for real time applications.Neural networks are not rely on model. They only need learn from input and output data in actual and storage the mapping relationship by means of weights' codes that is very difficult to comprehended. We can not understand the contents which networks have memorized, and also track the associate processes in logical. While carrying out classification of faults using traditional multi-layer perception neural networks, the numbers of output nodes equal the number of categories. During the networks training, the output value is 1 which corresponding with the fault mode and the others are 0, which is a modality of double value output. Since the existence of morbidity data in actual, the fault coupling phenomenon may be emerging which lead to fuzzy on border of classification, and the modes using for networks training usually have one or more values of non_zero. For the purpose of doing diagnosis correctly and improving the reliability of diagnosis, Diagnostic Reasoning of Mechanical Fault Via Fuzzy Neural Networks Method (DRMFFNNM) which combining the neural networks with fuzzy classification techniques is put forward in this paper. The main idea of this method is to achieve the membership which is expected during the procedure of training and output the membership which corresponding with the situation of faults during the procedure of diagnosis. In accordance with different fault, the networks' output different membership replacing the traditional double-value outputs, which is more efficiency for fault diagnosis and decision to the problem of fault feature data with fuzzy border. In addition, in order to solve the problem of convergence local minimum by using BP algorithm and achieve the optimization weights direct, the strategy for learning and training the networks using genetic algorithm is presented.

FEATURE EXTRACT FOR FAULT DETECTION AND FUZZY PROCESS FOR INPUT MODE OF NETWORKS

It is carried out by choice the appropriate membership function that the fuzzy processing of input vectors of fuzzy neural networks. For identifying the condition of a mechanical system, vibration monitoring is often effective [XBY93]. Vibration monitoring is based on the principle that components in engineering systems and plants produce vibration during operation. When a machine is operating properly, vibration levels are generally small and constant, However,when fault arises and some of the dynamic processes in the machine change, the vibration

signature also changes. The approach proposed here involves collecting vibration data and analyzing detailed the features which reflect the machinery operational state. The purpose is to detect the faults and analyze their causes. This procedure makes it possible that application of preventive maintenance[XY95] is effective.

The amplitude distribution of each frequency section on mechanical vibration spectrum are sensitivity to the operational condition of the system, so the features are chosen by this way. Because the difference in distribution of each frequency section is great, the member function is established by using the amplitude ratio which is the each frequency section with the whole feature frequency section selected. It is described as

$$\mu(x_i) = x_i / \sum_{j=1}^{n} x_j \quad i=1,2,\cdots n \quad [1]$$

where x_i, x_j is respectively the maximum value of the frequency spectrum in i-th and j-th feature frequency section, n is the number of feature frequency section.

DESCRIPTION OF FUZZY MEMBERSHIP FOR DIAGNOSTIC REASONING

The output of traditional perception is a double-value form (i.e. 0,1) while being used for the purpose of classification. When it is applied to classify the unknown data, the results of classification is decided by selecting the maximum output value. This method is relative simple and definite, but it is difficult for processing the data of fuzzy border. The output of fuzzy neural networks ,whereas, reflects the membership of some data belong to some categories, so this method can effective process the data which contains the fuzzy border.

Differing from resemble degree and close degree which are often used in fuzzy classification ,the angle cosine function is selected for calculating the classification membership. For one category mode problem, the node number of network's output is 1, the cluster center of the j-th category mode is described by using n dimensions vector c_j, thus the membership function of the i-th category mode to the j-th category mode is

$$\mu_{ij} = \frac{\sum_{k=1}^{n} x_{ik} \cdot c_{jk}}{\sqrt{\sum x_{ik}^2} \sqrt{\sum_{k=1}^{n} c_{jk}^2}} \qquad [2]$$

where x_{ik} , c_{jk} are respectively the *k-th* feature element of the *i-th* mode x_i and the *j-th* mode cluster center c_j , and c_{jk} is confirmed by following formula

$$c_{jk} = \sum_{m=1}^{n_j} x_{mk} / n_j \qquad k=1,2,---n \qquad [3]$$

where x_{mk} is the *k-th* feature element of the *m-th* mode , n_j is mode number of the *j-th* category.

When two kinds of fault modes show the absolute different signature(especially double-value signature) in output vector, the membership between the two kinds modes is zero ,which concords with decision of human brain.

GENETIC ALGORITHM FOR FUZZY NEURAL NETWORKS TRAINING

In genetic algorithms (GA),we represent the network with an array of floating point numbers, each of which corresponds to weights and biases.At the initialization, these floating point numbers are randomly generated in the range of ± 3.0. This distribution of initial values is decided based upon heuristic observation of weights and biases in the final state of the networks, and adjusted for each network. It is important to assure that the optimal value of weights and biases are covered by the range of initial distribution. To evaluate the fitness of each chromosome, feedforward computation is performed by presenting patterns which consist of one epoch. Total Sum Square (*TSS*) error is used as an error measure. The fitness of a chromosome is: fitness=$1/TSS^2$. Reproduction strategy is a proportional which normalizes fitness, and assigns higher reproduction probability for higher fitness chromosomes. In addition, we introduced

an elitist reproduction which always chooses the two best chromosomes and simple copies them, without crossover or mutation, to the population of the next generation. Crossover is either a single crossover or multiple crossover (two point or more) with a probability of 0.5 . Mutation probability is 0.05 for a static mutation unless otherwise stated. For an adaptive mutation, the value changes depending upon a similarity of paired chromosomes. For this experiment, we used two-point crossover, adaptive mutation, proportional reproduction based on a fitness measure, and elitist copy strategy.

EXPERIMENTAL RESEARCH AND RESULTS ANALYSIS

The experiment of fault detecting and diagnostic reasoning is carried out on the main gear couples of NF125 type motor engine using the method put forward in this paper. and two points vibration signals are collected using 4371 acceleration transducers. The signals having been amplified and filtered are sampled using 5000 Hz sampling frequency. To process the original data and construct the condition mode vector, the method presented in the paper is adopted. and then perform the fuzzed process so as to achieve the network input mode vector expressed by membership.During the stage of networks' learning and training, the expectation membership values of output nodes which all kinds of training examples to all kinds of fault is respective calculated reference the formulation [2] . Having been trained directed by GA, the fault diagnostic reasoning model for main gear couples is gained.

In experiment the fuzzy neural networks include one hidden layer, The number of nodes in input layer is 9 and output layer is 3. the number the nodes in hidden layer is 7 which decided by experimental comparison.There are fifty gear couples acted as samples for analysis and diagnostic reasoning, which include three kinds of fault mode types: normal, slight fault and serious fault. There are twenty samples for networks' learning and training, the others are used for testing the effects of the diagnostic reasoning model. The results of diagnostic reasoning for partly samples using fuzzy neural networks are shown in Table 1. The results of diagnostic reasoning using traditional neural networks are shown in Table 2, The

Table 1　The Results Of Diagnostic Reasoning Using Fuzzy Neural Networks

	O_1	O_2	O_3	Results
Sample No.1	0.575	0.934	0.245	√
Sample No.2	0.075	0.893	0.379	√
Sample No.3	0.954	0.750	0.664	√
Sample No.4	0.758	0.388	0.955	√
Sample No.5	0.675	0.374	0.897	√

Table 2　The Results Of Diagnostic Reasoning Using Traditional Neural Networks

	O_1	O_2	O_3	Results
Sample No.1	0.01	0.758	0.431	√
Sample No.2	0.035	0.844	0.214	√
Sample No.3	0.785	0.331	0.014	√
Sample No.4	0.693	0.028	0.155	×
Sample No.5	0.079	0.186	0.691	√

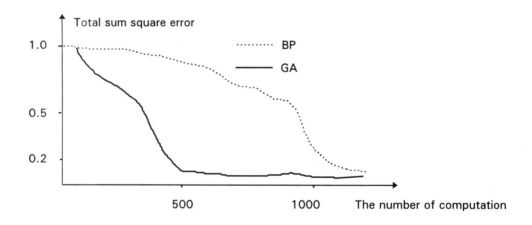

FIGURE 1　The convergence of GA-based and BP-based training

convergence of GA-based and BP-based training are shown Figure 1.

It is demonstrated by experiment that the fuzzy neural networks(FNN) has good ability for diagnostic reasoning correctly than traditional neural networks(NN). In this experiment the success ratio is ninety percent by using FNN and seventy-eight percent by using NN.

CONCLUSION

(1) The system of fuzzy neural network based on GA for learning has good effects to mechanical fault diagnosis;

(2) The effects is obvious that the networks weights optimization for all scale by using genetic algorithm;

(3) The diagnostic reasoning method presented in this paper is simple and utility, especially it is effective to process the problem which has the feature date with fuzzy border.

REFERENCES

[Guo95] Shang Guoqin,etc ,Research on the Method of Fault Diagnosis Based on Fuzzy Neural Networks, Journal of Vibration, Measurement & Diagnosis, Vol.15, No.s,Sep.1995

[SD93]   Yang Shuzi, Ding Hong,   Knowledge- Based Diagnostic Reasoning , Huazhong University of Science and Technology Press, 1993.

[XBY93] Wen Xisen, Tang Bingyang, Li Yue , Implementation of Operational Condition Monitoring and Fault Diagnosis System for Power Devices , 5th ICCOMADEM'93, England.

[XY95] Wen Xisen, Li Yue , Study of Fuzzy Classification Method on Condition Monitoring for Turbogenerator Groups. Journal of Vibration, Measurement & Diagnosis, Vol.15, No.s,Sep.1995

PORSEL: A PORTFOLIO SELECTION SYSTEM

Mehdi R. Zargham and Wei Xu
Department of Computer Science
Southern Illinois University
Carbondale, Illinois 62901
USA
Email: mehdi@cs.siu.edu

ABSTRACT

We have developed an expert system, called PORSEL (PORtfolio SELection system), which uses a small set of rules to select stocks. Each year, PORSEL selects twenty stocks from the listed stocks to form an investment portfolio for that year.

This paper improves the PORSEL by incorporating a better method of constructing portfolios. The new PORSEL now consists of two components: the Fuzzy Stock Selector and the Portfolio Constructor. The purpose of the Fuzzy Stock Selector is to evaluate the listed stocks and then assign a composite score for each stock. This component is essentially the old version of PORSEL [ZH96]. The second component, Portfolio Constructor, is a new component designed to generate the optimal portfolios for the selected stocks. It is based on the Markowitz portfolio selection model.

The results of simulation show that our new version of PORSEL outperforms the market every year during the testing period. For 14 years, the average excess return of our system over the S&P 500 Index is 121.38 percent.
Key Words: Fuzzy Logic, Stock Market, Securities Analysis, Inductive Learning.

INTRODUCTION

Investing to earn consistent and high returns is the ultimate goal strived for by all investors. People have tried many different techniques to obtain abnormal returns on their investments. Some use technical analysis, a technique in which historical data on prices, volume, leading indexes, and market-derived statistical measures are used to predict future stock prices. Others employ the technique of fundamental analysis in which a stock is evaluated based on the company's characteristics such as financial ratios and performance measures.

Despite increased efforts by investors, to earn consistent and high returns becomes more difficult today than ever before. According to the Efficient Market Theory, the market is efficient and people can not beat the market because stock prices have already reflected available information. Although the Efficient Market Theory is widely accepted by the academics, many began to doubt its validity because it failed to explain many events which occurred in recent years. One such event occurred on October 19, 1987, when the stock market dropped more than 500 points in one day. Thus, many believe that the market is not perfectly efficient and that there is inefficiency in the market which people can exploit.

To exploit market inefficiency, computer technology will play a more important role. Investors compete on the timeliness of getting the useful information to make their decisions. With the improvement of the computer and communication technology, those who are equipped with the right technology will have better information and better tools to analyze information and ultimately have a competitive advantage.

FUZZY STOCK SELECTOR

To exploit market inefficiency, we have developed an expert system called PORSEL (PORtfolio SELection system). This system, which is based on fuzzy logic [Z65], provides useful information for investors to make their decisions. The main objective of PORSEL is to find undervalued stocks. Undervalued stocks are those stocks whose prices are significantly lower than their intrinsic value, thus they have a large potential to grow in price. At the present implementation, one of the main function of PORSEL is to select a set of stocks based on a set of given rules. Each year, PORSEL selects twenty stocks from the listed stocks to form an investment portfolio for that year.

The rational behind using of fuzzy logic is that: in the real world, an ordinary investor usually evaluates stocks based on some rules of thumb. For instance, people usually choose a stock with a stable earning record, low price-to-earning ratio, low debt, and/or high dividend yield. Without doubt there are some principle rules associated with choosing the right stocks, although the rules differ from one person to the other. Therefore, it makes sense that we develop a model into which those rules of thumb are incorporated. Fuzzy logic is appropriate to construct the model since it is well known to have the ability to deal with linguistic variables, which are used a lot in the rules people state.

The rules are the true spirit of any expert system. However, it is not true that "the more rules, the better the system." Instead, a few golden rules can beat a system with a lot of unsatisfactory rules. At present our system contains 50 rules. These rules are derived from some of the well-know rules. For example, consider the following rule (known as Graham's rule [B77]):

"If a stock has price-to-earning ratio less than 40 percent of the highest price-earning the stock had over the past five years, then the rating for this stock is good".

In our system, the above rule is represented by few rules such as:

"If X is BIG for a stock, then the rating for this stock is GOOD"
"If X is VERY SMALL for a stock, then the rating for this stock is VERY BAD"
where:

$$X = \frac{\text{highest price-to-earning ratio over past 5 years}}{\text{current price-earning ratio}}$$

PORSEL uses a weight function to assign a weight to each rule. Assigning different weights can differentiates the importance of each rule. The proper weight for a rule is determined by the performance of this rule in the previous years by using past data. By training the system on past data, we can identify the comparatively more important rules to enhance the performance of the system.

In [ZH96], we have proposed a function for calculating the weight. We assume that each rule has just one linguistic variable in the consequent part; that is each rule is represented as:

Rule R_i: if X_1 is A_{1i} ... and X_n is A_{ni} then Y is B_i. where X_j and Y are linguistic variables with A_{ji} and B_i as their corresponding linguistic values. Also, let's assume that the data set contains a total of h data pairs; each data pair represented as $(x_{1k},..., x_{nk}, y_k)$, where k=1,...,h. Based on these assumptions, the weight $W(R_i)$ for rule R_i is obtained as:

$$W(R_i) = \frac{\sum_k AC_k * CC_k}{\sum_k AC_k} \text{, where}$$

AC_k = the compatibility degree of the antecedent and the k-th data pair. It is defined as:
$$AC_k = Bel_k(A) = \mu_{A_{1i}}(x_{1k}) * ... * \mu_{A_{ni}}(x_{nk})$$
CC_k = the compatibility degree of the consequence and the k-th data pair. It is defined as:
$$CC_k = Bel_k(B) = \mu_B(y_k)$$

$\overline{CC}_k$ = the incompatibility degree of the consequence and the k-th data pair. It is defined as:
$$\overline{CC}_k = Bel_k(\overline{B}) = 1 - \mu_B(y_k)$$

Bel(A) denotes the belief measure. (Details about fuzzy measures can be found in [ZH96].)

Using fuzzy logic, PORSEL applies each rule to the data of a given stock to obtain a rating. This rating represents the degree to which the stock's data matches the original golden rule. The closer the stock matches the rule, the higher is its rating. Combining the rating of all rules for a given stock together provides a final score for the stock. The stocks with the highest scores are selected to form portfolio.

PORTFOLIO CONSTRUCTOR

Although PORSEL can perform well, it needs some improvements, especially in the area of forming the final portfolios. In the original PORSEL, portfolios were constructed by subjectively investing an equal amount in each of the twenty stocks. To invest in twenty stocks instead of a single stock is to reduce the risk of the overall investment. However, portfolios with equal investment may not be optimal. To be optimal, a portfolio must be the least risky among all the portfolios that can achieve the specified return. Most of the time, however, optimal portfolios do not have an equal proportion in each stock.

Portfolio Constructor, is a new component to our system which generates the optimal portfolios for the selected stocks. It is based on the Markowitz portfolio selection model [M91, LTC90]. The Markowitz Model, also called the Mean-Variance Optimization Model, is an optimization approach to obtain optimal portfolios. Its goal is to minimize the risk of the entire portfolio while achieving the required expected return. Its focus is on the risk-return relationship of a stock and a portfolio.

Risk means uncertainty about future rates of return. It is generally represented by the standard deviation of the expected returns. The square of the standard deviation, called variance, is the expected value of the squared deviations from the expected return. Symbolically,

$$\sigma^2 = \sum_s Pr(s)[r(s) - E(r)]^2$$

where   Pr(s) is the probability of scenario s;

$r(s)$ is the return in scenario s;

$E(r)$ is the expected return of the stock. It is defined as the probability-weighted average of a stock's return in all scenarios, or $E(r) = \sum_s Pr(s)r(s)$.

A portfolio's expected return is a weighted average of the expected rate of return on each stock. It can be represented as the following formula:

$$E(r_p) = \sum_s x(s)E(r_s)$$

where $x(s)$ is the proportion of investment in stock s;

$E(r_s)$ is the expected return of stock s.

The variance of a portfolio can be calculated as follows:

$$V(r_p) = \sum_i^n \sum_j^n x_i x_j Cov(r_i, r_j)$$

where x_i is the proportion of investment in stock i;

x_j is the proportion of investment in stock j;

$Cov(r_i, r_j)$ is the covariance (variance if i is equal to j) between stock i and stock j.

PORSEL   minimizes   the   risk   of   the   entire portfolio, $\sqrt{V(r_p)}$,   by   carefully   constructing   the portfolio. This is done by solving the following problem:

Minimize

$$V(r_p) = \sum_i^n \sum_j^n x_i x_j Cov(r_i, r_j)$$

for all

$$\sum_{i=1}^n x_i E(r_i) = E(r_p)$$

Subject to

$$\sum_{i=1}^n x_i = 1 \qquad x_i \geq 0, \ i=1,...,n,$$

where   n is the number of the securities; x_i is the fraction of the portfolio held in stock i; $E(r_i)$ is the expected value of return on stock i; $E(r_p)$ is the target level of return on the portfolio; $Cov(r_i, r_j)$ is the covariance of returns of securities i and j; and $V(r_p)$ is the variance of the portfolio's return.

SIMULATION RESULTS

At present, there are two versions of PORSEL, Old-Version and New-Version. In the portfolio selected by Old-Version, every stock is invested in equal dollar amount. New-Version consists of Old-Version and a new component called Portfolio Constructor. Portfolio Constructor, is a component designed to generate the optimal portfolios for the selected stocks. (Optimal portfolios do not have equal proportion of fund in each stock.) The selected stocks (from Old-Version) are carefully analyzed; their expected returns and covariance matrix are estimated based on the historical data. A final optimal portfolio is constructed by using the Markowitz Model.

Both versions of PORSELs were tested on real stock market data from 1981 to 1994. The data were obtained from Standard & Poor's Compustat PC Plus database. The total number of companies downloaded was 6,344.

The investment horizon was set to be one year for the simulation. It was assumed that all transactions occur at each year's end. At each year's end, the system sells one year's portfolio and buys the next year's portfolio. Before the buying of the next year's portfolio, the system evaluates the current year's characteristics of the companies and selects twenty stocks as the candidates to be included in next year's portfolio. The returns on the portfolio are calculated based on the stock prices at the end of the year.

The benchmark used for testing the performance of our system is the returns on the S&P's Composite Index of common stocks. S&P 500 Index is a value-weighted portfolio consisting of the stocks from the largest 500 corporations in the United States. It is widely regarded as an approximation of the whole market. It is a good benchmark because beating the market is the goal of investors today. If our system can outperform the market most of the time, then it can be regarded as a good system. The returns on S&P 500 Index from 1981 to 1994 are shown in Table 1.

The performance of   Old-Version and New-Version are shown in Table 2. In this table, the second column lists the returns on the portfolios of New-Version; the third column shows the returns on the portfolios of the Old-Version; the fourth column shows the returns on the S&P 500 Index.

In Table 2, the portfolios of the Old-Version are only compared to the portfolios of the New-Version with identical risk. The results show that the New-Version outperforms the market every year during the testing period. For 14 years, the average excess return of our system over the S&P 500 Index is 121.38 percent. On a year-by-year comparison, the smallest excess returns come from the first six years, 1981 to 1986. After 1986, the excess returns become much bigger and the biggest occurs in 1990.

New-Version performs better than Old-Version in ten of the 14 years tested. The average excess return for these 14 years is 68.34 percent. The under-performance of New-Version occurs in 1981, 1983, 1986, and 1987. Even for these years, the differences between the returns of New-Version and Old-Version are small. The only big difference occurs in 1986 when New-Version yields 21.56 percent and Old-Version, 108.84 percent.

The new PORSEL is based on the idea of using Fuzzy Logic to select stocks and using the Markowitz Model to construct portfolios. The simulation results show that this new system has its potential in the financial market. The portfolios constructed by the new system consistently outperform the S&P 500 Index and those constructed by the old version of PORSEL.

REFERENCES

[B77]   P. Blustein, Ben Graham's last will and testament, Forbes, August 1, 1977, p. 43-45.

[CMB88]   S. Cottle, R.F. Murray, and F.E. Block, Graham and Dodd's Security Analysis, 5th Ed., New York: McGraw-Hill Book Company, 1988.

[LTC90]   J.K. Lee, R.R. Trippi, S.I. Chu, and H. Kim, "K-FOLIO: Integrating the Markowitz Model with a Knowledge-Based System," Journal of Portfolio Management. vol 16, pp. 89-93, Fall 1990.

[M91]   H. M. Markowitz, "Portfolio Selection-- Efficient Diversification of Investments," Second Edition, Basil Blackwell, Inc., Cambridge, Massachusetts, 1991.

[Z65]   L. A. Zadeh, Fuzzy sets. Information and Control, 8:338-353, 1965.

[ZH96]   M. Zargham and L. Hu, "Assigning Weights to Rules of an Expert System Based on Fuzzy Logic," proceedings of IEA/AIE-96, 1996, pp. 189-193.

TABLE 1
ANNUAL RETURNS ON STANDARD AND POOR'S 500 COMPOSITE INDEX (1981-1994)

Year	Return	Year	Return
1981	-4.91	1988	16.81
1982	21.41	1989	31.49
1983	22.51	1990	-3.17
1984	6.27	1991	30.55
1985	32.16	1992	7.67
1986	18.47	1993	9.99
1987	5.23	1994	-1.50

CONCLUSIONS

In this paper, a new version of PORSEL was proposed.

TABLE 2
RETURNS ON PORTFOLIOS OF THE NEW-VERSION, THE OLD-VERSION, AND S&P 500 INDEX

Year	New-Version	Old-Version	S&P 500 Index
1981	14.91	22.17	-4.91
1982	49.26	36.76	21.41
1983	53.20	74.91	22.51
1984	30.05	4.35	6.27
1985	66.76	37.10	32.16
1986	21.56	108.84	18.47
1987	46.38	49.30	5.23
1988	182.42	81.74	16.81
1989	67.87	63.66	31.49
1990	529.49	75.85	-3.17
1991	105.78	99.49	30.55
1992	33.77	6.42	7.67
1993	641.75	255.54	9.99
1994	49.08	19.39	-1.50
Average	135.16	66.82	13.78

USING THE SELF GENERATING NEURON-FUZZY MODEL FOR MACHINERY CONDITION MONITORING

Wen XiSen

National University of Defense Technology , Changsha, Hunan, P.R.China, 410073,
Email: WXS@nudt.edu.cn

Hu Niaoqing,   Qing Guojun

Department of Mechatronics Engineering and Instruments, National University of
Defense Technology, Changsha, Hunan, P.R.China, 410073
Email: S801@nudt.edu.cn

ABSTRACT

In this paper, we discuss the structure of Self Generating Neuron-Fuzzy Model, a self generating algorithm and its application to machinery condition monitoring. First, we analyze the neuron-fuzzy model based on radial basis function (RBF). Then, we discuss the concrete algorithm of this model, in particular, we improve the algorithm of the model for the application of machinery condition monitoring. At last, we use the model to identify modes of bearing operating condition. The results are shown that the model is adaptive according to the specified model error with less number of RBFs than the other methods by which only coefficients of the RBFs are tuned. So the model satisfies the basic requirements of machinery condition monitoring, such as real-time property, success rate, sensitivity and robustness.

KEY WORDS

radial basis function(RBF), neuron-fuzzy model, self generating algorithm, machinery condition monitoring

INTRODUCTION

In machinery condition monitoring, we usually have difficulties in obtaining sufficient samples of operating state, or it is impossible to obtain any samples of various operating states. In this case, decision-making strategy faces an austere challenge. It is necessary to have samples of each state at one's disposal in order to either learn the parameters of the decision rule or to find the neighbors of the observed vector [XBN93]. The more representative it is, the more pertinent the decision rule [DM93].

In real applications, one does not know the number of state classes or cannot do any recording of data from some classes. This case often appears in condition monitoring problems. For example, because of security problems, it is unsafe to run the system under dangerous states, therefore it is impossible to record data corresponding to these states. As another example, a man in charge of the system does not want his process to work in an abnormal state because of bad production quality, so, in most of the cases, it is impossible to obtain any samples from some states: the learning (or training) set is never complete.

Another comment concerns the number of states: this number is usually very important in real cases. Think, for example, of a car for which you want to build an on-board condition monitoring system. A car can run (or cannot run) under a lot of different states: some of them are normal ones (you can drive the car, but its power is not the optimal condition,   and,   as an example,   you cannot go at the speed you want), some of them are abnormal (it is impossible, under these states to drive the car), some of them are dangerous (for example, your brakes do not work ···). The number of states being very important, the discrimination problem becomes very difficult because you need samples from each class in order to learn the parameters of your discrimination rule.

In conclusion, a condition monitoring problem can be solved using a pattern recognition approach but, because the number of system states or classes and/or because it is impossible to record observations on some states, the learning set (i.e., the set of observations on each possible class) is never complete. It is impossible to discriminate using the usual rules. Hence, the usual decision model can not discriminate the problem real-timely and accurately. Although some modified Bayesian decision rules have been proposed, their adaptation is not sufficient. Considered these properties of condition monitoring problem, it is necessary to find a new distinguishing model to classify machine operating state.

To be discriminative model of condition monitoring, it is required to have the following requirements: (a) real-time processing ability, (b) adaptive ability, (c) fast convergence speed of learning, (d) avoidance using classifiers trained via supervised learning for monitoring. So the discriminative model is asked to be Simple, adaptive, practicable, affordable and fast computationally - inherently parallel.

In recent years, intelligent condition monitoring systems are widely and intensively developed. Some models being used as monitoring methods include pattern recognition, fuzzy systems, decision trees, expert systems and neural networks. Fuzzy logic, neural network, and neuro&fuzzy technology which integrates these approaches are now regarded as an effective method to realize such intelligent features.

For fuzzy technology, for example, fuzzy control is very useful for designing and controlling complex systems whose mathematical model cannot be easily obtained. In this case, fuzzy if-then rules are used to express skilled operator's experience, control engineer's knowledge and maintenance engineer's experience. Although this linguistic approach of fuzzy theory enables the fusion of symbol processing and numerical computation, it has not learning ability itself and exists the limitation of adaptation.

Multi-layered neural networks are widely used for a variety of applications such as diagnosis, pattern recognition, control, identification and prediction of nonlinear dynamical systems, as a universal function approximator. Moreover, the learning ability of neural network is used to generate fuzzy if-then rules from observed data. However, slow convergence speed by the commonly used backpropagation algorithm makes it difficult to put these applications into practice. Furthermore, the neurons are arranged according to a predefined structure. The structure is determined by the number of layers and the number of neurons in each layer. In a paper[GWG89], a criterion is introduced to estimate the number of neurons in two-layer neural networks. However, there have been no general results for designing a neural network structure. Some empirical rules for the application of condition monitoring include: (a) similar inputs (samples from a sample process condition) should be given similar representation in the network; (b) classes (different process conditions) to be separated should be given widely different representations in the network; If a particular class is important, then a large number of neurons should be used to represent it in the network, and (c) whenever possible, as much preprocessing as possible should be done, so that the learning and adaptive parts of the network would be required to do as little as possible.

In literature[Jyh93], adaptive-network-based fuzzy inference system (ANFIS) has been proposed. ANFIS can serve as a basis for constructing a set of fuzzy if-then rules with appropriate membership functions to generate the stipulated input-output pairs. Since its structure is predefined. Using ANFIS for machinery condition monitoring has no distinct predominance.

In this paper, we use neuron-fuzzy model based on RBF to classify modes of bearing operating condition which can satisfy characteristics and requirements of monitoring problem.

RBF algorithm proposed by paper [Pow87] has very fast convergence property compared to the backpropagation type algorithm, since an arbitrary can be approximated by the linear combination of locally tuned factorizable basis functions. This RBF model can also be regarded as multi-layered neural network or fuzzy model.

To identify the nonlinear systems by RBF, several algorithms such as k-means clustering method or Gram-Schmidt orthogonalization procedure have been proposed. However, in these methods, relatively large number of the basis functions is required since the tuning parameters are limited to only the coefficients of RBF.

In this paper, we improve the self generating algorithm proposed by paper [Kat93a] [at93b] for RBF to satisfy the specified model error with relatively small number of basis functions compared to the conventional methods, where not only the coefficients, but also the center values and widths of RBF are tuned. The algorithm adapts the case of condition monitoring where the real-time processing is needed. It consists of the following three processes: (a) model parameter tuning process by gradient method for fixed number of the basis functions, (b) RBF generation procedure, which is invoked when the effect of parameter tuning is diminished, by which new RBF is generated in such a way that whose center is located at the point where maximal inference error occurs in the input space, (c) if all the errors between each RBF value corresponding to any element of input condition feature vectors are least than any given small positive real number, the element can be considered to be redundant and cab be eliminated from the vectors in order to reduce the dimension number of vectors and improve the speed of learning and classification.

The second section discusses mainly the aspects of RBF model[Kat93a].

CONDITION MONITORING PROBLEM

Let $x=(x_1,\cdots,x_i,\cdots,x_m)\in R^m$ be inputs and $y=(y_1,\cdots,y_j,\cdots,y_s)\in R^s$ be outputs. The problem is identify the relationship between states and feature vectors $y=f(x)$: $R^m \to R^s$ with the given N input/output data.

$$(x^1, y^1),\ \cdots,\ (x^p, y^p),\ \cdots,\ (x^N, y^N) \quad (1)$$

Also if $s \geq 2$, output y_j $(j=1,\cdots,s)$ are assumed to be normalized such that $y_j \in \{y_j \mid -1 \leq y_j \leq 1\}$, $j=1,\cdots,s$, by dividing original y_j by the $\max(|y_j|)$, $j=1,\cdots,s$, respectively。

In RBF model, output y_j is given by (2):

$$y_j = \sum_{k=1}^{n} w_{jk} \cdot \mu_k(x, a_k, b_k), j=1,\cdots,s \quad (2)$$

where n is a number of RBF, $a_k \in R^m$, $b_k \in R^m (k=1,\cdots,)$, are the center value vector and the width value vector of RBF respectively, and $w_{jk}(k=1,\cdots,n, j=1,\cdots,s)$, are coefficients of RBF. Moreover

$$\mu_k(x, a_k, b_k) = \prod_{i=1}^{m} A_{ik}(x_i, a_{ik}, b_{ik}) \quad (3)$$

and $A_{ik}(x_i, a_{ik}, b_{ik})$ is a RBF for input x_i where a_{ik}, b_{ik} are the center value and the width of RBF, respectively. Several functions are used as A_{ik}, and in this paper we use Gaussian type RBF which is given by(4):

$$A_{ik}(x_i, a_{ik}, b_{ik}) = \exp\left(-\frac{(x_i - a_{ik})^2}{b_{ik}}\right) \quad (4)$$

In this case, the model is equivalent to generalized RBF or fuzzy model of class C^∞. If we regard this model as a fuzzy model, the k-th fuzzy rule is expressed by (5):
Rule k:

If $x_1 = A_{1k}$ and $\cdots$ and $x_i = A_{ik}$ and $\cdots$ and $x_m = A_{mk}$,
then $y_1 = w_{1k}, y_2 = w_{2k}, \cdots, y_s = w_{sk}, k=1,\cdots,n$ (5)

Let us introduce the following parameter vectors $a^n \in R^{mn}, b^n \in R^{mn}, w^n \in R^{sn}$ defined by (6) when the number of RBF is n :

$$\begin{cases} a_k = (a_{1k}, a_{2k}, \cdots, a_{ik}, \cdots, a_{mk}) \in R^m, & k=1,\cdots,n \\ a^n = (a_1, a_2, \cdots, a_k, \cdots, a_n) \in R^{mn} \\ b_k = (b_{1k}, b_{2k}, \cdots, b_{ik}, \cdots, b_{mk}) \in R^{mn}, & k=1,\cdots,n \\ b^n = (b_1, b_2, \cdots, b_k, \cdots, b_n) \in R^{mn} \\ w_k = (w_{1k}, w_{2k}, \cdots, w_{jk}, \cdots, w_{sk}) \in R^s, & k=1,\cdots,n \\ w^n = (w_1, w_2, \cdots, w_k, \cdots, w_n) \in R^{sn} \end{cases} \quad (6)$$

Then, given the N input/output data (1), the condition monitoring problem $P(n)$ with n basis functions is defined as (7):
Condition Monitoring Problem $P(n)$：

$$\min_{(a^n, b^n, w^n)} E_n(a^n, b^n, w^n) = \frac{1}{2}\sum_{p=1}^{N}\sum_{j=1}^{s}\left(y_j^p - {}^*y_j^p\right)^2 \quad (7)$$

subject to (2), (3), (4)

where y_j^p, ${}^*y_j^p$ are the j-th output and the j-th inference output for p-th input x^p, respectively. Let optimal solution for Problem $P(n)$ be $({}^*a^n, {}^*b^n, {}^*w^n)$ and the specified model error value be $\varepsilon > 0$, then the overall condition monitoring problem is formulated as follows.
Overall Condition Monitoring Problem:

Given the N input/output data (1) and the specified model error $\varepsilon > 0$, obtain the minimal number n of RBF and optimal solution $\left({}^{*}a^{n}, {}^{*}b^{n}, {}^{*}w^{n}\right)$ for Problem $P(n)$ which satisfies the inequality (8):

$$E_n\left({}^{*}a^{n}, {}^{*}b^{n}, {}^{*}w^{n}\right) < \varepsilon \tag{8}$$

ALGORITHM GENERATING

In order to solve the identification problem $P(n)$, we derive the gradients of E_n with respect to a_{ik}, b_{ik}, w_{jk},($i=1,\cdots,m$, $j=1,\cdots,s$, $k=1,\cdots,n$), under the equality constraints (2)~(4) as (9):

$$\frac{\partial E_n}{\partial a_{ik}} = \frac{-2}{b_{ik}} \times \sum_{p=1}^{N}\sum_{j=1}^{s}\left[w_{jk} \cdot \mu_k\left(x^{p}, a_k, b_k\right) \right.$$
$$\left. \cdot \left(y_j^{p} - {}^{*}y_j^{p}\right)\left(x_i^{p} - a_{ik}\right)\right], \tag{9a}$$
$$\left(i=1,\cdots,m,\ k=1,\cdots,n\right)$$

$$\frac{\partial E_n}{\partial b_{ik}} = \frac{-1}{bb_{ik}^{2}} \times \sum_{p=1}^{N}\sum_{j=1}^{s}\left[w_{jk} \cdot \mu_k\left(x^{p}, a_k, b_k\right) \right.$$
$$\left. \cdot \left(y_j^{p} - {}^{*}y_j^{p}\right) \cdot \left(x_i^{p} - a_{ik}\right)^{2}\right], \tag{9b}$$
$$\left(i=1,\cdots,m,\ k=1,\cdots,n\right)$$

$$\frac{\partial E_n}{\partial w_{ik}} = -\sum_{p=1}^{N}\left[\mu_k\left(x^{p}, a_k, b_k\right) \cdot \left(y_j^{p} - {}^{*}y_j^{p}\right)\right], \tag{9c}$$
$$\left(j=1,\cdots,s,\ k=1,\cdots,n\right)$$

By using the gradients (9), we can solve the identification problem $P(n)$ by appropriate gradient methods such as steepest descent method, conjugate gradient method, and quasi-Newton method. The learning algorithm by steepest descent method is given by (10), where h is iteration number and α is an optimal step size obtained by solving a one-dimensional search problem.

$$a_{ik}\left(h+1\right) = a_{ik}\left(h\right) - \alpha \cdot \partial E_n / \partial a_{ik} \tag{10a}$$
$$b_{ik}\left(h+1\right) = b_{ik}\left(h\right) - \alpha \cdot \partial E_n / \partial b_{ik} \tag{10b}$$
$$w_{jk}\left(h+1\right) = w_{jk}\left(h\right) - \alpha \cdot \partial E_n / \partial w_{jk} \tag{10c}$$
$$i=1,\cdots,m,\quad j=1,\cdots,s,\quad k=1,\cdots,n$$

When the decreasing rate of E_n becomes relatively small during the tuning process for fixed number of RBF, we generate new basis function. Although several strategies may be possible, we adopt such a strategy that a new basis function is generated in such a way that those center is located at the point where maximum of absolute inference error occurs in the input space. The self generating algorithm based on this idea is described as follows.

(1) Set the model error ε to be satisfied and stop criteria constant ε_1 for the parameter tuning process for the fixed number of RBF. Set ε_2 to be an appropriate small positive real number.

(2) Let h be the iteration number for the fixed number of RBF, and st be the total iteration number for the entire procedure. Let $h=0$, $st=0$.

(3) For N input/output data (1), compute (a^n (h+1), b^n (h+1), w^n (h+1)) from (a^n (h), b^n (h), w^n (h)) by learning rule (10).

(4) Let $E_n(h) \equiv E_n\left(a^{n}(h), b^{n}(h), w^{n}(h)\right)$ for simplicity. Then if inequality
$$E_n\left(h+1\right) < \varepsilon \tag{11}$$
is satisfied, n is the least number of RBF which satisfies the inequality (8) for overall condition monitoring problem. In this case, let (a^n(h+1), b^n(h+1), w^n(h+1)) be an optimal solution $\left({}^{*}a^{n}, {}^{*}b^{n}, {}^{*}w^{n}\right)$, and go to step (7); if inequality is not satisfied, go to step (5).

(5) Let us define the decreasing rate of E_n by (12).

$$D_n\left(h+1\right) = \left|\frac{E_n\left(h+1\right) - E_n\left(h\right)}{E_n\left(h\right)}\right| \tag{12}$$

If inequality (13)
$$D_n\left(h+1\right) < \varepsilon_1 \tag{13}$$
is satisfied, go to step (6) to generate a new basis function. If
$$D_n\left(h+1\right) \geq \varepsilon_1 \tag{14}$$
holds, let $=h+1$, $st=st+1$, and go back to step (3).

(6) Let (x^q, y^q) be an input/output vector such that the absolute inference error takes maximum at this point among the N input/output data $(x^p, y^p)(p=1,\cdots,N)$, that is to say, find (x^q, y^q) satisfying (15):

$$\left|y_r^{q} - {}^{*}y_r^{q}\right| = \max_{1 \leq p \leq N}\sum_{j=1}^{s}\left|y_j^{p} - {}^{*}y_j^{p}\right| \tag{15}$$

Then generating a (n+1)-th new radial basis function according to (16):

$$a_{i,n+1} = x_i^q, \qquad i = 1, \cdots, m \qquad (16a)$$

$$b_{i,n+1} = b_0, \qquad i = 1, \cdots, m \qquad (16b)$$

$$w_{j,n+1} = -y_j^q + {}^*y_j^q, \quad j = 1, \cdots, s \qquad (16c)$$

where b_0 is a given constant for the initial width of RBF. Let $n=n+1$, $h=0$, and go back to step (3).

(7) For any element x_i ($i = 1, \cdots, m$) of each vector x^p 对 of learning set $\{x^p\}$, if

$$\max_{\forall i,j} |A_{ik} - A_{jk}| \le \varepsilon_2 \quad (i = 1, \cdots, m, \ j = 1, \cdots, m) \qquad (17)$$

holds, then eliminate x_i from each x^p of $\{x^p\}$, go to step (3) and continue to learn, otherwise, terminate learning procedure.

APPLICATION TO BEARING OPERATING CONDITION MONITORING

We have used self generating neuron-fuzzy model to identify bearing operating condition. The test rig is shown in Fig. 1. Bearing belongs to rolling element type 207. The test bearing shaft was driven at 3000 rpm by an electric motor. B&K 4371 type accelerometer was used to sense bearing running vibration signal whose data acquisition flow diagram has been shown in Fig. 2. and test parameters were selected to maintain ingredients of signal which can reflect bearing running conditions.

We have tested and analyzed twenty bearings whose conditions are different. The possible conditions include three types: normal state, slight abnormal state and serious state. Time domain waveforms of different states have been shown in Fig.3. Obtaining condition feature samples in experiment have been shown in Table 1 [Nia92].

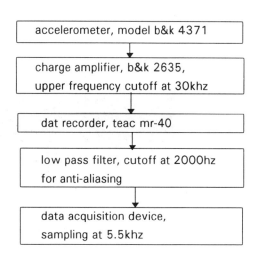

FIGURE 2   Data acquisition flow diagram

Samples whose number is 1, 2, 3, 9, 10, 11, 12, 13, 14, 15, 19, 20 have been incorporated into the learning (or training) sample set in order to train self generating neuron-fuzzy model., specified model error $\varepsilon = 0.3$, and stop criteria $\varepsilon_1 = 0.003$, $\varepsilon_2 = 0.1$ for parameter tuning for fixed number of RBF. Initial number of RBF $n=0$, and we apply the proposed algorithm with the case (a^n, b^n, w^n). Fig.4 and Fig.6 show the convergence curve. RBFs obtained after learning are shown in Fig.5. Learning process of the algorithm is visibly faster than that of BP network algorithm used by literature[NX97]. RBFs obtained from learning procedure have been used to identify the other spare samples and the results which have been identical with the real situation have been shown in Table 2. All analysis is run within the MATLAB software environment. Some general conclusions are given in the next section.

By using BP neural network model, the same learning set is trained. The trained results are shown in Table 3 compared with that of self generating algorithm. From Table 3, we have found that: (a) Learning speed of neuron-fuzzy model is faster than that of BP model. (b) Learned model error of the front is far least than the back. (c) The network structure of the front is simper than the back. In a word, the neuron-fuzzy model which satisfies the requirements of machinery condition monitoring problem has significant advantages over the BP model.

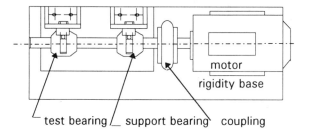

FIGURE 1   Test rig plan schematic

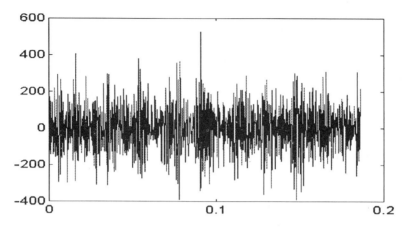

(a) Time domain waveform of normal state

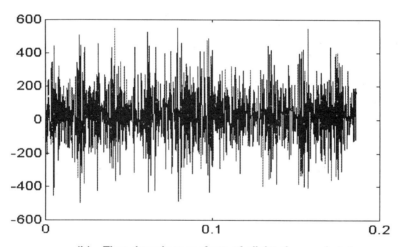

(b)   Time domain waveform of slight abnormal state

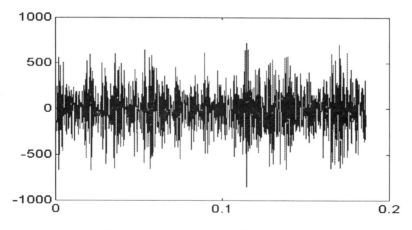

(c) Time domain waveform of serious abnormal state

FIGURE 3   Time domain waveforms of different states

Table 1.   Feature Data Of Bearing Running Condition

Number	Input feature elements of sample vectors								States
1	-1.7817	-0.2786	-0.2954	-0.2394	-0.1842	-0.1572	-0.1584	-0.1998	1
2	-1.8710	-0.2957	-0.3494	-0.2904	-0.1460	-0.1387	-0.1492	-0.2228	1
3	-1.8347	-0.2817	-0.3566	-0.3476	-0.1820	-0.1435	-0.1778	-0.1849	1
4	-1.4151	-0.2282	-0.2124	-0.2147	-0.1271	-0.0680	-0.0872	-0.1684	2
5	-1.8809	-0.2467	-0.2316	-0.2419	-0.1938	-0.2103	-0.2010	- 0.2533	1
6	-1.2879	-0.2252	-0.2012	-0.1298	-0.0245	-0.0390	-0.0762	-0.1672	2
7	-1.5239	-0.1979	-0.1094	-0.1402	-0.0994	-0.1394	-0.1673	-0.2810	2
8	-1.6781	-0.2047	-0.1180	-0.1532	-0.1732	-0.1716	-0.1851	-0.2006	2
9	-1.3479	-0.2883	-0.2479	-0.1534	-0.0689	-0.0463	-0.0326	-0.1464	2
10	-1.1476	-0.2026	-0.1948	-0.2173	-0.0808	0.0173	-0.0800	-0.0801	2
11	-1.4087	-0.2773	-0.2759	-0.2181	-0.0575	-0.0829	-0.0592	-0.1240	2
12	-1.7443	-0.1766	-0.1506	-0.1944	-0.1533	-0.1672	-0.2147	-0.2792	2
13	-1.8069	-0.2408	-0.1498	-0.2088	-0.1966	-0.1723	-0.1935	-0.2507	2
14	-1.7259	-0.1943	-0.1373	-0.2172	-0.1952	-0.1776	-0.1632	-0.2550	3
15	-0.5147	-0.1839	-0.1432	-0.0694	0.0285	0.0991	0.1326	0.0592	3
16	0.2741	0.1442	0.1916	0.1662	0.2120	0.1631	0.0318	0.0337	3
17	0.2045	0.1078	0.2246	0.2031	0.2428	0.2050	0.0704	0.0403	3
18	0.1605	-0.0920	-0.0106	0.1246	0.1802	0.2087	0.2234	0.1003	3
19	-0.7915	-0.1018	-0.0737	-0.0945	-0.0955	0.0044	0.0467	0.0719	3
20	-1.0242	-0.1461	-0.1018	-0.0778	-0.0363	-0.0476	0.0160	-0.0253	3

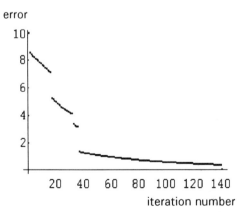

FIGURE 4   Model error vs total iteration number
(Not by features optimization)

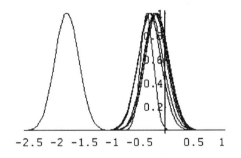

(a)   8 column RBFs of 1th row

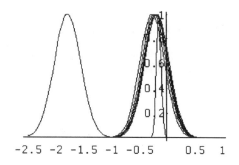

(b) 8 column RBFs of 2th row

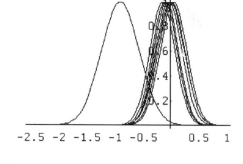

(d) 8 column RBFs of 4th row

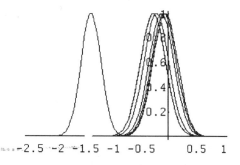

(c) 8 column RBFs of 3th row

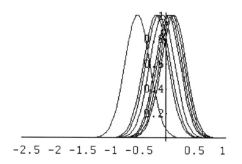

(e) 8 column RBFs of 3th row

FIGURE 5   RBFs from learning procedure
(including (a),(b),(c),(d),(e))

Table 2. Classification Results For The Spare Samples

Number	output net values			states
1	0.9307	0.0597	0.0018	1
2	0.9517	0.0125	-0.0004	1
5	0.9414	0.0194	-0.0006	1
4	0.0128	0.9365	0.1042	2
6	-0.0065	0.9748	0.0041	2
7	0.0396	1.0048	0.0019	2
8	-0.0224	0.9974	0.0058	2
16	0.0001	0.0001	1.0000	3
17	0.0067	-0.0598	1.0694	3
18	0.0020	-0.0172	0.9425	3

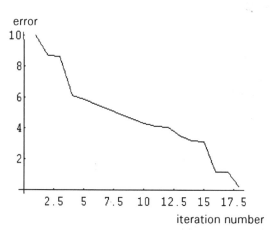

FIGURE 6   Model error vs total iteration number
(Not by features optimization)

Table 3   Comparative Results Between Neuron-Fuzzy Model And BP Model

	neuron-fuzzy model	BP model
network structure (input vectors are the same)	RBFs is self-generated by the algorithm, RBFs number is 5 after learning	The structure is predefined with 1 hidden layer of 9 units
learning algorithm	self generating iteration algorithm	BP network algorithm
predetermined max. total error	0.3	0.3
stop criteria or individual error	0.003	0.003
iteration number	141	161
learned max. total error	0.044629	0.298125

CONCLUSIONS

The aim of this paper is not to propose a new discriminative model, but to improve the self generating algorithm of neuron-fuzzy model in order to solve some specific application, such as real-time processing and accurate identification in machinery condition monitoring problems.   With the research of this paper, we have drawn the following general conclusions.

(1) Because the membership functions in RBF model are given by the Gaussian, the outputs of identification have been smoothed, which cannot be obtained by the fuzzy model with triangular or trapezoid shaped membership functions which are usually used.

(2) The self generating and tuning algorithm for RBF model can determine the minimum number of basis functions (fuzzy rules or hidden units) automatically to realize the specified model accuracy. We have gained satisfactory results by using RBF model for classification of bearing operating states.

(3) RBF model is also used to extract and optimize condition features.

(4) The algorithm discussed in this paper is useful not only for machinery condition monitoring and fault diagnosis, but also for the nonlinear prediction, dimension estimation of chaotic time series, nonlinear signal processing such as adaptive equalizers and portable telephone noise canceller, learning inverse kinematics and dynamics of robotics manipulator, adaptive controller, image processing and spatial filtering, etc.

(5) As shown in Fig. 4, the time when a new RBF is generated is represented. From Fig.4, monotonous decrease of E_n with respect to the total iteration number is

almost guaranteed for every st This because the learning results with n RBFs are effectively used as initial parameter values for the learning process of $(n+1)$ RBFs, and this works well for factorizable RBF model. In general, however, this initial value setting scheme may not work well for conventional neural network with sigmoid transfer functions.

(6) Self generating neuron-fuzzy model makes a compromise between real-time processing and accurate discrimination.

(7) Practical and efficient algorithm of self generating neuron-fuzzy model is necessary to be developed further.

REFERENCES

[DEW95]   Du,R. ,   Elbestawi, M. A. , and Wu,S. M. , Automated monitoring of   manufacturing processes,Part 1: Monitoring   methods, *ASME Journal of Engineering for Industry*, Vol.117, No.2, pp.121-132, January 1995.

[DM93]   Bernard Dubuisson,   Mylene Masson,   A Statistical Decision Rule with   Incomplete Knowledge about Classes, *Pattern Recognition*,   Vol.26,   No.1, pp.155-165, 1993.

[GWG89]   Gutierrez,H., Wang, J.,   and   Grondin, R. O.,   Estimating   Hidden   Units   for Two - Layer   Perceptons, Neural   Network Theory, Haykin, S.,ed., pp.120-123, 1989.

[Jyh93]   Jyh-Shing Roger Jang, ANFIS: Adaptive Network Based Fuzzy Inference System, *IEEE Trans. on SMC,* Vol.23, No.3, May/June 1993.

[Kat93a] R.Katayama,et.al. Self Generating Radial Basis Functions as Neuron-fuzzy Model and Its Application to Nonlinear Prediction of Chaotic Time Series. *Second Int. Conf. on Fuzzy Systems* (FUZZ - IEEE 93), San Francisco, Calif., pp.407-414(1993).

[Kat93b] R. Katayama, et al. Developing Tools and Methods for Applications Incorporating Neuron, Fuzzy and Chaos Technology, *Computers ind. Engng ,* Vol.24, No.4, pp.579-592,1993.

[Nia92]   Hu Niaoqing, Study for Operating Condition Monitoring and Fault Diagnosis System of Power Machine, *thesis for the Master in National University of Defense Technology,* 1992.1 (in Chinese).

[NX97]   Hu Niaoqing, Wen Xisen, A New Method of Feature Extracting Techniques Using Fractal Information for Machine Condition Monitoring, *6th COMADEM'97,* Xi'an,China.

[Pow87]   M.J. D. Powell, Radial Basis Function Aproximations to Polynomials, Proc.12th *Biennial Numerical Analysis Conference,* pp.223- 241,1987.

[XBN93]   Wen Xisen, Tang Bingyang, Hu Niaoqing, An Approach of Condition Model Recognition for Condition Monitoring System of Manufacturing Cell, *5th COMADEM'93,* Bristol, English.

ANSWERING NULL QUERIES IN SIMILARITY-RELATION-BASED FUZZY RELATIONAL DATABASES

Shyue-Liang Wang
Department of Information Management
Kaohsiung Polytechnic Institute, Taiwan
email: slwang@csa500.kpi.edu.tw

Ta-Jung Huang
Institute of Information Engineering
Kaohsiung Polytechnic Institute, Taiwan
m843202m@csa500.kpi.edu.tw

ABSTRACT

Null queries are queries that elicit a null answer from the database. In the context of fuzzy relational databases, obtaining approximate answers for null queries can enhance the user-friendliness of the system. Analogical reasoning utilizing fuzzy functional dependency to answer null queries on possibility-distribution-based relational database does not address the issue of redundancy. In this work, we present an approach to obtain approximate answers for null queries on similarity-relation-based fuzzy relational databases. Based on the concept of contexts on domain attributes, our approach is a generalization of the former model of analogy and provide the ability of handling the redundancy.

Keyword: Fuzzy Relational Database, Similarity Relation, Null Query, Analogical Reasoning, Fuzzy Functional Dependence.

INTRODUCTION

Fuzzy relational databases to accommodate incomplete, imprecise or uncertain information have been studied extensively in recent years[1-7,12-14,16-17,19]. To represent fuzzy information in the data model, two basic approaches can be classified: model through similarity relations or proximity relations [3-7,16] and model through possibility distribution[1,12-14,17,19].

Null queries are queries that elicit a null answer from the database. They could arise from missing data in the database, explicit null values, or no data items completely satisfying the conditions specified in the query. Since the answer of a query on fuzzy database is uncertain in nature and following the principle that something is better than nothing, it would be advantageous to have the database system to produce an approximate answer to the query instead of only a null answer.

Reasoning by analogy is an important inference tool in AI research. It consists of recognizing certain similarities between a source object/tuple and a target object/source and derive certain properties of the target on the basis of its observed similarity with the source.

A theoretical model utilizing analogical reasoning to answer null queries for possibility-distribution-based data model has been developed [8]. However, the model does not consider data with analogy over discrete domain[12], where data are better modeled through similarity relations. In addition, the issues of equivalency and redundancy between tuples in a relation is not addressed. In this work, we develop a generalized model of analogy to answer null queries on similarity-relation-based fuzzy relational data model. In fact, it can also accommodate possibility-distribution-based data types. Furthermore, the issue of redundancy is circumvented by adopting the concept of "contexts" on domain attributes.

The next section reviews the fuzzy relational data model based on similarity relations. Section 3 illustrates the inference mechanism of analogical reasoning for answering null queries. Section 4 defines the similarity measure used to measure the similarities between source and target objects/tuples. Section 5 reviews and defines fuzzy functional dependencies between attributes. Section 6 formally presents the problem model with proposed solution algorithm and an example is given to illustrate the idea of the approach. Finally, a short conclusion is reached at the end of the paper.

SIMILARITY RELATION AND THE FUZZY RELATIONAL DATA MODEL

For the following sections we will be utilizing the notion of similarity relation[18]. For each domain D, a similarity relation s is defined over its domain elements: $s : D \times D \rightarrow [0, 1]$. A similarity relation is a generalization of an equivalence relation in that if a, b, c $\in$ D, then s is

reflexive: $s(a, a) = 1$
symmetric: $s(a, b) = s(b, a)$
transitive: $s(a, c) \geq \max[\min(s(a, b), s(b, c))]$
        for all $b \in D$

A similarity-relation-based fuzzy relational database [3-7] is defined as a set of relations consists of tuples. Let t_i represent the i-th tuples of a relation R, it has the form $(t_{i1}, t_{i2}, ..., t_{im})$ where t_{ij} is defined on the domain set D_j, $1 \leq j \leq m$. Unlike the ordinary relational database, two simple but important extensions are defined. The first extension is that the tuple component t_{ij} selected from D_j is not constrained to be singleton, instead, $t_{ij} \subseteq D_j$ and $t_{ij} \neq \varnothing$. Allowing tuple component t_{ij} to be a subset of the domain D_j means that fuzzy information can be represented. If t_{ij} consists of a single element, it represents the most precise information; whereas if t_{ij} is the domain D_j itself, it corresponds to the fuzziest information. This leads to the definition:

DEFINITION [7] A fuzzy database relation R is a subset of the set cross product

$$2^{D_1} \times 2^{D_2} \times \ ... \ \times 2^{D_m},$$

where $2^{D_j} = 2^{D_j} - \varnothing$, 2^{D_j} represents the power set of D_j.

The second extension of the fuzzy relational database is that, for each domain set D_j of a relation, a similarity relation S_j is defined over the set elements:

$$S_j : D_j \times D_j \to [0, 1].$$

The similarity relation introduces different degrees of similarity to the elements in each domain and this is another mechanism for the representation of "fuzziness" in this fuzzy relational data model.

A simple illustration of the fuzzy database relation is shown in the Table 1 below representing the JOB, HOBBY and SALARY of seven EMPLOYEES. The similarity relations for domain attributes JOB and HOBBY are shown in Fig. 1.

EMPLOYEE	JOB	HOBBY	SALARY
Amy	Secretary	Music	25K
Bill	Software Engineer	{Swimming,Sports}	35K
Carol	Accountant	Sports	28K
David	System Engineer	Reading	39K
Ed	Software Engineer	{Music,Reading}	38K
Frank	VP	Sports	55K
Grace	Design Engineer	Sports	?

Table 1   A Fuzzy Database Relation

	Sec	SW Eng	Acct	Sys Eng	VP	D Eng
Sec	1	0.6	0.9	0.6	0.5	0.6
SW Eng	0.6	1	0.6	0.9	0.5	0.9
Acct	0.9	0.6	1	0.6	0.5	0.6
Sys Eng	0.6	0.9	0.6	1	0.5	0.9
VP	0.5	0.5	0.5	0.5	1	0.5
D Eng	0.6	0.9	0.6	0.9	0.5	1

JOB={Sec, SW Eng, Acct, Sys Eng, VP, D Eng}

	Reading	Music	Swimming	Sports
Reading	1	0.9	0.7	0.7
Music	0.9	1	0.7	0.7
Swimming	0.7	0.7	1	0.9
Sports	0.7	0.7	0.9	1

HOBBY={Reading, Music, Swimming, Sports}

Fig. 1 Similarity relations for domain sets

ANALOGICAL REASONING TO ANSWER NULL QUERIES

A query is a null query if it elicits a null answer from the database. Let R be a database, Q be a query posed to the database, and Q(R) be the answer generated by R in response to Q. For a null query Q, Q(R) is the empty set. These situations occur when no data items satisfy the conditions specified in the query or there are missing data in the database. Our goal here is to present an approach for obtaining an approximate answer Q'(R), based on analogical reasoning, where Q'(R) is an approximation to Q(R).

As an example to illustrate the obtaining of approximate answer for null query using analogical reasoning. Consider the relation shown on Table 1. Suppose that the attribute Salary of tuple Ed is missing and a query, Salary(Ed)?, is presented to the system. A null answer is produced. It is reasonable to assume that an approximate answer for this null query is that Ed's salary should be around 35K. This is because Ed's job is a software engineer and from the data present in other tuples of the relation we find that the salary of the other software engineer, Bill, is 35K (Ed and Bill share similar property -- software engineer, a determining factor of salary). Of course, this approximate answer may be incorrect due to other factors such as experience and ranks of employees. Nevertheless, producing an approximate answer is better than no answer. In addition, similarity-relation-based fuzzy relational data model assumes that no tuple elements can be empty. Without providing answers for missing data items, the fuzzy relational database can not function.

There are two issues to be resolved in this inferencing process. One is that which attribute P can determine the unknown attribute Q, e.g. P is *Job* and Q is *Salary*. This is referred to as the problem of fuzzy functional dependency (fuzzy determination) between attributes. The other one is that, with respect to attribute P, which tuple S is "most similar" to the tuple T with unknown attribute Q. This is referred to the problem of measuring the similarity between two tuple attribute values. The following two sections will provide some approaches for handling these two issues respectively.

SIMILARITY MEASURE

A similarity measure of two fuzzy sets, on the same domain of universe, is a measure that describes the similarity between fuzzy sets. In the context of fuzzy relational data model based on similarity relations, a similarity measure for the tuple components t_{ij} and t_{kj} can be defined to satisfy the axiom definition of Liu[14].

DEFINITION   Let t_{ij} and t_{kj} be tuple components defined on domain D_j with given similarity relation. The similarity measure $\mu(t_{ij}, t_{kj})$ is defined as

$$\mu(t_{ij}, t_{kj}) = \max S_j(x, y) \text{ for all } x \in t_{ij} \text{ and } y \in t_{kj}.$$

To illustrate this similarity measure, consider the two tuples Bill and Ed in the relation shown in Table 1. On attribute JOB, both tuples have attribute value Software Engineer which implies the similarity measure has value of one. On attribute HOBBY, tuple Bill's value is {Swimming, Reading} and tuple Ed's value is {Music, Sports}, which has $\mu(t_{ij}, t_{kj})$=0.9.

FUZZY FUNCTIONAL DEPENDENCY BETWEEN ATTRIBUTES

In this section, we will first review the fuzzy functional dependency based on the concept of "context", which is a generalization of the ffd defined on possibility-distribution-based fuzzy relational data model. A specialization of this ffd will then be applied on the calculation of the degree of relevance for the fuzzy determination on the similarity-relation-based fuzzy relational data model.

A context C is a partition on a set $\hat{D}$ generated by an equivalence relation ρ (or similarity relation) on $\hat{D}$ [9].

$\hat{D}$ is a subset of the underlying database domain D. The equivalence classes in C are set of "closely related" (indistinguishable) elements. For example, {[0k,20k), [20k,50k), [50k,80k), [80k+)} is a context for attribute *Salary* with four equivalence classes. The set of all equivalence relations on subsets of D is denoted by R_D; the corresponding context set is C_D.

DEFINITION[9] Let $(t)_\alpha$ be the α-cut of fuzzy set t. Fuzzy sets t and t' are equivalent at level α with respect to a context C, denoted by $t \sim_{(c,\alpha)} t'$, whenever $(t)_\alpha$ and $(t')_\alpha$ are non-empty subsets of the same equivalence class in C.

The precise salary {60k} and the imprecise salary [65k,75k] are equivalent with respect to the context

$\{[0k,20k),[20k,50k),[50k,80k),[80k+)\}$ at $\alpha = 1$.

DEFINITION[9]   Two tuples t and t' are redundant with respect to contexts C=$(C_1, C_2, ..., C_n)$, i.e., $t \sim_c t'$, whenever $t_i \sim_{(c_i, \alpha=1)} t_i'$ for each component i.

DEFINITION[9]   A database relation r with underlying scheme $R(A,C)$, where attributes A=$(A_1, A_2, ..., A_n)$ and associated contexts C=$(C_1, C_2, ..., C_n)$, is a set of non-redundant tuples with respect to the contexts in C.

The existence of a functional dependency (FD) implies that if some tuple components satisfy certain equivalences, then other tuple components must exist and their values must be equivalent. The following definition of ffd generalizes other versions of ffd in literature[8,15].

DEFINITION[9]   A fuzzy FD : $X(C_x) \rightarrow Y(C_Y)$ holds in scheme $R(A,C)$ if for all tuples t, t' in every extension r of $R(A,C)$, $t \sim_{C_X} t'$ implies $t \sim_{C_Y} t'$.

The context C_Y on attribute Y for a ffd is in fact induced by the context C_X on X from a relation. The induced context, $I_{X \rightarrow Y}(C_X)$, is the finest context for Y that yields a fuzzy FD from X to Y with context C_X for X. The induced context is computed as the greatest lower bound (glb) of equivalence classes merged in Y according to values in X.

An example of induced context is as following. Table 2 shows a precise relation of three tuples with attributes X and Y. It has precise contexts C_X={{a}, {b}} and C_Y={{1}, {2}} on the attributes X and Y respectively. The precise FD : X(C_X) $\rightarrow$ Y(C_Y) does not hold in the relation. However a ffd can be induced from X to Y by suitably coarsening C_Y.   The induced context, $I_{X \rightarrow Y}(C_X)$={{1}, {2, 3}} is the finest context for Y that "induces" the fd in the relation.

Precise Relation	
X	Y
a	1
b	2
b	3

Table 2

With the above ffd established, we define here a degree of relevance for a fuzzy determination with respect to contexts.

DEFINITION   For attributes X and Y in a relation r with context C_X and induced context $I_{X \rightarrow Y}(C_X)$, a fuzzy determination X $\rightarrow$ Y holds with a degree of relevance σ if for all equivalence classes p_1, p_2 in C_X and q_1, q_2 in $I_{X \rightarrow Y}(C_X)$, the following relation holds:

$$\mu(p_1, p_2) \le \mu(q_1, q_2)^\sigma$$

Then

$$R_{(x,C_X) \rightarrow (Y,C_Y)} = \sigma$$

Where $R_{(x,C_X) \rightarrow (Y,C_Y)}$ represents the degree of relevance for X determining Y with respect to context C_X.

ALGORITHM AND EXAMPLE

In this section, we describe formally the problem model and propose a solution algorithm for answering null queries.

Given a relations $r(t_1, t_2, ..., t_n)$ such that $(t_1, t_2, ..., t_{n-1})[A_1, A_2, ..., A_m]$ and $(t_n)[A_1, A_2, ..., A_{m-1}]$ are defined (i.e., data values exist) and $t_n[A_m]$ is not defined (i.e., data values is missing).   Find by analogy approximate values for $(t_n)[A_m]$.

The solution algorithm is

I   Calculate the degree of relevance σ_j for each $A_j \rightarrow A_m$, $\quad 1 \le j \le m-1$, with respect to contexts C_{Aj} and $I_{Aj \rightarrow Am}(C_{Aj})$, Where C_{Aj} consists of equivalence classes at level α_j.

II   Measure the similarity , β_j, of $t_n[A_j]$ to the closest equivalence class in C_{Aj}.

III   Find the equivalence class such that it has the maximum value of $\alpha_j * \dfrac{\sigma_j}{\sigma_{max}} * \beta_j$, $1 \le j \le$ m-1. Let it be q in context C_{Ak} for attribute A_k.

IV   Project the induced equivalence class (induced from q in C_{Ak}) in $I_{Ak \rightarrow Am}(C_{Ak})$ to $t_n[A_m]$.

The degree of relevance for the ffd: $A_j \rightarrow A_m$ is determined by σ_j

$$\mu(p_1, p_2) \leq \mu(q_1, q_2)^{\sigma_j}$$

where p_1, p_2 are equivalence classes of C_{Aj} and q_1, q_2 are corresponding equivalence classes of the induced context $I_{Aj \rightarrow Am}(C_{Aj})$. The selection of contexts depends on the redundancies and consistency at level α_j of the tuple components in each attribute A_j.

In addition, the "truth" of the whole analogical reasoning process can be conservatively measured by

$$\alpha_k * \frac{\sigma_k}{\sigma_{max}} * \beta_k = \max_{1 \leq j \leq m-1} \left(\alpha_j * \frac{\sigma_j}{\sigma_{max}} * \beta_j \right)$$

where α_k is the α-cut level for the equivalence class q in context C_{Ak}, σ_k is the degree of relevance for $A_k \rightarrow A_m$ and β_k is the similarity of $t_n[A_k]$ to the closest equivalence class q in C_{Ak}.

We now present a simple example to illustrate some of the ideas presented in the paper. Using the relation *Employee* shown in Table 1, we assume that the salary of tuple Grace is missing and would like to obtain an approximate answer.

First, we compute the degree of relevances for fuzzy determinations Job $\rightarrow$ Salary and Hobby $\rightarrow$ Salary with respect to contexts. The corresponding contexts are C_{Job}={{Secretary}, {Accountant}, {Software Engineer}, {System engineer}, {VP}} at level α_1=1, and C_{Hobby}={{Music, Reading}, {Sports, Swimming}} at level α_2=0.9. The induced contexts are $I_{Job \rightarrow Salary}(C_{Job})$={{25k}, {28k}, {35k,38k}, {39k}, {55k}}, and $I_{Hobby \rightarrow Salary}(C_{Hobby})$={{25k,38k,39k}, {28k,35k,55k}}.

The degree of relevances are

$$R_{(Job, C_{Job}) \rightarrow (Salary, I(C_{Job}))} = \sigma_1 = 0.754$$

$$R_{(Hobby, C_{Hobby}) \rightarrow (Salary, I(C_{Hobby}))} = \sigma_2 = 0.607$$

where the measure of similarity between precise numbers is calculated using $\mu(a,b) = \dfrac{1}{1 + \gamma|a - b|}, \gamma$ is

domain dependent [8]. We choose γ=0.05 in this example.

Since Grace's Job is a design engineer and her *Hobby* is *Sports*, the similarity for these tuple components to their closest equivalence classes are β_1=0.9 to software engineer as well as system engineer, and β_2=1 to sports, respectively. Accordingly, the equivalence classes {software engineer} and {system engineer} both have value

$$1 * \frac{0.754}{0.754} * 0.9 = 0.9,$$

and {sports} have value

$$0.9 * \frac{0.607}{0.754} * 1 = 0.7245.$$

Therefore, we project induced equivalence classes (induced from {software engineer} and {system engineering}) {35k,38k} and {39k} to the salary of Grace, i.e., Grace's salary is {35k,38k,39k}. The "truth" value for this approximate answer is then 0.9 as calculated in step 3 in the algorithm.

CONCLUSION

We have presented an approach for answering null queries using analogical reasoning in the context of similarity-relation-based fuzzy relational data model. The approach developed here is a generalization of model of analogy developed for possibility-distribution-based fuzzy relational data model. It utilizes the concept of contexts on domain attributes to deal with the issue of redundancy among tuples and successfully obtain approximate answers for missing data values in an incomplete database. This kind of facility will certainly improve the cooperative nature of databases and enhance the user-friendliness of the database systems.

REFERENCES

1. Anvari, M. and Rose, G.F., Fuzzy relational databases, in Bezdek, Ed., Analysis of Fuzzy Information, Vol. II (CRC Press, Boca Raton, FL, 1987).

2. Bosc, P. and Pivert, O. Fuzzy querying in conventional databases. In Fuzzy Logic for the Management of Uncertainty, Zadeh, L. and Kacprzyk, J. Eds, John Wiley, New York, 1992, 645-671.

3. Buckles, B.P. and Petry, F.E.   A fuzzy representation of data for relational databases, Fuzzy Sets and Systems, 7, 1982, 213-226.

4. Buckles, B.P. and Petry, F.E. Fuzzy databases and their applications, in Fuzzy Information and Decision Process, Gupta, M. and Sanchez, E., Eds, North-Holland, New York, 1982, 361-371.

5. Buckles, B.P. and Petry, F.E. Information-theoretic characterization of fuzzy relational databases, IEEE Trans. Systems Man Cybernetics, 13, 1983, 74-77.

6. Buckles, B.P. and Petry, F.E. Extending the fuzzy database with fuzzy numbers, Inform. Sci. 34, 1984, 145-155.

7. Buckles, B.P. and Petry, F.E. Query languages for fuzzy databases, in Management Decision Support Systems using Fuzzy Sets and Possibility Theory, Kacprzyk, J. and Yager, R.R., Eds,Verlag TUV Rheinland, Cologne, 1985, 241-252.

8. Dutta, S., Approximate reasoning by analogy to answer null queries, Int. J. of Appr. Reasoning, 5, 1991, 373-398.

9. Hale, J. and Shenoi, S., Analyzing FD inference in relational databases, Data & Knowledge Engineering, 18, 1996, 167-183.

10. Hale, J., Threet, J. and Shenoi, S., A practical formalism for imprecise inference control, in J. Biskup et al, eds., Database Security, VIII-Status and Prospects (Elsevier Science, Amsterdam, 1994) 139-156.

11. Liu, X. Entropy, distance measure and similarity measure of fuzzy sets and their relations. Fuzzy Sets and Systems, 52, 1992, 305-318.

12. Medina, J.M., Vila, M.A., Cubero, J.C. and Pons, O., Towards the implementation of a generalized fuzzy relational database model, Fuzzy Sets and Systems, 75, 1995, 273-289.

13. Prade, H. Lipski's approach to incomplete information databases restated and generalized in the setting of Zadeh's possibility theory, Inform. Systems, 9, 1984, 27-42.

14. Prade, H. and Testmale, C. Generalizing database relational algebra for the treatment of incomplete or uncertain information and vague queries, Inform. Sci. 34, 1984, 115-143.

15. Raju, K.V.S.V.N. and Majumdar, A.K., Fuzzy Functional Dependencies and Lossless Join Decomposition of Fuzzy Relational Database Systems, ACM Transactions on Database Systems, Vol. 13, No. 2, June 1988, 127-166.

16. Shenoi, S. and Melton, A., Proximity relations in the fuzzy relational database model, Fuzzy Sets and System, 31, 1989, 285-296.

17. Umano, M., Freedom-O: A fuzzy database system, in Gupta-Sanchez, Ed., Fuzzy Information and Decision Processes (North-Holland, Amsterdam, 1982).

18. Zadeh, L.A.   Similarity relations and fuzzy orderings. Inform. Sci., vol 3, no. 1, Mar. 1971, 177-200.

19. Zemankova-Leech, M. and Kandel, A.   Fuzzy Relational Databases - A Key to Expert Systems, Verlag TUV Rheinland, Cologne, 1985.

AUTHOR INDEX

SUBJECT INDEX

This index is produced from the titles, abstracts, and keyword lists provided by authors. Reference numbers are to the first pages of papers.